RESTRICTED

TORPEDO
INSTRUCTION PAMPHLET

TS-5

Visual Education Department
FLEET SERVICE SCHOOLS
U. S. DESTROYER BASE
San Diego, California

©2014 Periscope Film LLC
All Rights Reserved
ISBN #9781940453293
www.PeriscopeFilm.com

This document reproduces the text of a manual first published by the Department of the Navy, Washington DC. All source material contained in the reproduced document has been approved for public release and unlimited distribution by an agency of the U.S. Government. Any U.S. Government markings in this reproduction that indicate limited distribution or classified material have been superseded by downgrading instructions that were promulgated by an agency of the U.S. government after the original publication of the document. No U.S. government agency is associated with the publishing of this reproduction.

©2014 Periscope Film LLC
All Rights Reserved
ISBN #9781940453293
www.PeriscopeFilm.com

RESTRICTED SERIAL No._____

TS-5

MK 13-1

AIRCRAFT TORPEDO

★

INSTRUCTION PAMPHLET

FLEET TORPEDO SCHOOL
SAN DIEGO — CALIFORNIA

ILLUSTRATED & MULTILITHED BY VISUAL EDUCATION DEPT...FLEET SCHOOLS
U.S. DESTROYER BASE SAN DIEGO, CALIFORNIA

REG. NO._____

TABLE OF CONTENTS

FLEET TORPEDO SCHOOL PAMPHLET

MARK 13 MODS. TORPEDOES

CHAPTER	CONTENTS
I	History of the Torpedo.
II	General Description.
III	Exercise Heads & Attachments.
IV	War Head and Attachments.
V	Air Flask, Midship Section.
VI	Reducing, Superheating System and Starting Gear.
VII	Propulsion Mechanism.
VIII	Depth Control Mechanism.
IX	Gyro Mechanism.
X	Gyro, Mark 12-1.
XI	Adjustments and Tests.
XII	Loading. Air Trajectory Stabilizers. O.D. and T Letters.

LIST OF EFFECTIVE PAGES

REG. NO. _____

PAGES	ISSUE	TOTAL
Table of contents page 1,2,3,4		
I-1,2,3,4,5,6,7,8,9,10,11,12,13, 14,15,16,17,18,19,20,21.	Original	21
II-1,2,3,4,5,6,6A,6B,6C,6D,6E, 6F,6G.	Original	13
III-1,2,2A,2B,3,4,4A,4B,4C,5,6, 6A,7,8,9,10,11,12,13,13A, 13B,14,15,16.	Original	24
IV-1,2,3,4,5,5A,5B,6,7,7A,7B,7C, 8,9,9A,9B,9C,9D,9E,10,11,11A, 12,13,14,14A,14B,14C,15,15A, 15B,16,16A,16B,16C,17,17A,18, 18A,18B,18C,18D,19,20,21,22,23, 24,24A,24B,24C,24D,24E,24F,24G, 24H,24I,24J,24K,24L,24M,24N.	Original	62
V-1,2,3,4,4A,4B,5,5A,6,6A,6B,6C,6D, 6E,6F,6G,6H,7,8,8A,9,10,10A,10B, 11,11A,12,12A,12B,13,14,15,16,17, 18,18A,19,19A,19B,20,20A,20B,21, 21A,21B,21C,21D,21E,21F,21G,21H, 21I,22,22A,23,23A,24	Original	56
VI-1,2,2A,3,3A,4,4A,5,6,6A,7,7A,8, 8A,9,9A,10,10A,11,12,12A,13,13A, 13B,14,15,15A,15B,15C,16,17,18, 18A,19,19A,19B,20,21,21A.	Original	39
VII-1,2,2A,2B,2C,3,3A,3B,4,4A,4B, 4C,4D,4E,5,6,7,8,9,9A,9B,10, 10A,11,12,13,13A,14,15,16,16A, 16B,17,17A,18,19,19A,19B,19C, 20,20A,20B,20C,21,21A,22,22A, 22B,23,23A,23B,23C,23D,23E,23F, 24,25,25A,25B,25C,25D,25E,25F, 25G,25H,26,27,28,28A,29,30.	Original	71

TS-5

LIST OF EFFECTIVE PAGES

REG. NO._____

PAGES	ISSUE	TOTAL
VIII-1,2,2A,2B,2C,2D,2E,3,3A,4,5, 6,6A,6B,7,8,8A,9,10,10A,11, 12,13,14,15,15A,16,17,17A,17B, 18,18A,18B,18C,19,19A,19B,20, 20A,21,21A,21B,20C,21D,21E,21F, 22,22A,22B,22C,22D,23,24,25,26, 27,28.	Original	57
IX-1,2,2A,2B,3,3A,3B,4,4A,4B,4C,5, 5A,5B,5C,6,7,8,9,10,11,12,13, 14,14A,15,16,17,17A,18,18A,19, 19A,20,20A,21,22,23,23A,23B,24, 24A,24B,24C,25,25A,26,26A,26B, 26C,27,27A,27B,28.	Original	55
X-1,2,3,3A,3B,3C,4,5,6,7,7A,8,9,10, 11,12,12A,13,13A,13B,13C,14,14A, 15,15A,15B,15C,16,16A,17,18,19, 20,21,21A,22,22A,23,23A,23B,23C, 23D,23E,24,24A,25,25A,25B,26.	Original	49
XI-1,2,3,4,5,6,7,8,9,9A,10,11,12, 13,14,15,16,17,18,19,20,21,22, 23,24,25,26,27,28,29,30,31.	Original	32
XII-1,2,3,3A,3B,3C,3D,3E,3F,3G,3H, 3I,3J,3K,3L,3M,3N,3O,3P,3Q,3R, 3S,3T,3U,3V,3W,3X,3Y,3Z,3A',3B' 3C',3D',4,4A,5,6,7,8,9,9A,10, 11,12,13,14,15,16,17,18,19,19A, 19B,20,20A,20B,20C,21,21A,21B, 21C,21D,21E,22,22A,22B,22C,22D, 23,24,25,25A,26,26A,26B,27,27A, 28,28A,29,29A,30,31,32,33,34,35, 36,37,38.	Original	90

TS-5 REG. NO. _____

RECORD OF CORRECTIONS

Change No.	Date Entered	Signature	Change No.	Date Entered	Signature

Original Page.

CHAPTER I

A HISTORY OF THE DEVELOPMENT OF THE

TORPEDO

A. EARLY HISTORY

1001. Ancient history tells us of the use of fire ships and powder ships against fleets of hostile vessels. It remained for Captain David Bushnell, U.S. Navy, to make use of this idea shortly before the American Revolution in the invention of a floating keg-mine. Captain Bushnell's "keg" was loaded with a small charge of gun powder, fired by the pulling of a lanyard led from the keg to shore. The keg was floated down stream until it fouled the target, when the lanyard was pulled and the explosive charge set off. This crude device was the forerunner of the delicate and deadly torpedo of today.

1002. The name "torpedo" was not used in connection with explosive agents until about 1800. Robert Fulton, the inventor of the steamboat, developed a floating mine and named it the torpedo. Fulton's torpedo was merely a buoyant mine that was designed to be floated against the hull of a vessel, and there be exploded by the clockwork firing mechanism. He later improved his mine so that it could be attached to the underwater hull of a vessel from a crude submarine boat.

1003. From 1805 to 1865 there were many forms of floating mines devised, called torpedoes, and resembling each other very closely. They all depended upon fouling the target by drifting downstream, and then being exploded by various means. As a weapon they were ineffective and not dependable. It is interesting to note that this style of warfare was universally considered inhuman and barbarous. These devices, though they mark the beginning of torpedo warfare, were called torpedoes but were in reality floating mines.

1004. The development of the torpedo really began with the Civil War in the United States. From then on the growth of the torpedo can be clearly traced through three distinct classes, viz., the portable, the controllable and the uncontrollable.

B. SPAR TORPEDO

1005. The strict blockade imposed upon ports of the Confederacy by the Federal Navy imperiled the success of the Confederate cause. Ordinary efforts to break the blockade were unavailing. So the Confederate Government set up a Torpedo Bureau in 1862. The principal work of this Bureau was the development of the spar torpedo. The Federal

Navy Department likewise worked upon the idea, and several forms of the spar torpedo appeared. In principle, the spar torpedo consisted of a buoyant casing, loaded with about 40 lbs. of gunpowder and equipped with a percussion firing device. The torpedo was mounted on the end of a long spar or boom rigged out over the bow of a launch or torpedo boat. The attacking boat approached the target until the torpedo boom could be lowered beneath the hull of the target. The torpedo was then released from the spar, rose by its buoyancy to the bottom of the vessel, and was fired by a lanyard. It was with this device that Lieutenant Cushing, U.S.Navy, sank the rebel ram Albemarle. The spar torpedo developed to a high state of efficiency by the U.S.Navy, was mounted on booms rigged out from bows, beams, and quarters of capital ships. The latest spar torpedo carried as much as 100 pounds of gunpowder. The earlier torpedoes were designed to be released from the spar under the target. The later types were rigidly secured to the spar, and were exploded electrically by current drawn from batteries or dynamos. Some trouble was experienced with spars carrying away when vessels attempted to steam at high speeds. To overcome this difficulty, spars were so rigged that they could be run out through the side, under water, when needed. This device was the forerunner of the present day submerged torpedo tube.

C. THE OTTER

1006. It was soon found that the spar torpedo could rarely be used by vessels underway during action because of its unwieldiness. This brought about the "otter", or towing torpedo. The otter was a cigar-shaped buoyant container, fitted with fins and adjustable rudders. The otter was towed by a line led out from the stern or quarter. The rudders and fins were set at such angles that the otter towed submerged at the end of the line which tended about $45°$ with the keel line. The British adopted the otter about 1870 and retained it for about ten years. The otter carried about 100 lbs., of gun powder, exploded by a fulminate of mercury detonator. In 1880 the otter was abandoned as no longer useful. It is interesting to note that this device was revived during the World War, and employed as a defense against submerged mines.

D. THE PROPELLED TORPEDO

1007. The abandonment of the portable, i.e., the spar and towed torpedoes marks the beginning of the development of the propelled torpedo. This development was of two distinct types, each involving many different designs which may be grouped as follows:

UNCONTROLLABLE	CONTROLLABLE
1. Projectile.	1. Dirigible.
2. Rocket.	2. Locomotive.
3. Automobile.	3. Automobile.

E. THE PROJECTILE TORPEDO

1008. The controllable and uncontrollable torpedoes were rapidly developed, and work on all types was practically contemporaneous. The projectile torpedo was merely a container, (Ericcson type), 25 feet long, 16 inches in diameter, weighing 1500 lbs., carrying 300 pounds of explosive. It was fired from a smooth bore gun, through the hull, below the water line, had a range of 50 yards, and maintained its direction by rotation resulting from four vanes placed at the tail. This device was, for a time, promising but finally proved a failure, owing to the uncertainty of the trajectory and the short range.

1009. The projectile torpedo suggested the rocket torpedo, of which there were many types, their development extending over many years. An example of this type, the Weeks, was experimented with in the United States in 1884. The largest type weighed as much as 600 pounds, was 33 feet long, was boat-shaped with a V-bottom, carried 75 pounds of dynamite, and was propelled by two rockets. The torpedo was intended to be kept in a straight line by two stationary vertical rudders and ran on the surface. This torpedo developed speeds as high as 45 knots for about 300 feet. The rocket torpedoes, all types, proved inaccurate and undependable and were soon abandoned.

F. THE AUTOMOBILE TORPEDO

1010. The automobile torpedoes now remained as the last of the uncontrollable type to find favor. The first of the automobile or "fish" torpedoes, and the one that finally survived all others, was the Whitehead.

G. THE WHITEHEAD

1011. In 1860, an Austrian Naval officer conceived the idea of a small surface boat propelled by a small steam or hot air engine, steered by long tiller ropes from shore, carrying in the bow a charge of explosive. This unknown officer died before his plans could be executed. They were taken up by another Austrian Naval Officer, a Captain Luppis, who made a model. With this model, Captain Luppis, in 1864, enrolled the service of Mr. Robert Whitehead, who was the English manager of an iron works at Fiume, Austria. Two years later, Mr. Whitehead produced the first "fish" torpedo. The meager description of this weapon that survives, describes it as "made of boiler plate, carrying a charge of 18 pounds of dynamite, with a speed of 6 knots for a short distance". In 1867, the Whitehead torpedo had improved to the extent of having a horizontal rudder actuated by the "secret mechanism", i.e., the hydrostatic piston. This piston was, to all intents and purposes, exactly the same as the piston of today. The torpedo was propelled by compressed air, actuating a motor of peculiar, yet simple, type. It consisted of two cylinders, eccentrically placed, one inside the other. The air entered the outer cylinder, controlled by a valve so that the air acted always upon one side of the inner cylinder and in one direction. The cylinders were thus made to revolve, the revolutions of the outer cylinder being transmitted to the propeller through a propeller shaft.

1012. In 1868 the Whitehead torpedo was introduced to the world in trials conducted by the Austrian government. The torpedo was 14 feet long, 16 inches in diameter, weighed 650 pounds, and carried an explosive charge of 50 pounds of gun cotton. The torpedo was ejected from a submerged tube by compressed air, attained a speed of 6 to 7 knots for 700 yards, maintained approximately a desired depth, and was kept in a straight course by a stationary vertical rudder. The secret of the Whitehead was purchased in turn by Great Britain, France, Italy, Germany, and Denmark, and by 1875 its supremacy seemed undisputed. By 1887 the Whitehead was greatly improved. It had acquired a speed of 27 knots up to 400 yards with an initial air pressure of 1500 pounds per square inch. The mechanical features of the torpedo may be summarized as follows:

1. A warhead, exploded by a war nose with "whiskers".
2. A combined hydrostatic piston and pendulum actuating.
3. The horizontal rudder engine, air controlled.
4. A three cylinder Brotherhood engine, operating under a reducing valve pressure of 450 pounds, developing 48 H.P. and a propeller speed of 1100 R.P.M.
5. A sinking valve.
6. Bevel gears by which the revolutions of one engine shaft were transmitted to two opposite turning propellers.
7. A distance gear.
8. A stationary vertical rudder for compensating the torpedoes natural deflection.

This was the condition of the Whitehead torpedo in 1887, and with the exception of the gyro control in deflection and the superheating of the air, it will be seen to include the major mechanical features of the modern torpedo.

H. THE HOWELL TORPEDO.

1013. The United States Government was offered the Whitehead torpedo at a price of 20,000 pounds, but the offer was refused. The price was apparently a minor consideration; the real reason for the refusal being the development of two American torpedoes, the Howell and Newport Torpedo Station fish torpedo.

1014. Captain J. H. Howell, U. S. Army, a mathematician of high attainment, invented his automobile torpedo in 1870. For 15 years, considerable effort was made to perfect his invention. The torpedo was a spindle of revolution, having a cylindrical central portion, a sharply conical forward end, with a rounded, tapering after end. The forward end carried the propelling mechanism, a flywheel weighing 100 pounds. The after section carried the steering mechanism and propellers. The torpedo was 11 feet long, 14 inches in diameter and weighed 518 pounds. It could be launched from an above water or submerged tube, and attained a speed of 26 knots for 800 yards, <u>with great accuracy</u>. It could be set to maintain a desired depth and explode upon contact with its target.

1015. The motive plant of the Howell was unique. A heavy flywheel was placed with its axis horizontal and athwart the torpedo. By means of a steam engine, or electric motor, temporarily connected by gears and shafts through the shell of the torpedo to the flywheel, the wheel was spun until it attained a speed of 10,000 R.P.M. The torpedo was then launched, the spinning shaft being automatically disconnected and the momentum of the spinning flywheel provided the propelling power, transmitted through gears and shafting to the twin propellers. As the speed of revolution of the wheel decreased, the pitch of the propellers was automatically changed, thus keeping the torpedo at a constant speed. The remarkable accuracy of the Howell torpedo, possessed by no other torpedo of its time, was a great factor in its adoption by the United States Navy. This accuracy was obtained by use of the gyroscopic effect of the spinning flywheel. The wheel was mounted in the vertical plane of the torpedoes longitudinal axis. Any force tending to deflect the torpedo from the plane of the flywheel, that is, to change course of the torpedo, resulted in a tilting of the gyroscopic wheel, therefore of the torpedo. This canting of the torpedo set in motion a system of levers connected to a vertical pendulum, which in turn threw a vertical rudder to one side, bringing the torpedo back to its original course. This control, in addition to the depth control, gave the Howell torpedo a marked superiority over the Whitehead torpedo.

I. THE NEWPORT TORPEDO STATION.

1016. The success of the Whitehead torpedo lead to many efforts to produce other torpedoes equally efficient and similar in design. The United States Naval Torpedo Station at Newport, R. I., had been established in 1869, to coordinate the torpedo work of the U. S. Navy. Its principal activities were in the development of the old spar torpedo. But in 1874, the Torpedo Station proposed a "fish" torpedo made along the supposed lines of the Whitehead. Instead of bronze, which the Whitehead used, this fish torpedo was to be made of steel, or deposited copper. Two types were proposed, one driven by air at 900 pounds pressure, the other by carbonic acid gas. The two types were otherwise alike. A three cylinder Brotherhood engine was designed that would drive the torpedo 900 yards at 20 knots. The torpedo was to be about 13 feet long, 16 inches in diameter, and weigh about 600 pounds. The torpedo had the usual hydrostatic piston for depth seeking and control. The explosive charge of guncotton varied from 160 up to 210 lbs. The weight of the consumed air or gas was compensated for by admitting sea water to a compensating tank.

J. OTHER UNCONTROLLED TYPES.

1017. Other interesting torpedoes of the uncontrollable type were the Hall, the Peck and the Paulson. The Hall steam torpedo carried a quantity of steam and water at 550°, stored in a flask. The torpedo was driven by two two-bladed propellers of large disk area. The depth control was by means of two fins forward and a drag on the tail, the depth thus depending upon the buoyancy and the speed. The fins were automatically adjusted as the buoyancy increased. Rolling of the torpedo was corrected by automatic righting fins.

1018. The Peck torpedo also used steam for the motive power, was 14 feet long, 14 inches in diameter and carried 100 lbs. of guncotton. Except for the use of steam, this torpedo was otherwise a Whitehead. It was claimed that a speed of 32 knots could be maintained for 1000 yards. The Paulson torpedo had for its motive power compressed carbonic acid, at a pressure of 1500 pounds. As this pressure was gradually released, the liquid volatilized and was made to syphon a stream of sea water which was expelled according to the well-known ejector theory. This torpedo embodied the first real attempt to steer in any given direction.

In the torpedo was a mariner's compass. Mounted over the needle was a magnetic plate carrying two insulated silver studs. Each stud was wired through batteries to its own electro-magnetic, the circuit including the shell of the torpedo as the ground. Rotating with the compass needle, insulated from the needle, and grounded on the shell of the torpedo was a course indicator. The course indicator and studded dial were set to the course it was desired that the torpedo steer. Should the torpedo deviate from this course, the course indicator, riding on the compass needle, would touch one of the studs. This completed the circuit from the battery to stud, through indicator to ground and back to the battery. The magnet, being energized, would draw it back on its course. Contact with the other stud gave the opposite rudder. This device was first proposed by Gunner Burdett, U.S.N. in 1874 for the Torpedo Station fish torpedo but did not find favor.

K. THE CONTROLLABLE TORPEDO.

1019. All of the uncontrollable torpedoes had their respective defects, but common to all were lack of accuracy at "long" range, and liability of blowing up one's own ship by lack of control over the torpedo after it was discharged. These two features led to the development of the third class, or controllable torpedoes. The principal examples of this class were:

1. Borden.
2. Sims-Edison.
3. Brennan.
4. Maxim.
5. Lay.
6. Lay-Haught.
7. Patrick.
8. Nordenfeldt.

1020. The Borden consisted of a surface boat 31 feet long, 31 inches in diameter, and carried 220 pounds of dynamite. This torpedo was steered from the launching vessel, or the shore, by tiller ropes up to a distance of one mile. It was turbine driven, the energy being derived from gases liberated from 12 rockets in the torpedo. The towing feature was intended to defeat the torpedo net. Upon the surface boat becoming fouled in the net, the smaller towed torpedo dove under the net and fired upon contact with the hull.

L. THE LOCOMOTIVE TORPEDO

1021. The Sims-Edison, Brennan and Maxim torpedoes were locomotive torpedoes, that is, their propelling power was supplied from a source outside the torpedo.

1022. The Sims-Edison was electrically driven by a motor in the torpedo. As the torpedo ran ahead, the cable unreeled, until the maximum range of two miles was reached. The torpedo was steered by electro-magnets operated from ashore through small cables. The torpedo was 28 feet long, weighed 4000 lbs., carried an explosive charge of 400 lbs., and attained a speed of 10 knots.

1023. The Brennan torpedo carried two reels of wire lead to two drums ashore. The torpedo was launched and the shore drums started. As the wire was unwound from its reels, the rotation thus imparted to the reels propelled the torpedo through suitable gearing and shafting. The torpedo was steered by varying the rate of unwinding of one drum or the other. While at first glance, the winding of the wire upon the shore drums would seem to tend to haul the torpedo in shore, the fact is that this operation actually resulted in a forward propulsion of the torpedo. A speed of 20 miles per hour for $1\frac{1}{2}$ miles at a depth of ten feet was obtained by this torpedo with a limited directional control.

1024. The torpedo devised by Hudson Maxim in 1886 was an improvement on the Brennan, and incorporated some new features. The depth of run of torpedo could be varied to enable launching in shallow water and submerging to greater depth as the depth of water increased. The exploding mechanism was not in firing condition until the torpedo had run a certain number of revolutions of the propellers. When the end of the run was reached the torpedo could be brought to the surface and its exploding mechanism rendered safe before the torpedo was hauled back to shore.

M. THE AUTOMOBILE CONTROLLABLE TORPEDO.

1025. The locomotive torpedoes proved impractical for ships use, were never adopted, and gave way to the automobile or self-propelling controllable torpedo. The Lay was in this class. This was a surface torpedo, 23 feet long, 18 inches in diameter, weighed $1\frac{1}{4}$ tons and carried 200 lbs. of explosive in the forward cone. The propelling mechanism was a series of reciprocating engines operated by superheated carbonic acid gas, driving a pair of two bladed propellers in a cavity in the under side of the torpedo about 3 feet from the forward end. A speed of 16 knots was claimed for this weapon, for a range of 2 miles. The engine exhausted through a pipe in the stern through which two cables were lead to the controlling station ashore. One cable was connected to the electrically controlled steering engine operated by gas pressure, and the other operated the stop and starting mechanism. This torpedo failed to give satisfactory results and was superseded by other more successful torpedoes.

1026. The Lay-Haight and the Patrick torpedoes were practically the same, and a brief of one, the Patrick, will serve for both. The torpedo was suspended from a float of sheet copper 41 feet long, 12 inches in diameter carrying two guide rods with flags.

1027. The torpedo, made of copper, was 36 feet long, 22 inches in diameter, and was rigidly connected to the float. Propulsion was effected by gas actuating 6 double acting cylinders, driving a two bladed propeller astern. The torpedo was started, stopped, steered, and exploded electrically by means of a double cable connecting the torpedo with a battery of Bunson cells at the controlling station. The total weight of the torpedo was 4700 pounds including 200 pounds of dynamite. A speed of $16\frac{1}{2}$ to 17 knots for one mile was attained in tests in 1886, conducted in the United States.

1028. The last of the automobile, controllable torpedoes was the Nordenfeldt, differing from all others heretofore developed in that it was driven by electricity supplied from within the torpedo. Sets of storage batteries furnished the current for propelling the torpedo at a speed of 16 knots for a range of two miles. The torpedo was 35 feet long, 29 inches in diameter, weighed 5000 lbs. and carried a charge of 300 lbs. of dynamite. It ran submerged and was controlled electrically from the firing station.

1029. Up to 1890, the decision as to the best type of torpedo had not been made. Experiments were being conducted with all the types described herein, with varying results. The locomotive torpedo was finally abandoned. Next went the automobile, controllable torpedo because its use was limited to harbor defense, and even then proved unreliable. This cleared the field for the development of the automobile, uncontrollable torpedo. In this class, the Whitehead, owing to its superiority over the Howell, became the standard torpedo of all navies.

N. THE WHITEHEAD TORPEDO.

1030. Improvement of the Whitehead was rapid. From the early type described above was finally evolved the present day Whitehead. This torpedo consists essentially of four main sections: (a) the Head carrying an explosive charge of over 500 lbs. of guncotton or T.N.T. exploded upon contact with the target; (b) the Air Chamber made of high grade steel, carrying a charge of air compressed to 2650 lbs. per square inch; (c) the Immersion Chamber carrying the depth control mechanism; (d) the Afterbody carrying the propelling and steering mechanism, the propellers and the rudders. The engine, weighing but little more than 300 lbs., is a four cylinder reciprocating engine, driven by superheated compressed air, generating from 75 to 100 H.P. By means of twin propellers, the little engine drives the 2500 lb. torpedo at a speed of 30 knots for a distance of five miles. Constant direction is maintained by means of an air actuated steering engine controlled by a tiny gyroscope. Various speed adjustments make possible a greater range at reduced speeds. The torpedo will run at any desired depth, for any desired distance up to the limit of its range, and will sink or float at the end of its run as desired. Great accuracy has been obtained with this torpedo, and so satisfactorily has it performed that its adoption became universal. It was apparently the last word in torpedo design.

O. THE BLISS-LEAVITT TORPEDO.

1031. But just as the United States was first in the development of torpedoes for war use so has it been the first to improve upon the Whitehead. Although the Whitehead was adopted by the United States, it was soon superseded by a totally new and superior design, known as the Bliss-Leavitt torpedo.

1032. This new torpedo is driven by two turbine wheels, operated by a combination of compressed air and steam, superheated to give high efficiency. The torpedo possesses all the advantages of the Whitehead coupled with greater efficiency and ease of operation. Air compressed to nearly 3000 lbs. in the specially designed air chamber provides the motive power for the two turbines, and the auxiliary machinery. The gyroscope guides the torpedo along the desired course, at the depth for which it is set to run and the great explosive charge carried in the forward end, is sufficient to pierce the hull of the strongest vessel afloat. The range of the torpedo exceeds that of any torpedo heretofore developed, and it is deadly in its accuracy.

1033. The Bliss-Leavitt torpedo, like the Whitehead, may be launched either from a tube on deck, from an airplane, or from a submerged tube, by means of a small charge of gunpowder or compressed air. Once speeding on its way, unseen, unheard, the torpedo becomes a weapon of terrible destruction, before which even the proudest ship is humbled. With such a weapon, its possessor could have made himself master of the situation at Jutland, and perhaps brought the World War to an early end.

P. RECENT DEVELOPMENTS.

1034. Following characteristics of torpedoes of foreign navies have been reported:

- Great Britain: 21" and 24" torpedoes,
    maximum speed 47 knots with
    3000 yard range.
    Maximum range 16,400 yards at
    24 knots.
- Italy: 21" torpedoes.
    Maximum speed 54 knots at 3000 yard range.
    Maximum range 21,800 yards at 24 knot speed.
    Reported to use reciprocating engines in
    the Lombardi torpedo.
- Japan: 21" torpedo.
    Maximum speed 45 knots at 4400 yard range.
    Maximum range 22,000 yards at 22 knot speed.
    Reported to be experimenting with 24"
    torpedoes.

1035. The weight of explosive carried by the foreign torpedoes averages about 600 lbs., which is considerably more than that of American torpedoes.

1036. Developments which may be expected by way of increasing speed and range of torpedoes are:

    (a) Reduction in the number of speed changes in present multispeed torpedoes.
    (b) Increase in diameter and possibly in weight.
    (c) Increase of turbine speed.
    (d) Increased concentration of oxygen in air flask.

1037. Q. DEVELOPMENT OF THE AIRCRAFT TORPEDO
AIR STABILIZATION.

1. The value of the aircraft torpedo as an offensive weapon was recognized a number of years ago, but for many reasons it was not considered a primary means of attack until rather recently. All ideas concerning the use of Naval aircraft have undergone a gradual change since the first planes were used by the Navy and the use of the torpedo plane is no exception. At first the aircraft torpedo was considered secondary to dive bombing and horizontal bombing. However, with improvements in aircraft torpedoes, methods of launching them, and the greater damaging effect on ships of underwater explosions, they are now considered a primary aircraft weapon for attack upon surface craft.

2. The development of the aircraft torpedo and methods of launching covers a considerable period of time. Only a brief summary of this development work will be made and many experiments and changes will be omitted. The summary is not presented in chronological order since some of the early problems are still unsolved for certain Marks of torpedoes.

3. The Mark 7 torpedo and some of its modifications were the first torpedoes used for aircraft work. The Mark 7 is an 18 inch torpedo originally designed for use by submarines. It had two distinct disadvantages for use as an aircraft torpedo: First, since it was originally designed for submarines it did not have the strength to stand water impact when dropped from a plane and, second, due to its shape it was hard to stabilize in the air.

4. Other torpedoes of the long, thin variety have been tried with fair success but they are not entirely satisfactory for aircraft use. The Mark 14, a submarine torpedo, and the Mark 15, a destroyer torpedo, are being used at the present time with moderate success for a limited range of speed and altitude. The Mark 13 torpedo was designed especially for aircraft use. In shape it is a radical departure from the conventional long, thin torpedo and it is considerably stronger than the other

types. The Mark 13 torpedo has the rail tail, that is, the rudders are carried on rail type outriggers and are aft of the propellers. The rail tail has some advantages but since the drag of the rails reduces the speed, the straight tail was adopted. The Mark 13 Mod. 1 has the straight tail and the Mark 13 Mod. 2 is the same as the Mod. 1 except that the speed is increased to about 40 knots. The Mark 13 torpedo and its modifications are satisfactory aircraft torpedoes and can be launched with good results over a fairly wide range of speeds and altitudes.

5. The development of the Mark 13 aircraft torpedo covers a period of several years and improvements to the torpedo were constantly made. Many changes were made in tail surfaces, rudders, rudder throw, head shapes and so forth. During the early years of experiments, progress was slowed to a considerable extent by the lack of adequate air stabilization and it was difficult to distinguish between failure due to poor design and failure due to poor entry into the water. In the past two years the air stabilization of the Mark 13 torpedo has been improved to the point where good entry is now the rule rather than the exception. The Mark 13 is now a fairly reliable aircraft torpedo when dropped at speeds up to 125 knots and from altitudes up to 200 feet. Recent drops at higher speeds indicate that certain parts of the torpedo may have to be strengthened to withstand the impact forces at ground speeds in excess of 150 knots.

1038. R. <u>DEVELOPMENT OF LAUNCHING METHODS.</u>

1. To make a successful run, the torpedo must enter the water at, or very close to, the proper angle, or in other words along the torpedo flight path or trajectory. There are a number of means available which will cause the torpedo to enter at the proper angle, and many of them have been tried with varying degrees of success. The various ways of obtaining the desired entrance angles may be divided into two general classes. The first includes those which depend upon certain launching technique or attitude of the torpedo or plane at release, and the second which provides air stabilization by the use of auxiliary stabilizing surfaces attached to the torpedo which may be either fixed or controlled.

2. A number of methods of launching the torpedo without auxiliary stabilizers have been tried. Some of these methods which are described below have been fairly successful and others have been entirely unsuccessful. Most of them depended a great deal upon the skill of the pilot and a set of very rigid rules for launching had to be followed in order to obtain good results.

3. A torpedo without auxiliary stabilizing surfaces is aerodynamically unstable and therefore does not tend to remain tangent to the trajectory. If the movement of the torpedo away from the trajectory is not violent, that is, if the angular velocity is not much greater than that which the torpedo would have if it were following a constant trajectory, and if the torpedo enters the water when it is tangent to the trajectory a good entrance should result. By carrying the torpedo so that it is nose down to the relative wind at release, it will gain angular velocity any may approach the trajectory soon after release. For any given rack angle, there are several combinations of speed and altitude of release which will produce the desired entrance angle, or the rack angle may be varied to suit the speed and altitude of release. This method of changing rack angles has been tried and has produced fairly good results, but for any given rack angle, the range of speeds and altitudes are rather limited and any departure from either of the three conditions results in poor entrance angles. This method is limited to low altitudes since it is desirable to have the torpedo enter the water when it first approaches the trajectory and the air speeds must be low so that high angular velocities are not gained.

4. In order to follow the vacuum trajectory, the torpedo must have angular velocity at the instant of release and in order to follow any path which approaches the parabolic trajectory the torpedo must acquire angular velocity shortly after release. One method of imparting this angular velocity was to nose the plane over sharply just before release and release the torpedo while the plane was still turning downward. If properly executed, this maneuver should give the torpedo the correct angular velocity and at low altitudes, about 50 feet or less, the torpedo should enter the water at about the right angle. This method also was limited to rather low airspeeds, since at higher air speeds the oscillations are more violent. Since it is difficult to give the plane and thereby the torpedo the proper angular velocity at the instant of release, this method was not very successful except for pilots who had a great deal of training for this type of release.

5. If the torpedo could be launched directly into the relative wind it should remain headed into the relative wind for a short period. Therefore, if the flight path of the plane approached the desired entrance angle and the torpedo is released at low altitude, it should enter the water at the proper angle. This method was tried, but the natural tendency to pull up as the plane approached the water resulted in the torpedo entering the water at angles other than those desired.

6. A number of other methods were tried but none of them were very successful. Changing the rack angle or changing the attitude of the plane at the instant of release gave fair results for very limited ranges of altitude and speeds. The altitudes were about 50 feet maximum and maximum speed about 110 knots. Much depended upon the skill of the pilot and 50 feet or less is an uncomfortably low altitude especially under combat conditions.

1039. S. DEVELOPMENT OF AIR STABILIZATION.

1. The results of dropping torpedoes without auxiliary stabilizers and a study of the theory of stabilized flight indicated that some form of auxiliary stabilizing surfaces was required if the torpedo was to follow any constant trajectory. Many forms of stabilization were proposed and a number of them were tested with various degrees of success. The fixed auxiliary stabilizers now in use on Mark 13 torpedoes are reasonably successful and are the gradual development of a type proposed some time ago. They are satisfactory and simple to make, but are not perfect and cannot be successfully used with the Mark 14 and Mark 15 torpedoes. Other type stabilizers now under development may solve the problem of entrance angle for all marks of torpedoes.

2. Various forms of extensions to the horizontal vanes of the torpedo have been tried. Some were made of metal and several kinds of wooden vanes were used. Two types of monoplane plywood extensions were developed which gave fairly good results. One was developed at the Naval Torpedo Station and was issued to the service. It was quite satisfactory for the TBD air plane at speeds of less than 110 knots and altitudes less than 50 feet. The other was developed by Patrol Wing One and was superior for PBY airplanes to the one developed at the Naval Torpedo Station. These auxiliary stabilizers were made of quarter inch plywood and were bolted to the horizontal

vanes of the torpedo. Although they produced fair results they had several serious disadvantages. Since they were in effect monoplanes, it was very difficult to make them rigid and it was difficult to secure them rigidly to the torpedo. If the vanes were not rigidly secured, one or both vanes dropped off in the air, especially from the PBY, and the air performance of the torpedo was then poor. If the vanes were rigidly secured, part of the vane frequently remained on the torpedo after entry into the water and a poor water run resulted. Since these stabilizers were of a monoplane construction, part or all of the stabilizing effect was lost if the torpedo rolled in the air. There was nothing but inertia keeping the torpedo from rolling. This type of stabilizer was limited to low altitudes where the time of fall was not sufficient to allow the torpedo to roll much more than $30°$.

3. Controlled monoplane tails have been used with good results. In this type a fairly large monoplane elevator is mounted aft of the torpedo and is controlled through the horizontal steering engine of the torpedo by a small wind vane mounted on the side. When the torpedo axis departs from the relative wind, the wind vane moves and through a control valve causes the elevator to operate to bring the torpedo back to the trajectory. The torpedo thus oscillates about the trajectory but never stays on for any length of time. If the torpedo once starts to roll, this stabilizer is not effective and unless an anti-roll device is used in conjunction with it, the torpedo must be dropped from low altitudes where roll cannot be effective. Although this stabilizer produces good results, so far as the Mark 13 torpedo is concerned it is considered less satisfactory than the biplane stabilizer now in use, and is much more complicated to make, install, and service.

4. Small parachutes (Pilot Chutes) and small sleeves similar to target sleeves have been used with fair results. The parachute, or sleeve, was attached to the tail shaft securing nut by a steel wire and then was stuffed into the tail shaft. When the torpedo started the chute or sleeve was expelled by the exhaust. The stabilization of the torpedo was mediocre since the chute was free to oscillate, and it frequently produced considerable yaw and the entrance angles were not consistent. However, it indicated that drag may be made an important factor in the air stabilization of the aircraft torpedo.

5. Several types of biplane stabilizers have been used. They have several advantages over the monoplane stabilizers. Since the span is less, the rolling moment which may be imparted by the stabilizer is smaller. Also they are easier to attach rigidly to the torpedo and come completely off upon impact with the water. One of the first types used had an air-foil section, and the two airfoils were mounted so that the lift on each acted toward the torpedo. In other words, they were mounted on the upper and lower vertical tail fins of the torpedo with the cambered surfaces facing each other.

Several combinations of span and chord were used, with the airfoils set at various angles of incidence. A number of methods of securing the stabilizer to the torpedo were also tried. These stabilizers were fairly successful but had the following disadvantages:

    (a) The air foil section was rather difficult to construct and would not be readily available to units operating from advance bases.

    (b) It was rather difficult to attach the airfoils with the cambered surface toward the torpedo.

    (c) It was difficult to set the airfoils at the correct angles.

6. Although the biplane airfoil stabilizers were superiod to the monoplane extensions, it appeared desirable to develop a simpler type. Several forms of flat board stabilizers were used and a number of modifications of these were tried before the present form was finally adopted. Chapter IV contains a detailed description of the biplane stabilizers now in use.

7. Since the TBD airplane carries the torpedo partly within the fuselage, the span of the stabilizers that can be used with this airplane is limited to about 23 inches. The span of the stabilizers used on the PBY airplane is 32 inches and it therefore cannot be used on the TBD. Several types of non-symmetrical biplane stabilizers have been tried with fair success, but the wedge type symmetrical stabilizers now recommended for the TBD gave better results. This type works well on the TBD or the PBY but, since it is a little harder to construct, usually it is not used on the PBY airplane.

8. Although the fixed type biplane stabilizers are very good on the Mark 13 torpedo, they are not as effective on the Mark 7, 14 and 15 torpedoes. For these Marks of torpedoes a biplane stabilizer with a two position flap (or trailing edge) has been developed. This type is moderately successful for a limited range of speeds and altitudes only. It appears that for torpedoes of conventional shape it is not sufficient to obtain symmetrical air stabilization, but that other conditions must be obtained.

9. Projects are now underway for the development of auxiliary stabilizers of several different types for all marks of torpedoes that may be carried by aircraft. As these stabilizers reach the stage where they can be used by operating units, information about them will be issued either as additions to this pamphlet or in separate pamphlets.

T. <u>NOTES ON CONTEMPORARY U.S. NAVY TORPEDOES</u>.

1040. <u>MARK VII</u>.

12' x 18" - Range 4000 yards, speed 32 knots. This was the first torpedo to use water spray in conjunction with alcohol spray in the superheater. Combined diving gear with gyro mechanism as one unit. Turbines placed in a horizontal plane. - Obsolete.

1041. <u>MARK VII-1</u>.

Similar to the Mark VII except that it contained a compound regulator (two reducing valves). The first stage reduced flask pressure to 850 lbs., and the second stage reduced to 450 lbs. A more uniform speed was obtained. Explosive charge 245 lbs. - Obsolete.

1042. MARK VII-2. Same as the Mark VII-1 except designed to run 6000 yards at a speed of 27 knots. - Obsolete.

1043. MARK VII-2A - 18" diameter, 18' long. Weight of explosive charge - 319 lbs. Range 6000 yards at 30 knots. Afterbody and tail strengthened for aircraft use. Dropped from height of 25 feet. 2 stage reducer. - Obsolete.

1044. MARK VII-2B. Same as VII-2A except improved gyro,

Uhlan immersion mechanism, could be dropped from height of 75 feet. Sylphon type reducer. - Obsolete.

1045. MARK VIII-3. 21' by 21". Range 13,000 yards at 27 knots. This torpedo was not strong enough to withstand the shock of firing. - Obsolete.

1046. MARK VIII-3A. Similar to VIII-3 except heavier parts and tail cone than the Mark VIII-3. Satisfactory for destroyers but still not strong enough for cruiser use. - Obsolete.

1047. MARK VIII-3B. Similar to VIII-3 except steel braces to brace guide vanes. - Obsolete.

1048. Mk.VIII-3C. Similar to the VIII-3A, except depth and deflection performance improved by use of air sustained gyro; Uhlan depth gear. Premixer top added. 384 lb. explosive charge. Range 13,500 yards at 27 knots.

1049. MARK VIII-3D. Similar to Mark VIII-3C except stronger tail construction to permit use from cruisers.

1050. MARK VIII-4. Similar to Mark VIII-3 except heavier afterbody. Air and water preheaters. Air sustained gyro. Mechanical control of igniter and air, fuel and water, range 16,000 yards at 26 knots. Length 21½ feet. Explosive charge 466 lbs. Exercise head *not* water ballasted.

1051. MARK VIII-5. Similar to Mark VIII-4 except higher turbine speed. Range 14,000 yards at 29 knots. Obsolete.

1052. MARK VIII-8. Afterbody strengthened and tail vanes and cone strengthened. Air blowing head. Uhlan Depth Mechanism. Air sustained non-tumble gyro. Explosive charge 478 lbs. 98 shaft H.P. Range 14,000 yards at 29 knots. Spring sylphon reducer. Premixer top. Air preheater. Overhead check valves.

1053. MARK X. 16' x 21" - speed 30 knots. Obsolete.

1054. MARK X-1. Similar to Mark X except speed of 34 knots. Obsolete.

1055. MARK X-2. Similar to Mark X except speed of 36 knots at 3500 yards range. Explosive charge 497 lbs.

1056. MARK X-3. Similar to Mark X-2 except improved depth mechanism (Uhlan Gear). Explosive charge 508 lbs. Premixer top, overhead vented check valves. Mark XIII air sustained, non-tumble gyro spring sylphon reducer S.H.P. 158.

1057. MARK XI. First of the multispeed torpedoes. Shape of head streamlined. Air flask solid afterhead water compartment - separate from air flask and bolted to it. Speed change mechanism similar to Mark XV. Balanced reducer with control valve. Compound Gear Oil Pump. Four Exhaust tubes. Explosive charge 500 lbs. Developed 340 H.P. in high power
Range 15,000 yards at 27 knots.
" 10,000 yards at 34 knots.
" 6,000 yards at 45 knots.

1058. MARK XII. Second Multispeed Torpedo. Similar to Mark XI except removable after bulkhead, screwed in place water compartment forged integral with air flask. Gear oil pump not compound. Two exhaust tubes but four exhaust valves. Chromium plating on various parts introduced for the first time.

Range 15,000 yards at 27 knots.
" 10,000 yards at 35 knots.
" 6,000 yards at 44 knots.

1059 MARK XIII. First aircraft torpedo originally designed as such 22½" x 14'. Explosive charge 400 lbs. 98 horse power Range 4,500 yards at 30 knots. Center of gravity of explosive charge below centerline. Exercise head has air compartment at after end and head is rugged. Air flask solid after bulkhead.

1060 MARK XIV. The third multi-speed torpedo manufactured by the Bureau of Ordnance. Primarily for submarines. Shorter than Mark XI and XII and weighs 500 lbs. less.

Range 4,500 yards at 46 knots.
" 9,000 yards at 31 knots.

No gear shift to change speeds

Movable top plate type gyro mechanism.

CHAPTER TWO

		ARTICLES
A.	General Description of Mark 13, 13-1, 13-2 torpedo	2001-2003
B.	Principal dimensions, weights and characteristics	2004
C.	Interchangeability	2005
D.	Parts carrying numbers	2006
E.	Characteristics of construction	2007
F.	Buoyancy	2008
G.	Lubricants	2009
H.	Mixing depressant with hot running torpedo oil	2010

A. GENERAL DESCRIPTION.

2001. The complete torpedo is composed of the following assembled units: War head, exercise head, air flask, afterbody, tail, gyroscope.

2002. The designation "Mark 13", "13-1" and "13-2" applies to the complete torpedo and to the air flask, afterbody and tail. The war head, exercise head and gyroscope have individual designations which are given over their respective descriptions elsewhere in this pamphlet.

2003. The Mark 13 torpedo is the first torpedo originally designed for launching from aircraft only. The Mark 13-1 and 13-2 torpedoes are similar to the Mark 13 except for a new design of tail and other minor improvements. Mark 13 and modification torpedoes include increased structural and launching strength without sacrificing torpedo control or effectiveness of war head.

B. PRINCIPAL DIMENSIONS, WEIGHT AND CHARACTERISTICS.

2004.

	Mk. 13	Mk. 13-1	Mk. 13-2
Torpedo Mark	Mk. 13	Mk. 13-1	Mk. 13-2
War Head Mark	13	13	13-1
Exercise Head Mark	26	26-1 or 26-2	26-3
Gyro Mark	12-1	12-1	12-1
Diameter	22"42	22"42	22"42
Length overall, with war head	13'8"55	13'5"0	13'5"0
Length overall, with exercise head	13'8"55	13'5"0	13'5"0
Length of war head to joint line	53"99	53"99	53"99
Length of exercise head to joint line with towing eye	53"99	53"99	53"99
Length of air flask, joint line to joint line	52"890	52"890	52"89
Length of afterbody, joint line to joint line	37"32	37"32	37"32
Length of tail, end to joint line	20"35	16"80	16"80
Weight, lbs.			
Explosive charge	400	400	600
War head empty, without attachments	195.9	195.9	202.5
War head, loaded, with exploder	625	625	836.0
Exercise head, empty	312	312	
Exercise head, ready for firing	625 ± 3	625 ± 3	
Air Flask section	622	622	622

	Mk.13	Mk.13-1	Mk.13-2
Afterbody, complete with gyro and tail........................	499	477	477
Air charge, total, 62° F., 2800#............................	130.4	130.4	130.4
Fuel................................	16.8	16.8	16.8
Water...............................	48.0	48.0	48.0
Oil.................................	7.5	7.5	7.5
Ballast, lead in war head........	12.0	12.0	12.0
Ballast, lead, in exercise head.	93.6	93.6	
Ballast, water, in exercise head.	319.5	320.0	
Torpedo, empty, with war head and attachments and gyro......	1746	1724	
Torpedo, ready for war shot......	1949 ± 20	1927 ± 20	2127
Torpedo, empty, with exercise head full and gyro............	1746	1724	
Torpedo, ready, for exercise shot................................	1949 ± 20	1927 ± 20	
Buoyancy, Trim and Stability.			
Displacement (water 1.026 sp.gr.)............................	1709 lbs.	1703 lbs.	1703 lbs.
Buoyancy, ready for war shot, negative...........................	-240	-224	-424
Buoyancy, ready for exercise shot, negative....................	-240	-224	
Buoyancy, empty, with war head.	-37	-21	
Buoyancy, empty, with exercise head not blown................	-37	-21	
Buoyancy, exercise head blown, 350 lbs. air pressure, 5% fuel, 10% water, 100% oil...........	±253	±270	
Center of buoyancy to end of tail................................	94.01±.25	90.64±.25	
Center of gravity from center of buoyancy:			
(a) Ready for war shot, aft....	3.96	2.77	
(b) End of war shot run, aft...	3.19	1.86	
(c) Ready for exercise run, aft.	3.96	2.77	
(d) End of exercise run, head not blown, aft..................	3.19	1.86	
Pull around (minimum)..........	77 ft.lb.	77 ft.lb.	
Capacity, cu.ft. air flask Charged..............	9.16	9.16	9.16
" fuel flask, pints.....	20	20	20
" water compartments, pints...............	46	46	46
" fuel spray, seconds to deliver 6 pts. water 35 lbs. pressure...............	115 sec.	115 sec.	75 sec.

	Mk.13	Mk.13-1	Mk.13-2
Capacity, water, spray (same standard)	46 sec.	46 sec.	20 sec.
Sprays, type	Whirl	Whirl	Whirl
Fuse and pistol	Igniter	Igniter	Igniter
Turbine clearance, nozzle and rotors	.060	.060	.060
Differential ring, dia	.280	.280	.307
Nozzles, conical, throat diameter	3-".21875	3-".21875	3-".256
Nozzle working pressure, lbs gauge	395	395	458
Nozzle working temp. Max. F	1550°	1500°	(1450 av.
Exhaust temperature at exhaust valves	550°	550°	(1500 max. 584°
Turbine speed	10983	10983	13050
Gear ratios, turbines to propellers	9.566±1	9.566±1	9.566±1
Shaft H.P. developed in tank	93-98	93-98	170±2½
Forward propeller, diameter, pitch, L.H.	16"-30".0	16"-30".0	16"-30".0
After propeller, diameter, pitch, R.H.	14".38-29".5	14".38-29".5	14".38-29".5
Propeller R.P.M.	1150	1150	1170±20
Pressure, air flask test, lbs. per sq. in	4000	4000	
Pressure, air flask, working	2800	2800	2800
Pressure, regulator working lbs. per sq. in	420	420	513
Pressure, nozzle working lbs. per sq. in	395	395	458
Pressure, gyro and depth engines	420	420	
Pressure, gyro nozzle, continuous spin	125±5	125±5	
Range, acceptance yards	4500	6000	4000
Range, service yards	4500	6000	4000
Speed, acceptance and service knots	29.8±.5	33.5±.5	40
Distance head blows about, yds	5700	6700	5088
Total distance of run, yards	6400	7000	

NOTE: The data in above tables are average values. Dimensional values vary by tolerances on the drawings. Torpedo weights and buoyancy will vary plus or minus 20 lbs; capacities, flow rates of air, fuel and water, pressures and temperatures will vary in different torpedoes, but these variations are held within such limits that service and test requirements are not impaired.

C. <u>INTERCHANGEABILITY</u>.

2005. All assembled units and mechanisms are interchangeable as such, in the Mark and Modifications to which assigned; and in general all detail parts are also interchangeable except for special assembling operations such as lapping of pistons, doweling, etc.

D. **PARTS CARRYING NUMBERS.**

2006. The register number is the torpedo identifying number. All other numbers are serial numbers, which identify some part or unit of assembly. The register number of a torpedo is stamped in three places: on the air flask near the forward joint line; on the afterbody near the forward joint line; and on the tail. Serial numbers are assigned to each of the following units: the Exercise Head, the War Head, the Air Flask, the Afterbody, the Gyro and the Tail. Component units or mechanisms of the above units are also given the serial number of the unit to which they belong for assembly and identification as follows: Engine, Reducer, Nozzle Unit, Starting Gear, Oil Pump, Depth and Gyro Steering Engine.

E. **CHARACTERISTICS OF CONSTRUCTION - GENERAL DESCRIPTION.**

2007. These torpedoes are constructed in four major exterior sections, each detachable as a unit from its adjoining sections, viz: (a) The War Head or the Exercise Head; (b) The Air Flask; (c) The Afterbody; (d) The Tail.

F. **BUOYANCY (EXERCISE CONDITIONS).**

2008. The three factors affecting buoyancy are the displacement, the fixed (not expendable) weight of the torpedo, and the expendable weight. These factors are given in the table of characteristics and are as follows:

	Mk.13	Mk.13-1	Mk.13-2
Displacement (sea water 1.026 sp.gr.)	1709 lbs.	1703 lbs.	
Fixed weights, all air, all fuel, 36 pints water expended, exercise head blown, 10 lbs. water remaining in head	1454 lbs.	1432 lbs.	
Maximum expendable weights - 13, 13-1 and 13-2			
Air	130.4 lbs.	130.4	130.4
Fuel	16.8 "	16.8	16.8
Water, superheater	48. "	48.	48.
Water, ballast	319.5 "		
Total expendable weight	514.7 "		
Maximum weight, exercise run	1969 "	1947 lbs.	
Maximum buoyancy, end of exercise run	±253 lbs.	±270 lbs.	

Buoyancy is a matter of expendable weight which may be modified by salt water leakage into the afterbody. The air, water and fuel are expended in direct proportion to length of run. The water ballast is expended at end of run, serving the function of causing the torpedo to finish its run on the surface.

G. LUBRICANTS.

2009. The following lubricants are used in torpedo work and will frequently be designated only by the letter symbols in this pamphlet:

A.......... Gyro oil, blended by NTS, Newport, R. I., and obtainable only by requisition to tenders, bases or stations.

*B.......... Hot running torpedo oil, purchased under annual contract by the Bureau of Supplies and Accounts.

C.......... Light lubricating oil; forced feed and motor cylinder oil, light, Purchased under annual contract by Bureau S&A.

D.......... Compound steam cylinder oil; symbol 6135. Purchased under annual contract by the Bureau of S&A (600-W).

E.......... Petrolatum. Navy specifications 14P1. Fed. Stock Catalog #14-P-100 in 5 lb. cans and 14-P-110 for 10 lb.

F.......... Grease, mineral, lubricating (cup), medium, Fed. Spec. 14G1. Fed. Stock Cat. #14-G-1680 for 10 lb. cans.

G.......... Naval Torpedo Station Tail Packing Compound. Obtainable only by requisition to tenders, bases or stations.

*. In the event it is not possible to obtain hot running torpedo oil for use in torpedoes, the Bureau of Ordnance authorizes the use of Navy contract oils, Symbols 1100 and 3100 as substitutes therefore. Information concerning these oils is contained in N.Eng.#31 "Lubricating oil - General Information; Requirements; and Methods of test. The above oils can be used in all ball bearing engines such as the Mark 14-1, 15-1. These oils can also be used in the case of bushing engines, but without quite the same reliability, however they are dependable enough to be used as substitutes.

H. INSTRUCTIONS FOR MIXING POUR POINT DEPRESSANT WITH HOT RUNNING TORPEDO OIL FOR TORPEDOES (Santopour, Monsanto Chemical Co.)

2010. Six (6) oz. of Pour Point Depressant should be used for each five (5) gallons of hot running torpedo oil. The hot running torpedo oil must be heated to between 140-150° F, the Pour Point Depressant added and the mixture agitated until thorough blending is assured.

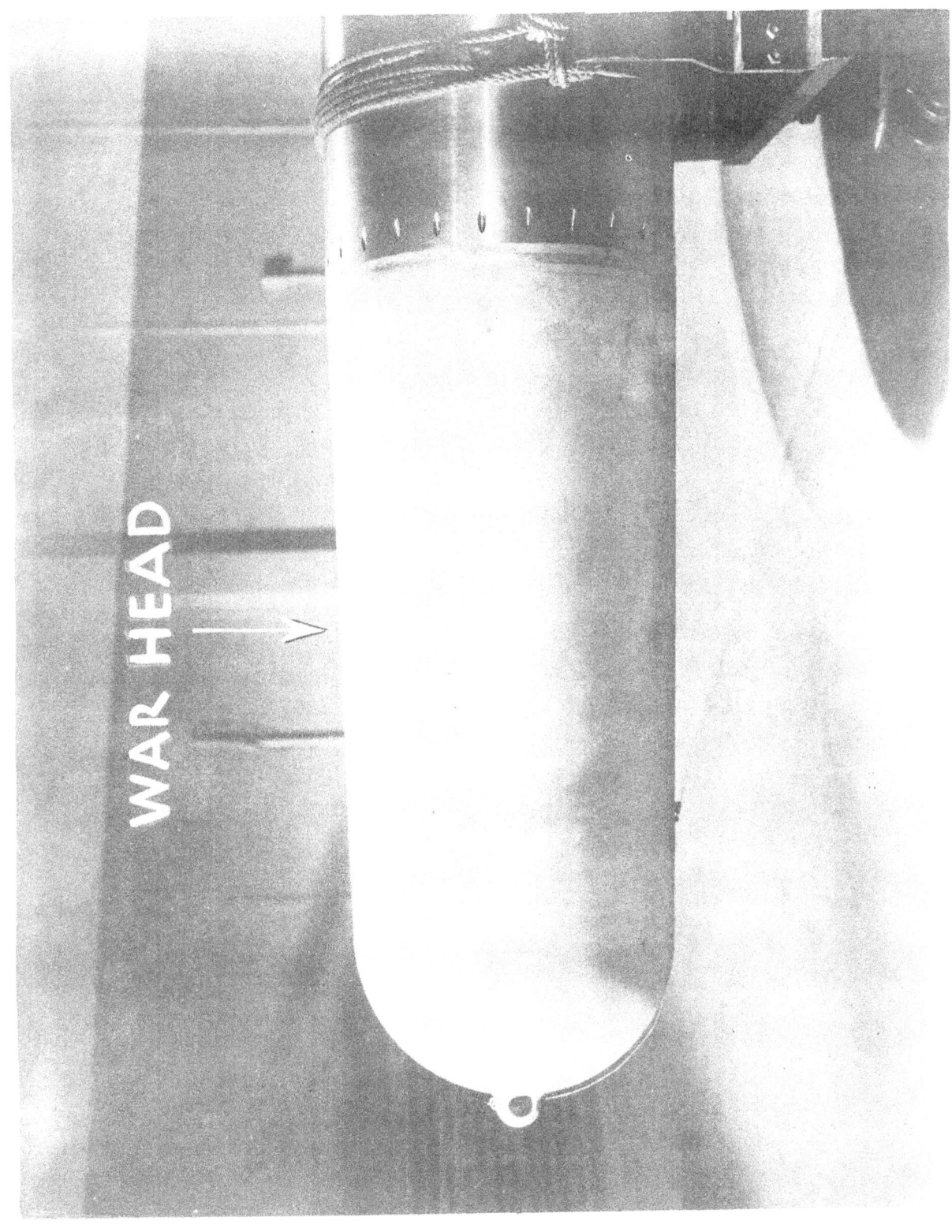

TS-5 II-6A

AIR FLASK SECTION

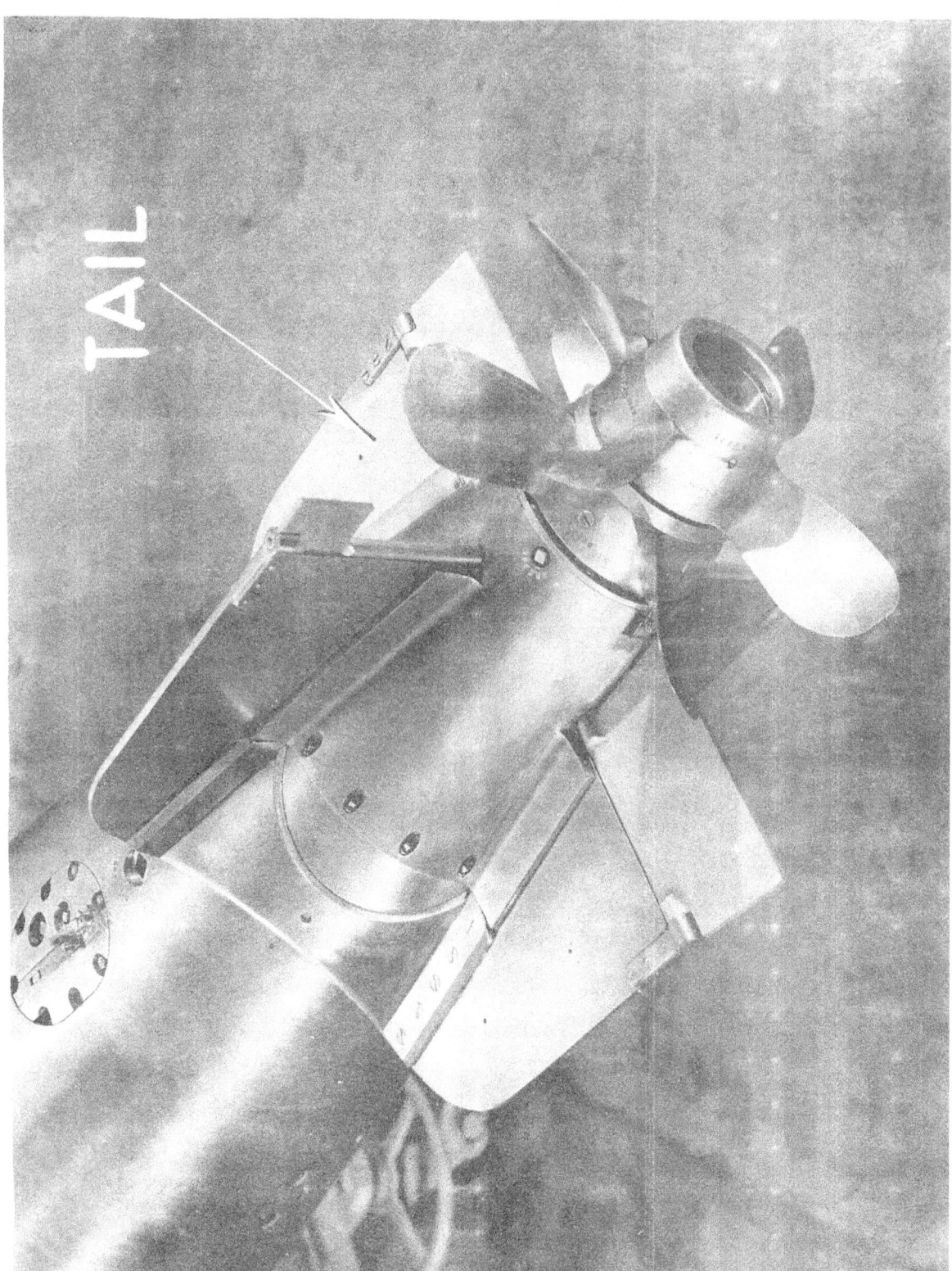

TORPEDO - TOP VIEW

TORPEDO – Left Side View

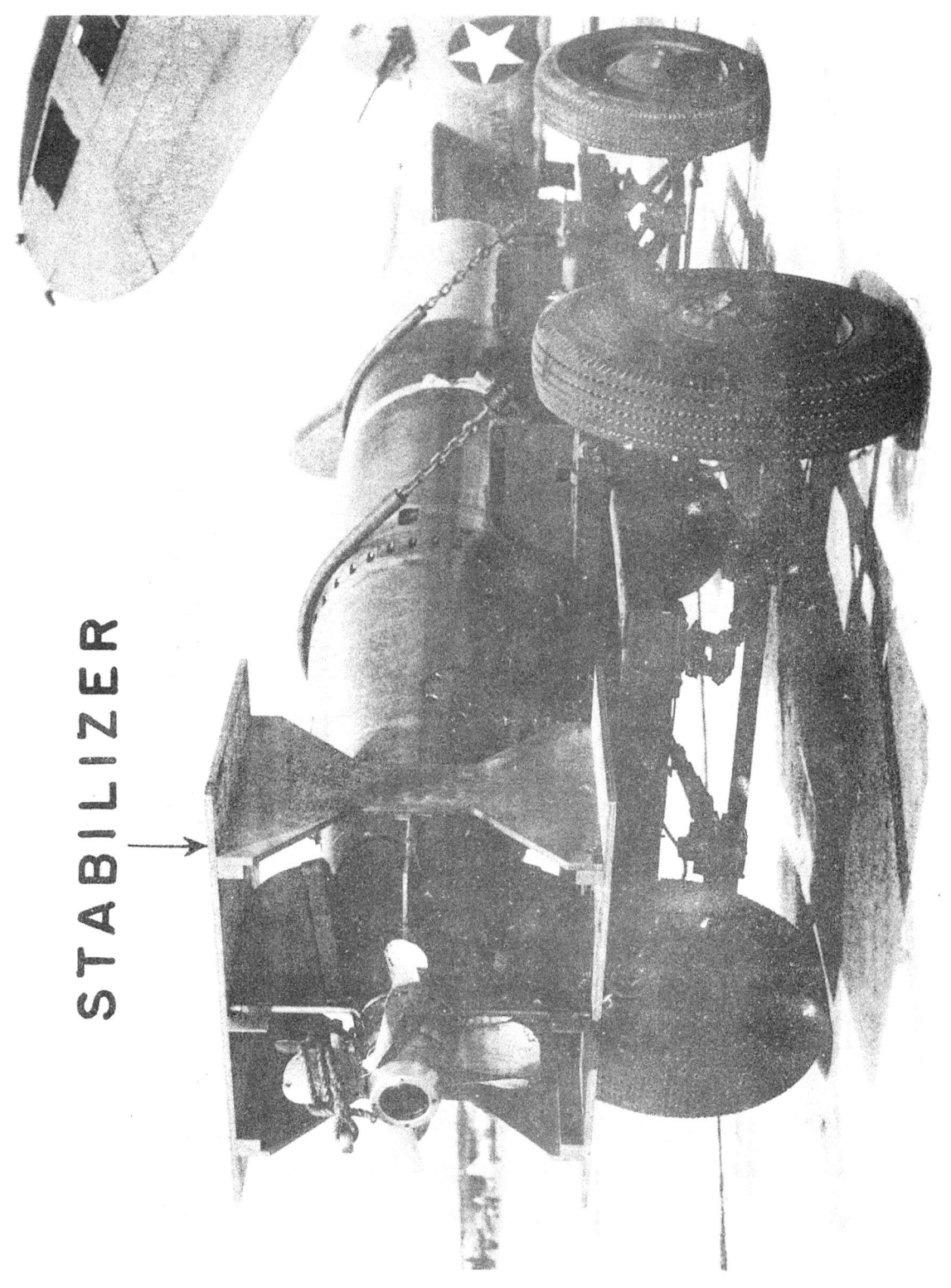

CHAPTER THREE

		ARTICLES
A.	Mark 26, 26-1, 26-2 Exercise Head.............	3001-3005
B.	Mark 2 and 3 Air Releasing Valve.............	3006-3010
C.	Water Discharge Valve........................	3011
D.	Disassembly of Exercise Head.................	3012-3020
E.	Overhaul, Tests and Assembly.................	3021-3029
F.	Notes on Exercise Heads......................	3030-3036
G.	Torpedo Headlight............................	3037-3057

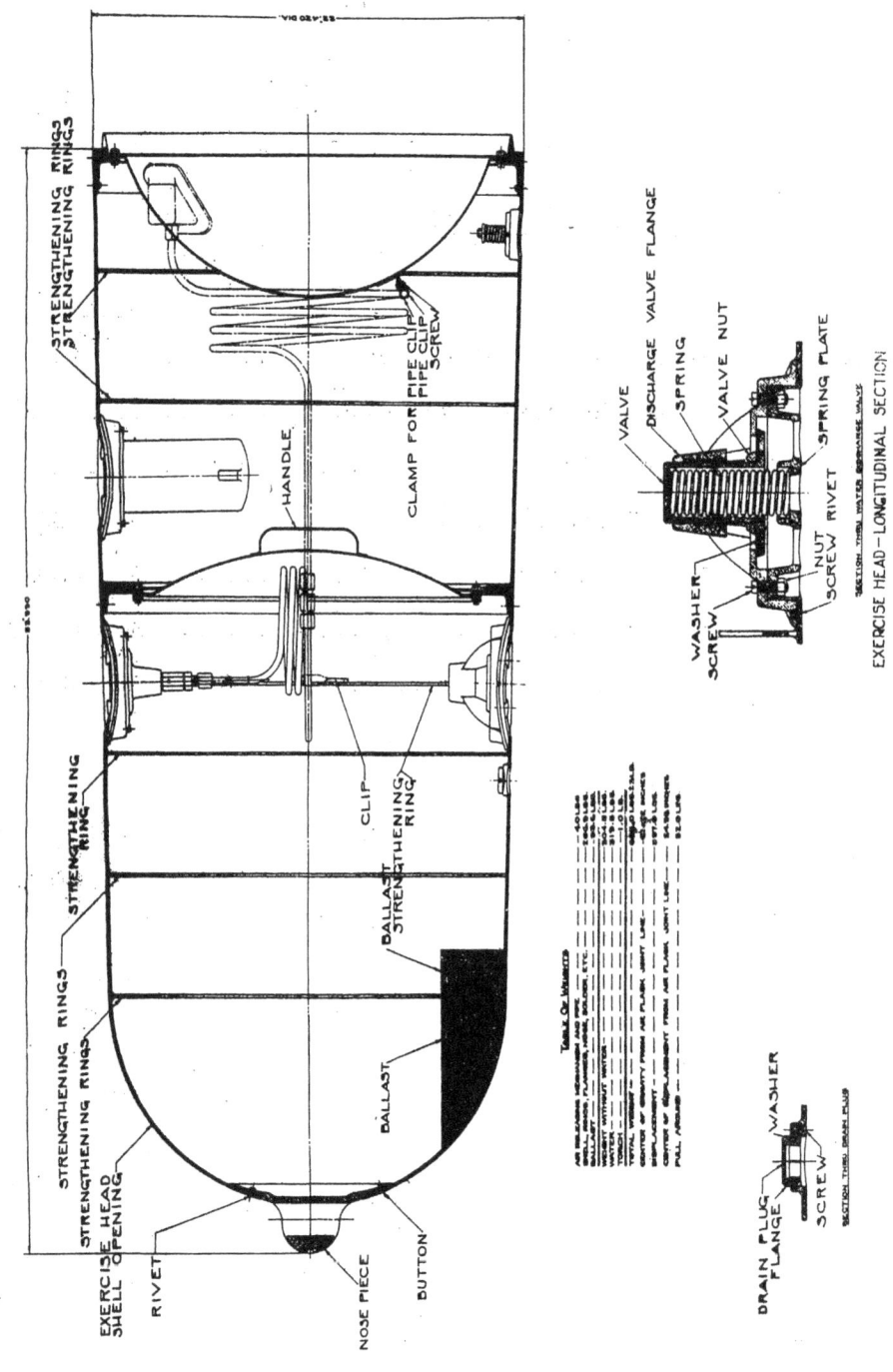

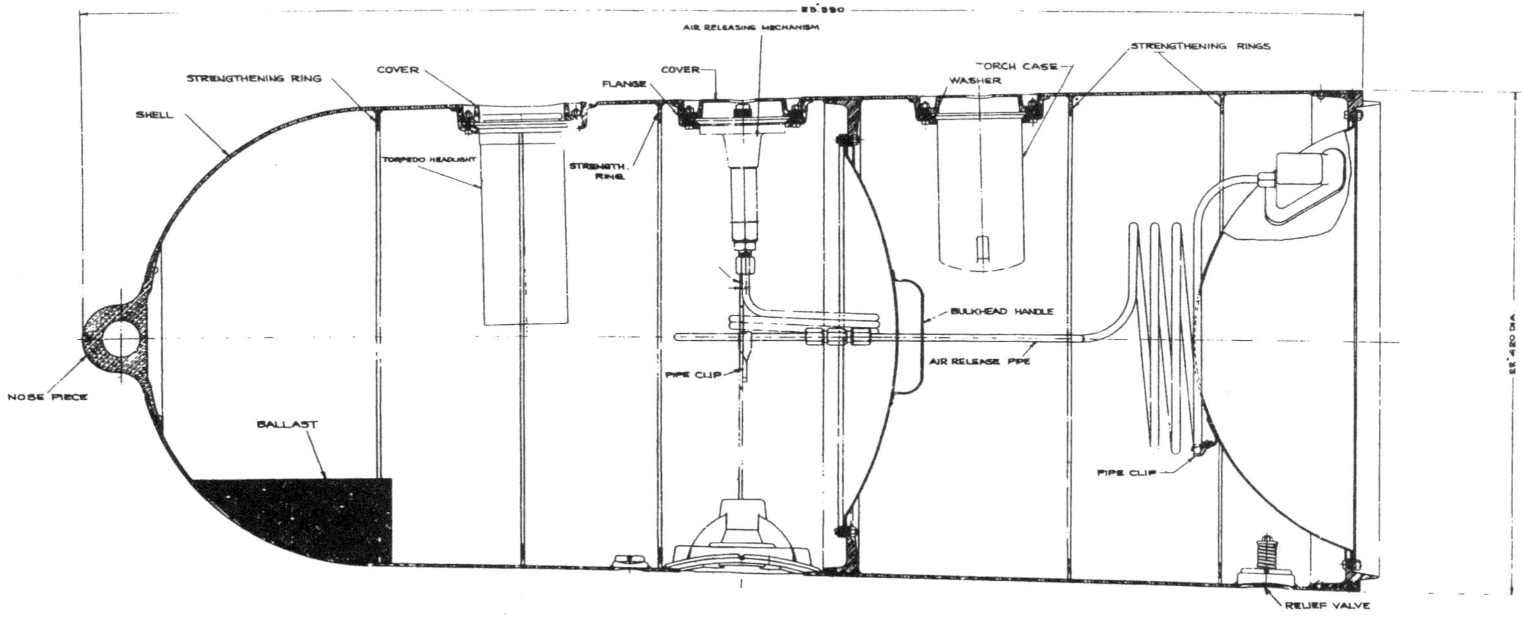

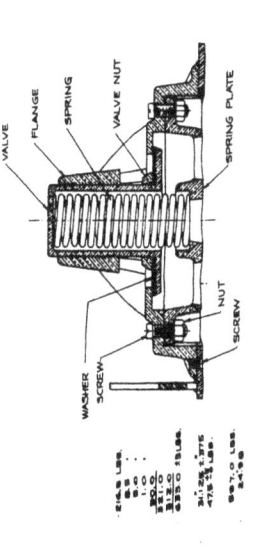

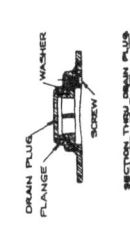

MK. 26-1 EXERCISE HEAD

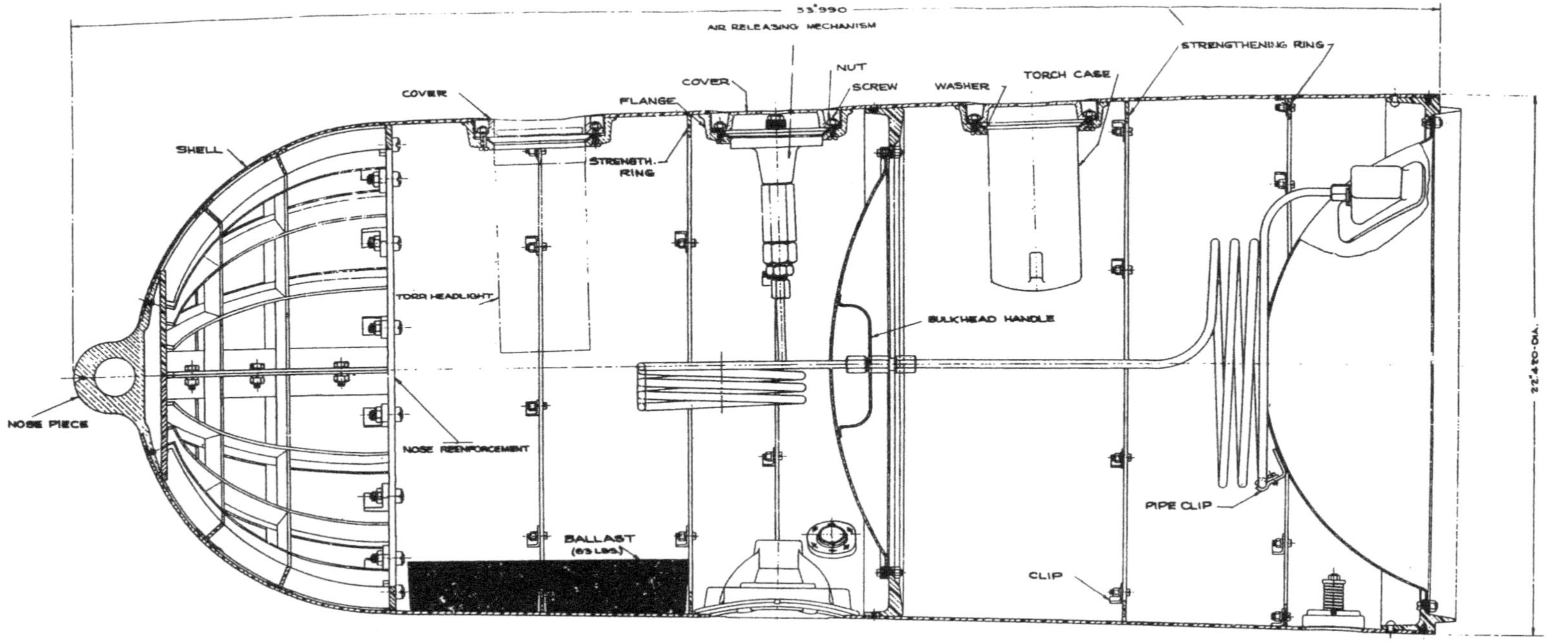

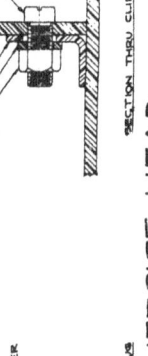

TABLE OF CHARACTERISTICS

AIR RELEASING MECHANISM AND PIPE	5.0 LBS.
SHELL, DOME REINFORCEMENT, RINGS, FLANGES, COVERS, SOLDER, ETC.	240.5 "
BALLAST FOR PULL AROUND	83.0 "
WEIGHT WITHOUT WATER	328.5 LBS.
WATER	287.0 "
TORCH	1.0 "
HEADLIGHT	8.5 "
TOTAL WEIGHT	625 ± 3 LBS
CENTER OF GRAVITY FROM JOINT LINE	31".42 ± .375
PULL AROUND	32.0 ± 2 LBS.
DISPLACEMENT	897.0 LBS.
CENTER OF DISPLACEMENT FROM JOINT LINE	24".98

SECTION THRU WATER DISCHARGE VALVE

SECTION THRU DRAIN PLUG

SECTION THRU CLIPS

MK 26-2 EXERCISE HEAD

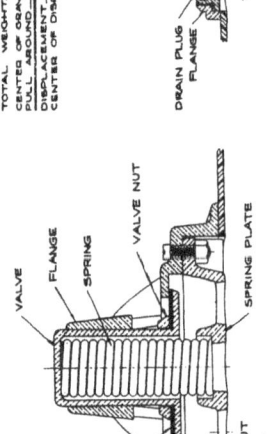

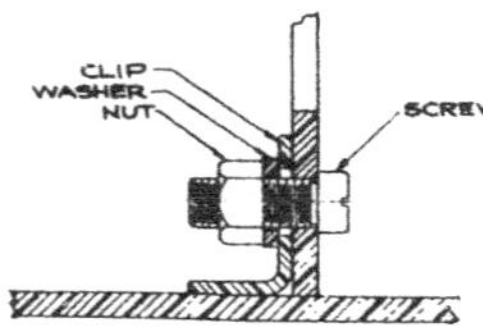

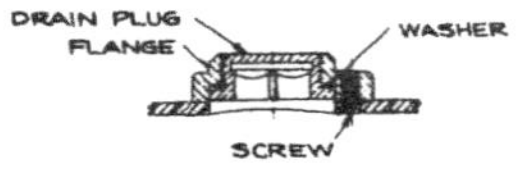

A. THE EXERCISE HEAD.

3001. The exercise head used with this torpedo is of the air blowing type and is designated Mark 26, 26-1 and 26-2. This design is fitted with an air chamber at the after end of the head and the head is considered more rugged in construction to withstand the high drop when launching from aircraft.

3002. The head shell is of steel, identical with that of the war head. The joint ring for the after bulkhead is a steel forging and the six (6) steel strengthening rings form a structural barrier which greatly increases the strength of the shell to withstand crushing or denting. In the Mark 26-2 exercise head the hemispherical portion of the forward end is reenforced by a basket shaped spider made in two sections connected together by bolts and nuts and held in place by bolting to the forward strengthening ring. The exercise heads are fitted with an air chamber at the after section of the head. This chamber is an air-tight section sealed on the after end by the exercise head bulkhead, and on the forward end by a smaller bulkhead which is seated against No. 4 strengthening ring, this ring being designed so as to withstand the strain incidental to pressure from air blow valve. The torch case and torch case flange are located on the upper centerline in this casing. Both of these bulkheads are made watertight by being seated against rubber gaskets. The air pipe from the after bulkhead to the air releasing mechanism passes through the air chamber where it is coiled for flexibility; nipples brazed on the ends of this pipe pass through holes drilled in an air inlet pad in the after bulkhead and the reenforced strengthening ring for the forward bulkhead, and are locked against their seats by lock nuts, copper washers being interposed to prevent the possibility of leakage in the air chamber. The after coil of this pipe is supported by a clamp secured to the bulkhead.

3003. The possibility of air leakage in the air pipe passing through the air chamber is a condition which must be guarded against and in order that the operator may be warned in case there be an air leak in this pipe, the drain plug is designed with a small relief valve integral therewith. This relief valve opens outboard against spring pressure and will open to relieve any pressure accumulating in the air chamber. If air escapes through the relief valve when opening the blow valve, the pipe connections are leaky, a condition which must be corrected before the head can be used.

3004. The water ballast is carried in the forward compartment of the head. The Mark 26 exercise heads have two flanges, the air releasing mechanism being located in the forward flange and connected to a pipe coiled for sufficient flexibility to permit withdrawal and installation. The inner end of the air releasing valve pipe is connected to a nipple extending through the reenforced bulkhead strengthening ring from the pipe section passing through the air chamber just forward of the inner bulkhead. An additional flange is installed in the forward portion of the Mark 26-1 and 26-2 exercise head for use with torpedo head light. In the bottom center line of the head is located one discharge valve of sufficient capacity to properly discharge the water ballast from the head. This valve, while somewhat larger, is similar in design to discharge valves installed in previous types of exercise heads.

3005. In the Mark 26 and 26-1 exercise heads a drain plug is located just forward of the water discharge valve for draining the interior of the water chamber of the head. In the Mark 26-2 exercise head this drain is located on the after side of the discharge valve and slightly to the right of the center line. In the Mark 26 and 26-1 exercise heads a lead ballast weight (90 lbs.) is located in the forward bottom end of the water chamber for stability and one ballast weight is located on each side for center of gravity. The stability ballast weight of the Mark 26 Mod. 2 exercise head is located between first and second strengthening ring in the bottom. The three heads being ballasted with this lead and water to approximate war head weights and moments.

B. MARK 2 AIR RELEASING MECHANISM.

3006. The air releasing valve is contained in the central portion of a circular body, the circumference of which is machined with an angular seat for seating in the flange provided in the exercise head. A leather gasket is interposed between these seats to make the joint water and air tight. The upper central section of this body is suitably machined for the insertion of the air releasing spring with spring support, and threaded for the adjusting nut and its lock screw. The lower end of the body is drilled and tapped to receive the valve guide which is screwed up and sweated to the body. A hole is drilled through the center of the valve guide extension and is suitably machined for a push fit bushing which carries the valve seat, and acts as a guide for the valve stem. Two radial escape ports are drilled above the valve seat for passage of air into the head. The lower end of valve guide extension is threaded for insertion on the air inlet and restriction nipple. A copper washer is interposed between the valve guide extension and restriction nipple. The air delivery pipe is attached to the restriction nipple. The air releasing valve is machined on the lower end of the valve stem. The diameter of this stem is reduced between the valve

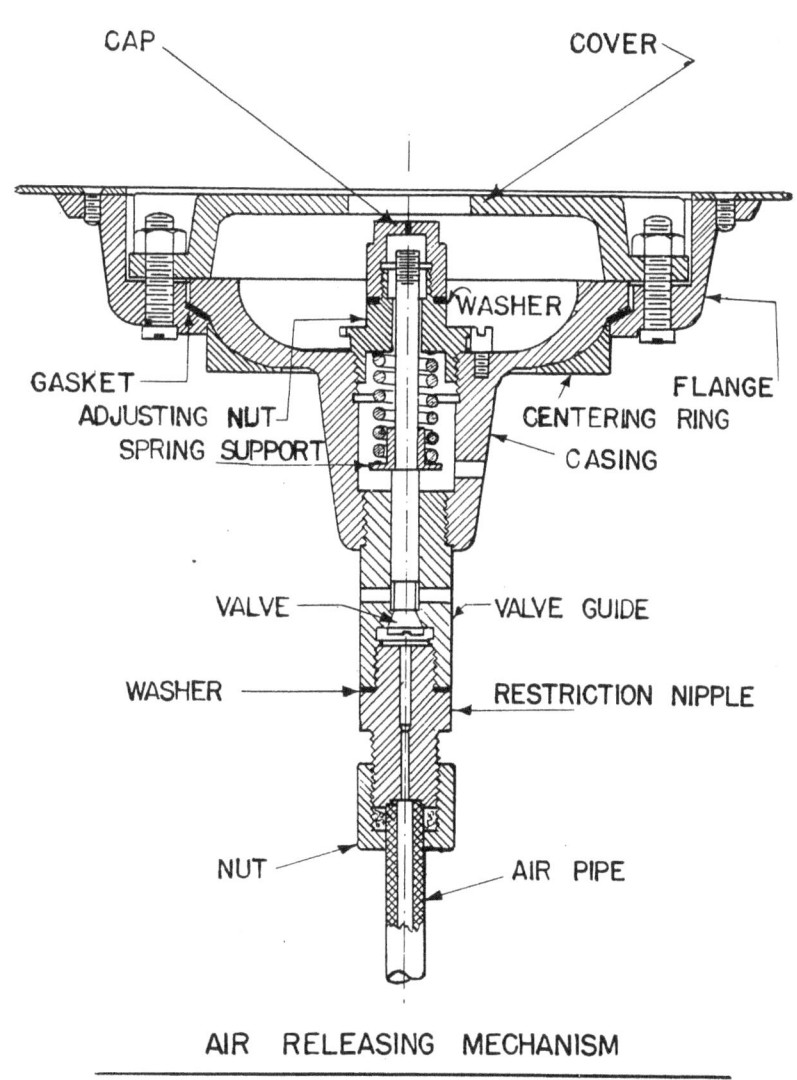

AIR RELEASING MECHANISM

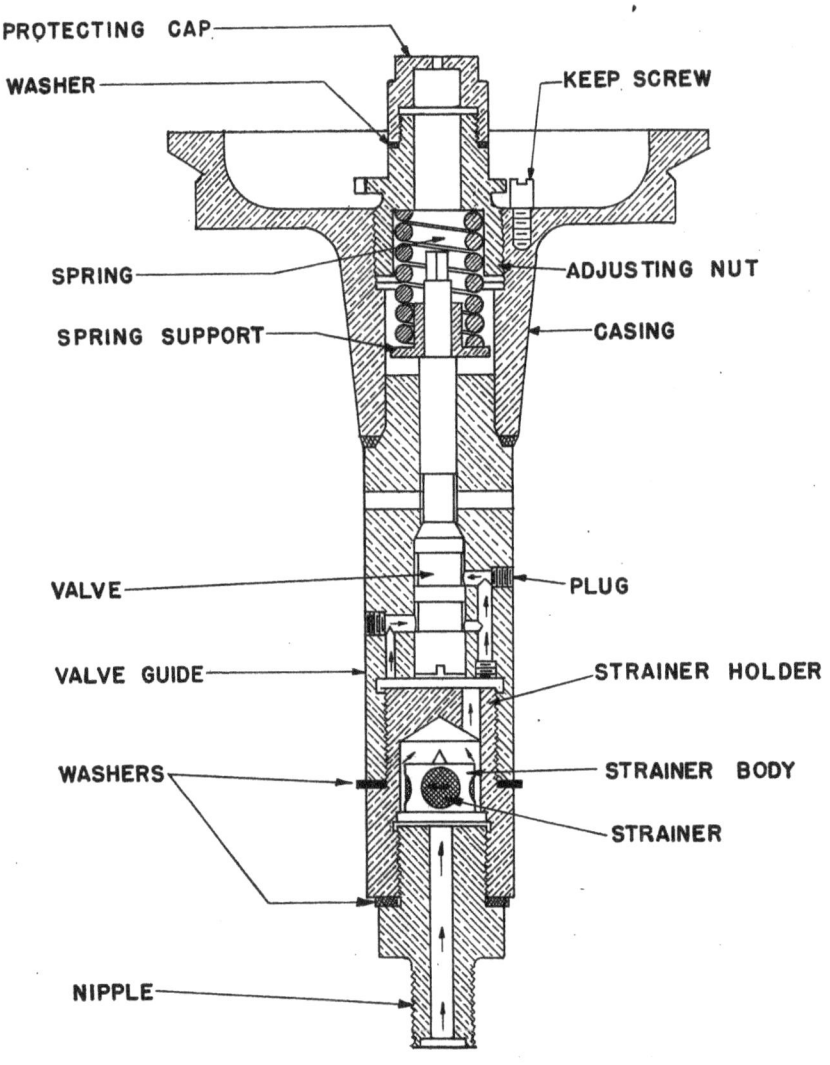

SECTION SHOWING MECANISM IN COCKED POSITION

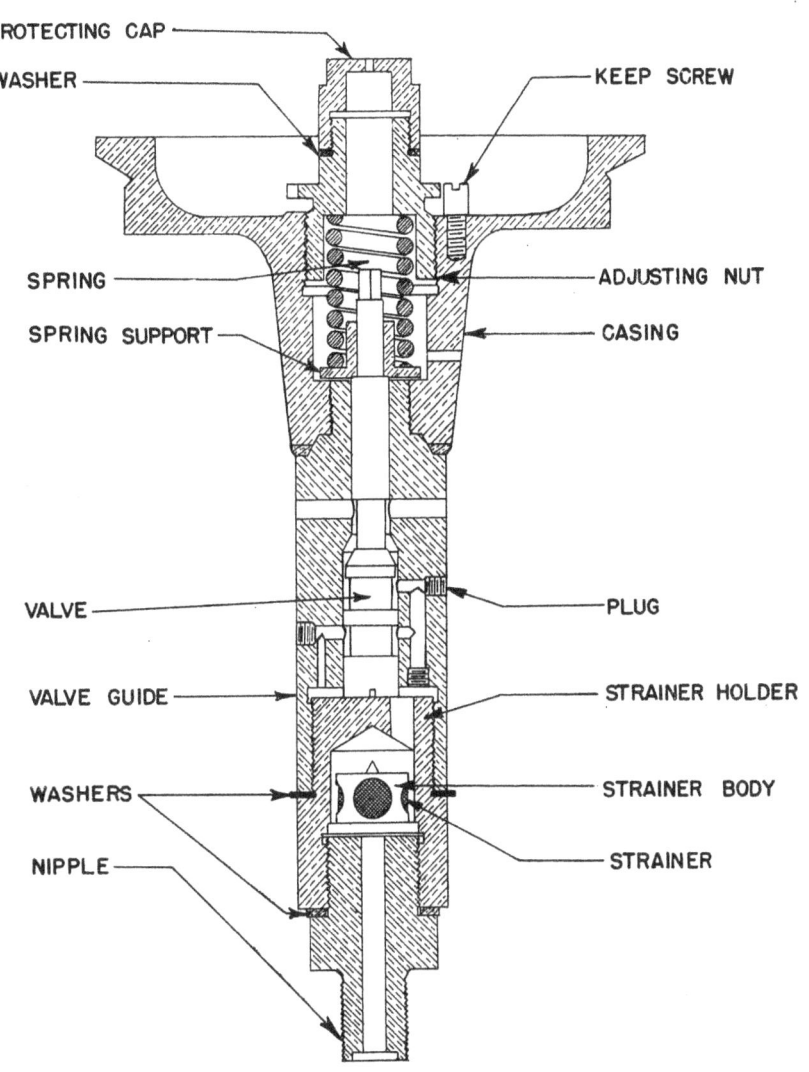

SECTION SHOWING MECHANISM IN UNCOCKED POSITION

and guide to give clearance for the passage of air into the escape ports, the diameter of the upper end being reduced to form a shoulder for the spring support. A threaded portion of the upper end of the stem permits the attachment of the cocking tool. The valve spring is inserted between the spring support and the spring seat in the adjusting nut. The spring support rests on a shoulder of the valve stem. Compression is regulated by the turning of this nut. A replacement cap, screwed up against a washer on the upper end of the adjusting nut protects the valve stem for accidental handling. A 0"020 hole is drilled through this cap to prevent any air that may leak by the valve stem from backing up.

3007. The air release mechanism is adjustable for blowing pressures of 300 to 600 lbs. It operates when the falling air flask pressure reaches the pressure for which it is set. The head then blows and expels its water ballast. The torpedo comes to the surface and finishes its run, usually on a deviating or circular course. As there is no distance gear cut off in this torpedo, the air will be completely exhausted from the air flask at the end of each run.

3008. <u>Operation</u>. The valve is first cocked by removing the adjusting nut cap, screwing the cocking tool to the top of the valve stem, and pulling up on the tool at the same time that the flask blow valve is opened. The valve is now held on its seat by flask pressure. As the flask pressure drops below that pressure for which the spring is set, the spring takes charge, forcing down on the spring support, which in turn forces the valve down and open. Air banked on the under side of the valve then forces its way upward around the valve through the restriction and out the holes in the valve casing to the inside of the exercise head, building up pressure in the head and expelling water ballast through the water discharge valve. The cut away collar on the valve stem permits the passage of air through the escape ports.

3009. A centering ring has been added to the air releasing mechanism to insure proper seating on the flange in the exercise head.

3010. The Mark 3 self-cocking air releasing valve incorporated for present and future use is contained in a circular casing with angular seat. The upper central portion of this casing is machined for the insertion of the air releasing spring and the spring support and threaded for the adjusting nut and its locking screw. The lower end of the casing is drilled and tapped to receive the valve guide which is screwed and sweated to the casing. A hole is drilled through the center of the valve guide and is suitably machined for pistons, valve and guide for the valve stem. Two radial escape ports are drilled above the valve seat for passage of air into the head. A suitable restriction and passages are drilled to allow the air to pass from the space below the lower cocking piston to the annular groove between the upper and lower piston and from there to the annular groove between the upper piston and valve. The The lower end of the valve guide is drilled and tapped to receive the strainer holder which is screwed into the valve guide. A hole is drilled through the strainer holder to the space below the lower cocking piston. The strainer is held against a machined seat in the strainer holder by an air inlet nipple, a washer being interposed between this nipple and the strainer body to prevent air leaking past. The air delivery pipe is connected to this nipple. The diameter of the valve stem is reduced between the valve and guide to give clearance for the passage of air into the escape ports. The upper end of the stem is reduced to form a shoulder for the spring support. The valve spring is inserted between the spring support attached to the valve stem above its guide and the spring seat in the adjusting nut. Compression is regulated by turning this nut. An adjusting nut cap with a vent hole to prevent banking up of air protects the valve from accidental handling.

C. WATER DISCHARGE VALVE

301. The water discharge valve is similar in design to previous types installed in exercise heads. It is larger, however, to provide sufficient capacity to properly discharge the water ballast from the head. It has a cast bronze housing, riveted and sweated to the shell. The upper part of the housing forms a guide for the valve proper and the lower flanged outboard section is surfaced for seats for both the discharge valve and spring plate. The valve is a naval brass hollow cylinder closed at the top, and with a flange machined at its lower end. A portion of the cylinder just above the flange is threaded to receive the valve nut which fits over the cylinder and screws down against the flange with a flat leather washer interposed between the two. The cylinder rides in the valve guide with the flange and washer seating upward on the machined seat in the under side of the housing. The brass spring plate with seven perforations is secured on its seat in the under side of the housing by four studs and four nuts. A bronze spring is housed within the hollow valve with its upper end bearing on the under side of the valve top and its lower end bearing in a guide on the inside of the spring plate. The spring is under compression, and holds the valve on its seat with about 8-10 lbs. pressure above the depth head. Sections of the valve housing are cut away so that water within the head may pass between the valve guide and flange through the valve opening and out through the perforations in the clamp plate when the valve is open.

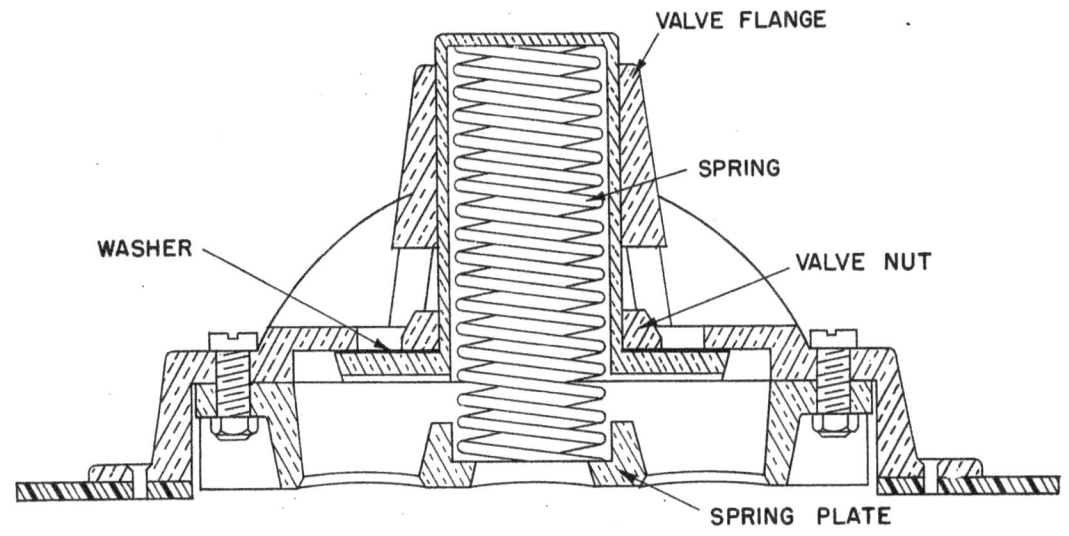

WATER DISCHARGE VALVE

D. **DISASSEMBLY OF EXERCISE HEAD.**   Tool No.

3012. **Remove head from air flask:**
   (a) Drain, sling head, and remove joint screws. 184,13-14
   (b) Push head just clear of air flask and disconnect air pipe from blow valve.......... 141A

3013. **Remove outer bulkhead:**
   (a) Unscrew clamp nut of air releasing pipe nipple................................ 141A
   (b) Remove nuts holding bulkhead. Break bulkhead joint and withdraw bulkhead sufficient to permit removal of clamp securing air releasing pipe to bulkhead................ 48
   (c) Remove clamp and push after nipple of air releasing pipe clear of bulkhead.......... 41
   (d) Remove bulkhead and gasket.

3014. **Remove torch case:**
   (a) Remove nuts, cover, case and gasket........ 48

3015. **Remove air releasing mechanism:**
   (a) Remove holding nuts and cover for releasing mechanism and lift mechanism clear of head.................................... 48
   (b) Disconnect air pipe, remove mechanism and gasket and push pipe back into head....... 141A

3016. **Remove forward bulkhead:**
   (a) Remove bulkhead nuts...................... 48
   (b) Push bulkhead forward to break joint.
   (c) Turn flats on periphery of bulkhead into line with slots in bulkhead seat and withdraw bulkhead.
   NOTE: When removing bulkhead through slot in seat, exercise great care not to damage holding screws.

3017. **Remove air releasing pipes:**
   (a) Disconnect and remove pipe (forward section)................................... 141A
   (b) Unscrew clamp nut holding air releasing pipe nipple in place on forward bulkhead ring and remove after section of pipe with nipple from head........................ 141A

3018. **Remove relief valve:**
   (a) Remove relief valve and body assembled..... 391A
   (b) Remove nut for stem, washer for spring, spring and valve with washer.............. 48

                                                            TOOL NO.
3019. <u>Remove water discharge valve</u>:
   (a) Remove holding nuts and water discharge
       valve spring plate........................ 48
   (b) Remove water discharge valve and spring.
   (c) Unscrew clamping nut for discharge valve   220,
       washer.................................... 448
       (Clamp discharge valve in tool 448 in
        vise and remove nut for washer with
        tool 220).

3020. <u>Disassemble air releasing mechanism</u>:
   (a) Unscrew restriction nipple with washer.... 18,141A
   (b) Unscrew adjusting nut cap with washer,
       adjusting keep screw and adjusting nut.... 406,407,
                                                   41
   (c) Remove air releasing valve spring, spring
       support and valve. (Removal of valve
       guide seat should not be done except when
       removal is necessary).

With the above steps completed the exercise head with
attachments is disassembled, ready for cleaning and inspection of parts.

E. <u>OVERHAUL TESTS AND ASSEMBLY</u>.

3021 Replace or resolder studs for bulkhead and
      flanges where necessary:
   (a) Apply heat to stud until stud will unscrew
       with screw driver......................... 41
   (b) Tin new stud and screw in place while hot,
       wipe excess solder off threads with rag
       and chase thread on screw with $\frac{1}{4}$" by 20
       finger die................................ 442,WE19

3022 <u>Treatment of interior of head</u>:
   (a) Clean interior of head thoroughly.
   (b) Treat exposed surfaces with a protective
       coating of bitumastic solution (allow to
       dry at least 12 hours prior to testing
       with air).

3023 <u>Lead ballast</u>:
   (a) Should necessity arise for the removal and
       re-installation of lead ballast, due to
       corrosion under ballast, or removal of
       dents in the shell, the ballast must be re-
       placed in exact location from which re-
       moved.
   (b) Shell having been cleaned and given a
       heavy coating of tin, replace lead weight
       and solder in place using soldering iron.

3024. <u>Torch case</u>:   TOOL NO.

    (a) Clean, inspect and test torch case in accordance with instructions contained in O.D. 750 (leakage through the torch case may result in the sinking of the torpedo.). Torch cases which do not satisfactorily pass a hydrostatic test of 85 lbs. per sq. inch should be scrapped and replaced.

3025. <u>Air releasing mechanism pipes</u>:
    (a) Anneal pipes and bend to shape in accordance with drawings.
    (b) Reseat seats on pipe collars if necessary.. WE86
    (c) Reseat nipple on inside of bulkhead........ WE85
    (d) Connect pipe to nipple on bulkhead and secure to supporting clip.................. 141A

3026. <u>Test exercise head for leaks.</u>
    (a) Oil (D) and replace washer on water discharge valve. Secure washer with lock nut.. 220,448
    (b) Clean, inspect and oil (C) valve guide, seat and studs in discharge valve flange.
    (c) Insert discharge valve in flange, place gag across the studs and secure with nuts, setting up until valve is held firmly on seat. 48
    (d) Replace air release pipe (after section) with nipple in forward bulkhead ring. Note that plated copper washer is in place against shoulder of nipple.
    (e) Replace clamp nut on nipple and secure, holding after end of pipe in alignment for attaching to after bulkhead................ 141A
    (f) Connect forward section of air release pipe to nipple, holding forward end of pipe in position for connecting to air release mechanism..................................... 141A
    (g) Replace after bulkhead gasket on its seat.
    (h) Note that plated copper washer is in place against the shoulder of nipple on the after end of air release pipe and guide nipple into hole in pocket on after bulkhead.
    (i) Replace after bulkhead and secure, tightening up evenly on holding nuts.............. 48
    (j) Replace and tighten clamp nut on nipple for air release pipe on after side of outer bulkhead.................................... 141A
    (k) Blank off nipple on after end of air release pipe.................................. 141A
    (l) Install air releasing mechanism blanking off plate (gauge attached) and gasket...... 48

		TOOL NO.

(m) Install torch case, cover and gasket................ 48
(n) Insert test connection in relief valve flange and connect low pressure pipe............... 391A, 141A
(o) Crack air valve and build up pressure in head to 15 lbs. per square inch. Close valve, submerge head in water and examine for leaks.
(p) If leaks occur, mark same and remedy.
(q) When satisfactory test has been made, unclamp nipple and remove after bulkhead............. 141A, 48
(r) Remove torch case to provide clearance for the installation of forward bulkhead............. 48
(s) Place a new gasket on the forward bulkhead and install the bulkhead on its seat, being careful not to damage threads on holding screws when passing through the slots in bulkhead ring. Set up evenly on nuts for holding screws..... 48
(t) Connect low pressure line to after end of air release pipe. Crack air valve and build up 15 lbs. per square inch pressure in forward (water) compartment........................ 141A
(u) Close valve, place head nose down and pour water around forward bulkhead and examine for leaks.
(v) If leaks occur, mark same and remedy.
(w) When satisfactory test has been made, wipe dry and clean after compartment inside. Reassemble torch case and after bulkhead on head, securing pipe to clip and clamp nut on nipple on after side of head........................ 48 41, 141A
(x) Remove test connection from after compartment and blanking off plate from water compartment reassemble and replace relief valve.......... 491A, 48
(y) Remove gag from water discharge valve and replace spring plate...................... 48

NOTE: Water discharge valve spring should be left out until head is to be used.

3027. <u>Air releasing mechanism</u>:

(a) If necessary to replace air releasing valve guide and seat, unscrew old valve guide, applying heat around threads. Screw and solder new valve guide and seat into body. Reseat valve seat and lap valve to seat (using cocking tool for turning valve when lapping into guide).... 41, 441. WE24 18, 141A
(b) When finished lapping, wash parts thoroughly in spirits, oil (C) and assemble.
(c) Check size of hole in restriction nipple. Should be ".0625 (use drill #52).
(d) Screw nipple into valve guide and seat against a copper washer........................ 18, 141A

                                                                    TOOL NO.
   (e) Oil (C) and replace spring support around valve
       stem, replace spring, and screw adjusting nut
       into body over spring........................... 406
   (f) Connect mechanism to air pipe from test panel
       or test set..................................... 141A
   (g) Hold valve on seat by pulling out on cocking
       tool, turn air on and test for leaks by sub-
       merging in water with valve seated. Pressure
       release tests should be made in accordance with
       procedure outlined in O.D. 750. Mechanism should
       be set to release at 500 lbs. pressure.......... 441
   (h) With the above tests completed and desired re-
       lease pressure obtained, insert lock screw for
       adjusting nut................................... 41
   (i) See that ".020 hole in top of adjusting nut
       cap is clear. If this hole is clogged a small
       air leak in the head may cause a premature
       blowing of the head.
   (j) Screw cap in place with washer under seat, thus
       completing assembling of air releasing mechan-
       ism... 407

3028. **Air releasing mechanism Mk. 3.**

   (a) Reseat valve seat, and lap valve to seat....... 41,WE236
       NOTE: If necessary to lap in a new valve lap
             hole in body with Lap Tool WE 237 and
             new valve with Lap Tool WE 238.
   (b) When finished lapping, wash parts thoroughly
       in spirits, oil (C) before assembling.
   (c) Blow out restriction hole in valve guide with
       air, and check size using a #52 drill. Blow
       out air channels with air.
   (d) Oil (A) and assemble valve in guide. Replace
       spring support over valve stem and insert spring.
   (e) Replace spring adjusting nut................... 406
   (f) Clean and replace strainer body with washer
       in valve guide.................................. 451
   (g) Clean and replace strainer in valve guide.
   (h) Remove burr from threads if necessary and re-
       place air inlet (restriction) nipple in strain-
       er body... 229
   (i) Connect mechanism to air pipe from test panel
       or test set.(BuOrd Dwg.No.44322,79646).......... 141A
   (j) Turn on air and test for leaks by submerging
       in water.
   (k) With the valve tight, set valve to release for
       pressure required in accordance with procedure
       outlined in O.D. #750........................... 406
   (l) With the above test completed and desired re-
       lease pressure obtained, insert lock screw for
       adjusting nut................................... 41

TOOL NO.

   (m) See that #020 hole on top of adjusting nut cap is clear. If this hole is clogged, a small air leak by the valve may cause premature blowing of the head.

   (n) Screw cap in place with washer under seat, thus completing assembly of Mark 3 air releasing mechanism.................................... 407

3029. <u>Install air releasing mechanism in head</u>:

   (a) Put washer in place on beveled flange. Pull out air releasing pipe connection through pocket and connect to mechanism............. 141A

   (b) Drop mechanism down to its seat. Replace cover and secure with holding nuts........... 48

The above steps complete overhaul, assembly and test of exercise head up to assembly on air flask.

F. NOTES ON EXERCISE HEADS.

3030. <u>Insure proper operation of the air release mechanism.</u> Immediately prior to firing this mechanism should be thoroughly overhauled, cleaned, lubricated and tested on board a tender or firing ship, under the direct observation of firing ship personnel. Special attention should be given to condition of spring and valve, since a broken or weak spring or a bent valve may cause failure of the mechanism. The hole in the restriction nipple should be examined to insure that it is not obstructed. Before installing the mechanism, the air pipe from the air flask must be thoroughly blown out. Several pressure release tests should be made prior to connecting the release mechanism to its air pipe on the torpedo.

3031. <u>Insure that water discharge valve functions properly.</u> This valve should be thoroughly cleaned, overhauled and lubricated prior to firing. The discharge valve washer requires special attention. Its periphery is not secured and consequently the washer may readily become so folded, wrinkled or otherwise distorted that it will not seat tightly. The valve body must slide freely in its flange, the working surfaces of both being free from corrosion, grit or other foreign matter which might prevent such free operation. The valve spring should be examined for weak or broken condition.

3032. <u>The torch case must be properly seated.</u> This fitting has a bevel seat similar to that of the air releasing mechanism and may therefore be leaky due to an improperly seated gasket. Furthermore, since the case is not always removed in preparation for firing, the gasket may become so deteriorated and defective that the pressure during head blow will cause its failure.

3033. The torch case itself must be in good condition and should be tested by hydrostatic pressure at least once each year.

3034. <u>The bulkheads must be seated properly</u>. The shock of launching or the internal pressure during head blow may loosen an improperly secured bulkhead which has appeared tight prior to firing. The gasket should be in good condition and properly located. All loose studs should be resoldered. All nuts must be set uniformly and tightly.

3035. <u>Insure that flask blow valve is completely open</u>. A partial opening of this valve may be sufficient to hold the air release mechanism valve on its seat prior to firing but insufficient to blow the head at the end of a run before the torpedo has sunk to a crushing depth.

3036. <u>After firing care</u>:

    (a) Completely clean and dry the interior of the head, remove rust spots; coat unplated spots with oil.
    (b) Disassemble and overhaul air release mechanism.
    (c) Disassemble and overhaul water discharge valve.
        <u>NOTE</u>: Spring should be left out of head until head is to be used again.
    (d) Inspect ballast weight to insure that it is properly secure.

G. <u>TORPEDO HEADLIGHT</u>.

3037. The torpedo headlight is designed for use in night torpedo practices. It is carried in a sealed container in the forward torch case flange of the exercise head.

3038. The torpedo headlight consists of a cylindrical brass case $9\frac{1}{4}$" long and $3\frac{1}{2}$" in diameter. This case contains the light socket with light, reflector, lens, switch and switch operating mechanism in the upper section, and 8 flashlight dry cells in the lower section. The weight of the headlight assembled, complete, is $8\frac{1}{4}$ pounds.

3039. The upper section (headlight body) is suitably machined to fit the flange in the exercise head. The outer end of this body is closed by a $\frac{1}{4}$" lens (headlight lens) made watertight by 2 rubber gaskets, and secured to the body by a retainer ring. A highly polished reflector, secured by 3 holding clips directly under the lens, reflects the light upward through the water. An elongated slot is machined near the lower end of the headlight body for access to manual operation of the headlight switch.

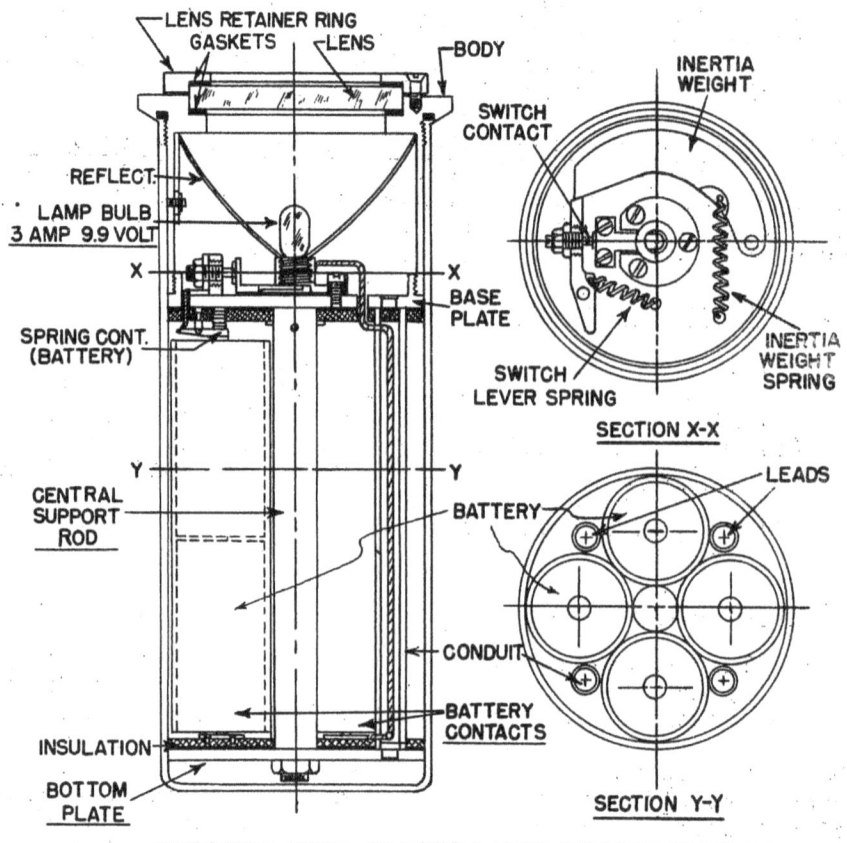

·· TORPEDO HEAD TRACER LIGHT FOR NIGHT FIRING ··

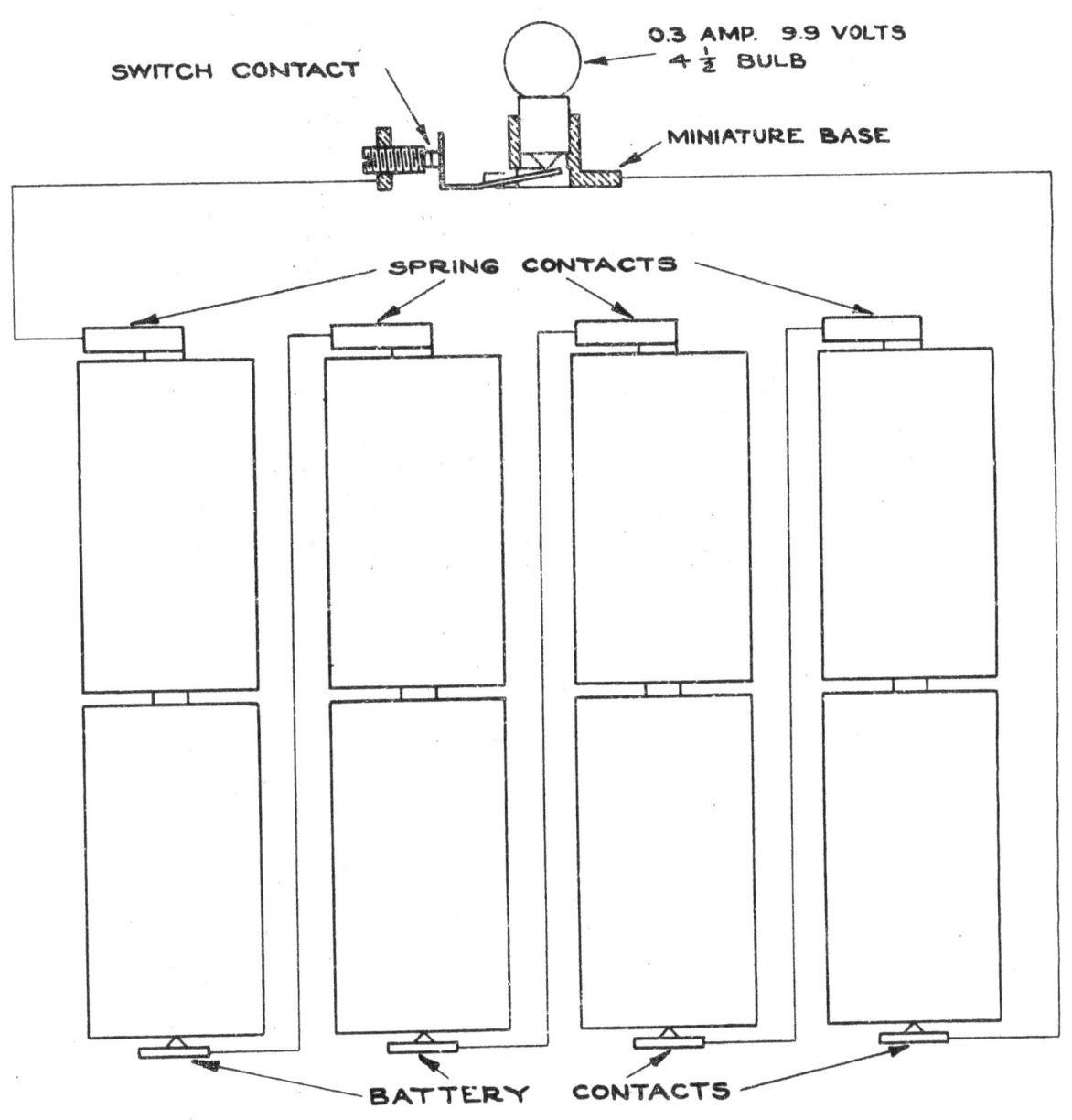

3040. The lower end of the headlight body is closed by a brass base plate, on which is assembled the light socket base, switch and switch operating mechanism. The light socket base is secured in the center of this base plate against a fibre insulating washer by three screws, which pass through insulating bushings.

3041. The switch is composed of a contact extending from the central terminal of the light socket base and a similar insulated contact secured to the free end of a lever (switch lever) pivoted on the base plate. A tension spring between this lever, and a poppet on the base plate, tends to keep the switch contacts closed.

3042. The switch is operated by a brass inertia weight, one end of which is pivoted on the base plate. The free end, in contact with the free end of the switch lever, keeps the switch open until actuated by the inertia of the torpedo upon launching. At that time the inertia weight is swung clear of the switch lever, and the spring closes the switch, thereby lighting the headlight. A tension spring, connected between the inertia weight and a poppet on the base plate, tends to keep the inertia weight in contact with the switch lever.

3043. The lower section is supported from the base plate by a central rod to which is attached a bottom plate. Four brass conduit pipes are equally disposed near the outer circumferences between the base plate and the bottom plate. These pipes are located by small buttons riveted to the base plate and bottom plate. Holes are drilled near the top and bottom of the pipes through which the battery connecting wires are conducted.

3044. The dry cells are of the commercial flash light type and are mounted in pairs in spaces between the conduit pipes. The positive terminals are placed upward against the insulated spring battery contacts pivoted in forked bearings attached to the upper insulating disc while the battery casings (negatives) are in contact with the insulated bottom battery contacts attached to the lower insulating disc. Insulating washers are interposed between the upper insulating disc and the base plate and the lower insulating disc and bottom head to insulate rivet heads and springs against contact with these parts. Wires are led from the bottom battery contacts through the conduit pipes to the battery spring contacts, switch and lamp socket terminals.

3045. Flashlight batteries are issued in sets of 3 cells. It will be necessary to remove one cell from each set before installation in headlight.

3046. The headlight assembly complete, including batteries, is enclosed in a brass casing, the upper end of which is threaded to fit similar threads on the headlight body. The joint between the body and casing is made watertight with a leather gasket.

3047. Special covers with the central hole enlarged to uncover the lens are furnished for securing the headlight in the torch case flange. The torch case flange being fitted with studs and nuts and using covers instead of follower nuts.

3048. Miniature Mazda Lamps .3 amp., 9.9 volts, are used in the headlight. These bulbs are furnished with the headlight when issued. Spare bulbs are furnished to tenders.

CARE AND HANDLING OF TORPEDO HEADLIGHT.

3049. Water-tightness is very necessary. The points to check for leakage are:
    (a) Through seat of headlight lens.
    (b) Through seat of headlight casing.

3050. It is important that the two rubber gaskets for the headlight lens are in good condition and that the screws for the lens retainer ring are set up even and tight. The headlight casing must be set up tight against its leather gasket before installation in the exercise head.

3051. <u>IT IS IMPORTANT THAT THE HEADLIGHT BE INSTALLED WITH SIDE MARKED 'AFT' ON THE AFTER SIDE, OTHERWISE THE INERTIA WEIGHT WILL NOT CLOSE THE SWITCH AND THE HEADLIGHT WILL NOT LIGHT.</u>

3052. When loading a torpedo, fitted with the headlight in the tube, care should be exercised not to permit the torpedo to strike the tube stop with sufficient force to close the light switch.

3053. To open the switch, remove casing, insert a finger in the elongated slot on the side of headlight body, and push the flat end of the switch lever back until the end of the inertia weight engages the end of this lever.

3054. To remove a headlight lamp bulb, it will be necessary to unscrew the casing and the headlight body, after which the bulb may be removed from the socket.

    To disassemble the headlight mechanism, proceed as follows:

    (a) Remove retainer ring, lens and washers.
    (b) Unscrew and remove headlight casing.

    (c) Remove batteries (apply force upward against spring contacts and lift out batteries lower end first).

    (d) Unscrew and remove headlight body from base plate. This will expose the entire mechanism for cleaning and repair if necessary.

    (e) To remove reflector from the headlight body, remove 3 screws and nuts, retainer clips and reflector will then drop out.

    (f) Wiring, contactors, switch mechanism, conduits, insulation and bottom plate should not be disassembled unless necessary through breakage of insulation, wiring or contactors, in which case it will be necessary to unsolder the wires from the spring contacts to the switch and unscrew nut holding the bottom plate to the central rod after which the bottom plate with conduits, pipes and wiring may be removed.

3055. To assemble, reverse procedure outlined in paragraph 3054

3056 When new batteries are installed, the switch mechanism should be tested by striking the side of the headlight marked 'aft' with the hand to see if lamp lights, after which switch lever is reset to off position by pressing with finger.

3057. The headlight is issued with batteries installed. A square piece of fibre is inserted between contact points on the switch to prevent possible contact during transportation. THIS FIBRE MUST BE REMOVED PRIOR TO INSTALLATION IN THE EXERCISE HEAD.

CHAPTER FOUR

		ARTICLES
A.	War Head Mark 13.....................................	4001-4007
B.	Mark 4-1 Exploder, Description...................	4008-4026
C.	Mark 4-1 Exploder, Operation.....................	4027-4030
D.	Disassembly of Mark 4-1 Exploder...............	4031-4032
E.	Overhaul and Assembly of Exploder..............	4033
F.	Notes on Mark 4-1 Exploder Firing Mechanism....	4034
G.	Tests of Mark 4-1 Exploder.......................	4035-4036
H.	The Anti-Counter Mining Device.................	4037
I.	Tests of the Anti-Countermining Device.........	4038-4039
J.	War Head Booster Mark 2...........................	4040-4044
K.	Detonator Mark 7.....................................	4045-4054
L.	Preparation of War Head for War Shot...........	4055
M.	Notes on War Heads...................................	4056-4065

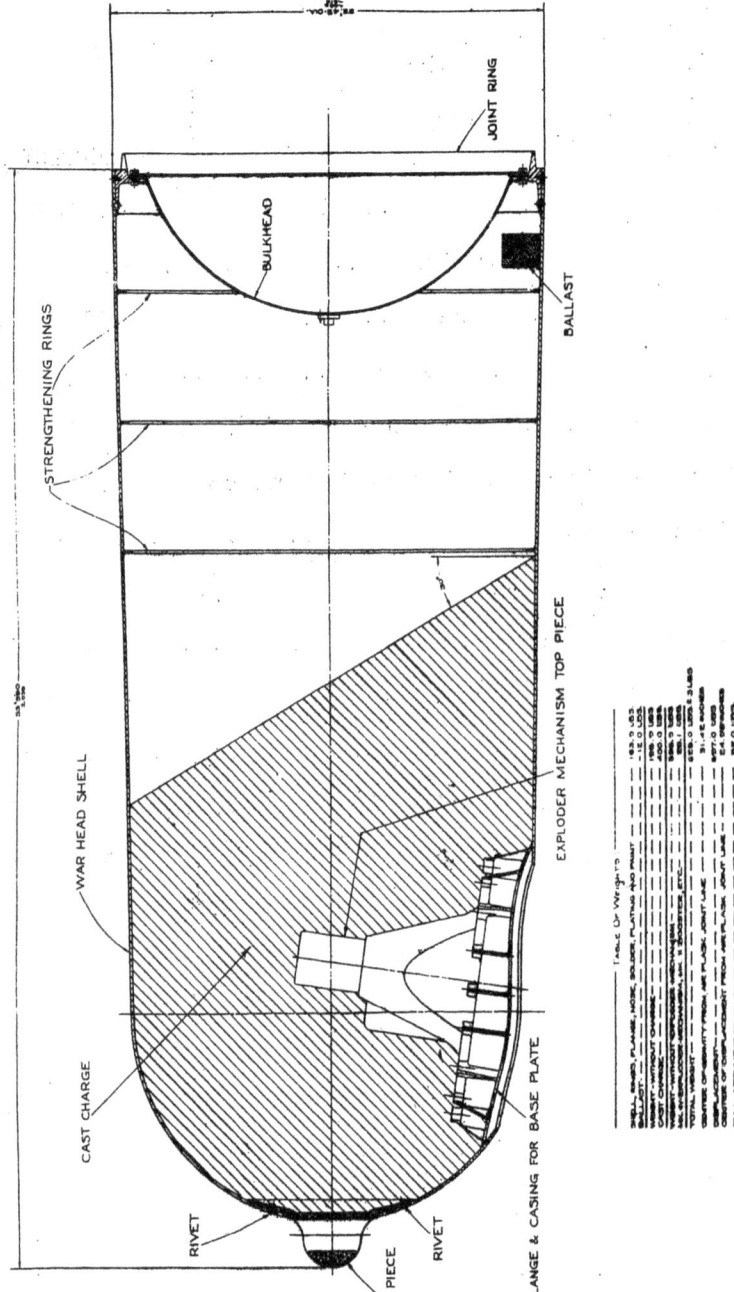

WAR HEAD

A. **WAR HEAD.**

4001. The war head carries the explosive charge and the firing device; it is the prime function of a torpedo to carry this charge to and explode it against an enemy vessel.

4002. The war head for the Mark 13 and 13-1 torpedoes is designated Mark 13. The shell is made of sheet steel, slightly conical in form at its after portion and hemispherical in form at its forward end. A steel joint ring is riveted and soldered to its after end, this ring being tapped to receive the joint screws for connecting the head to the air flask. The joint ring is also flanged to receive a steel bulkhead, dished in form, which is bolted to it and by means of a rubber gasket forms a watertight closure for the after end. A fitting for attaching an air pressure testing line is secured to the bulkhead. At the forward end is riveted and sweated a bronze nose piece. The after end of the head is reinforced with three strengthening rings.

4003. The shell contains 400 lbs. of cast T.N.T. which is loaded in the molten state with the head nosed down at an angle of 30 degrees. In this way when the T.N.T. solidifies the center of gravity of the cast charge is below the fore and aft axis. A lead ballast weight (12 lbs.) is soldered on the bottom center line, four inches forward of the joint ring for giving the torpedo its specified stability. The head is fitted to receive a Mark 4-1 exploder. A large flange is riveted and sweated on the bottom of the shell to which the exploder casing is attached. Extending from the top of this casing is a pocket for the Mark 2 tetryl booster. The bottom of the casing is closed by a large base plate on which the exploder mechanism is mounted.

4004. The joints of all parts attached to the shell are made watertight by soldering. The shell is cadmium plated inside and out to prevent corrosion. The inside of the shell is painted with projectile cavity paint or lacquer to separate the cast TNT from contact with the lead contained in the ballast and in the solder. This is a safeguard against the formation of lead picrate.

4005. All parts of the warhead and attachments are designed and tested to withstand an external pressure of 135 lbs. per square inch. Each empty war head is subjected to an internal pressure of 5 lbs. per square inch.

NOTE: The joint screws for the war heads and exercise heads are not interchangeable with the joint screws for the afterbody and tail, the former having 18 threads to the inch and the latter 24 threads to the inch. In order that no attempt will be made to use afterbody joint screws for the heads with consequent stripping of the threads, a slot is milled across the squared ends of the head joint screws. It is important that only screws so slotted be used when installing heads on air flasks in order to prevent serious damage to threaded joint screw holes in the bulkhead joint ring.

4006. TNT is an extremely stable, but when detonated, a very very powerful explosive. Its detonation is accomplished by the detonation of a small amount of tetryl, known as the booster, which is carried on the upper part of the exploder mechanism inside the exploder casing. The tetryl booster is located approximately at the center of the TNT charge. The sequence of explosions is initiated when the firing pins of the exploder strike the caps of the fulminate of mercury detonator. Fulminate of mercury is a very unstable explosive compound, requiring the utmost care in transportation and handling. It is desirable to use only the smallest amount of fulminate of mercury to produce the desired result. Between the fulminate and TNT stands the comparatively safe tetryl booster, which can be detonated by a small amount of the fulminate and has sufficient power to detonate the TNT.

4007. Except when the exploder is in the armed position, the fulminate detonator is withdrawn from the booster into a safety chamber, where it can explode without setting off the booster. For storage purposes, the tetryl booster and the fulminate detonator, together with the safety chamber, are removed from the war head and the exploder, and are stored in separate boxes.

B. THE MARK 4, MOD. 1 EXPLODER.

4008. In the Mark 4-1 exploder mechanism there are three features to be borne in mind:
    (a) The arming mechanism.
    (b) The firing mechanism.
    (c) The anti-countermining device.

4009. The function of the arming mechanism is to rearrange the component parts of the exploder mechanism so that the safety features are inoperative and the firing mechanism will be in a position to operate.

4010. The arming consists of:
    (a) Raising the safety lugs so that the two balls holding the firing pin body drop out and no longer prevent operation of the firing mechanism.
    (b) Placing compression on the firing spring.
    (c) Moving the detonator out of the safety chamber into the recess of the booster.

4011. The firing consists of:
  (a) Displacement of firing ring due to its own inertia upon impact of the torpedo with some outside object.
  (b) Resulting movement of trigger plate which raises the trigger cap and allows the two balls holding the firing pin body to go into a recess and release the firing mechanism.

4011A. Anti-Countermining Feature.
  (a) The function of this device is to prevent the firing of the war head, due to countermining or other nearby explosion, before it reaches its target. This is accomplished by interposing between the trigger cap and the underside of the top plate, a locking rack, which prevents the trigger cap from moving.

4012. The exploder mechanism is mounted on the base plate of the war head, which is a bronze casting shaped to fit into place in the casing, to which it is secured against a gasket with 18 joint screws. A channel and chamber is cast in the base plate for the impeller, the recess for the anti-countermining device, and the bosses for the exploder case, upon which the mechanism is built up. A guard is fitted in the impeller channel; it protects the blades of the impeller and helps to direct the flow of water on to them.

4013. The impeller is a die casting of brass, having fifteen buckets on its periphery and a square central hole for its shaft which runs in bearings in the base plate. A centrally drilled hole from the outer end of the shaft, with lateral holes at the bearings, provides for lubrication. A nut screwed into the base plate holds the impeller shaft in place. The inner end has a hexagonal hole broached in it to receive one end of the worm shaft. A gland prevents leakage. The other end of the worm shaft is carried in a ball bearing in the bearing bracket, which is held in place on the base plate by two screws and two dowel pins. A retainer keeps the ball race in place. The worm pinned to the worm shaft, drives the exploder mechanism through a vertical shaft.

4014. The arming and firing elements of the exploder mechanism are carried on a base of naval brass forging. Five projections, two for dowel pins and three for screws, match the bosses cast on the base plate of war head. The exploder base has four supports raised on its upper surface, suitably machined with slots for holding case hardened steel fingers. These supports are also tapped for the lower ends of four screws which support the top plate. The arming and firing elements are assembled on the frame thus formed.

4015. The top plate is suitably shaped and dimensioned to carry the assembly of parts of the gear train, and arming

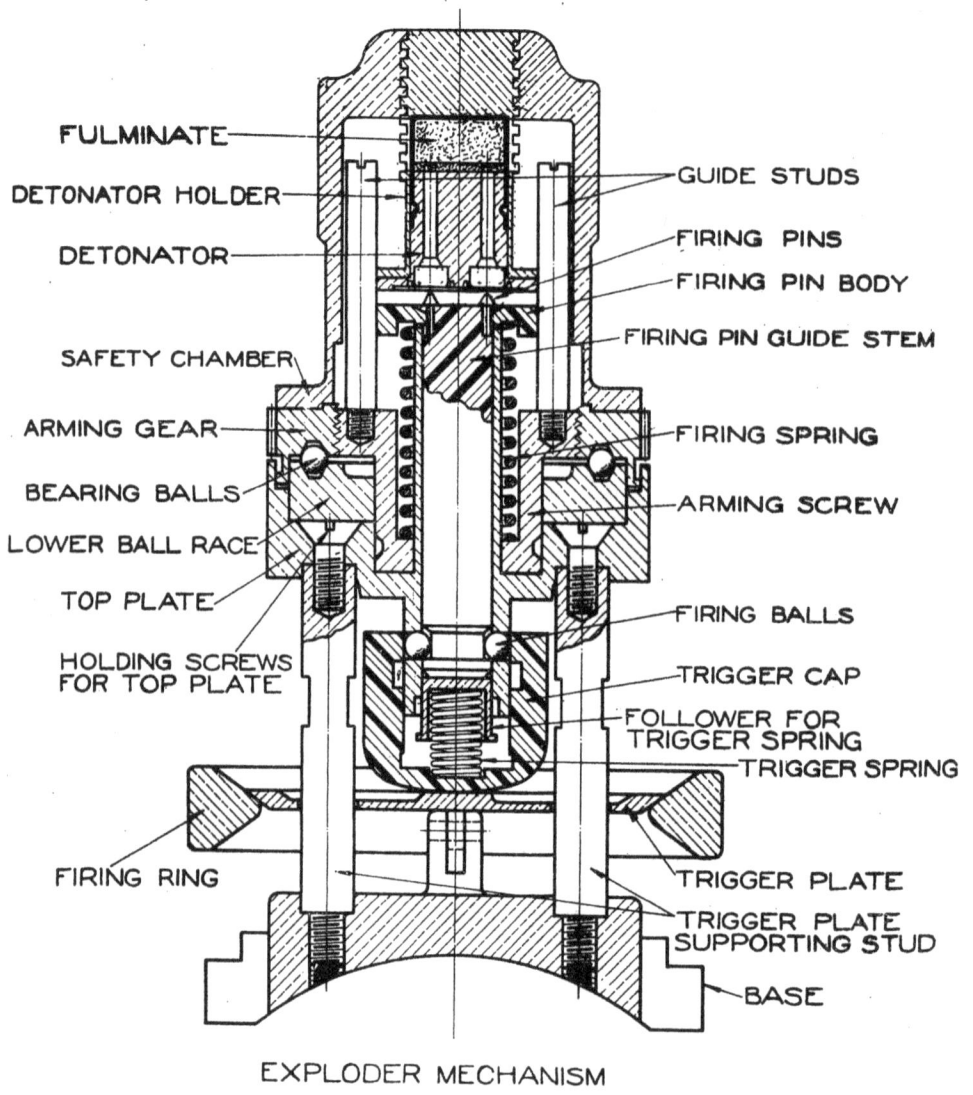

EXPLODER MECHANISM
INITIAL POSITION

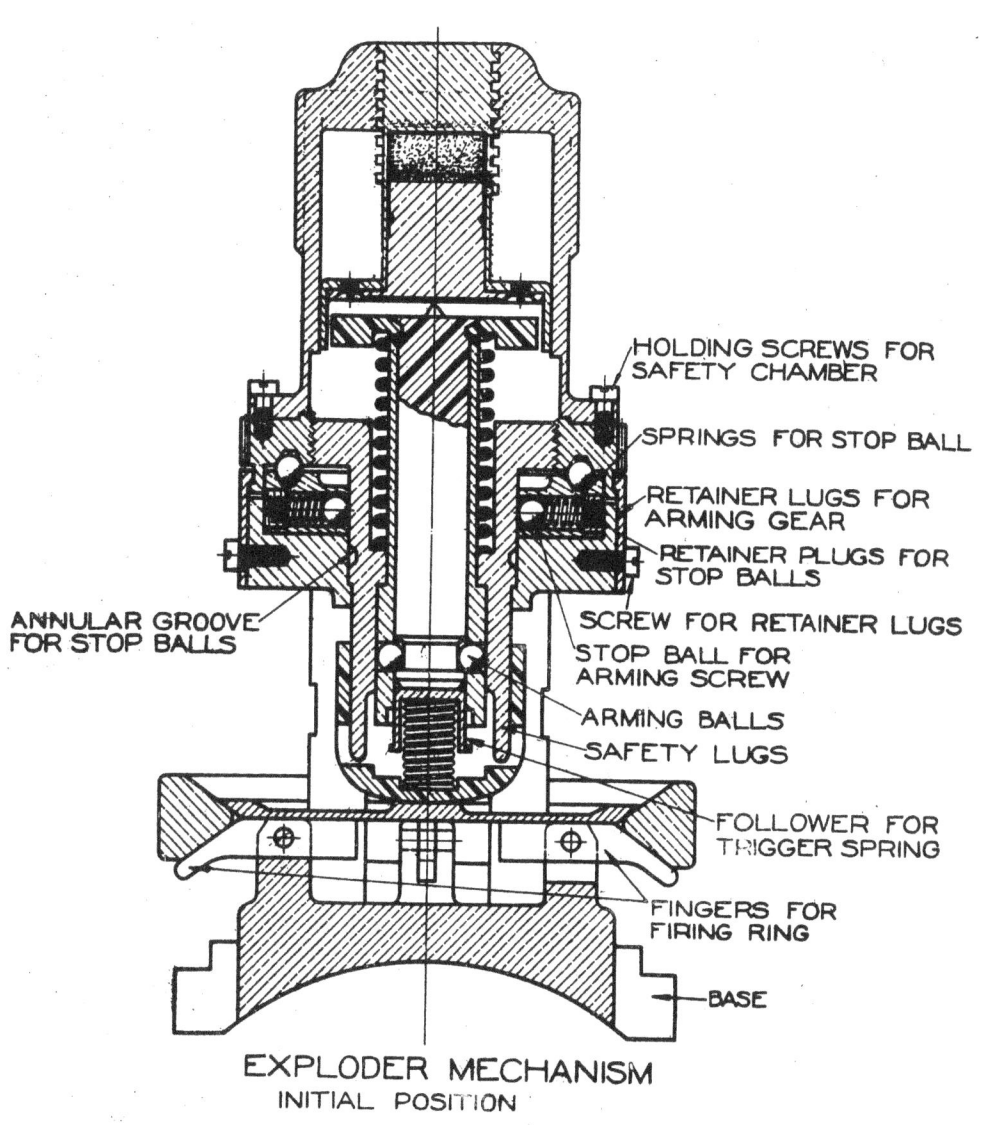

EXPLODER MECHANISM
INITIAL POSITION

and firing mechanism. The bottom of the top plate is drilled and recessed for four supporting studs to which it is attached and secured with screws. The upper and lower central portions of the top plate are extended. The upper portion forms a guide sleeve for the firing pin body and the lower portion is suitably machined for the assembly of the trigger element. A hole is drilled through the entire central portion for the passage of the firing pin body guide. Four holes are drilled, radially and equally disposed, in the lower portion into which fit the arming and firing balls. In drilling these holes a small shoulder is left to prevent the balls from passing through, and after insertion of the balls the outer ends of the holes are peened over to prevent the balls from falling out. A recess is machined in the end of the lower extended portion to afford a seat for the trigger spring follower and to permit passage of the safety lugs, which extend downward from the arming screw. The upper side of the top plate is suitably machined for the arming screw and the lower ball race.

4016. The arming element is composed of the arming gear which, by receiving rotating motion from the impeller through the gear train, transmits a traversing upward movement to the arming screw, and by this movement compresses the firing spring and unlocks the safety features from the trigger mechanism. The turning of the arming gear, which carries the safety chamber, also advances the detonator into the firing position.

4017. Eighty-four teeth are machined around the circumference of the arming gear of a suitable pitch to mesh with the idler gear pinion. The gear is threaded internally to fit similar threads on the arming screw and revolves upon a ball bearing containing thirty-three 3/16" brass balls. The upper race is machined in a separate piece which fits into a circular recess in the top plate. A shoulder is machined on the upper face of the arming gear to fit in the flange of the safety chamber in order that proper alignment may be obtained in assembly.

4018. Four radial holes are equally disposed in the lower ball race in which fit the locking balls with the springs. The outer ends of these holes are closed by screw plugs, affording sufficient compression on the springs to hold the balls against the arming screw and force them into the annular groove on this screw when in fully armed position. This locks the arming screw against any further upward movement, and relieves strain on the threads of the arming screw and arming gear when disengaged.

4019. The arming gear is restrained for upward movement by three retainer lugs equally disposed around the circumference. The upper ends of these clamps fit in an annular groove machined around the lower end of the arming gear, and the lugs are secured to the top plate with screws.

4020. The exterior diameter of the upper end of the arming screw is threaded to fit similar threads in the arming gear. Two extensions are machined 180 degrees apart on the lower end of the screw to form safety lugs. The inner sides of these lugs are chamfered to permit clearance for the arming balls when the mechanism is armed. An annular groove is machined around the lower diameter of the arming screw to permit entrance of four locking balls when the mechanism is fully armed. The interior diameter of the arming screw is suitably machined for a sliding fit over the sleeve for the firing pin guide in the top plate, and also for the lower seat of the firing spring. Two holes are drilled and tapped in the upper end of the arming screw 180 degrees apart, into which the guide posts are screwed.

4021. The firing pin guide and stem moves in the central sleeve of the top plate. Two notches are machined $180°$ apart on its upper end to fit between guide posts on the arming screw. An annular recess is machined on the under side of the top to form a seat for the firing spring. Two holes are drilled in the upper end in which the firing pins are inserted with a drive fit. An annular groove is machined around the lower end of the stem, affording a means for the arming and firing balls to restrain the upward movement until the mechanism is armed and fired.

4022. The trigger cap is a cylindrical case-hardened steel cap, the lower end of which is machined with a radius. The interior of the trigger cap is suitably machined to fit around the lower cylindrical portion of the top plate. An annular groove is machined in the inner circumference of the cap, into which the firing balls will roll when the cap is forced up a sufficient distance to bring the balls and groove into alignment and thus release the firing pins. A recess is machined in the interior bottom center to form a seat for the trigger spring. A large hole is drilled radially on the lower end, in wake of the safety lugs, for use as a sighting hole, and to prevent air cushioning in the cap. Square holes are cut away in the top of the cap for the safety lugs.

4023. The lower end of the trigger cap rests on a projection in the center of the trigger plate (the radius on the end permits uniform contact with this projection from any angle), and is held rigidly against the compression of the trigger spring when the mechanism is cocked.

4024. The trigger plate is of brass, the outer lower circumference of which is machined with a taper to fit in the tapered inner portion of the firing ring. Four holes are drilled through the trigger plate in wake of the top plate supporting studs to permit assembly of the trigger plate over these studs.

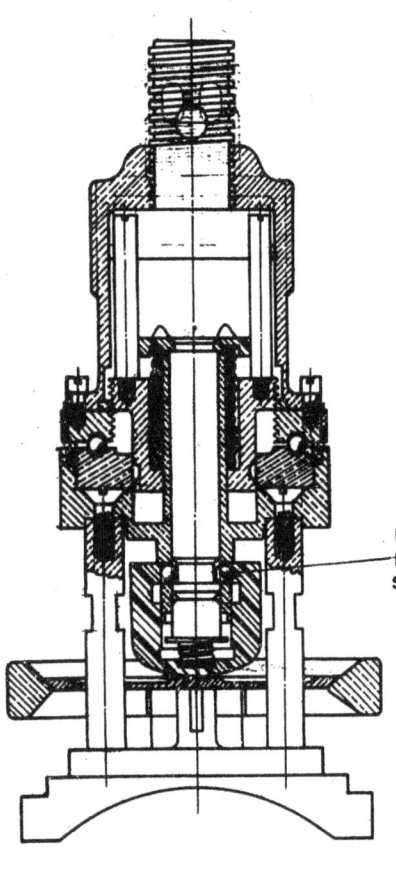

NOTE THAT FIRING BALLS ARE BEING HELD IN ANNULAR GROOVE OF FIRING PIN GUIDE STEM BY THE TRIGGER CAP.

ARMED POSITION

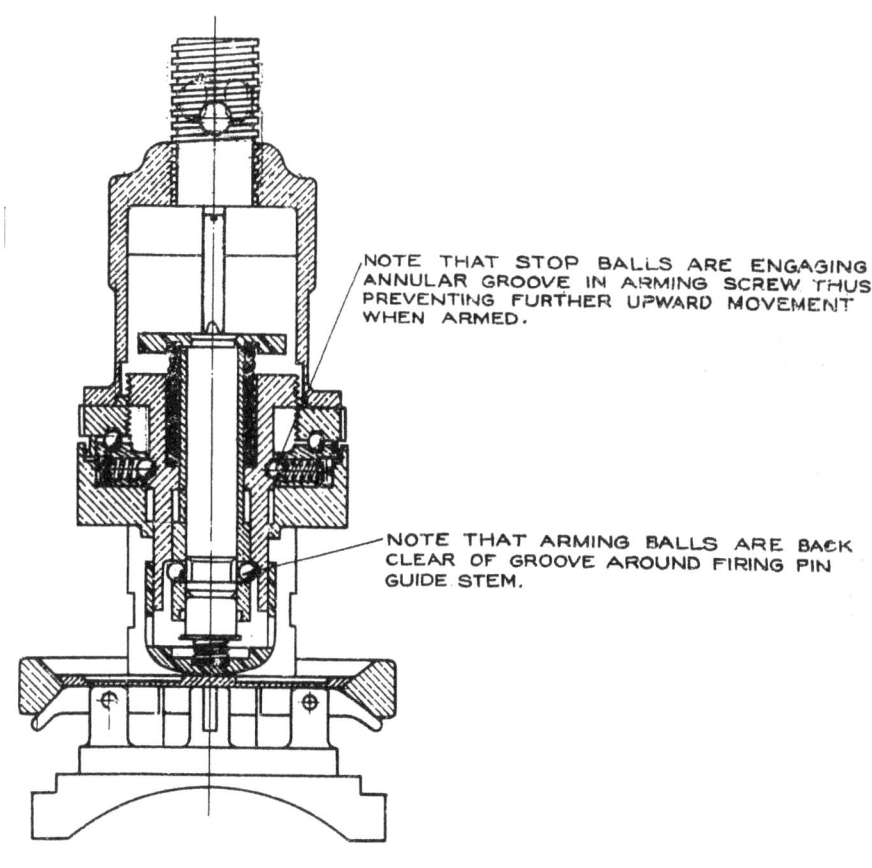

ARMED POSITION

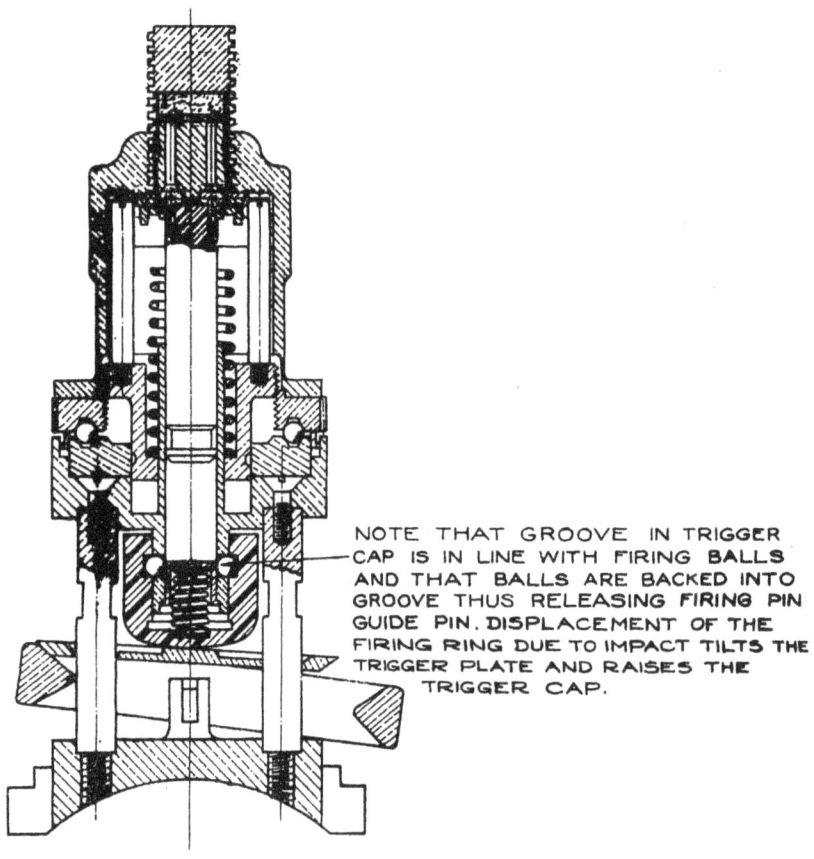

NOTE THAT GROOVE IN TRIGGER CAP IS IN LINE WITH FIRING BALLS AND THAT BALLS ARE BACKED INTO GROOVE THUS RELEASING FIRING PIN GUIDE PIN. DISPLACEMENT OF THE FIRING RING DUE TO IMPACT TILTS THE TRIGGER PLATE AND RAISES THE TRIGGER CAP.

FIRED POSITION

4025. The firing ring is of brass of sufficient weight (8 oz) so that, if given a severe shock, it will overcome factors normally holding it central, and through its displacement fire the mechanism. When assembled in the mechanism in the uncocked position, the ring is free to move in any direction within the limits permitted by the top plate supporting studs. Both sides of this ring are tapering toward its inner circumference and with the mechanism in the cocked position the ring is held central between four fingers (profiled to the same angle as the tapering inner circumference of the firing ring in relation to the under side of the trigger plate). Thus the trigger plate by the compression of the trigger spring exerts a downward pressure through the trigger cap to the trigger plate. The tapering under side of the trigger plate enters the tapered inner circumference of the firing ring, and at the same time the bottom side of the trigger plate contacts the inner ends of the four fingers, causing the outer ends of the fingers to center and hold the firing ring rigid in the central position. The firing ring is thus held central until a sufficient shock is received to overcome the compression of the trigger spring and release the firing element.

4026. The safety chamber is a cup-shaped container made of copper nickel alloy. In the open end is a circular recess in which fits a projecting lip on the arming gear to facilitate alignment during assembly. The safety chamber is secured to the arming gear with four screws. A single pitch thread is machined in the upper end for the reception of the detonator holder which is similarly threaded. The detonator holder when installed and in the unarmed position is entirely within this chamber and will require 781 turns of the impeller to move out to fully armed position and into the pocket (detonator end) of the tetryl booster in the war head. The safety chamber is considered as part of the detonator unit and is assembled with same in storage.

C. EXPLODER MECHANISM - MARK 4, MOD. 1.

4027. The operation of the exploder mechanism is as follows:
   (a) Upon impact with the water the impeller starts to revolve transmitting motion through the gear train to the arming gear, which in turn gives a traversing movement upward to the arming screw.
   (b) The firing pins are unlocked by the withdrawal upward of the two safety lugs which normally prevent the arming balls from moving out of the annular groove on the firing pin guide.
   (c) The firing spring is fully compressed.
   (d) **The threads on the arming gear are next disengaged** from the arming screw. This required about 370 turns of the impeller, after which time the screw is locked from further vertical movement by the four locking balls.

(e) The detonator is advanced out of the safety chamber and into firing position in the tetryl booster. This takes about 780 turns of the impeller to complete, after which the impeller continues to rotate idly. A water run of about 400 yards is normally required to completely arm the exploder.

4028. After steps (a) to (c) have been completed the exploder is completely armed. Upon contact with anything that will produce a deceleration or acceleration of 6 g, the taper portion of the firing ring will wedge in between the trigger plate and trigger fingers, causing the trigger plate to be tilted upwards. The central projection on the trigger plate will then apply an upward force to the trigger cap, moving the cap upward against the compression of the trigger spring a sufficient distance to permit the alignment of the annular groove with the holes in the top plate. The firing balls then move out into these holes under pressure of the firing spring, thereby releasing the firing pins and firing the detonator. The mechanism will operate at any angle of impact, provided the impact blow is sufficient (about 6 g.).

4029. The anti-countermining device prevents the armed exploder mechanism from firing when the torpedo passes through the pressure wave set up by another detonation close by. Without the anti-countermining device the torpedo when armed, might be set off by some other detonation. The device interposes a locking rack between the trigger cap and the under side of the top plate, so that the firing ring cannot operate under the impact of the detonation wave of another explosion. The operation is as follows: The force of the countermining explosion is transmitted through the diaphragm, lever and push rod to the bell crank, which pushes the locking rack in between the upper edge of the trigger cap and the under face of the top plate, thus preventing the exploder from firing. The action of the locking rack is faster than that of the firing ring, so that under the influence of the pressure wave, the mechanism is locked before the firing ring can act. The return of the locking rack is slowed down to guard against the secondary pressure wave which closely follows the primary wave.

4030. While the exploder can be fired before reaching the fully armed position, extreme care should be taken that the firing mechanism has reached that position before attempting to re-cock it. Attempts to re-cock the mechanism when it is not fully armed or attempts to un-arm when it is not cocked will result in sufficient damage to parts as to cause excessive friction, which, under certain conditions, may actually result in failure to fire.

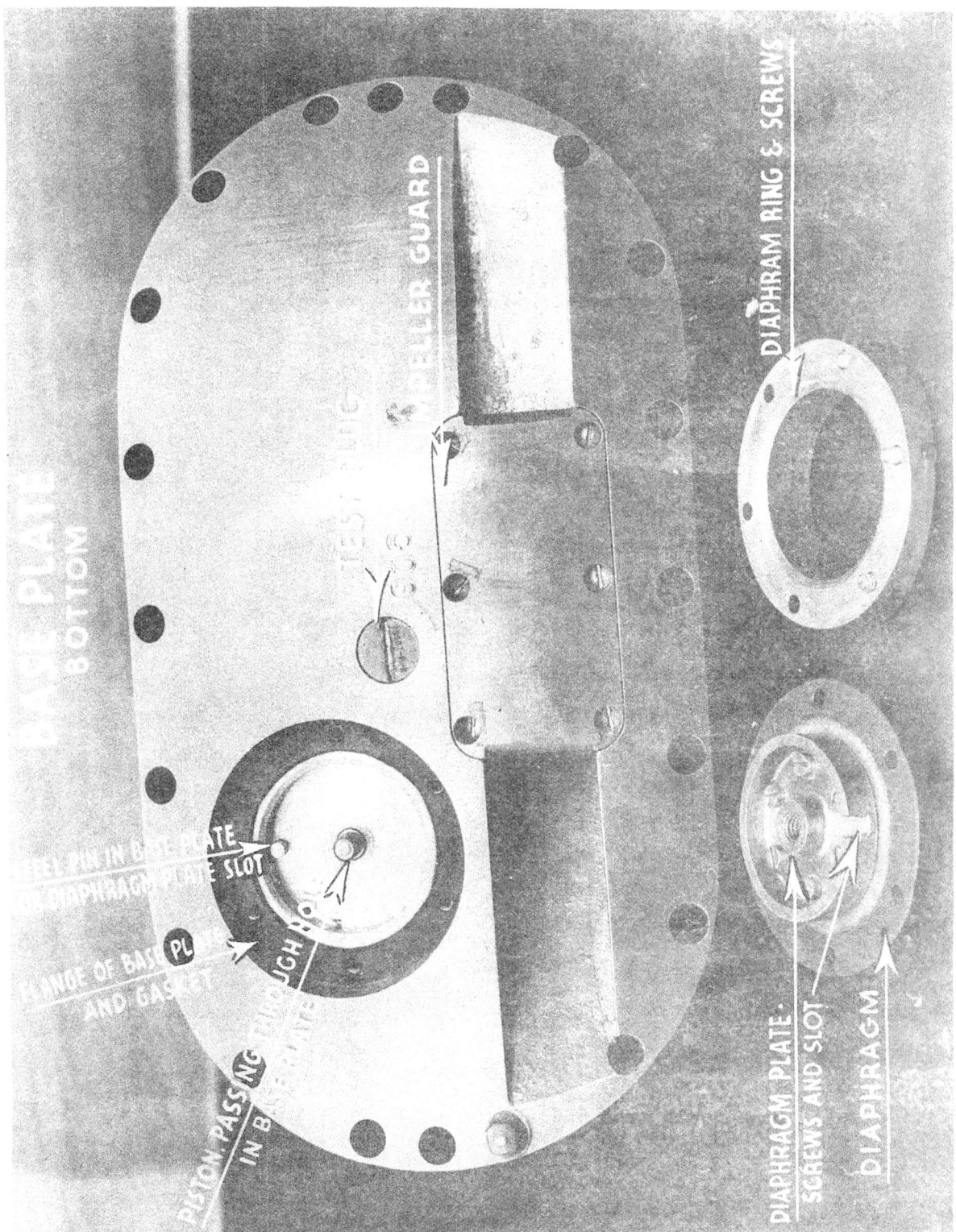

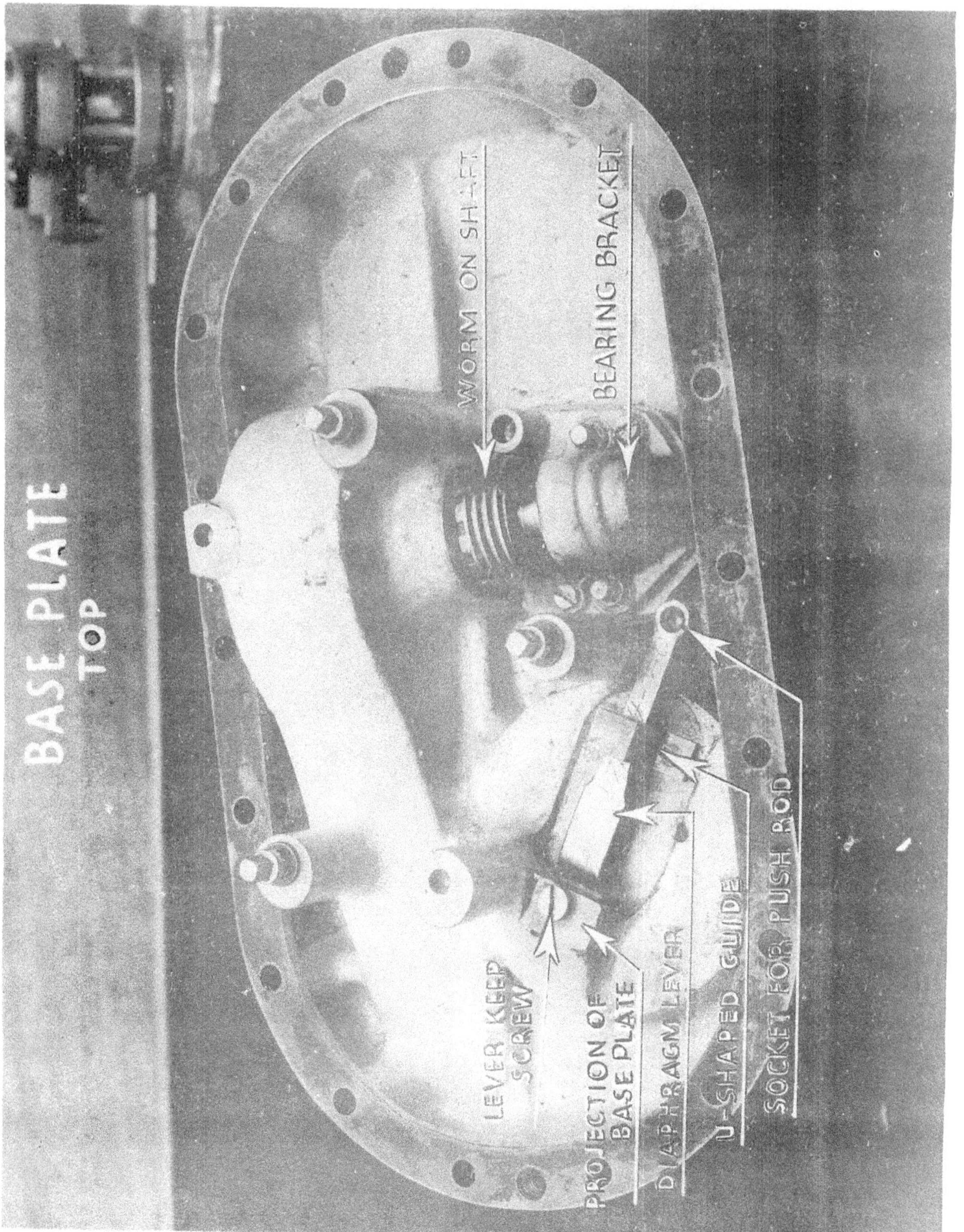

BASE PLATE
TOP

WORM ON SHAFT
BEARING BRACKET
LEVER KEEP SCREW
PROJECTION OF BASE PLATE
DIAPHRAGM LEVER
U-SHAPED GUIDE
SOCKET FOR PUSH ROD

TS-5 IV-9B

TS-5

IV-9D

SAFETY CHAMBER

VERTICAL SHAFT

TS-5

IV-9E

D. **DISASSEMBLY OF THE MARK 4-1 EXPLODER.**   TOOL NO.

4031. **Parts on base plate attached to war head:**
    (a) Remove 18 screws and base plate from war
        head.. 49,MF2
    (b) Remove screws and impeller guard........... 37
    (c) Remove nut, impeller shaft and impeller.... 39

    <u>Exploder Firing Mechanism.</u>
    (d) Remove firing pin and firing spring........
    (e) Remove holding screws and 3 retainer lugs.. 37
    (f) Turn arming gear to locked position and re-
        move arming gear assembly with lower ball
        race (it may be necessary to pry loose by
        the insertion of a small screw driver in
        opening for retainer lugs, in which case
        great care must be exercised not to mar the
        finished surfaces).
    (g) Unscrew arming gear out of arming screw
        and remove bearing balls.
    (h) Remove screw plugs, spring and balls in
        lower ball race and slide arming screw out
        of place... 37
    (i) Unscrew four holding screws and remove top
        plate from supporting studs on base........ 41
    (j) Loosen set screw for idler gear and remove
        idler gear.. 37
    (k) Remove pin on driven gear on top plate and
        remove gear and driving pinion.............. 166
    (l) Insert the point of a screw driver through
        top plate sleeve for firing pin guide and
        push follower for trigger spring clear of
        firing balls, remove trigger cap........... 37
    (m) Remove trigger spring and follower.
    (n) Remove trigger plate and firing ring.
    (o) Remove 4 fingers by driving pivot pins
        clear... 166

4032. Guide posts on upper end of arming screw and supporting studs on base should never be removed except for renewal. The supporting studs are peened in place and the four screws securing the top plate to these studs are locked in place. If these are removed for any purpose they must be again set up tight and secured in place as before. These precautions are necessary in order that once adjustment is made to the anti-countermining device it will not change by virtue of looseness between the top plate and the exploder base plate.

E. <u>OVERHAUL AND ASSEMBLY OF EXPLODER MECHANISM.</u>

4033. <u>Exploder Firing Mechanism.</u>
    (a) Clean, inspect and oil parts on base, note

TOOL NO.

particularly that fingers are working free on their pivot pins, working oil into bearings.
(b) Inspect firing ring for possible burrs on tapering portions. If found, remove by stoning out. Replace ring over fingers.
(c) Inspect trigger plate for burrs or for being bent out of alignment with taper seat in firing ring. If found out of alignment, trigger plate should be renewed. Replace trigger plate over firing ring.
(d) Clean and inspect trigger cap for burrs with particular attention to edges around annular groove and seat for trigger spring. Try fit of trigger cap over sleeve bearing on top plate which should be free, also note that safety lugs on arming screw move freely in square holes of trigger cap.
Apply a light coating of oil all over trigger cap.
(e) Check length of trigger spring. If less than 0"98, spring should be renewed. Oil and replace spring in trigger cap, being particular that spring is down on its seat.
(f) Inspect trigger spring follower for burrs, try for fit in sleeve on top plate; oil and replace over trigger spring in trigger cap.
(g) Note that firing and arming balls move freely in their recesses in sleeve of top plate. Oil balls.
(h) Assemble trigger cap with spring follower over sleeve on top plate with groove on cap in line with 3/16" hole in top plate. (To accomplish this it is necessary to insert a screw driver through firing pin guide sleeve to hold the trigger spring follower clear of firing balls until groove in trigger cap comes into alignment with balls and permits them to roll into same, thus locking the trigger cap to sleeve).
(i) Replace top plate with trigger cap assembled on the top plate on supporting studs with 3/16" holes in top plate and base in alignment. Secure top plate to supporting studs with four screws. These screws should be securely locked in place..37
(j) Clean, inspect, oil and replace arming screw retainer balls with spring and plugs in lower ball race. Apply a light coating of oil on ball race.37
(k) Replace 33 3/16" brass balls calipered to uniform size with tolerance of "001 in lower ball race.
(l) Clean and inspect arming screw for burred threads or teeth, with particular attention to groove for retainer lugs. Apply a light coating of oil on upper ball race.
(m) Clean and inspect arming screw for burred threads. Note that central portion and locking groove are clean and free of burrs, check guide posts and safety lugs for proper alignment.

TS-5   IV-11   Original Page.

                                                                TOOL NO.
(n) Oil threads on arming screw and screw into arming
    gear, note that threads are working freely through-
    out their length. Set arming gear and screw flush.
(o) Slip arming gear and arming screw assembly into
    lower ball race until balls contact races. Turn
    lower ball race around arming gear and note that
    balls roll freely.
(p) Note that groove on side of arming cap is in line
    with 3/16" hole in top plate and insert arming
    gear assembly in top plate with safety lugs pass-
    ing through square holes in firing cap. Press
    the assembly down until lower ball race bottoms
    in top plate.
(q) Replace the 3 retainer lugs for arming gear and
    secure to top plate with screws................ 37
(r) Clean and inspect pinion gear and driven gear
    for burrs, oil and reassemble in bearing in top
    plate. Secure driven gear to pinion with pin,
    and note that shaft turns free in its bearings.. 166
(s) Clean and inspect idler, oil and assemble on
    top plate. Tighten set screw for idler gear.... 37
(t) Insert vertical shaft with pinion meshing into
    driven gear, turn shaft to arm and unarm and note
    if moving freely and smoothly. If not, cause
    for stickiness must be removed.
(u) Turn arming gear to fully armed position (until
    arming screw stop balls enter groove on arming
    screw). Uncock firing mechanism by tilting trig-
    ger plate and note that firing cap moves up until
    ball rolls into annular groove of same.
(v) Clean and test firing spring. **New springs are** 2.4" long
    (free length) as against 1.688" for the old springs.
    **Number of actual coils is twelve in the case of new and
    10 in the case of the old springs. Use only** the new springs
(w) Clean, inspect and stone out any burrs found on
    firing pin guide stem, with particular attention
    to arming and firing ball groove on its lower end.
    Apply a thin coating of oil over firing pin body
    and stem. Insert stem through guide sleeve in
    top plate with notches on periphery of firing pin
    body engaging in the guide posts on arming screw,
    push down until arming and firing balls engage
    groove around stem. Turn arming gear until top
    face is flush with arming screw and scribe marks
    in line.
    NOTE: The mechanism should be stowed and shipped
          in this condition.
    <u>PARTS ON BASE PLATE ATTACHED TO WAR HEAD.</u>

(x) Clean and inspect impeller for distortion of
    blades. Try impeller shaft through hub for a
    free sliding fit.

TOOL NO.

- (y) Clean and inspect impeller shaft. Note that grease holes are clean and free.
- (z) Clean impeller shaft stuffing box; remove old packing and inspect impeller shaft bearing in base plate.
- (aa) Locate impeller in channel of base plate in line with shaft bearings and insert impeller shaft. Turn shaft and note that impeller blades do not bind or touch walls of the channel and that blades have concave surfaces forward in channel. Replace lock nut over outer end of shaft, setting up tight............... 39
- (bb) Repack impeller shaft stuffing box, using three rings, each about 1-5/8" long, of 3/32" dia. round packing. Oil packing and by turning impeller work the packing in gradually, tightening the gland until solid. In packing the shaft, care should be taken that an adjustment is made with minimum friction required to exclude leakage of water.
- (cc) Clean and inspect exploder casing and flange for base plate in war head. Renew gasket and install the mechanism base plate in war head. Secure with 18 holding screws............... 49,MF2

The above completes the overhaul and assembly of the exploder and impeller mechanism in preparation for tests.

NOTES: It will be noted that during manufacture, assembling and fitting of the exploder mechanism, a line is scribed across the top of arming gear and arming screw (below flush position). A similar line is scribed on safety chamber and detonator holder. This line-up should be adhered to in assembly.

Where oil is indicated for lubrication or preservation, the use of a light oil such as gyro oil is intended. A mineral sperm oil is used for dipping the packing before assembly.

A bronze dummy detonator and safety chamber are provided in each exploder box for the actual firing of caps for test purposes.

F. NOTES ON MARK 4-1 EXPLODER FIRING MECHANISM.

4034. The fits mentioned below between parts of the firing mechanism which are within drawing tolerances militate against complete interchangeability of the parts:

- (a) Arming screw and top plate. There must be an absolute fit of these two parts in order that the legs of the arming screw will pass through the holes in the top plate without binding. Even though these holes are broached in accordance with the drawing, they may be out sufficiently to cause the arming screw to bind.

(b) Trigger cap and top plate. This is an accurate job of assembly as no more than ".003 is allowed between the trigger cap and the lower sleeve of the top plate. This necessitates a selection of parts within the tolerance to fit properly.
(c) Guide posts and firing pin body. These parts must be selected because when the posts are assembled in the top plate they may be off sufficiently to cause binding between themselves and the firing pin body.
(d) Shaft from lower bearing on base plate of warhead to top plate of exploder. When this shaft is installed the worm wheel is pinned in place to suit the distance between the gear at the top and the worm at the bottom. This is necessary due to the differences which may be found in accumulative tolerances which may amount to as much as ".012 to ".015.

For storing or shipping, Mark 4-1 exploders should be in the following cocked, unarmed condition:

(a) Top face of arming screw flush with top plate of arming gear, with safety lugs extending into the trigger cap, preventing it and trigger plate from lifting
(b) The arming and firing balls held in the groove in the firing pin guide stem by the safety lugs and the trigger cap.
(c) The firing spring under only slight compression.
(d) Detonator and safety chamber removed for separate storage.

A light oil, preferably gyro oil, should be used for preservation and lubrication of the exploder.

G. **TESTS OF MARK 4-1 EXPLODER MECHANISM.**

FRICTION TEST.

4035 This test is conducted to ascertain that exploder mechanism is properly assembled with minimum friction consistent with its operation. Note particularly that:

(a) Guide posts are in alignment.
(b) Threads on arming gear and screw are free of burrs.
(c) Safety lugs are straight and free from burrs.
(d) Gears in gear train in proper mesh without binding.
(e) Impeller shaft stuffing box properly packed.
(f) Impeller and vertical shafts straight.

4036 The test for excessive friction of the Mark 4-1 exploder mechanism is conducted as follows:

(a) Remove impeller guard.
(b) Install 4 legs, MF-8, on mechanism base plate, in holes nearest longitudinal and transverse centers of mechanism base plate.

MF7 TEST FOR FRICTION ON IMPELLER GEAR TRAIN

NOTE: WITH PACKING GLAND NUT BACKED OFF AND SHAFT GREASED, IMPELLER MUST START TO MOVE AT NOT OVER 3 OUNCES

(c) Place mechanism base plate with legs resting on work bench
(d) Grease impeller shaft.
(e) Place vertical shaft in its lower bearing. It is necessary that thin tooth marked by arrow be inserted in worm.
(f) See that exploder mechanism is in the unarmed position, install mechanism on base plate, lining up dowels with holes in base, and upper end of vertical shaft with its bearing in top plate. Secure mechanism with screws.
(g) Turn impeller a few turns by hand and note that gears are meshing. See that arming gear and arming screw are flush and scribe marks are in line.
(h) With mechanism in unarmed position and impeller shaft stuffing boxes loose, place flat spring gauge tool MF-7 on impeller blade and note force required to rotate shaft and assembled mechanism. This should not exceed 3 ounces. If force required exceeds 3 ounces, excessive friction must be traced and eliminated. When eliminated, so that force is 3 ounces or less, turn mechanism back to unarmed position; set up stuffing box gland until force is not less than 12 nor more than 20 ounces. Turn mechanism again to unarmed position; with arming gear and arming screw flush and scribe marks in line.
(i) Mechanisms requiring more than 20 ounces push to initiate movement of impeller and gear train must be corrected until passing this test.
(j) The pressure on the firing ring required to fire the exploder in the fully armed position should not be less than $3\frac{1}{2}$ nor more than $5\frac{1}{2}$ lbs. This pressure should be applied on the outside of and in the plane of the ring, and between studs rather than in line with one of them. Use tool #98 or MF-10

H. THE ANTI-COUNTERMINING DEVICE.

4037. The anti-countermining diaphragm assembly is located in the base plate of the war head to the right and rear of the exploder mechanism. The diaphragm, which consists of two thin sheets of copper, is held between the diaphragm cap and the diaphragm plate by six screws, with a gasket between the cap and the diaphragm. The flange of the diaphragm is held against its seat in the base plate, with a gasket under it by the diaphragm cover and six screws. A steel pin, forced into the base plate engages a slot in the flange of the diaphragm plate, insuring proper assembly. The movement of the diaphragm is transmitted to the lever by a piston, which is in the form of a hexagon headed cap screw, screwed into the diaphragm plate and passing through a hole in the base plate. The lever hooks under a projection of the base plate, and is held in place by the point of a screw and a

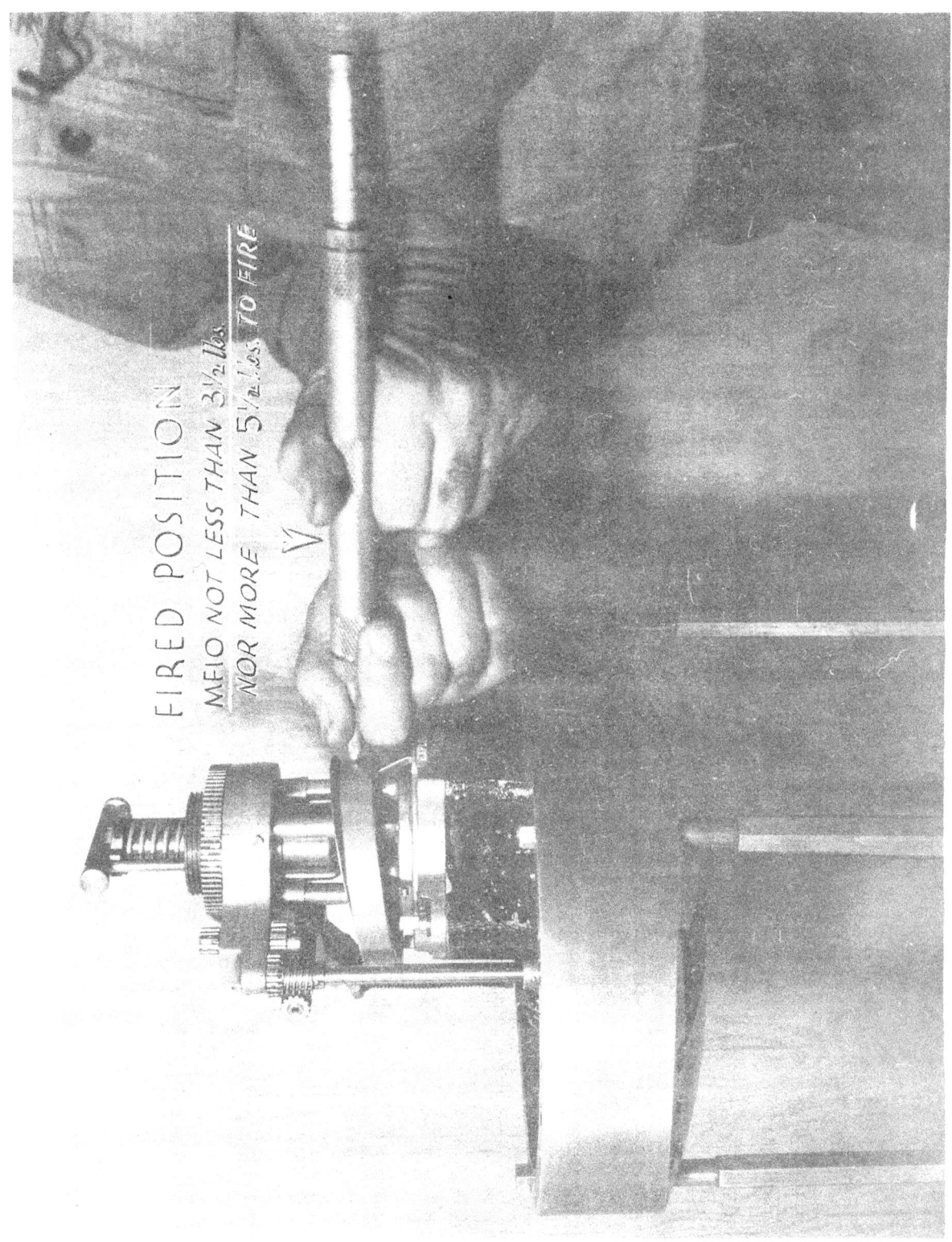

TS-5 IV-15B

U-shaped guide which is screwed to the base plate. A socket in the outer end of the lever takes the lower end of the push rod, through which the movement of the diaphragm is transmitted to the bell crank and rack.

I. **TEST OF THE ANTI-COUNTERMINING DEVICE.**

4038. The purpose of this test is to determine that the adjustment made between the locking fork of the anti-countermining device and the trigger cap of the firing mechanism is correct. When the mechanisms are assembled originally, the clearance between these two parts is at least ".005 when the fork is in the full retracted position. This is obtained with the trigger cap in the fired position and with a push rod clearance of from ".002 to ".005 when the fork is against the trigger cap, pressure to bring the fork against the cap being applied to the bell crank. This small clearance is necessary in order that the locking fork may engage before the firing ring can trip the firing mechanism.

4039. 1. Install on the base plate the bracket for the motor which is to drive the impeller shaft.
    2. Engage the connecting shaft of this motor with the slotted impeller shaft.
    3. Install the bracket which carries the mechanism for imposing a thrust on the diaphragm cap.
    4. Start the motor and hang the weight provided on the lever arm at the notch nearest the mechanism. With the weight at this position, it should be possible to fire the mechanism by tripping the firing ring after arming screw has disengaged from the arming gear. If the locking fork does engage, however, it will prevent the mechanism from being fired and the source of trouble will have to be located. This may be due to one or more of the following:

    (a) Dirt or other interference back of the shoulders of the locking fork between the locking fork and the bracket.
    (b) No clearance between the push rod and the bell crank.

If due to the former, it will be necessary to break down that part of the firing mechanism over the anti-countermining bracket in order to get at the fork. The clearance between the push rod and the bell crank lever should be between ".002 and ".005, when the locking fork is held by finger pressure against the trigger cap. If it is less than the minimum, then steps must be taken to correct this adjustment. When originally assembled, the rod is fitted between the socket of the diaphragm lever and the under side of the bell crank. This is necessary because of the accumulative tolerances in the firing mechanism and the anti-countermining assembly in the exploder mechanism base. The rod may thus fit this mechanism and no other. It may be that the spring which holds the rod in engagement with the diaphragm lever is too weak so that it does not keep the piston of the anti-

TS-5   IV-16A

countermining assembly in its full retracted position. This can be determined by measurement of the spring and weighing its load. If the spring is not of the proper strength, that is, too weak, a new spring should be fitted. The old spring should not be made to fit by stretching it, as it will soon take its original set. If the trouble is not at either of these two points, it is with the piston, probably as the result of dirt or grit between the piston and its seat in the exploder mechanism base, or on top of the piston.

    5. With the motor running, hang the weight on the lever in the outboard notch. If the anti-countermining device is in proper adjustment, this will engage the fork sufficiently to prevent the mechanism from firing. If it does not, proceed as follows:

    (a) With the motor running, push the fork into engagement by hand - it should have an extreme travel of about 9/32" and should return to the fully retracted position in from one to two seconds. The force required to push the fork into engagement is from 2 to 3 lbs. If this part of the mechanism functions in accordance with above, it is certain that the trouble is not in the bracket assembly and must be looked for in the anti-countermining assembly on the base plate, with the probability that the difficulty lies with a sticky piston or bent lever. In case the difficulty appears to be in the anti-countermining assembly in the base plate, this unit must be disassembled as follows:

      (1) Cast loose two screws securing the firing mechanism to the base plate and remove the firing mechanism.
      (2) Remove push rod and spring.
      (3) Remove guide over diaphragm lever.
      (4) Remove screw from projection of base plate over the after end of the diaphragm lever.
      (5) Withdraw lever.
      (6) With socket wrench remove hexagonal screw (piston) in the center of the diaphragm assembly.
      (7) Turn mechanism on side and remove fillister head screws in diaphragm ring.
      (8) Withdraw diaphragm assembly.

    6. It will be noted when this is done that the diaphragm screws in step (7) above have been white leaded at assembly. This has been done to prevent leaks and must be accomplished upon reassembly.

J. <u>WAR HEAD BOOSTER MARK 2.</u>

4040. The tetryl booster is contained in a copper cylinder container which fits snugly inside the war head housing, and is held in place by the safety chamber on top of the exploder.

TS-5 IV-17A

This snug fit insures proper alignment so that the fulminate detonator may move up into the recess of the booster when the exploder is armed.

4041. The booster consists of 8 ounces of compressed tetryl contained in a hollow cylindrical copper casing, the casing being about $2\frac{1}{2}$" in diameter by $2\frac{1}{2}$" long. The diameter of the booster is larger than the safety chamber and fits over it, making two recesses in the bottom, one to fit over the safety chamber and another for the detonator to fit into. The assembled booster is flat across one end but on the other end is a central indentation. This indentation is the detonator recess.

4042. All parts of the container are of thin sheet copper. In assembling, the detonator recess is first put in place inside the hollow cylindrical case and then the pellets are inserted, the ring pellet first. A copper disc closes the upper end and the top of the casing is crimped over to hold the assembly firmly together. Both the top and bottom joints are then shellacked for moisture proofing and the exterior surface painted with red pyroxylin lacquer. An identification label is then secured to the side of the casing.

4043. Tetryl is almost as stable as TNT but should be carefully handled as a matter of precaution. Tetryl boosters are supplied ships in sufficient quantity to provide one for each war head. They are enclosed in individually sealed metal containers, and issued in sealed metal boxes, each box containing six boosters. Seals of the individual containers or boxes shall be broken only when the boosters are actually required for installation in war heads. They are stowed in the war head locker in special racks provided therefor. No special tests are as yet prescribed for tetryl boosters.

4044. In assembling in the war head, the booster fits snugly in the brass booster housing brazed to the top of the exploder mechanism housing. The booster MUST be installed with the detonator recess outward next to the exploder mechanism.

K. <u>DETONATOR MARK 7</u>.

4045. The detonator is composed of a brass body, one end of which is flanged for assembly in the copper nickel alloy detonator holder with the notches machined to fit guide posts on the exploder mechanism. Two holes are drilled through the body to permit the flash from the primers into the ignition charge. Primer caps are inserted in recesses machined in these holes. The primer end is sealed by a thin copper disc soldered in place. Assembled on one end

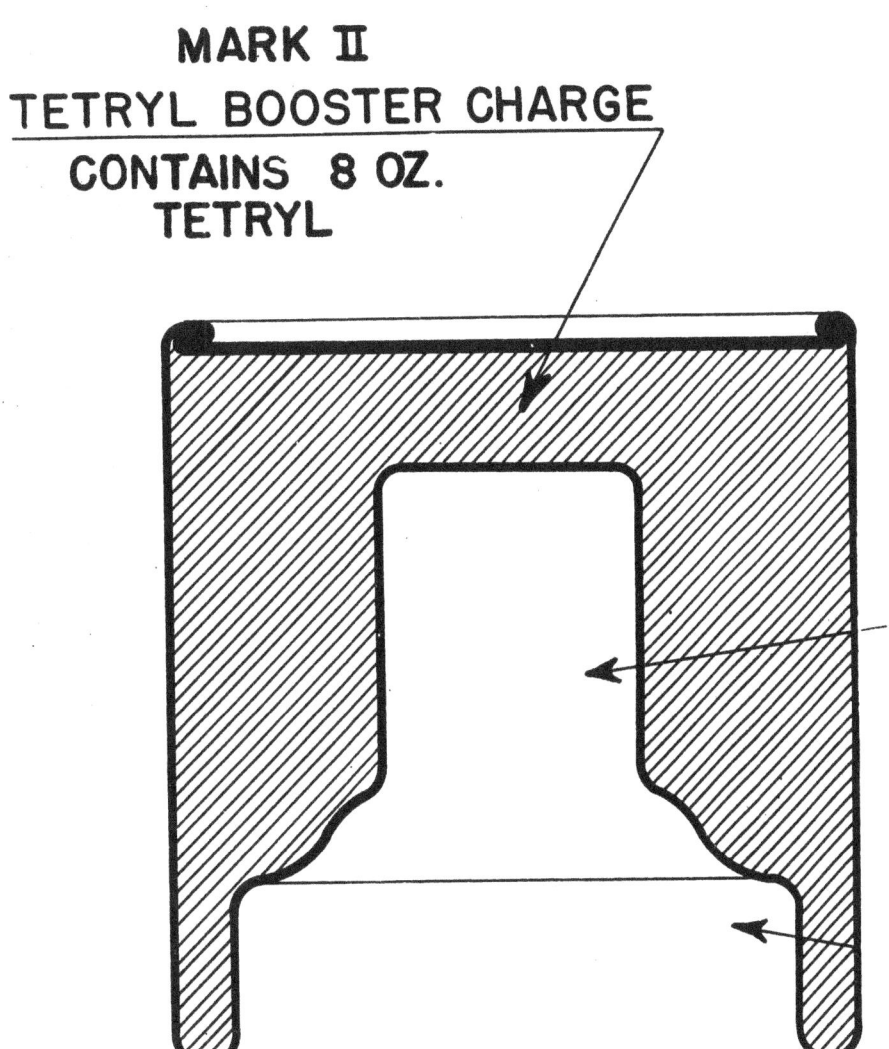

BOOSTER CONTAINER

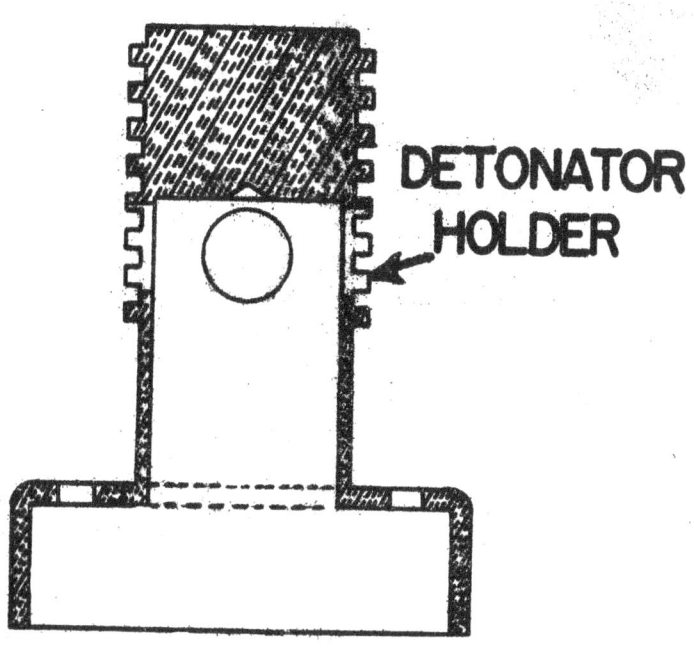

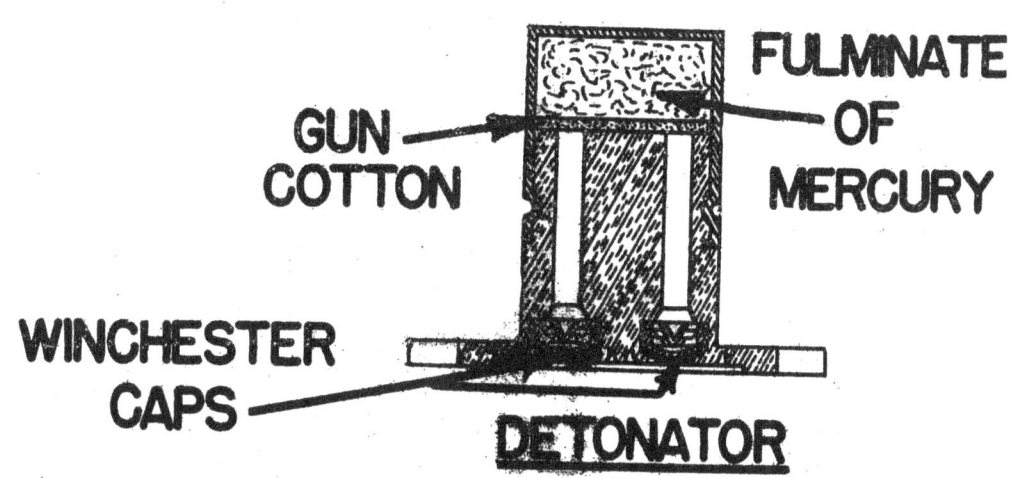

IV-42 A

TS-5 IV-18C

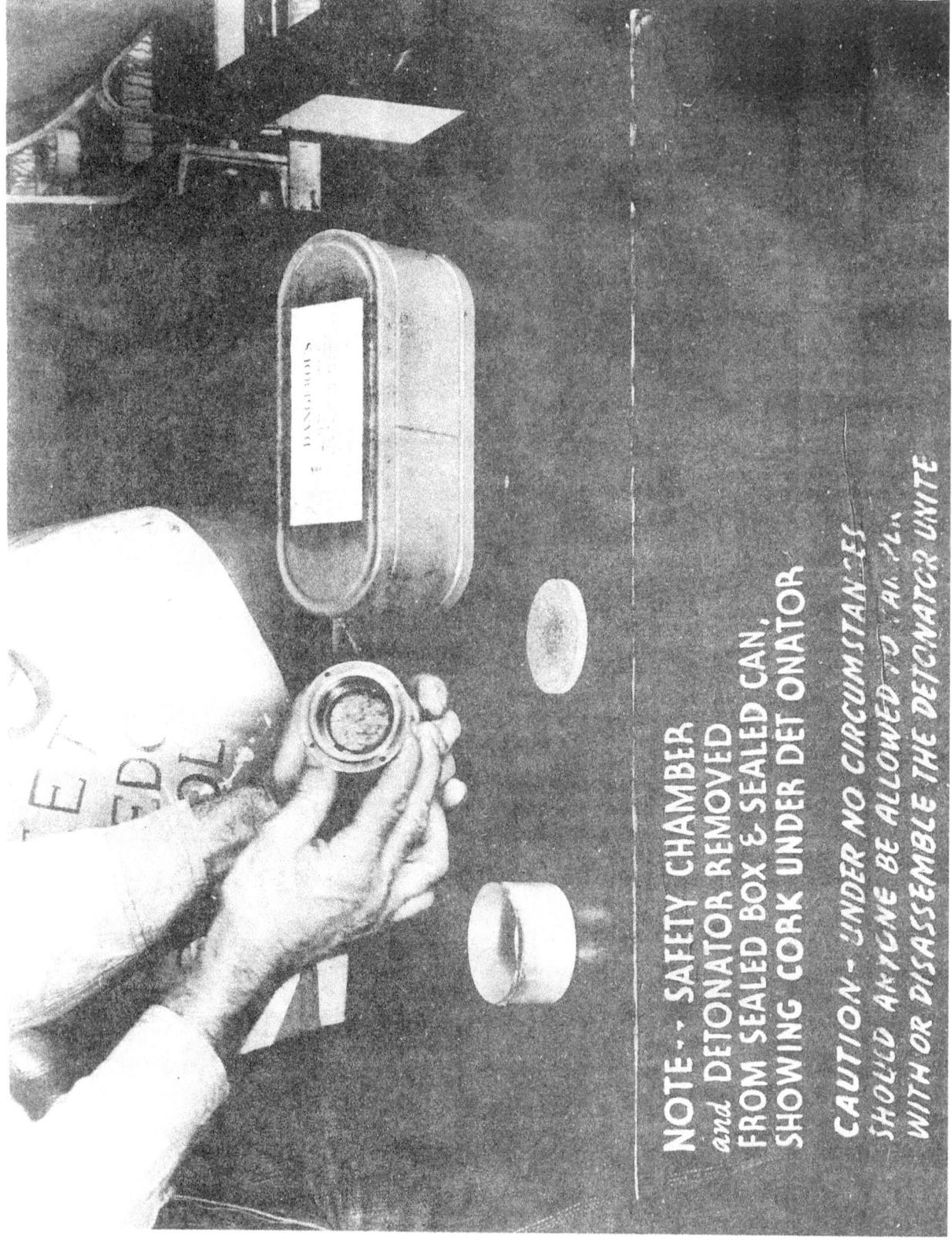

NOTE -- SAFETY CHAMBER and DETONATOR REMOVED FROM SEALED BOX & SEALED CAN, SHOWING CORK UNDER DETONATOR

CAUTION -- UNDER NO CIRCUMSTANCES SHOULD ANYONE BE ALLOWED TO TAMPER WITH OR DISASSEMBLE THE DETONATOR UNITE

of the body is a magazine containing a detonating charge of 65 grains of fulminate of mercury and an ignition charge of $\frac{1}{2}$ grain of gun cotton. The magazine is held in place on the body by a cover which is crimped in place and made waterproof by coating with white lead.

4046. Red lacquer is applied over the primer end of the detonator; mark of detonator, year of manufacture, lot letter and assemblers number are stenciled in white over the red lacquer.

4047. The assembled detonator is secured in a copper nickel alloy holder, the inner end of which is machined with four flash holes in line with the magazine on the detonator, and having a modified square single pitch thread. The outer end of the holder is flanged so as to guide the holder in the safety chamber during the arming of the exploder mechanism. The detonator is secured in its holder by two screws, during manufacture and should never be removed by the service.

4048. Detonator and holder are assembled with the safety chamber in storage. It will be noted that a line is scribed across the face of the safety chamber and one on the inner end of the detonator holder. During the assembly of the detonator in the mechanism of the war head, the inner end of the detonator holder should be flush with the face of the safety chamber and scribe marks in alignment. When safety chamber is installed on the arming gear those same scribe marks should be aligned with that of the arming gear.

4049. Under no circumstances should anyone be allowed to disassemble the detonator unit. While the detonator is designed with a view to eliminate all danger in handling, **it is nevertheless recommended that all care be enjoined.**

4050. Detonators are issued in standard detonator boxes. Each box contains three detonators. Each detonator, with its holder and safety chamber is enclosed in a sealed metal container. Mark 7 detonators are issued in copper boxes.

4051. Stowage of war head detonators in service is by means of metallic water-tight containers of a suitable size to hold one box of 3 detonators. Containers for surface vessels are furnished by the Bureau of Ships.

4052. Dummy detonators and safety chambers are furnished for use in testing exploder mechanisms. The dummy detonator is of the same dimensions as the live detonators, fitted with flash holes and recessed for the insertion of primers. The safety chambers for dummy detonators are manufactured of bronze. The outside form has been slightly changed from that of the live detonators for identification purposes.

4053. Primer caps are pressed into dummy detonators by the use of a capping tool and punch. Primers are first inserted in the capping tool after which the tool is installed with the lugs fitting into guide slots on the detonator, the capping punch being used to drive primers down through the capping tool to seats in the dummy detonator. It is important that primers be inserted from the side of capping tool marked "top", as the holes for guiding the primers through the tool are slightly tapered, and should the primer be inserted from the wrong end, there is a possibility of the primer being out of line with its seat in the dummy detonator. Spent primers may be removed by the use of a small brass rod.

4054. The Bureau of Ordnance regulations require that war head detonators be given a routine percentage test, 3 detonators being sent in annually from each ship from the oldest lot of detonators on board. The test consists of firing the detonator in a sand test bomb. The requirement is that the explosion of the detonator shall shatter not less than 50% of the sand in the bomb. The detonators will be forwarded to one of the following stations for annual test:

      Naval Torpedo Station, Keyport, Wash.
      Naval Torpedo Station, Newport, R. I.
      Naval Ammunition Depot, Cavite, P. I.
      Naval Ammunition Depot, Oahu, T. H.

The detonators turned in for test will be accompanied by invoices. It is not necessary to requisition replacements.

L. PREPARATION OF A WAR HEAD FOR A WAR SHOT.

4055.
1. Remove protecting ring from war head. REASON: To inspect joint ring and install war head.
2. Place torpedo on truck or chocks.
3. Place hoisting strap on war head at center of gravity, 31.42 inches from joint ring (for Mk.13); hoist war head and fit to air flask. Line up hole in joint ring with hole in air flask and secure head to flask with joint screws. Turn torpedo 180°. #49.
4. Put on propeller lock, secure lanyard. REASON: To keep propellers from turning; personnel safety.
5. Remove base plate from war head. MF2-40. REASON: To install exploder.
6. Inspect interior of exploder casing for dirt and corrosion; wipe dry. Inspect base plate gasket for tears and wear.
7. Remove impeller guard. Inspect impeller and anti-countermining plate for corrosion and burrs. REASON: Access to impeller for tests which follow.
8. Inspect inside of base plate. REASON: Corrosion and dirt. Removal of three holding screws and lock washers.

9. Insert grease gun, MF-1, into hollow impeller shaft and force in grease. REASON: To lubricate impeller shaft bearings.

10. Place tool MF-6 on castlelated nut and back off on impeller shaft packing gland nut. REASON: Easier to get to before the exploder is mounted. Check freeness of movement by hand.

11. Check serial numbers on base plate, exploder, vertical shaft, vertical shaft upper gear and push rod. Set base on legs. REASON: Same number on each unit.

12. With MF-11 compress push rod spring on push rod; insert rounded end of rod in free end of the diaphragm lever. Check worm gear on bottom of vertical shaft for arm which indicates thin tooth and engages with worm wheel on transverse shaft.

13. Engage removable gear of vertical shaft with gear train on top plate. Slots in gear should be placed parallel to pins on vertical shaft. REASON: To install exploder on base plate without damage to any part.

14. Place exploder on base plate, inserting vertical shaft through gear to top plate; push rod through hole in bracket in bell crank bracket. Line up dowel pins with base plate. If exploder does not seat itself readily, turn impeller until mechanism seats in line with base plate and removable gear falls in place over pins on vertical shaft.

15. Install three holding down screws on lock washers. Set up evenly with tool #64. REASON: So that exploder is not canted on base plate.

16. Remove MF-11. REASON: So that push rod spring will force rod down into cup of diaphragm lever.

17. Connect base plate to exploder test set No. 3. Turn on motor and arm exploder. Place step of plunger of tool MF-11 on anti-countermining rack and force in. Pressure required should be between 1 to 2 pounds; time for withdrawal, 1 to 2 seconds. REASON: To test for sensitivity and to ascertain that scribe on rack returns clear of bell crank bracket.

18. Fire exploder. Force forked rack in all the way until it contacts trigger cap. With feeler gauge between top of push rod and heel of bell crank. REASON: To gauge clearance between fork and trigger cap when exploder is fired. Clearance should be between "002 to "005. If proper clearance is not obtained check for corrosion; alignment.

19. Hang 3 pound weight in outer notch of lever arm bearing in anti-countermining diaphragm plate. Turn on motor and arm exploder. (Note: Dummy detonator is installed with Winchester caps.) REASON: Attempt is made to fire exploder and the rack should prevent this by inserting the forked rack between the exploder top plate and trigger cap.

20. Move the 8 pound weight to the inboard notch in the lever arm. Force the firing ring until the mechanism fires. Remove dummy safety chamber; cock and unarm exploder. Replace MF-9 on guide posts. REASON: The live caps explode and

indicates that the firing spring is of sufficient strength. The anti-countermining gear will not operate before a depth of 55 feet is reached.

21. Place flat spring gauge, MF-7, on edge of impeller blade. Impeller should rotate at not more than 3 ounces pressure. REASON: Friction test.

22. With MF-6 set up packing gland nut on impeller shaft until the pressure required to turn impeller lies between 12 to 20 ounces. REASON: Places enough drag on impeller to prevent the exploder to arm up to air speed of 270 miles or 234 knots.

23. Lock castlelated nut with safety wire. REASON: To lock the packing gland in place.

24. Remove idler gear from gear train in top plate. REASON: To facilitate the arming and unarming of the exploder.

25. Arm exploder by hand. REASON: To remove the safety devices so that the exploder can be fired.

26. Place MF-10 in center of firing ring between two supporting studs. Exploder must fire between $3\frac{1}{2}$ to $5\frac{1}{2}$ pounds. REASON: Sensitivity. The MF-10 can introduce a drag if not used correctly. Using the step on the rod on the top of the firing ring as the mechanism fires, the tool is usually caught by the edge of the trigger plate denting it. If mechanisms fail to fire or require more than $5\frac{1}{2}$ pounds, friction must be found and eliminated.

27. By hand, cock and place exploder in the unarmed position. Replace the idler gear and set up on lock screw. Note that scribe mark on arming screw and arming gear are flush. (NOTE: Disregard the raised ridge for centering the safety chamber). REASON: To rearrange the component parts of the exploder so that the safety devices are operative.

28. The detonator box, containing 3 detonators assembled in their safety chambers, is then opened. One detonator is removed from its own sealed container, protected by corrugated paper and cork. REASON: The detonators are usually stored above the water line in sealed containers and protected from heat. (Fulminate of mercury is a very unstable explosive and deteriorates rapidly at temperatures of 104° F or over).

29. Test the detonator holder for ease in operation. Wipe with lint free rag. REASON: If burrs are present, remove. NOTE: Never disengage detonator holder from safety chamber; to do so the safety chamber's purpose is defeated.

30. Oil the thread of detonator holder and the exploder mechanism. Oil (A). REASON: Lubrication.

31. Screw detonator holder in flush with safety chamber, scribe marks in line. REASON: It requires 781 turns of the impeller to fully arm the exploder, that is, to disengage the thread of the detonator holder from the thread of safety chamber.

32. Remove MF-9 from guide posts. REASON: To install the detonator, holder and safety chamber over guide posts.

33. Line up scribe marks of safety chamber with scribe mark on arming gear. REASON: To align detonator holder with guide posts and arming gear.

34. Secure safety chamber on arming gear. REASON: So that safety chamber will rotate with arming gear.

35. Remove tetryl booster from sealed metal container and unseal from its individual container. Inspect and install base plate gasket. REASON: Six boosters are stored in metal boxes and usually with war heads.

36. Install booster in booster recess with recess facing man installing. REASON: So that detonator holder can be inserted in recess.

37. Screw MF-2 in base plate and install assembled exploder in the cocked-unarmed position in the exploder casing on the base plate flange. The anti-countermining diaphragm plate is aft.

38. Secure base plate on base plate flanges with monel joint screws. Set up evenly. REASON: To prevent sea water from entering.

39. Remove testing plug from base plate. REASON: To test for tightness.

40. Install adapter in test plug hole. Pump interior to 5 pounds air pressure for 5 minutes. Test all joints with soapy water. Fill cavity around impeller housing above the level of impeller shaft and look for bubbles. REASON: If air can come out, sea water can enter and cause corrosion.

41. Thread arming wire through nose piece, impeller and plug in exploder base plate. Connect with two Fahnstock clips. Pressure to pull free not to exceed 20 pounds.

42. Replace impeller guard.

M. NOTES ON WAR HEADS.

4056. Never handle warheads without the cast iron protecting ring secured to the joint ring, otherwise the joint ring may become injured.

4057. Warheads should be stowed aboard ship unboxed, nose up, on metal protecting rings, in racks in the warhead locker.

4058. Warheads should never be rolled on their cylindrical surfaces since there is danger of damaging or distorting the shell in so doing.

4059. Warheads should be lifted monthly from their stowage positions, at which time examinations should be made and steps taken to prevent corrosion. When making this inspection, particular attention should be paid to condition of bulkhead studs, joint rings, joint screw holes, and to determine if exudate is leaking from them. Exploders should be fitted periodically to insure proper fit in their heads. See that the serial numbers correspond.

4060. It should never be necessary to remove the detonator holder from the safety chamber.

4061. The temperature in the detonator storage should never exceed 104 degrees, as fulminate of mercury deteriorates rapidly above that temperature.

4062. War heads should be inspected to insure that joint screw holes and close-fitted surfaces are free of TNT. A report should be made to the Bureau of Ordnance of any war heads in which contrary conditions are found.

4063. Since TNT can be ignited by tapping, as well as by other means, work other than routine inspection, cleaning and handling should not be undertaken on loaded war heads aboard ship.

4064. In order to insure proper function of the exploder mechanism, it is essential that damage to the base plate be avoided. This applies especially to the base plates of those exploders which have the anti-countermining device installed, as any lack of freedom of the diaphragm to return to its full out position will probably result in locking the exploder mechanism during the entire run of the torpedo.

4065. As the exploder mechanism base plates are carried installed in the Mark 13 war head, leaving the anti-countermining device diaphragm constantly exposed to damage, these heads must be handled with extreme care to **avoid** damage. A heavy blow on any part of the base plate, or damage to the diaphragm of an apparently minor nature, such as a light blow or any condition which might interfere with its freedom of movement, may render the exploder mechanism entirely inoperative.

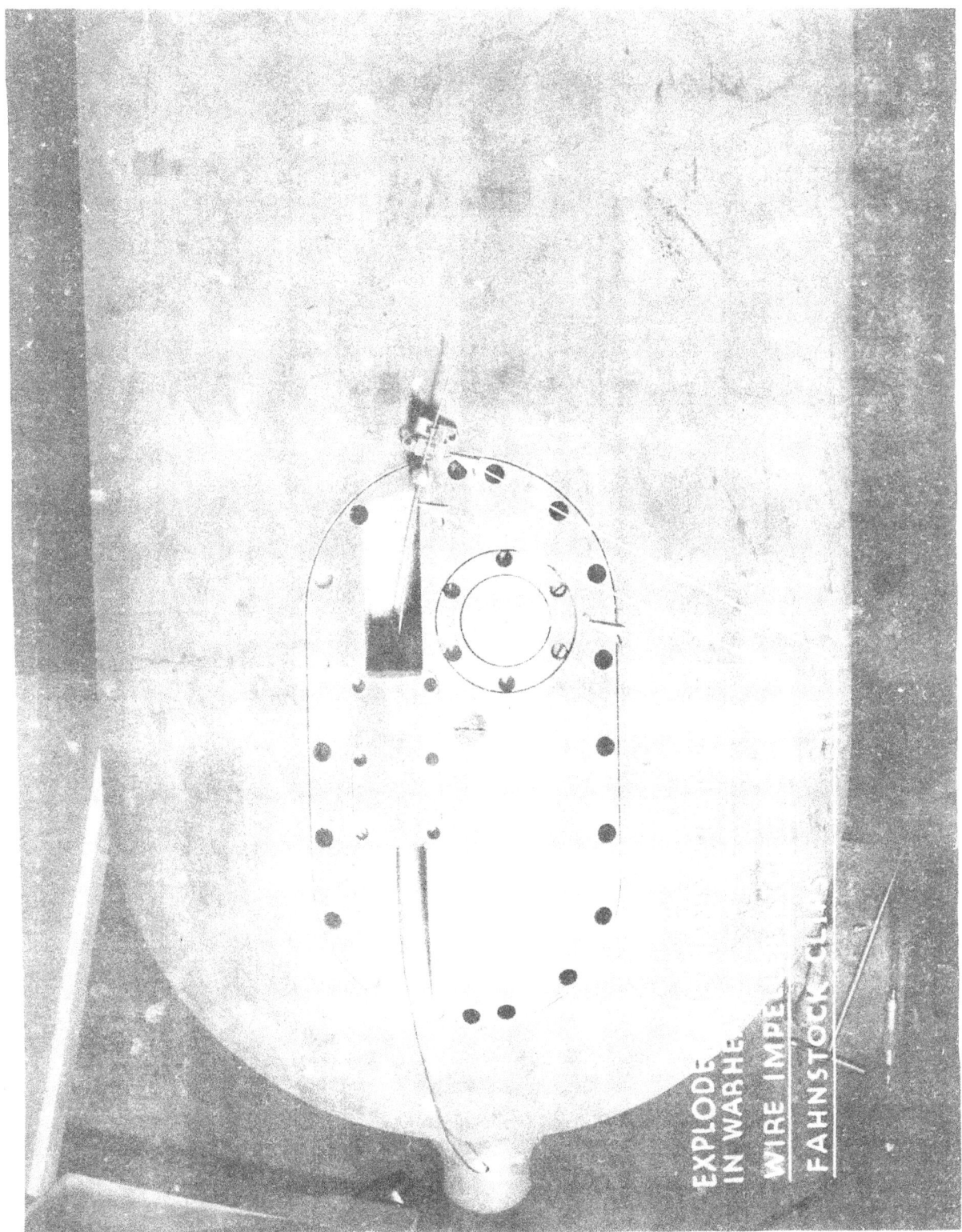

VIEW OF WIRE IMPELLER STOP INSTALLED TO PREVENT IMPELLER TURNING

TS-5　　　IV-24B

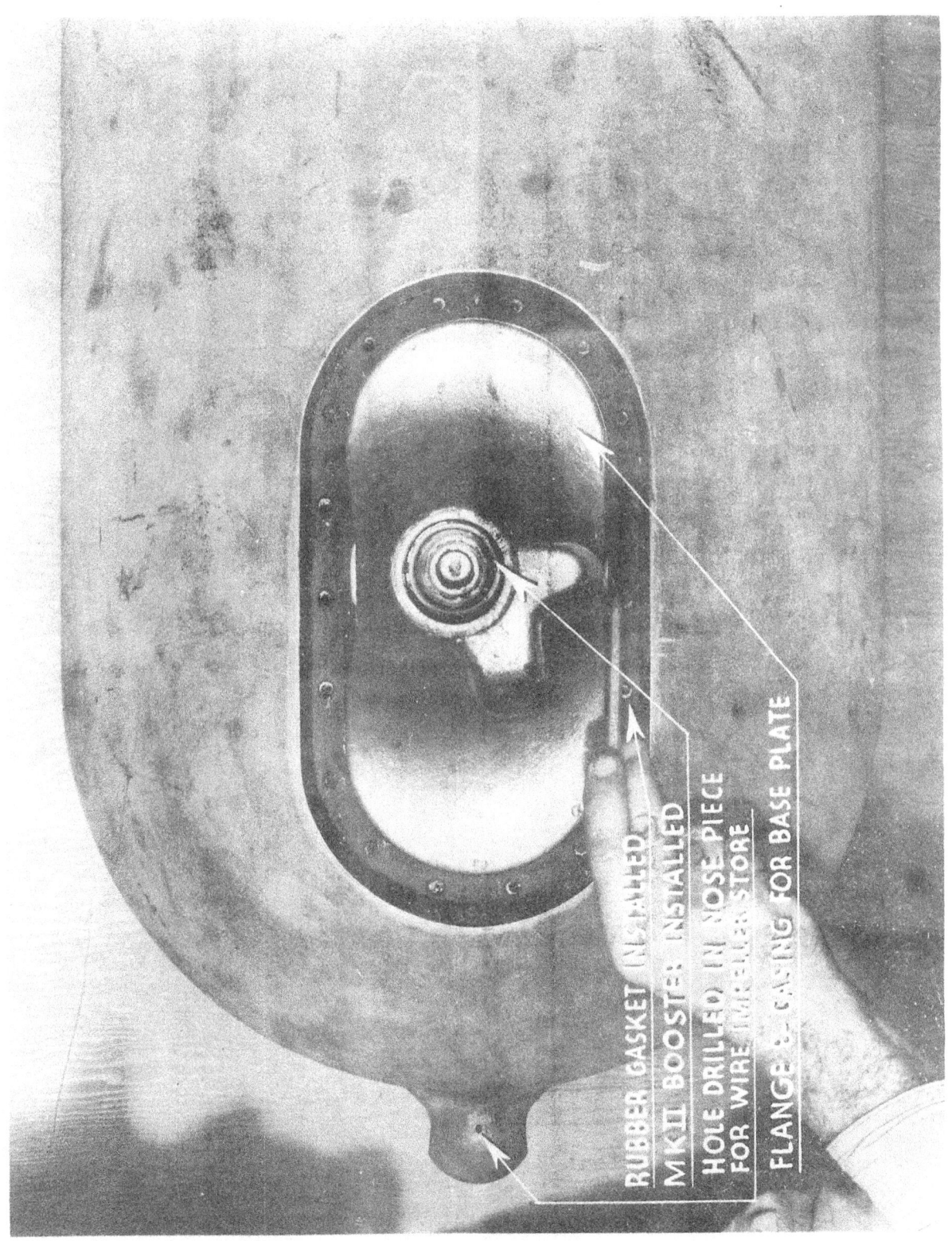

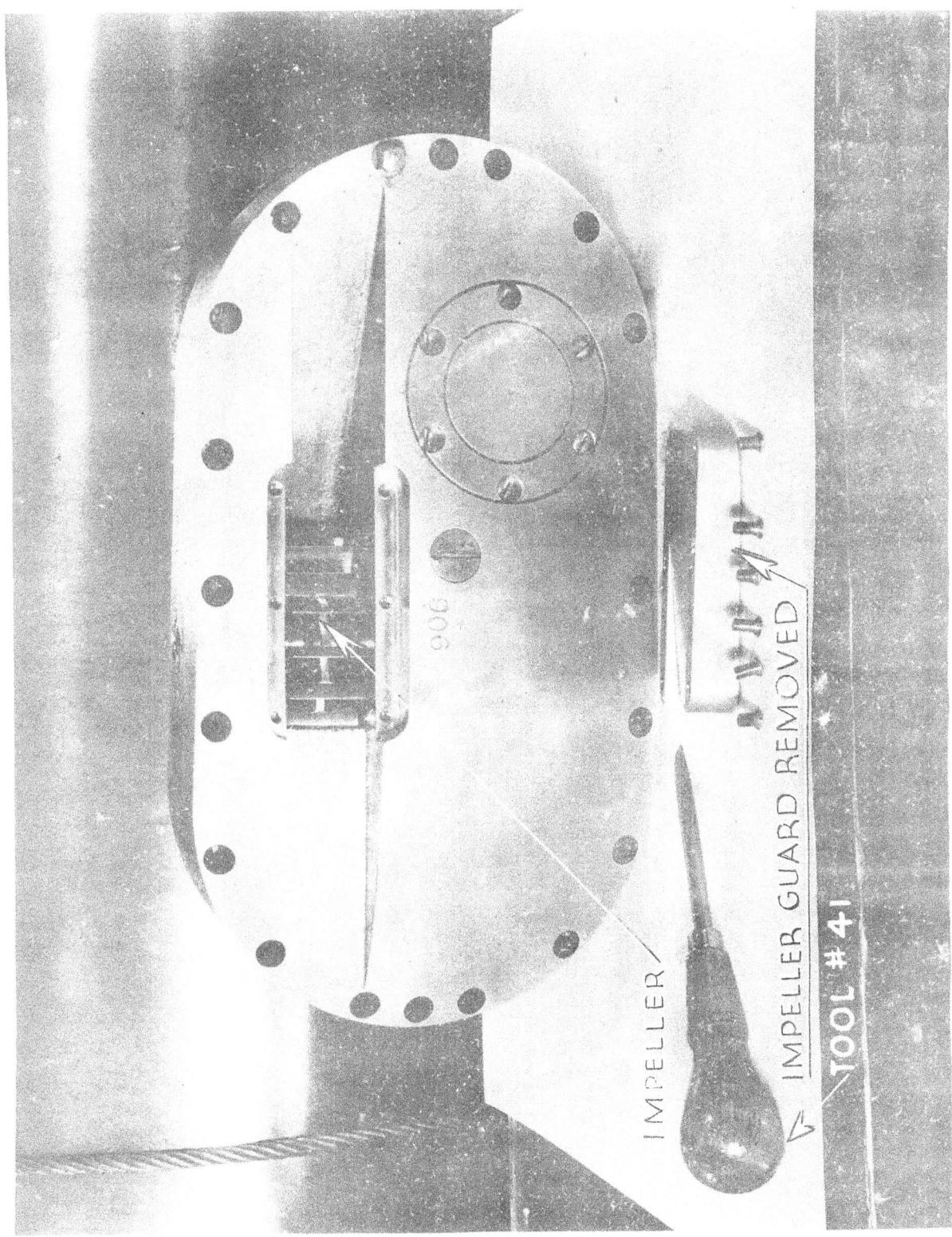

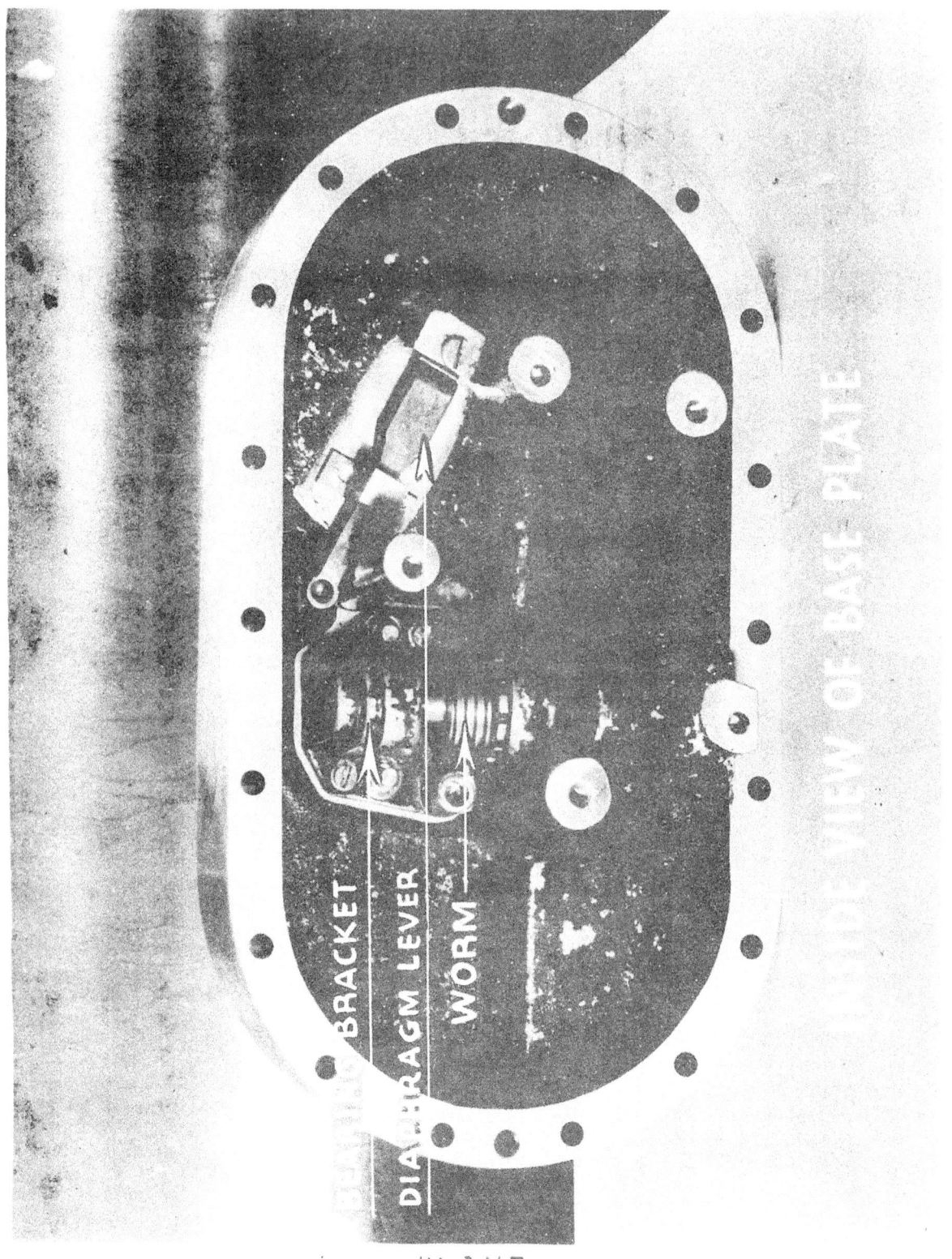

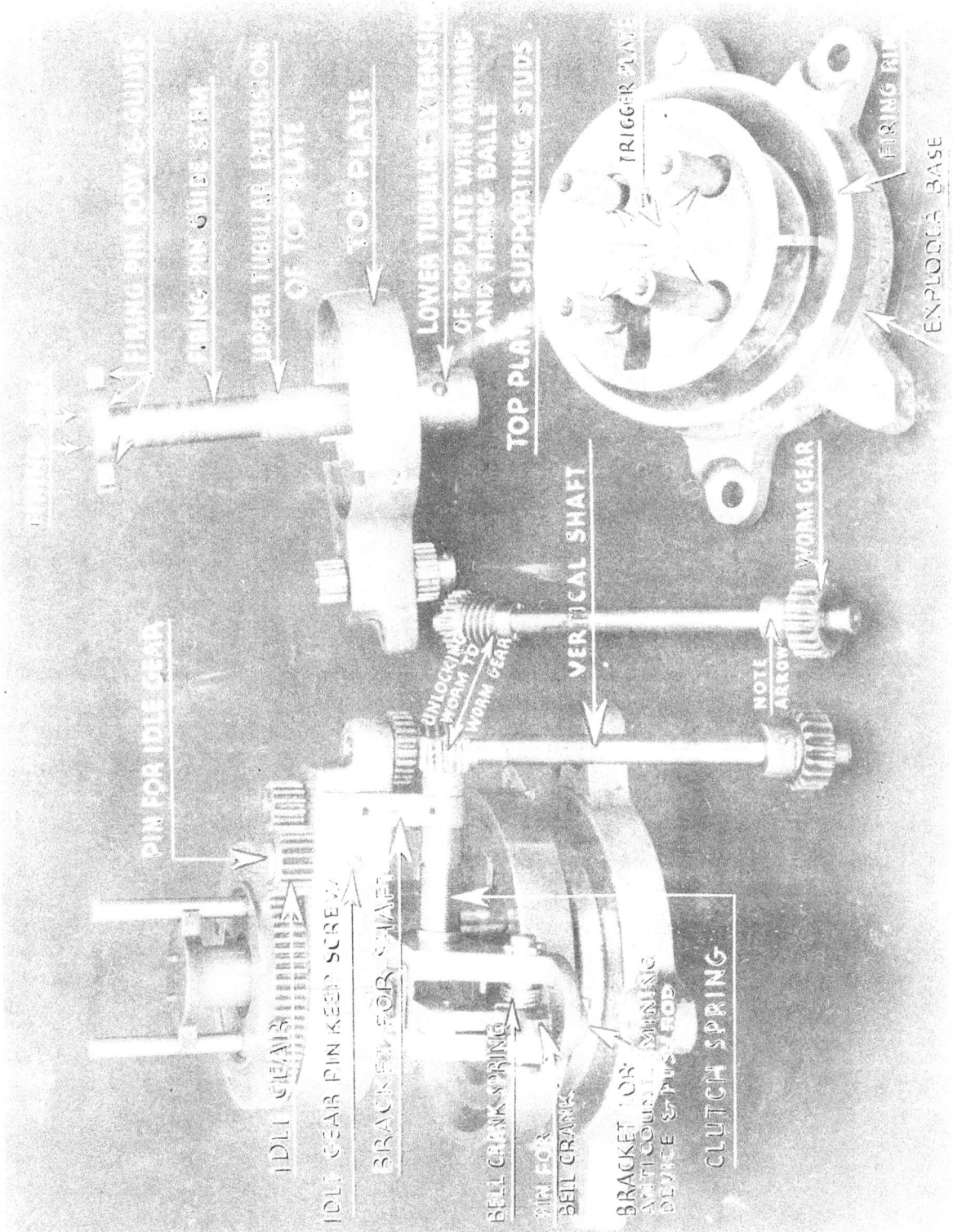

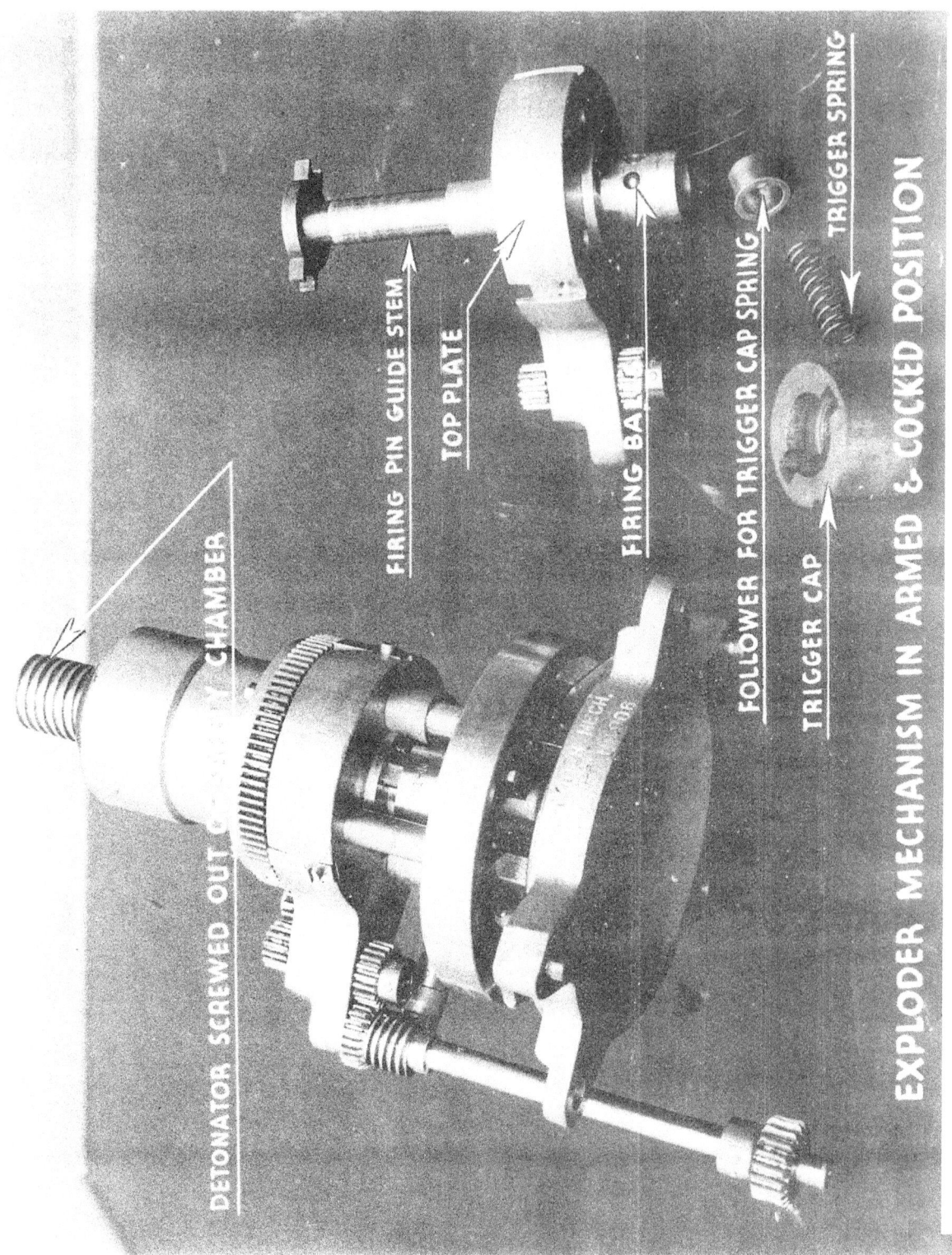

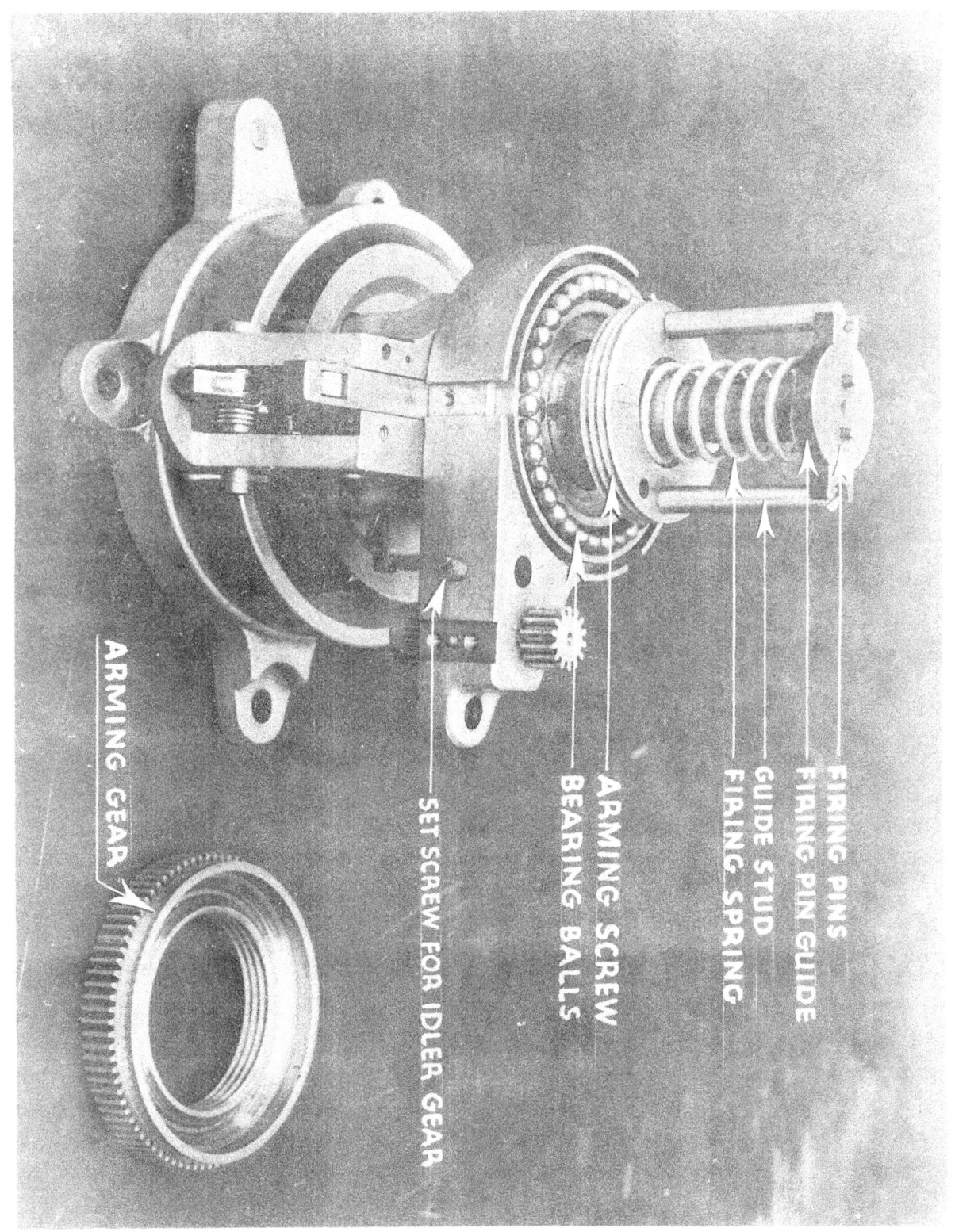

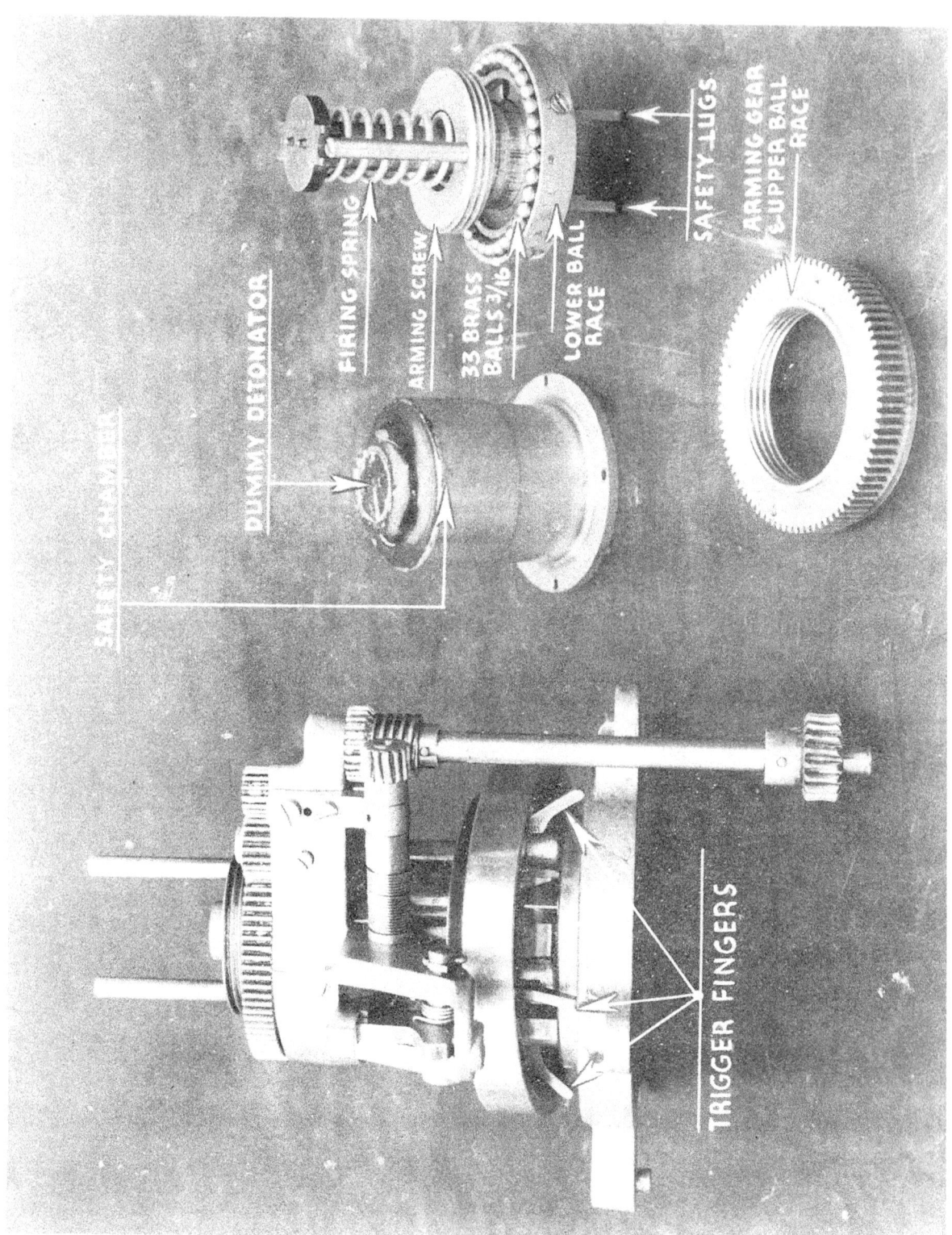

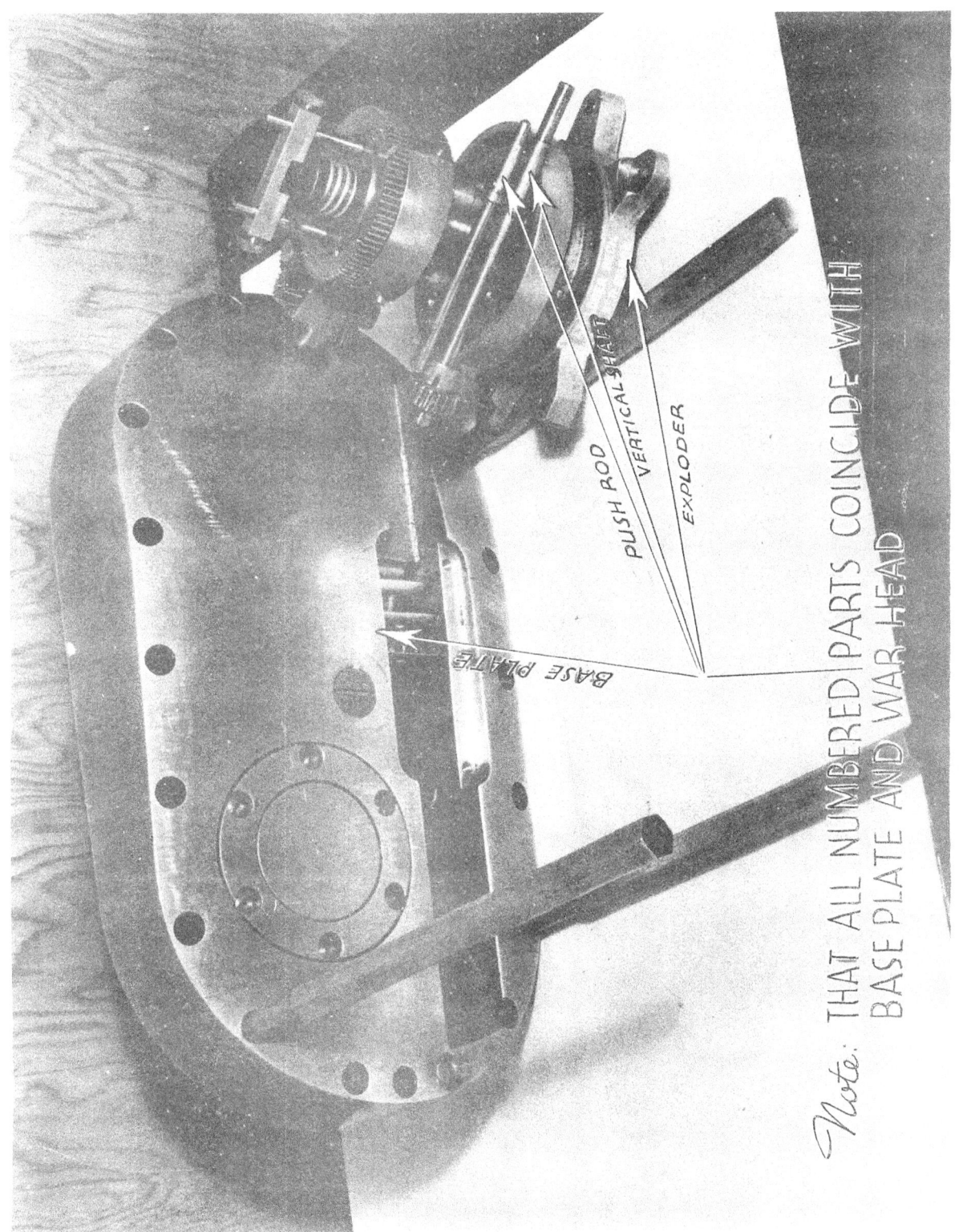

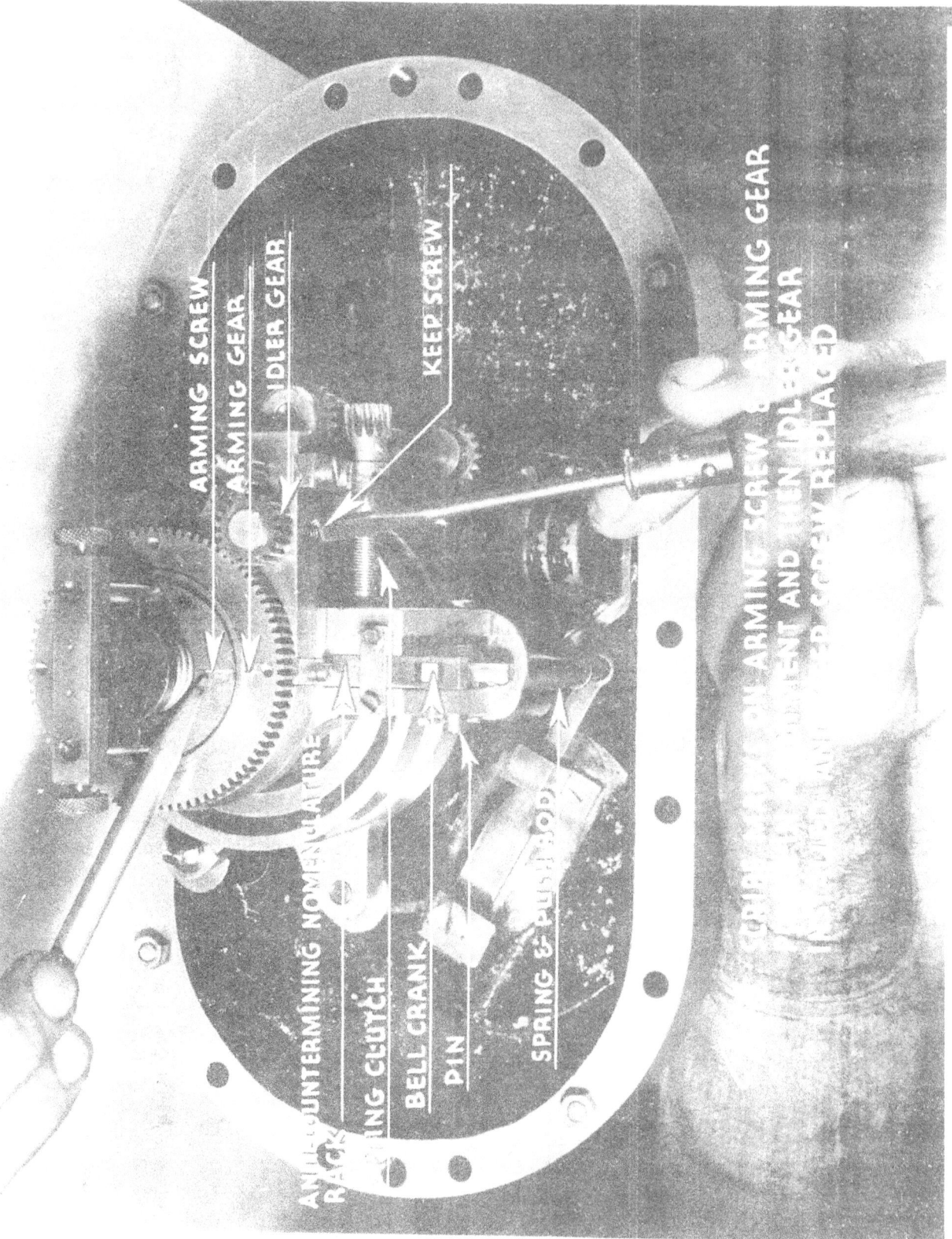

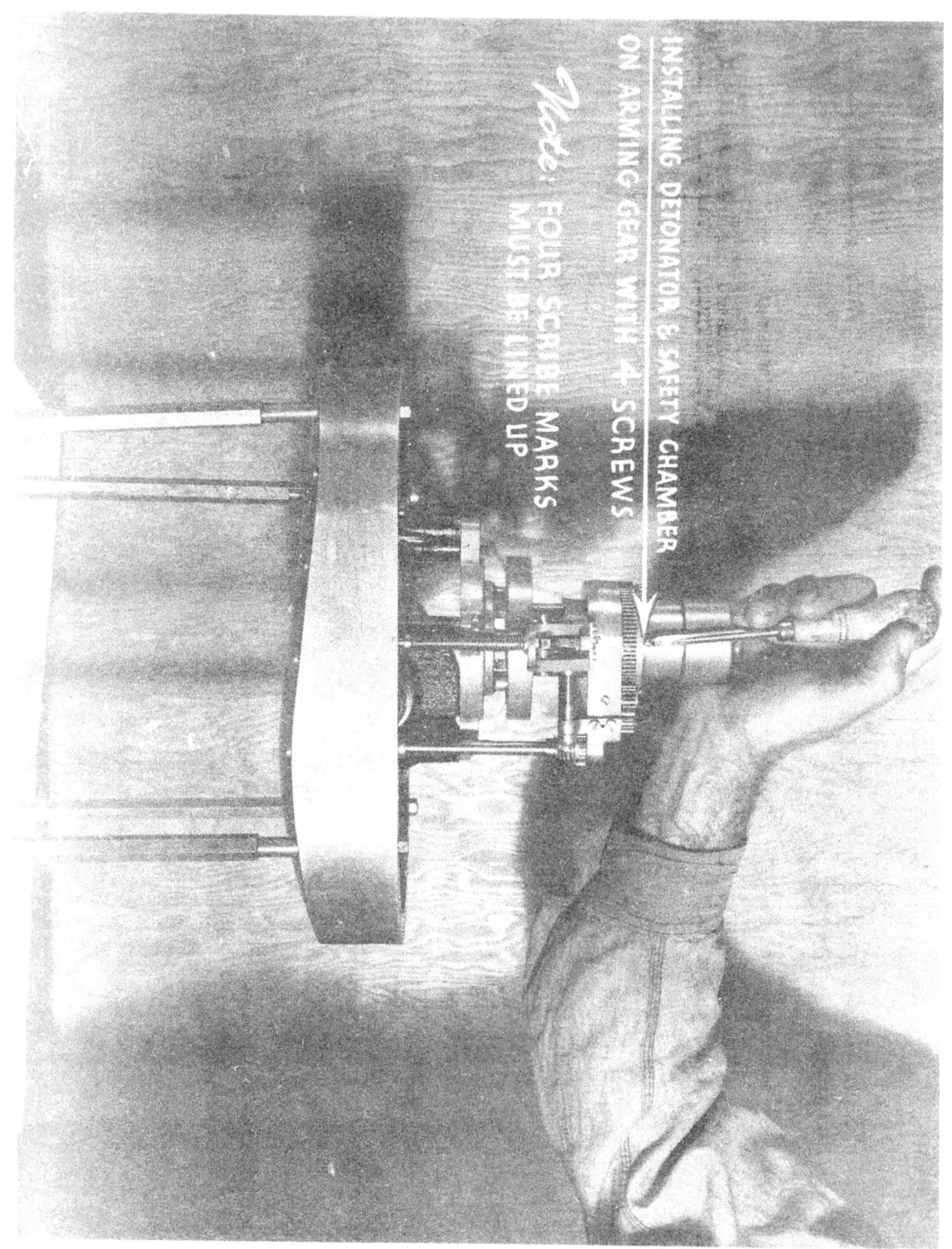

CHAPTER FIVE

		ARTICLES
A.	The Air Flask Section........................	5001
B.	The Air Flask...............................	5002-5007
C.	Water Compartment............................	5008-5019
D.	Stop and Charging Valve......................	5020-5023
E.	Air Check Valves.............................	5024-5032
F.	Fuel and Water Strainers and Check Valves....	5033-5038
G.	Disconnect and remove Afterbody from Air Flask..	5039
H.	Disassemble, Overhaul and Assembly of the Air Flask....................................	5040-5056
I.	Overhaul and Test Stop and Charging Valve....	5057-5059
J.	Overhaul and Test Fuel and Water Delivery Check Valves and Strainers...................	5060-5064
K.	Overhaul and Test of Air Check Valves........	5065-5068
L.	Charge Air Flask.............................	5069
M.	Safety Precautions...........................	5070
N.	Test of Air Flask for Tightness..............	5071
O.	Test of Flask Stop and Blow Valve............	5072
P.	Test of the Water Compartment................	5073
Q.	Inspection and Care of the Air Flask.........	5074

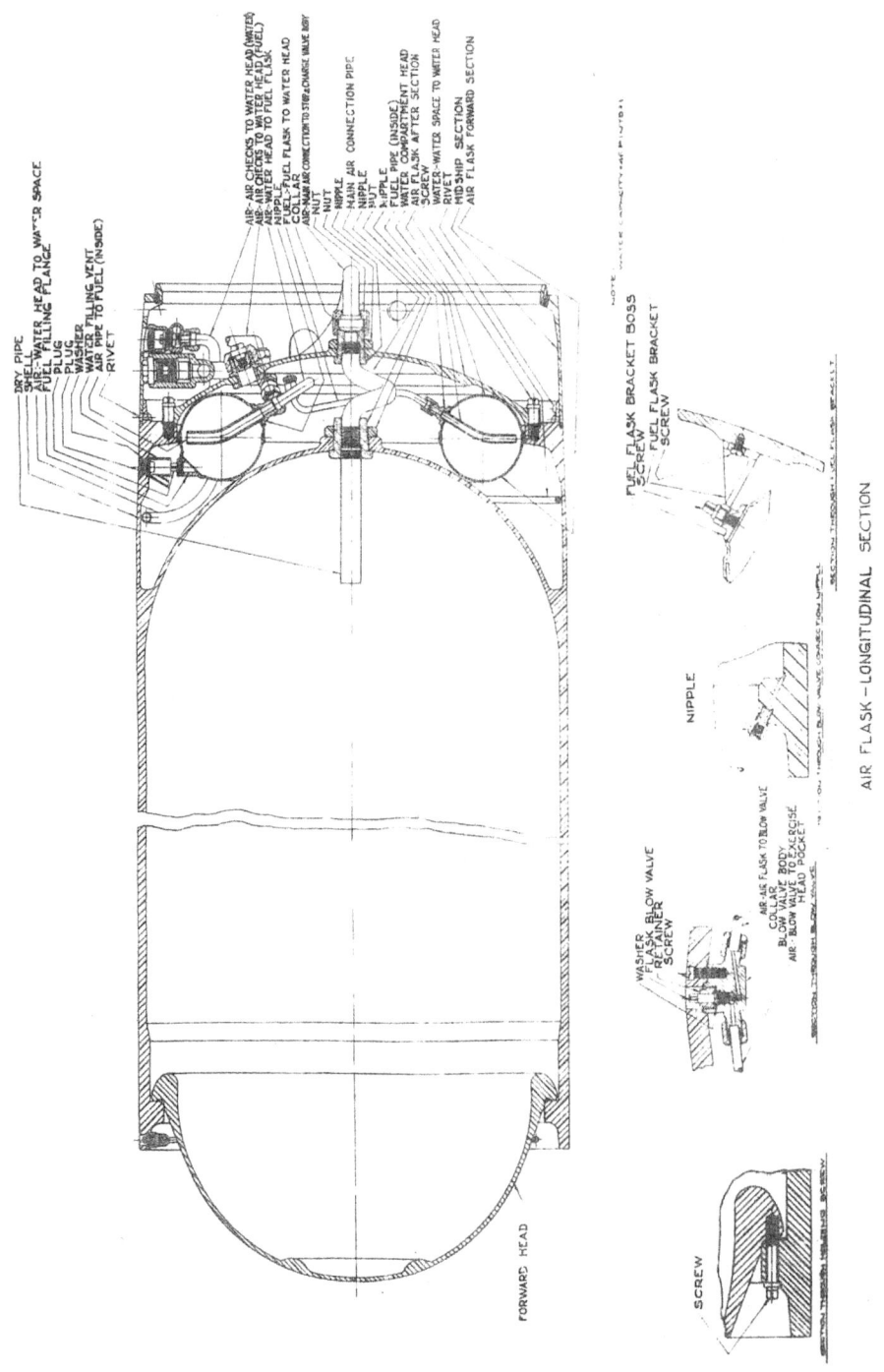

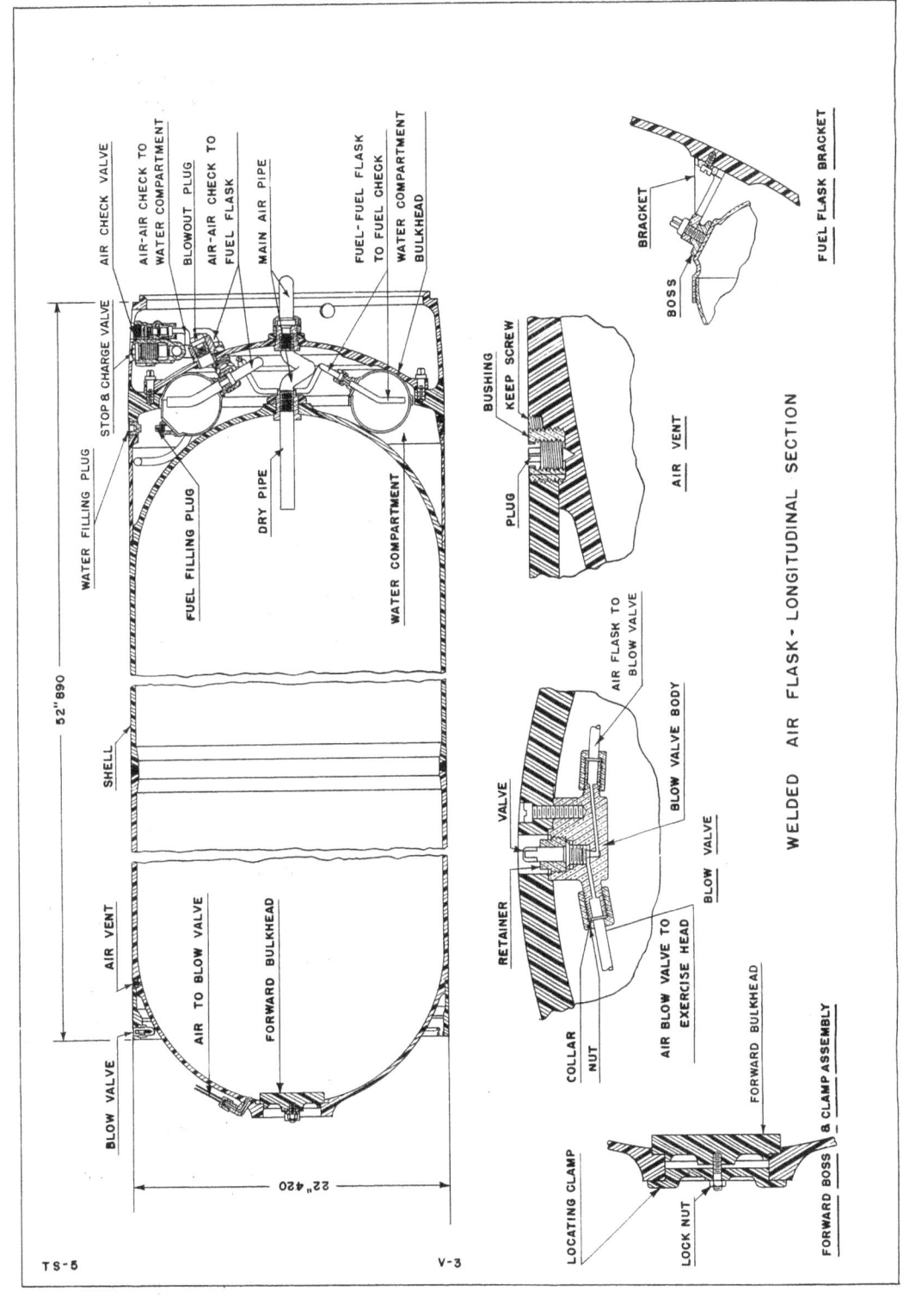

A. THE AIR FLASK SECTION.

5001. The air flask section consists of the air flask proper, the water compartment containing the fuel flask and the midship section. They will be described individually.

B. THE AIR FLASK.

5002. The air flask, consisting of the air chamber and water compartment forged integral therewith, is a drawn forging of alloy steel. The air flask is cylindrical in shape, the outside diameter tapering slightly at its after end. Both ends of the air chamber are closed by dome shaped heads, the after one of which is integral with the air flask. The forward head is removable and seats against a ground surface on a shoulder machined on the inside of the forward end of the air flask. This head is flattened at the ends of its horizontal diameter and two step slots are cut in the shoulder of the flask, which allow the reduced diameter of the head to pass through in the horizontal plane. The head is held in place on its seat with ten (10) steel screws. Pressure in the air flask tends more firmly to seat this head. Two shallow holes are tapped and threaded on a reinforced portion of the outer face of the head for the insertion of screw eyes for lifting and general manipulation. The flask is tested to 4000 lbs., hydrostatic pressure. In service, the use of air pressure in excess of 2800 lbs., is prohibited.

5003. The necessity for good ground joints on the forward bulkhead and its mating flange in forward end of air flask is of greatest importance. When making the joint, the surfaces must be absolutely clean and free from any particles of foreign matter; the surfaces should be wiped off with a piece of clean cloth. Waste should not be used for this purpose as the lint is disastrous to tight seats. The forward air flask bulkhead is removed and the interior of the flask inspected and cleaned periodically as prescribed in the Bureau of Ordnance Manual.

5004. In removing and repairing the forward air flask bulkhead, it is extremely important that the seat on bulkhead and flask be protected. A small burr caused by a tool being thrown against a seat may cause days of labor to remove by grinding, in order to stop leaks caused by same. If after installing bulkhead, a leak develops which does not disappear as the flask is charged, the head should be removed and the surface thoroughly cleaned and examined. In nearly all cases it will be found that the leak can be stopped by cleaning the seats. Occasionally it may be found necessary to apply a thin coating of strained white lead to seat. In extreme cases the bulkhead will have to be ground in place.

The grinding should be accomplished with a very fine carborundum and oil. This job requires a skilled mechanic and should generally be done only by experienced personnel.

5005. Air is admitted to and drained from the air chamber in the air flask through an opening in the center of the after bulkhead through which is assembled a section of the main air line extending through the water compartment to the stop and charging valve. A nipple on the inner end of this pipe is held against a lapped surface on the inside of the after air flask bulkhead by means of a lock nut. The inner end of this nipple is threaded to receive a dry pipe which extends into the air compartment about six inches. The dry pipe is used to prevent water or oil in the air flask being drawn into the air line when the torpedo is floating vertically at the end of the run. The interior of air flask and the after bulkhead is also electroplated with cadmium. **The forward bulkhead is also electroplated with cadmium except on the seat.**

5006. The flask stop and blow valve is located in the forward end of the flask on the top center line. The valve body is made of brass and is secured to the flask by one screw. The two opposite sides of the valve body are threaded to receive the inlet and outlet pipe connections. Two channels are drilled in the body; the inlet channel at a downward angle with a short vertical channel forming a seat for the valve. The outlet channel is drilled on an upward angle. The valve body is drilled and tapped to receive the valve and its retainer. A cup-shaped copper washer is interposed between a shoulder on valve stem and the retainer making the valve tight when fully open. The retainer is held in place by a small keep screw passing through the valve body, locking retainer in place. A channel is drilled in the flange of flask bulkhead seat approximately 30° left of the centerline. The forward end of this channel is greater in diameter and is tapped to receive the inlet pipe nipple, which is screwed and sweated in place. A short length of pipe from this nipple is connected to the inlet blow valve nipple. A length of pipe is similarly connected from the outlet nipple of blow valve body to a nipple on exercise head bulkhead when head is attached to air flask.

5007. The purpose of the flask stop and blow valve is to:
  (a) Afford a means for draining moisture from the flask after charging
  (b) Supply flask pressure to air releasing mechanism,
  (c) Bleed down air flask.

The operation is as follows: Using tool #49 and turning valve to left opens valve, allowing air to pass from flask through channels in valve body. When valve is fully open, the copper washer is jammed between shoulder on valve stem and retainer, thus forming a seal. Turning valve to the right will seat the valve.

NOTE: A new design of a welded air flask, as shown on page V-2A, is now in process of manufacture; pertinent supplementary instruction will be issued when complete information becomes available.

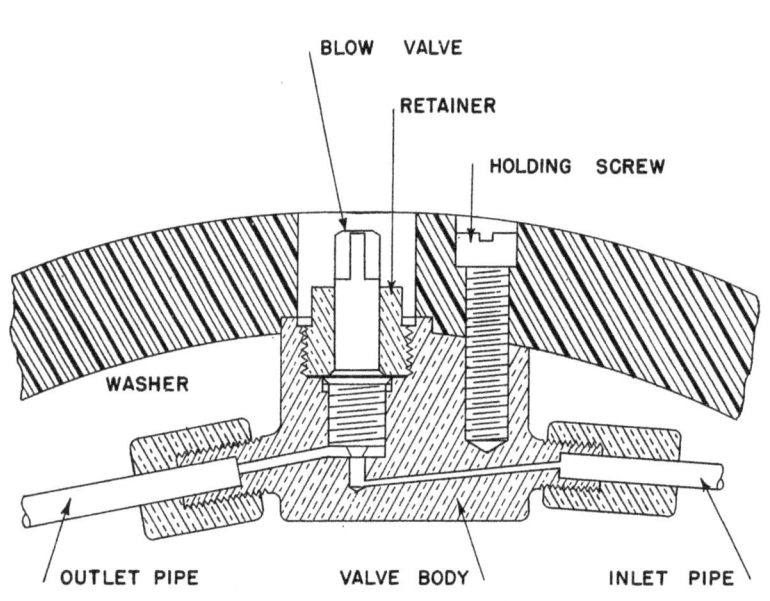

BLOW VALVE

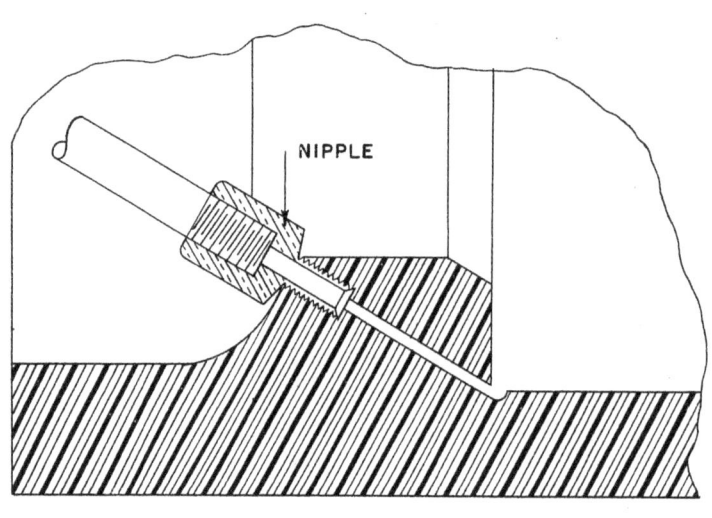

BLOW VALVE CONNECTION

TS-5 V-5A

C. **WATER COMPARTMENT.**

5008. The water compartment located abaft the air compartment is closed at its after end by a dome shaped head (water compartment head) which is held against a ground seat on a shoulder at the extreme end of the air flask section. In addition to sealing this compartment, the head sustains the fuel flask by three brackets attached to its inner surface, and provides an avenue for the main air line through a central opening which is made tight by the use of a copper washer and a hexagonal nut. Air from the fuel-air check to the fuel flask and fuel from the fuel flask to the delivery check are brought through reinforced portion of the head by means of nipples made pressure tight by soldering. Air to the water compartment is delivered through a nipple which replaces one of the holding screws; water from the water compartment to the delivery check passes through a pipe extending from the bottom of the water compartment through a nipple arranged similar to that of the air nipple for water.

5009. An opening in the top of the water compartment provides the filling hole for the water and access to the filling plug for the fuel flask.

5010. The complete inside of the air flask and water compartment, both sides of the heads and all the contents of the water compartment are electroplated with cadmium for protection against corrosion. The outer surface of the forward and water compartment heads (after assembly with nipples), are painted with one coat of iron oxide primer and one coat of navy gray paint.

5011. That portion of the main air pipe within the water compartment has a spirally wound coil which provides sufficient longitudinal flexibility to take up the varying expansion of the heads. Each end of this section of the main air pipe is screwed and soldered to a suitable fitting which is securely fastened by means of clamp nut, one to the after flask head and one to the water compartment head.

BLOW-OUT PLUG.

5012. To guard against destructive pressures being built up in the water compartment, due to the absolute failure of the reducing valve or a rupture in the main air line passing through the water compartment, a blow-out plug is provided. The body consists of a brass nipple secured by means of a nut and by being soldered to the water compartment bulkhead. It provides a 7/8" diameter opening in the bulkhead. It carries the blow-out plug which consists of a copper disc ".01 thick clamped between the plug nut and the plug screw, both of cylindrical form. The diameter of the cylindrical portion is ".015 less than the diameter of the hole in the fitting and is of sufficient length so that should the plug be blown, the parts cannot cant and bind in the fitting. In effect the blow-out

TS-5 V-6A

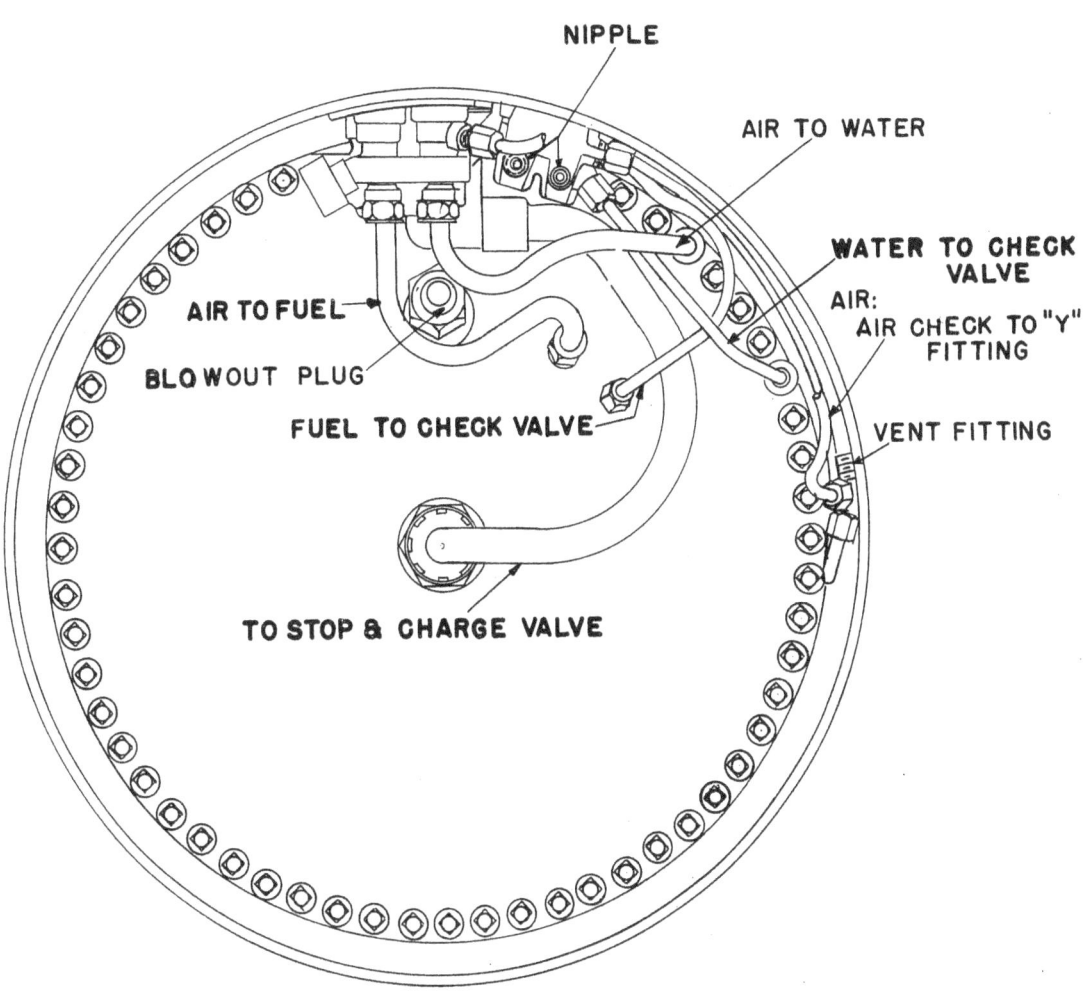

AIR FLASK
(END VIEW)

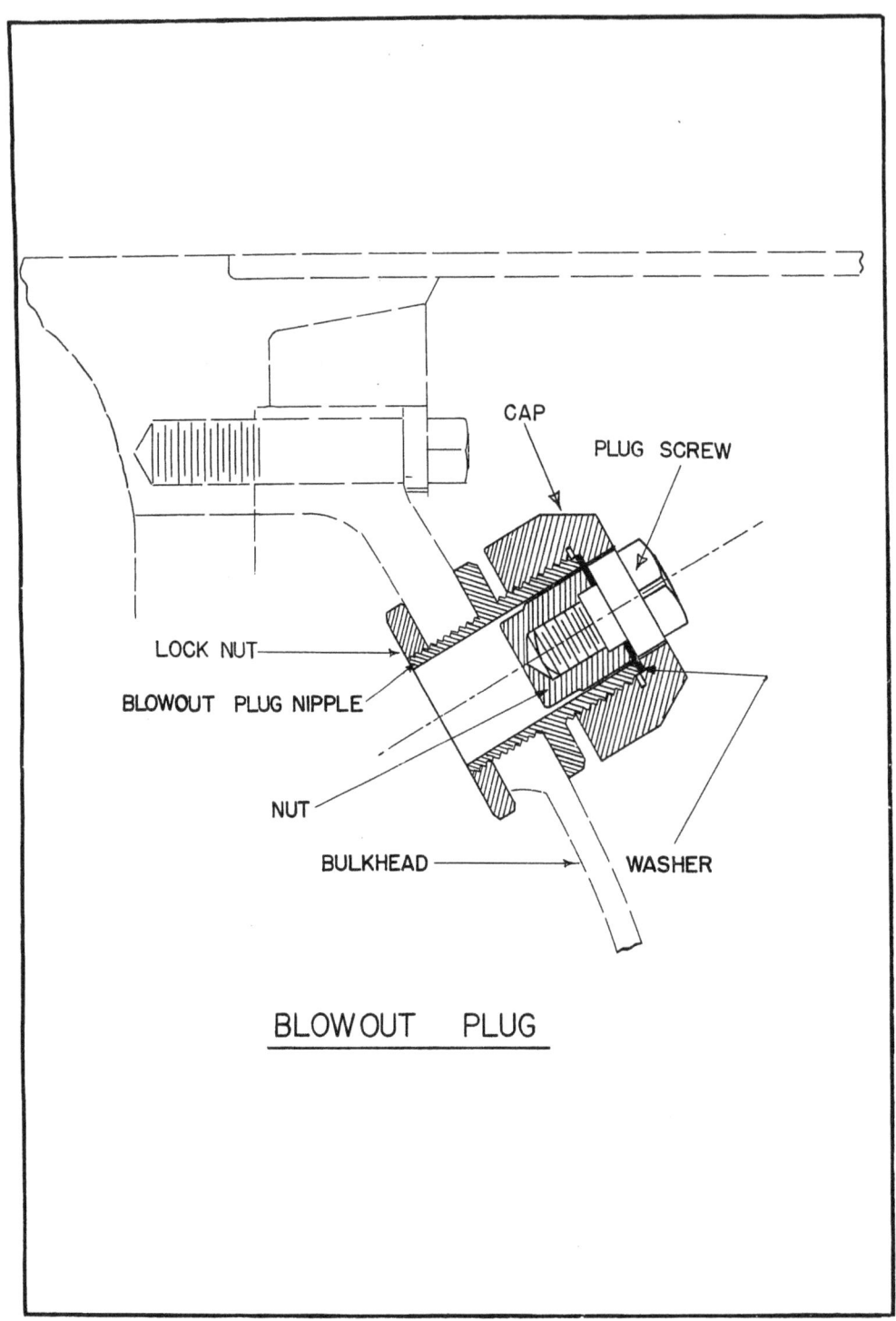

BLOWOUT PLUG

TS-5 V-6C

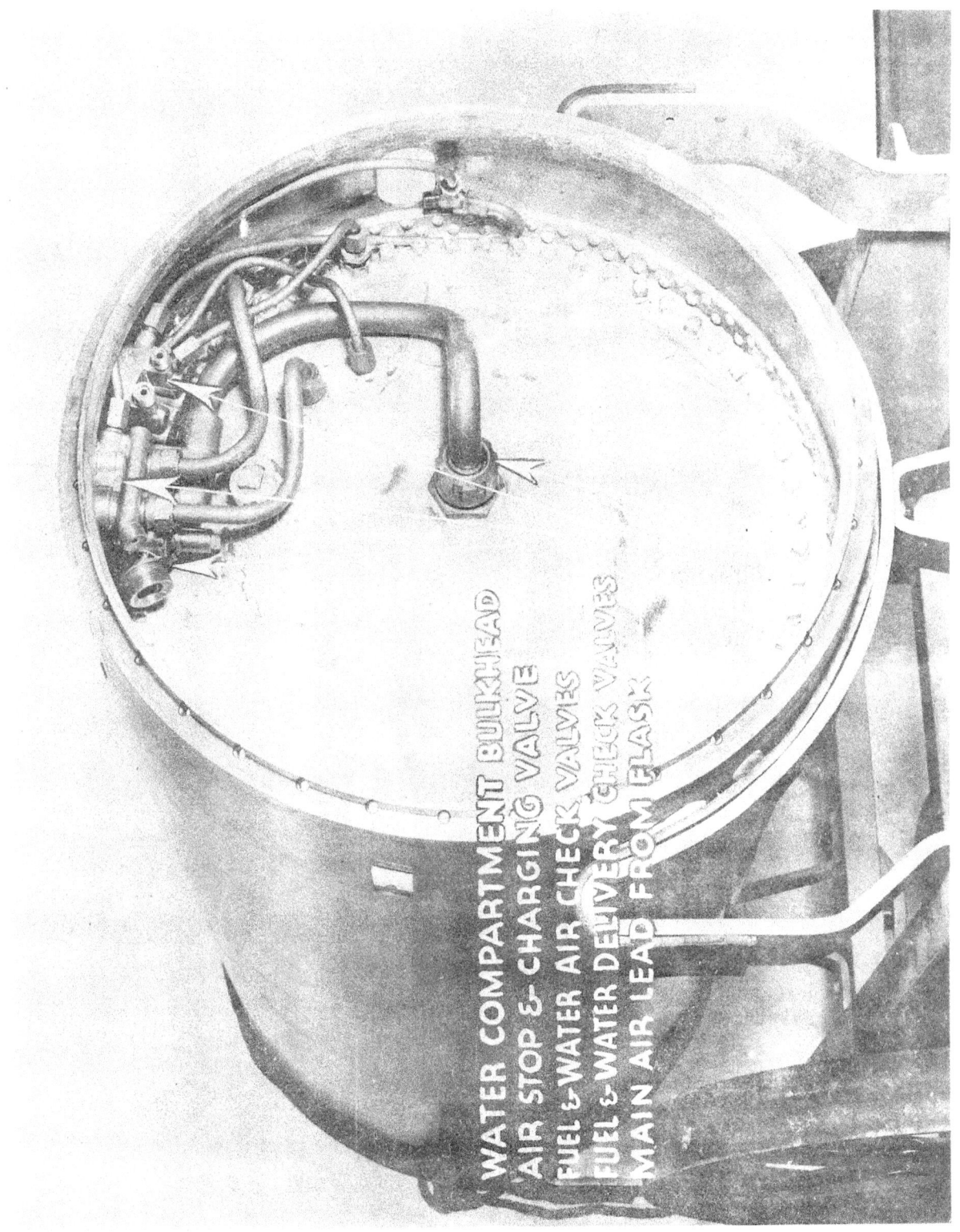

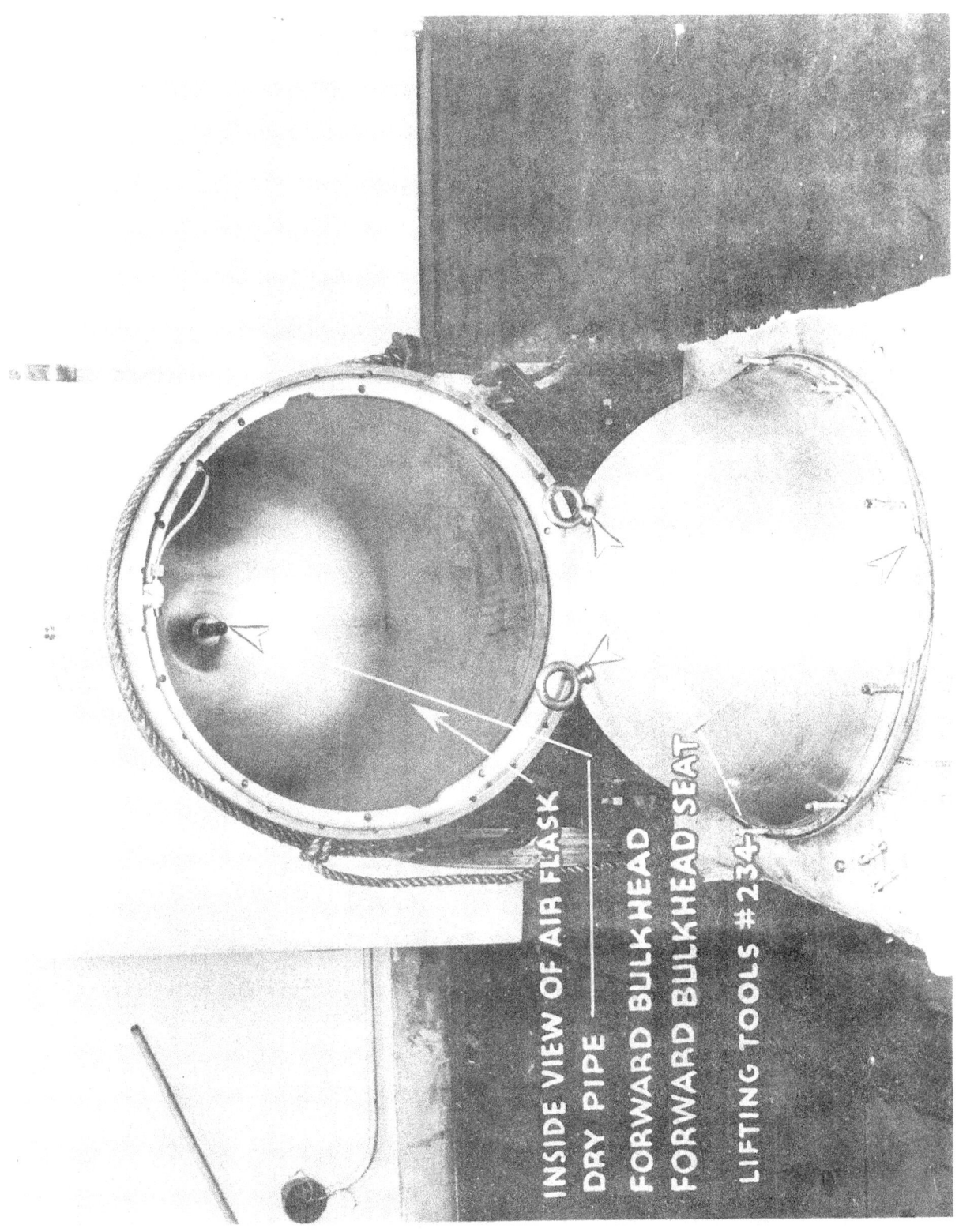

plug when assembled consists of a brass cylinder carrying a projecting copper flange ".01 thick. This is secured to the nipple in the bulkhead by means of a nut which securely clamps the copper flange to the face of the nipple. The whole arrangement is such that the stresses on the copper flange are those of shear only. Its strength is such that it will shear off at a pressure of about 1200 lbs. per sq.in. in the water compartment. Should this pressure be attained, the flange will shear and the plug will be blown out into the midship compartment, and thus a 7/8" clear opening will be provided for venting the pressure in the water compartment. As the ordinary running pressure is less than 500 lbs. per sq.in., there is sufficient margin of safety so that the shear flange will not be unduly strained by successive application of the working pressure.

5013. Its clear opening is such that it will pass water at a sufficient rate to prevent a rise of pressure in the water compartment in excess of 1500 lbs., even though the main air pipe should rupture. The construction of the blow-out plug makes the shear disc easily replaceable. It is important that copper sheet of uniform quality be used for these discs, and therefore only those furnished from the Torpedo Station should be used. It should never be the practice to make a disc from a sheet of copper selected at random.

FUEL FLASK.

5014. The fuel flask is designed in the form of a doughnut in order to permit its assembly in the limited space available in the water compartment; the flask is formed of brazing brass and tested to 50 pounds per sq.in. internal pressure after complete assembly with fittings.

5015. The filling flange on the fuel flask is located directly under the water filling flange, thus giving access through the water filling hole for manipulation of the fuel filling plug. The upper edge of the hole in the fuel filling flange is located about ".50 below the top of the fuel flask so that it cannot be completely filled with fuel. The air space thus provided, together with a similar air space in the water compartment, provides a margin of variation in the time for building up of pressure in the water compartment and the fuel flask when the torpedo is started, and tends to prevent a momentary unbalance of pressure on the fuel flask in excess of its strength.

5016. The air inlet and fuel outlet connections are located on the inner surface of the flask at an angle, which is most convenient for making the pipe connections. A short section of tubing is brazed on the inside of the air inlet connections extending through the fuel flask to within 3/16" of the top of the fuel flask. A similar pipe is attached to the fuel outlet connection, extending to near the bottom of fuel flask, for withdrawal of fuel from the flask. The air inlet and the fuel outlet nipples are connected by pipes to their respective nipples on the interior of water compartment head.

MIDSHIP SECTION.

5017. This section, which is riveted to and so becomes a permanent part of the air flask, is a forged steel cylinder appropriately machined for attaching to the air flask, the after end being machined to form a joint with the afterbody. The nature of the parts found in the midship section does not require the exclusion of water from their exterior surfaces. Further, certain parts conveying hot gases must be externally cooled in order that they may not be injured by the interior temperature, and access to certain other parts must be provided for adjustment or manipulation. Therefore, provision is made for free circulation of sea water in this section and for access by making openings in the shell. The shell is suitably marked adjacent each opening, indicating access to the following parts: main air connection; air charging valve; main stop valve; oil to reducer; reducing valve; strainers - fuel-water; fuel and water check valves; air to fuel and water; air check valve - fuel-water; igniter and water trip valve. A stop bolt hole is drilled through the top center near the joint for afterbody into which the stop bolt on the plane fuselage is inserted.

5018. The parts enclosed in this section and attached to the turbine bulkhead are known as the valve group and superheater.

5019. Attached to the shell of this section are the charging and stop valve body and fuel and water check valve and strainer body. Held in place by screws, they are readily removed in case it is necessary.

D. STOP AND CHARGING VALVE.

5020. The means provided for charging air into the air flask and for isolating the air in the air flask at will after charging until ready for use, thereby making it possible to disassemble and overhaul all the parts of the torpedo while the air flask is charged, is known as the Stop and Charging Valve.

5021. High pressure air leaves the air flask through the dry pipe in the after bulkhead and the spirally wound pipe in the water compartment to the stop and charging valve. The stop and charging valves are contained in one body located at the top of the midship section and are reached through openings in the shell, which are marked. The charging valve opening is closed by a plug which carries the charging valve. This plug is removed for charging, escape of air being prevented by a check valve. This check valve is guided in a check valve plug and held against its seat by a spring, a stop being provided in the bottom of the plug for limiting the opening of the check valve. A washer is interposed between the check valve plug and the body to prevent leakage around the seat. The charging valve seats against a leather washer as a final barrier to escape of air in case the check valve leaks.

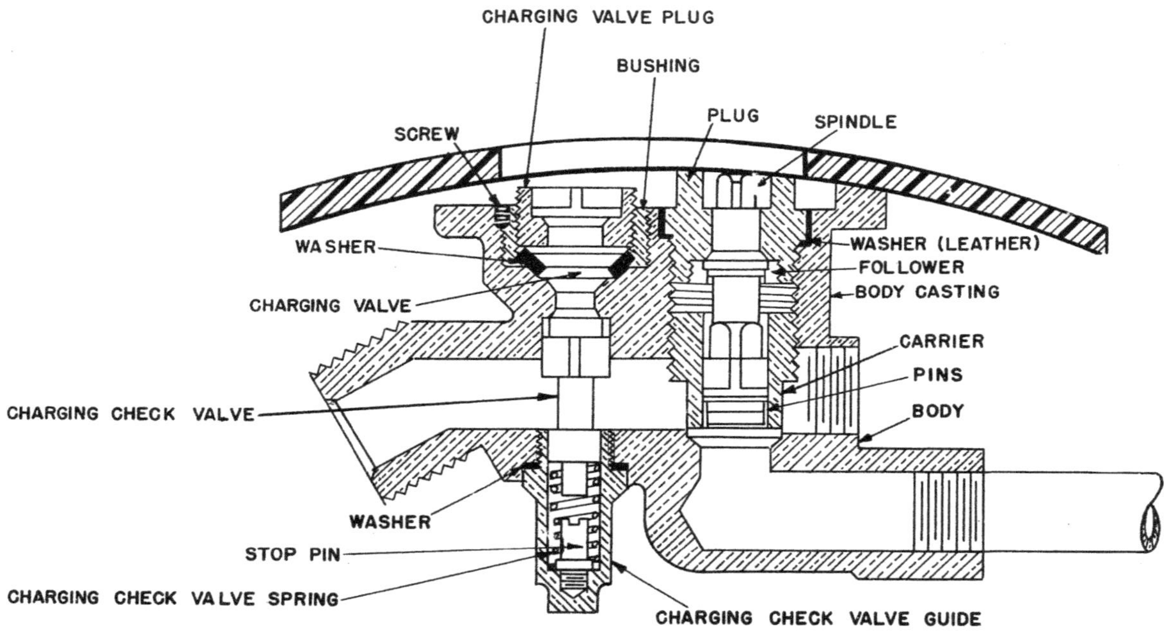

STOP & CHARGING VALVE

TS-5

V-8A

5022. For charging, the charging pipe connection is screwed into the threaded hole vacated by the charging valve plug. As this threaded hole is liable to be damaged due to its exposed position and the nature of its use, it is made in the form of a bushing screwed into the valve body, and is thus replaceable.

5023. The stop valve is of the compression type and closes airtight on its ground seat in the valve body. The stop valve carrier is threaded on the large diameter and at the outer end has an axial recess fitted to receive the square shank of the operating spindle. The stop valve proper is fitted to the bottom of the carrier in such a manner that it is free to rotate, but travels with the carrier; this is accomplished by means of a circular groove around the stem of the valve embraced by two small pins through the carrier. The operating spindle, by means of a collar machined about midway around its circumference, is held with freedom of revolution in the stop valve plug by a follower which in turn is held in place by a keep screw. A beveled seat machined on the operating spindle and matching a similar seat machined on the stop valve plug prevents leakage around the operating valve spindle. Turning the operating spindle screws the carrier up or down, lifting or seating the stop valve. The stop valve plug threaded the same as the carrier is set up hard against its seat, the joint being made tight by a leather washer.

E. **AIR CHECK VALVES.**

5024. The air check valves are interposed between the reducing valve and fuel and water compartments. Their purpose is to close those compartments against each other and into the reducing valve except when the torpedo is in normal operation and to vent pressure accumulations which may be caused by air leaks into the water compartment.

5025. The air check valves are contained in a naval brass body secured **by screws to the midship shell** on the after side of the stop and charging valve. This body is machined with the necessary inlet, outlet, and vent nipples, together with valve and plug seats for the fitting and assembly of the air check valves. In the inlet (offset) nipple is inserted a 3/16" restriction to check a sudden flow of air and permit the pressure to equalize on both sides of the fuel flask at the instant of opening, and thus prevent a possible rupture.

5026. The two air check valves are machined in the form of cylindrical plungers with the diameters of their inner ends increased to form the valve face, and the enlarged beveled area required for the air pressure to open the valves. The cylindrical portions of the valves are lapped into the bores of the valve plugs, thus forming guide sleeves for their movements in the plugs. Flat shoulders formed over the increased diameters of the valves are lapped to seats on the ends of the valve plugs to prevent leakage of air by the lapped guide sleeves with possibilities of cold runs. Bosses are machined in the bottom of the valve guide sleeves for centering the valve spring and to reinforce tapped holes used in connection with manipulation of check valves during overhaul and tests.

5027. As previously stated, the valve plugs are bored out and lap-fitted to the valve sleeves, thus forming a guide for the check valves. Holes are drilled through the outer ends of the valve plug into which bushings for centering the check valve spring are inserted. The diameters of the outer ends of the check valve plugs are increased to form flanges for seating the plugs in the valve body. The portion of the plugs under the flanges is threaded to fit similar threads in the check valve body and the outer ends of the plugs are machined with squares for insertion of a tool when overhauling.

5028. The check valves are held against their seats in the body by springs inserted in the check valve sleeve guides and compressed between the valves and their guide bushings in the check valve plugs.

5029. The outer ends of the spaces above the check valve plugs are closed by screw plugs and interconnected by an air vent passage, the entire space being vented into the afterbody with pipes passing through a common vent manifold fitting attached to the midship shell. Washers are interposed under the valve plug flanges and screw plugs to seal against leakage of air. NOTE: Venting of the air check valves is necessary to prevent pressure accumulation above the valve from cushioning the valve or preventing its opening.

5030. Reduced air pressure upon entering through the restriction will immediately fill the passage around the air check valves and will, due to the areas of the beveled surfaces of the valves above their seats being larger than the opposing area on the flat seats, open the valves against the pressure of their springs and hold the valves so open until the spring pressure again exceeds the forces tending to open the valves, at which time the valves will again close. Air pressure required to unseat the valves when acting against the beveled area above the valve seat is approximately 12 lbs.

5031. The air check valves may also be considered as relief valves in that any pressure accumulation in the fuel or water compartment in excess of 6 lbs. will act against the areas under the air check valve seats, unseating the valves and relieving such pressure through the restriction valve and combustion pot.

5032. The air is conveyed from the air check valves through pipes connecting through the water compartment bulkhead to the top of the fuel and water compartments.

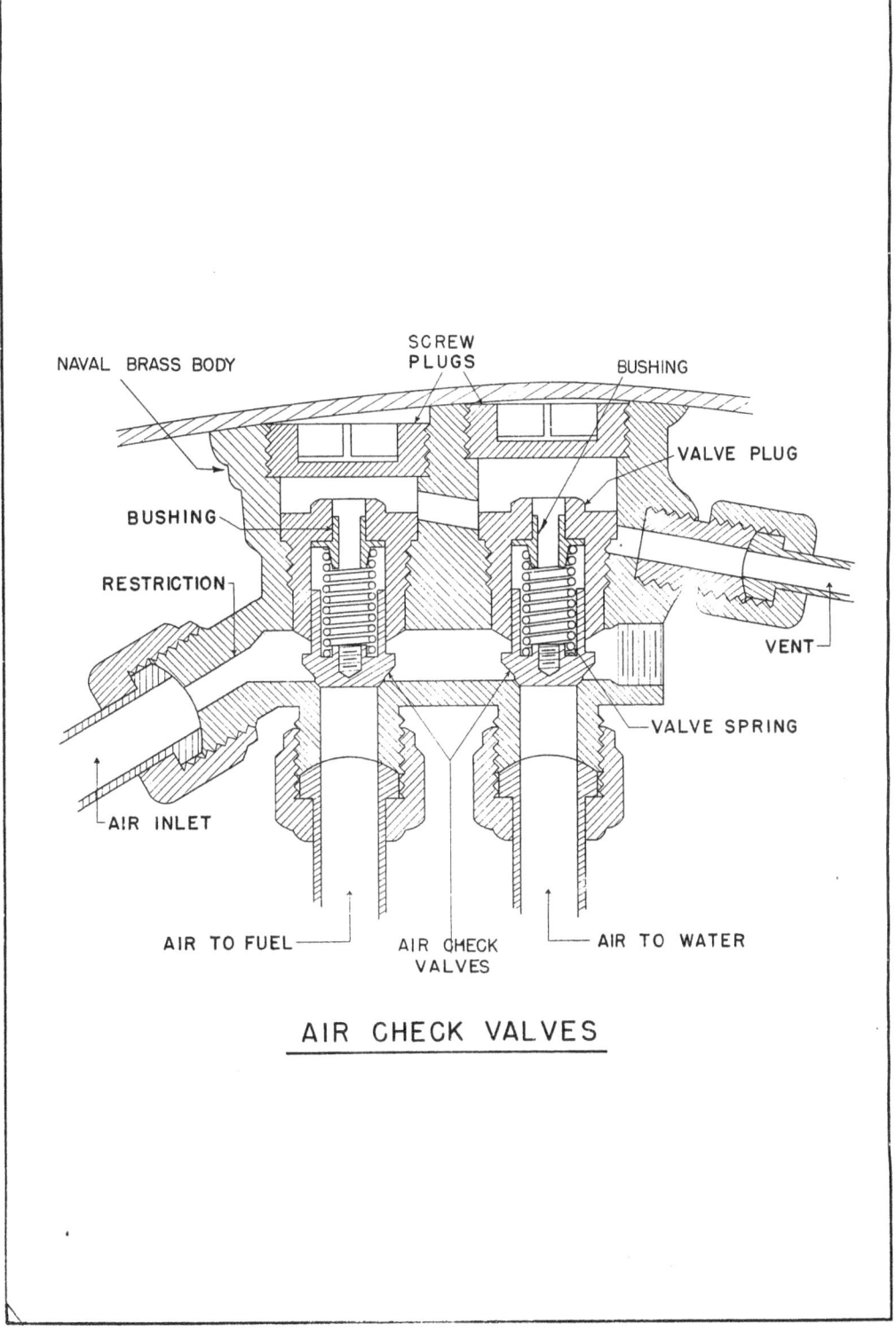

AIR CHECK VALVES

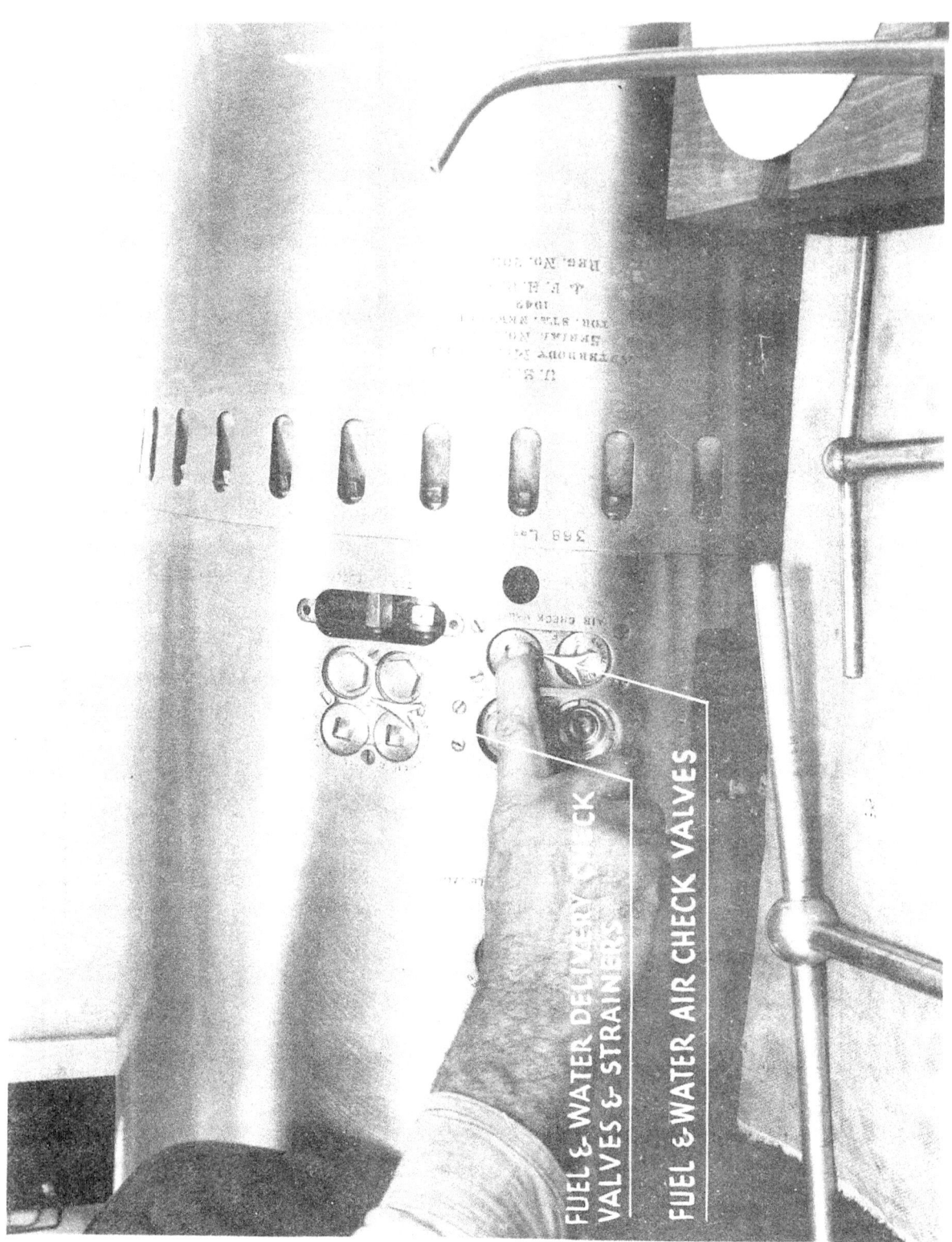

F. FUEL AND WATER STRAINERS AND CHECK VALVES.

5033. As previously stated the fuel and water are delivered to the sprays through strainers and check valves interposed between the fuel and water compartment and the combustion flask. The purpose of the strainers is to pick up any foreign matter which may be present in the fuel and water before reaching the sprays and thereby prevent possibility of clogging of the sprays. The fuel and water check valves automatically seal the fuel and water compartment against leakage of fuel and water through the sprays into the combustion flask until opened by reduced air pressure upon launching of torpedo.

5034. The fuel and water strainers and check valves are contained in a forged naval brass body secured to the midship shell adjacent to the stop valve. This body is suitably machined with four chambers in which are carried the fuel and water strainers and fuel and water check valves, together with the necessary air passages and nipples for the intake and discharge of fuel and water and for venting the outboard ends of the check valves.

5035. The fuel and water inlet ports are arranged so that the liquid enters through the top of the strainers, the spaces above these strainers being closed by screw plugs seated against washer and passing through, emerges at the bottom under the check valves and thence through these valves to their outlet nipples directly above the valve seats.

5036. The valves are of the poppet type and double acting, so that when closed fuel and water will not enter into the combustion flask and when open fuel and water are prevented from outboard leakage. The diameter above the valve face is increased and again decreased to fit the bore of the valve guide and thus form a shoulder which by seating against the inner end of the valve guide forms an effective seal against outboard leakage; the diameter is further decreased to form the valve which is lapped into a hole in the valve guide. The outer end of the valve stem is drilled and tapped for the insertion of a manipulating tool. A spring encircling the stem and compressed between its shoulder and the valve guide normally keeps the valve on its seat, 18 to 20 lbs. pressure being required to unseat both valves.

5037. The valve guides are threaded to fit similar threads in the check valve body, the outer end of the guides being machined with hexagonal heads for insertion of manipulating tool. The spaces above the valve guides are closed by screw plugs seated against washers and inter-connected by air venting passages to a common nipple which in turn connects with the vent manifold by a pipe.

5038. From the check valves the fuel and water passes through pipes to the sprays, fuel and water spray bodies and the combustion flask.

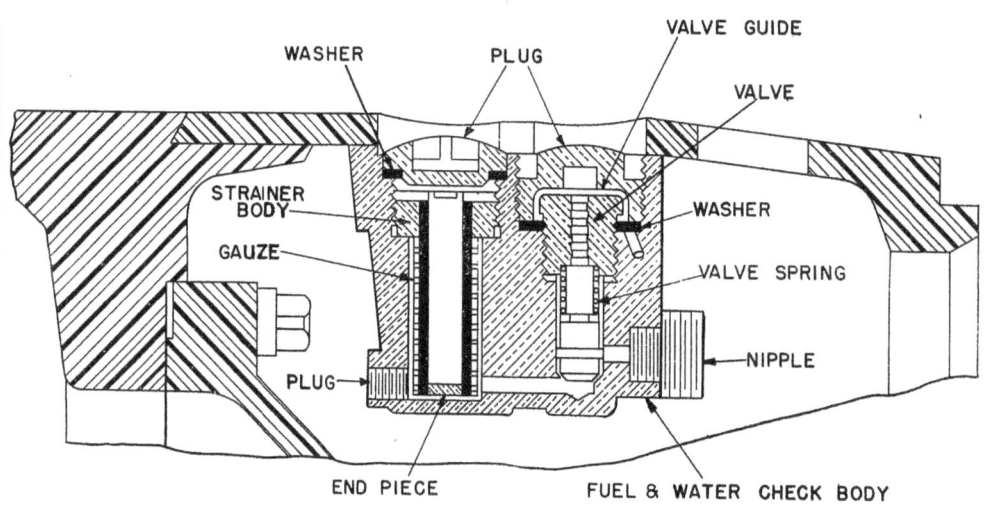

FUEL CHECK & STRAINER

NOTE: Steps preceded by an asterisk (*) are necessary only when repairing or replacing damaged parts.

	TOOL NO.

G. DISCONNECT AND REMOVE AFTERBODY FROM AIR FLASK.

5039.
  (1) Disconnect the following pipes:
    (a) Main air connection........................ 134
    (b) Air pipe to air check valves............... 229
    (c) Pipes to fuel and water strainers.......... 141A
    (d) Pipe from venting fitting to afterbody..... 141A
  (2) Install sling around afterbody and take the weight with chain fall...
  (3) Unscrew joint screws........................... 386
  (4) Lift tail slightly higher, rotate afterbody left to clear main air connection and pull clear of flask.
  (5) Place afterbody on stand.

H. DISASSEMBLE AIR FLASK.

5040. Remove air check valve assembly.
  (1) Disconnect pipes from check valve to water compartment bulkhead......................... 229
  (2) Disconnect pipe to venting fitting............ 141A
  (3) Remove holding screws and check valve body assembly.. 41

5041. Remove fuel and water check valve and strainer body.
  (1) Disconnect pipe from bulkhead to fuel strainer. 141A
  (2) Disconnect and remove pipe from check valve body to venting fitting......................... 141A
  (3) Disconnect pipe from bulkhead to water strainer. 141A
  (4) Remove holding screws and remove check valve and strainer body assembly..................... 41

5042. Remove stop and charging valve complete.
  (1) Disconnect main air pipe at water compartment bulkhead.. 134
  (2) Remove holding screws and stop and charging valve group..................................... 41

5043. Remove vent fitting.
  (1) Remove holding screws and fitting............. 41

5044. Remove water compartment bulkhead assembly.
  (1) Remove water and fuel filling plug............ 11,74,217
  (2) Unscrew clamp nut for main air connection through water compartment bulkhead............. 242A
  (3) Loosen clamp nut and unscrew nipple for air to water compartment............................... 144

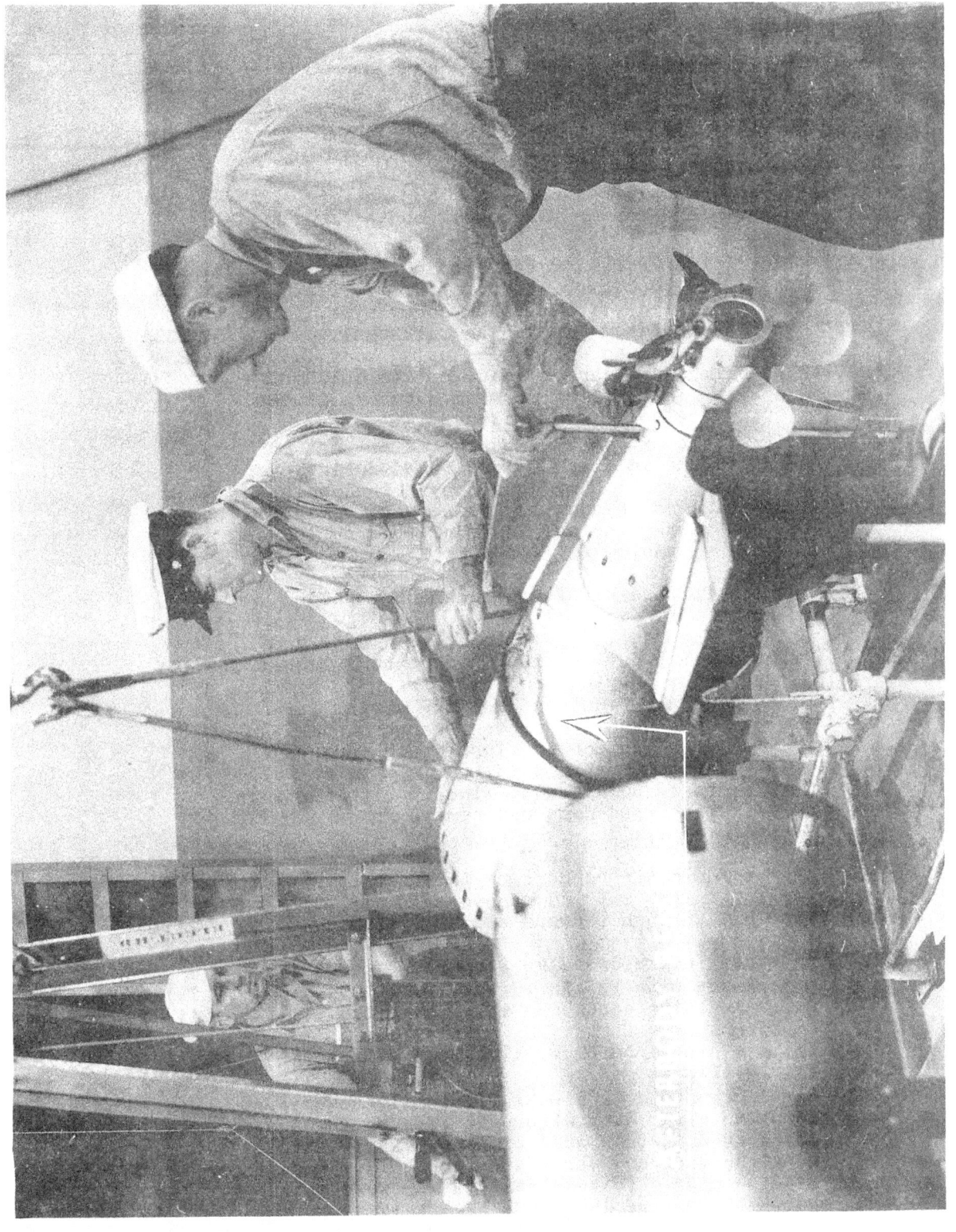

TS-5 V-12B

	TOOL NO.

(4) Loosen clamp nut and unscrew nipple for water from water compartment.......................... 144
(5) Remove water compartment holding screws......... 12
(6) Insert lifting handles (S.G.2868) and remove water compartment bulkhead assembly.
   NOTE: It may be necessary to tap around joint with lead hammer to loosen white lead joint.

5045. Remove forward air flask bulkhead - forged flask.
   (1) Remove pipe from blow valve body to exercise head.. 404
   (2) Remove replacement plugs for screw eyes........ 39
   (3) Remove 10 holding screws for bulkhead.......... 135A
   (4) Insert screw eyes for manipulating head........ 234
   (5) Push bulkhead clear of seat, rotate bulkhead and with flats lining up with cutaways in seat of flask slide head out clear.
   NOTE: It is extremely important that seat on bulkhead and in flask be protected during overhaul, as a small burr caused by a tool being thrown against a seat may cause days of labor to remove by grinding, in order to stop leaks caused by same.

5046. Remove forward air flask bulkhead - welded flask.
   (1) Loosen clamp nut, remove locating clamp, and replace clamp nut on stud in forward bulkhead... 404
   (2) Attach piece of wire or heavy string to clamp bolt under clamp nut and holding string (or wire) in one hand, to prevent forward bulkhead from falling inside air flask, push forward bulkhead aft to break seat.
   (3) Turn flats in bulkhead to line up with slots in air flask forward dome and remove forward bulkhead.

*5047. Remove main air connection in after bulkhead.
   (1) Back off clamp nut and remove main air pipe assembly through forward end of flask.......... 241A

5048. Remove flask blow valve assembly.
   (1) Disconnect pipe to air flask................... 141A
   (2) Remove holding screw and valve assembly........ 41

*5049. Remove pipe from blow valve to air flask - forged flask.
   (1) Straighten bend in pipe sufficiently to unscrew.
   (2) Saturate a rag with water and pack on forward flask head seat in wake of pipe. Apply sufficient heat around threads to run solder and unscrew pipe. (Excessive heat may warp bulkhead seat out of line and cause a leak).

The above completes disassembling of air flask.

OVERHAUL, ASSEMBLY.

                                                                                   TOOL NO.

*5050. Replace pipe from bulkhead flange to blow valve - forged flask.
  (1) Clean tapped hole in flange with a 5/16" - 20 thread bottoming tap.
  (2) Tin threaded end of pipe and tapped hole.
  (3) Apply enough heat on flange to run solder and screw pipe in place.
  (4) After resoldering see that hole in pipe is clear of solder.

*5051. Replace main air connection on after head.
  (1) Apply a 50-50 mixture of pure strained white lead and light lubricating oil (C) to lapped surface of nipple and after head.
  (2) Insert main air connection in bulkhead through forward end.
  (3) Slip clamp nut in place over after end of pipe and secure.................................... 241A

5052. Treat interior of air flask - forged flask.
  (1) Clean thoroughly of oil and examine for corrosion. If corrosion or discoloration is noted, the surface must be scrubbed with clean hot water, using a stiff bristle brush and thoroughly dried. Use of alkalies, gasoline (or other volatile liquids) for cleaning is forbidden. When flask is thoroughly dry, all electroplated surfaces shall be given a light coating of equal parts of steam cylinder oil and hot running torpedo oil
  NOTE: No oil should be applied to interior of welded flasks.

5053. Replace forward bulkhead - forged flask.
 *(1) Apply carborundum grinding compound mixed with oil (D) sparingly around seat on bulkhead.
 *(2) Insert bulkhead through slots in air flask seat and pull the bulkhead up to its seat in flask, catching one holding screw to hold in place when installing grinding tool................. 135A
 *(3) Install grinding tool and proceed to grind bulkhead to its seat, wiping and inspecting occasionally until a good seat is obtained all around.
 *(4) Remove grinding tool.
  (5) Clean and dry ground seats thoroughly (do not use lye or other alkali nor gasoline or volatile oils - hot water only should be used) and apply a 50-50 mixture of pure strained white lead and light lubricating oil (C) to both seats.

TOOL NO.

    (6) Install screw eyes, insert bulkhead through slots in air flask seats and pull bulkhead up on its seat with numbers stamped uppermost, line up to holes and proceed to secure with 10 holding screws coated in oil (D) (it is sometimes necessary to run a 5/6" x 20 die over the holding screws to secure a proper fit)...... 135A
    (7) Remove screw eyes and replace replacement plugs. 39

5054. Replace forward bulkhead - welded flask.
  *(1) Apply carborundum grinding compound mixed with oil (D) sparingly around seat on bulkhead.
  *(2) Insert bulkhead through slots in air flask seat and pull bulkhead up to its seat.
  *(3) Install grinding tool and proceed to grind bulkhead to its seat, wiping and inspecting occasionally until a good seat is obtained all around.
  *(4) Remove grinding tool.
  (5) Clean and dry ground seats thoroughly (do not use lye or other alkali nor gasoline or volatile oils - hot water only should be used) and apply a 50-50 mixture of pure strained white lead and light lubricating oil (C) to both seats.
  (6) Install bulkhead on its seat - hold in place by center boss and put on locating clamp - making sure that clamp seats in center slot in bulkhead and slots in air flask forward dome.
  (7) Assemble clamp nut with bellows part of nut away from locating clamp and secure............ 404

5055. Flask blow valve.
  (1) Clean and inspect blow valve body and parts. Remove burrs where necessary on nipples and in blow valve seat. Apply oil (B) to valve.
  (2) Replace new washer around blow valve stem and assemble blow valve in body. 49
  (3) Replace retainer over blow valve stem and secure with set screw............................. 37
  (4) Replace blow valve on air flask and secure with screw... 41
  (5) Clean out joint screw holes with a tap..........

5056. Water compartment bulkhead with parts assembled on.
  (1) Disconnect air pipe from bulkhead to fuel flask. 229
  (2) Disconnect fuel pipe from bulkhead to fuel flask.. 141A
  (3) Remove 3 screws for bracket and fuel flask, and remove fuel flask............................... 64,49A
  *(4) Disconnect and remove air pipe to fuel flask and fuel pipe from fuel flask at bulkhead....... 229

	TOOL NO.

*(5) Unscrew screws (2 each) and remove the three brackets for fuel flask.................... 40

*(6) To move nipples for fuel and fuel air leads, apply only sufficient heat to each nipple to run solder and unscrew nipples................. 141A

*(7) To replace nipples, clean tapped holes in bulkheads and tin threads on nipples and screw into holes in bulkhead, apply only sufficient heat to keep solder soft until screwed all the way up. After soldering in place, reseat seats on end of nipples and base burrs off thread.
                WE85
                WE83
                WE94
                WE95

*(8) Replace the 3 brackets for fuel flask on bulkhead and secure with holding screws............ 40

*(9) Clean, inspect, dress burrs off where necessary and connect pipes to bulkhead as follows:
  - (a) Air pipe from bulkhead to fuel flask at bulkhead................................... 229
  - (b) Fuel pipe from bulkhead to fuel flask at bulkhead................................... 141A

(10) Clean, inspect and test fuel flask for leaks as follows:
  - (a) Blank off air connection................... 229
  - (b) Connect test pipe from fuel outlet connection to valve on outlet pipe for spray and check valve testing outfit, described in O.D. #750. No water should be used in test tank when testing fuel flask............... 141A
  - (c) Open low pressure air valve and allow pressure to enter fuel flask (about 45 lbs).
  - (d) Submerge fuel tank in water and note if bubbles rise indicating a leak.
  - (e) Test completed, dry fuel flask, cover with light lubricating oil (C), and dress burrs of nipples on fuel flask if necessary...... WE82,WE83 WE94 WE95

(11) Replace fuel tank in bracket and secure with 3 screws.................................... 49 bent

(12) Connect air and fuel pipes to fuel flask....... 229,141A

(13) Clean interior of water compartment, clean out fitting and vent holes.

(14) Clean, inspect and install air pipe and nipple for air to water compartment................... 229

(15) Clean bulkhead seats including tapped screw holes and apply a coating of white lead and light lubricating oil (C) on seats.

(16) Install water compartment bulkhead assembly, guiding main air connection nipple through hole in bulkhead and secure with holding screws dipped in compound steam cylinder oil (D)............. 12

(17) Screw clamp nut over main air connection nipple and secure..................................... 241A,242A

I. OVERHAUL AND TEST STOP AND CHARGING VALVE GROUP.

5057.
                                                                                                           TOOL NO.

  (1) Remove charging valve plug and washer............ 13-14
  (2) Remove check valve guide, check valve and spring. 12
  (3) Remove stop valve plug and spindle............... 185
  (4) Remove stop valve and carrier.................... 405
*(5) Remove keep screw and bushing for charging
     valve... 81,420 with 377
*(6) Scribe assembly marks on main air pipe and stop
     valve body. Apply sufficient heat around threads
     to soften solder, unscrew main air pipe and
     wipe solder from valve body.
*(7) Try threads on main air pipe for fit in thread-
     ed hole in valve body, tin threads of both.
     Slip nut over end of pipe, apply sufficient
     heat around threads to soften solder and screw
     pipe in place in body, lining up scribe marks.
     NOTE: If a new pipe is fitted screw pipe into
     valve body without soldering. Install valve
     group on air flask and secure in place. Fit
     main air pipe to line up with main air nipple
     in center of water compartment bulkhead and
     scribe marks on pipe and stop valve body, remove
     and solder in place as described above.
*(8) Reseat main air connection nipple. Reseat seat   WE81
     for stop valve. Reseat stop valve plug seat in   WE80
     valve body...................................... WE41,WE42
*(9) Screw new charging valve bushing in place, mark
     place opposite keep screw hole, remove bushing
     and with small round file remove sufficient   420
     stock on bushing to correspond with half of   with
     hole drilled before tapping, replace bushing in   377
     valve body with hole for keep screw lining up,
     run a drill into hole and follow up with a tap.
     Insert keep screw for bushing................... 81
*(10) Reseat charging check valve seat................ WE46
*(11) Lap charging check valve to seat................ 41
 (12) Clean charging check valve and seat, replace
     valve and spring.
 (13) Clean, inspect, remove burrs where necessary
     and see that stop pin is screwed in tight on
     guide for check valve. Replace over spring
     and check valve in valve body, screwing up   12
     tight against a copper washer. Replace charg-
     ing valve washer and plug....................... 13-14
 (14) Grind stop valve to seat. Apply lapping com-
     pound sparingly, screw stop valve and carrier
     down to seat but not hard enough to take up.
     Insert a screw driver through carrier to slot   405
     in stop valve and grind until seated............ 41
 (15) Wash grinding compound from valve seat, clean,
     oil (C) and install stop valve carrier with
     valve... 180

	TOOL NO.

(16) Remove screw and stop valve spindle follower and stop valve spindle from stop valve plug. Clean, inspect, remove burrs, lap stop valve spindle to seat in plug, oil (C) and assemble spindle follower and keep screw in plug........ 181,81

(17) Replace washer and stop valve plug in valve body.

TEST OF STOP AND CHARGING VALVE.

5058. The assembled valve should be tested for tightness with 2300 lbs. pressure, as follows:
  (1) Connect air to stop valve side of valve body. Close stop valve tight.
  (2) Turn on H.P. air.
  (3) Submerge male end of valve body in water and look for bubbles.
  (4) Screw blank on male end.
  (5) Remove charging valve and plug.
  (6) Open stop valve wide.
  (7) Submerge body in water and look for bubbles at:
    (a) Stop valve spindle.
    (b) Body Plug.
    (c) Main air connection.
    (d) Stop valve plug.
    (e) Charging check valve guide.
    (f) Charging check valve.
    (g) Around body casting.
  (8) Close stop valve, turn off air, remove leads and blanks.

5059. Install stop and charging valve group on air flask as follows:
Catch threads on main air pipe nut to nipple on bulkhead, line up and secure valve body to midship shell with 4 screws. Turn main air connection nut until it is fully tightened........ 134 / 41 / 134

J. OVERHAUL AND TEST FUEL AND WATER DELIVERY CHECK VALVES AND STRAINER.

5060.
  (1) Remove plugs for strainers and remove strainers. 405,372A
  (2) Remove plugs and washers for fuel and water check valves.. 406
  (3) Remove valve guides with valves and springs..... 74,407
*(4) Remove burrs out of holes in body for check valves... WE119
*(5) Reseat counterbored nipples on body and remove burrs on threads........................ WE85 WE94A,WE95,WE87
*(6) Remove burrs in tapped hole for check valve guides.. WE117
  (7) Clean, inspect and oil (C) check valve body.

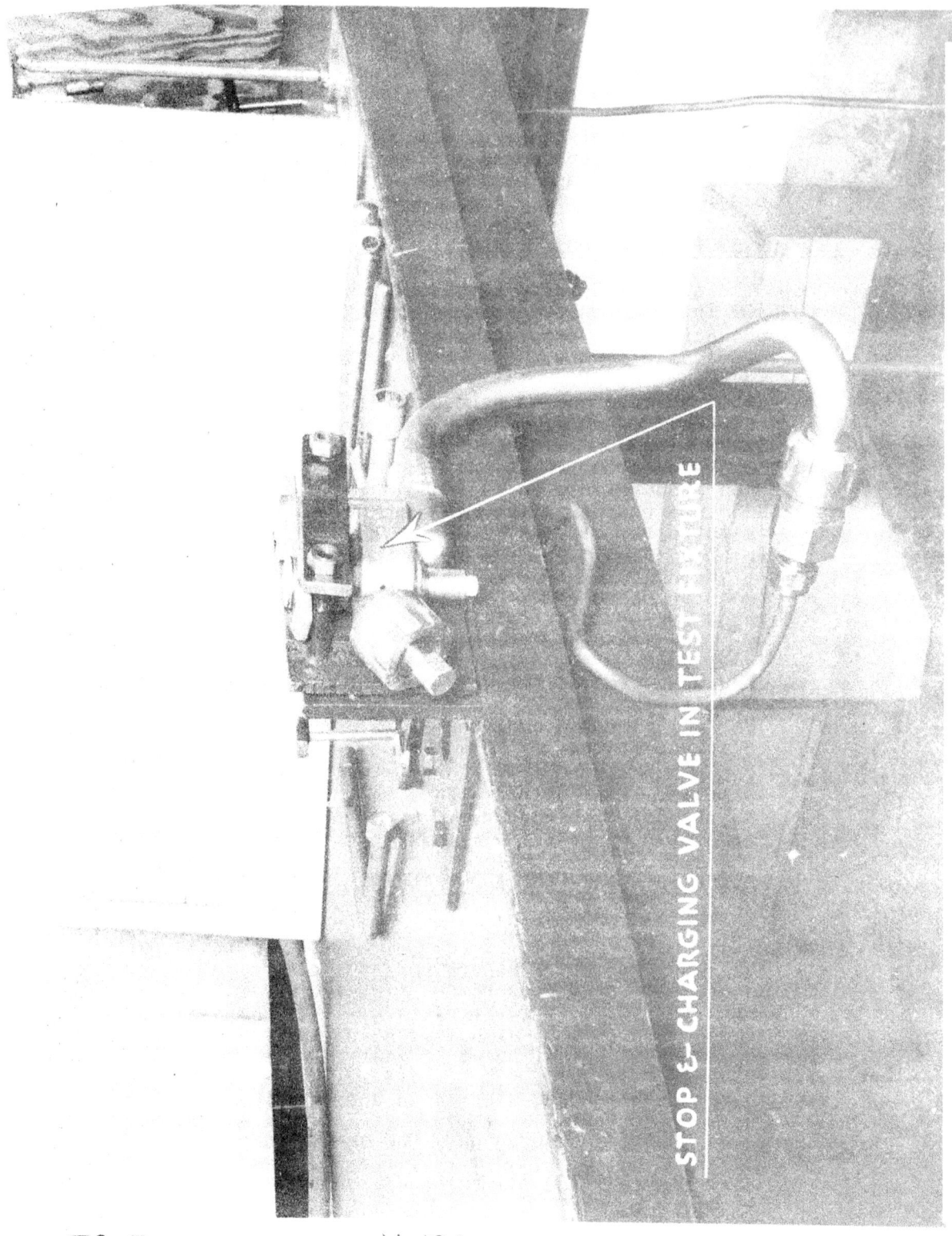

		TOOL NO.

*(8) Ream burrs out of small diameter hole in check valve guide.................................... WE120

*(9) Ream burrs out of large diameter hole in check valve guide.................................... WE124

*(10) Reseat outer seat for valve in check valve guide... WE121

*(11) Reseat outer seat on valve (a tool is provided for holding valve during this operation and when lapping valve stem into hole in guide)..... WE122 / WE127

*(12) Lap valve stems into guide.

*(13) Lap valves to their seats in body............... 74

(14) Clean, inspect, oil (C) and assemble spring and valve in guide and replace in check valve body.. 141A

(15) Clean, inspect and replace check valve plugs and washers..................................... 407

(16) Clean, inspect, oil (C) and replace strainers (do not set up strainers more than hand taut to prevent sticking on the next removal).

(17) Replace washers and strainer plugs.............. 405

TEST OF FUEL AND WATER STRAINERS AND DELIVERY CHECKS.

5061. Test of inboard seats.
   (1) Fill tank of testing set with water.
   (2) Connect inlet of check valve body to outlet pipe of testing tank beyond strainer. The inlet nipples to fuel and water sides of check valve body are not the same size. Appropriate fittings must be made.
   (3) Open valve in outlet line.
   (4) Open valve in inlet line slowly, admitting air pressure on top of water in tank.
   (5) Note that inner seats are tight with 3 to 5 lbs. pressure.
   (6) Build up pressure slowly and note when valves fully open. Should be 18 to 20 lbs. (Spring calibration).
   (7) Remove check valves, dry and oil.

   NOTE: Test of the inboard seats can be made, using air only.

5062. Test of outboard seats.
   (1) Blank off outlet nipples.
   (2) Remove outboard plugs.
   (3) Connect inlet nipple to test line.
   (4) Turn on 350 lbs. air.
   (5) Squirt oil between guide and stem, and around fuel strainer casting plug to determine leaks.
   (6) Turn off air and disconnect, remove blanks and replace plugs. Oil and work valves.

5063. Strainers may also be tested with the check valve testing outfits as follows:
   (1) Remove strainer in testing outfit.
   (2) Substitute the strainer to be tested in its place.

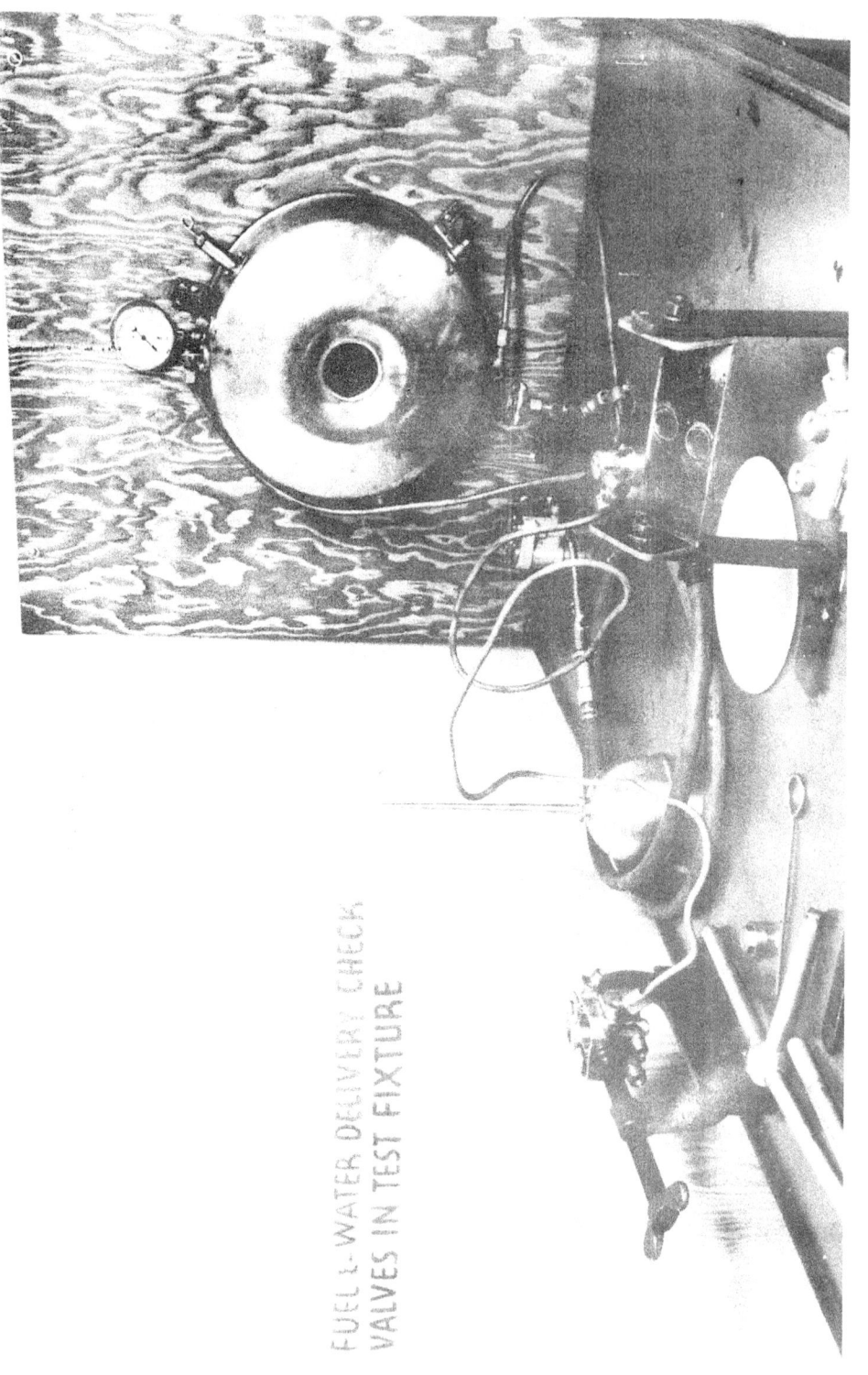

FUEL & WATER DELIVERY CHECK VALVES IN TEST FIXTURE

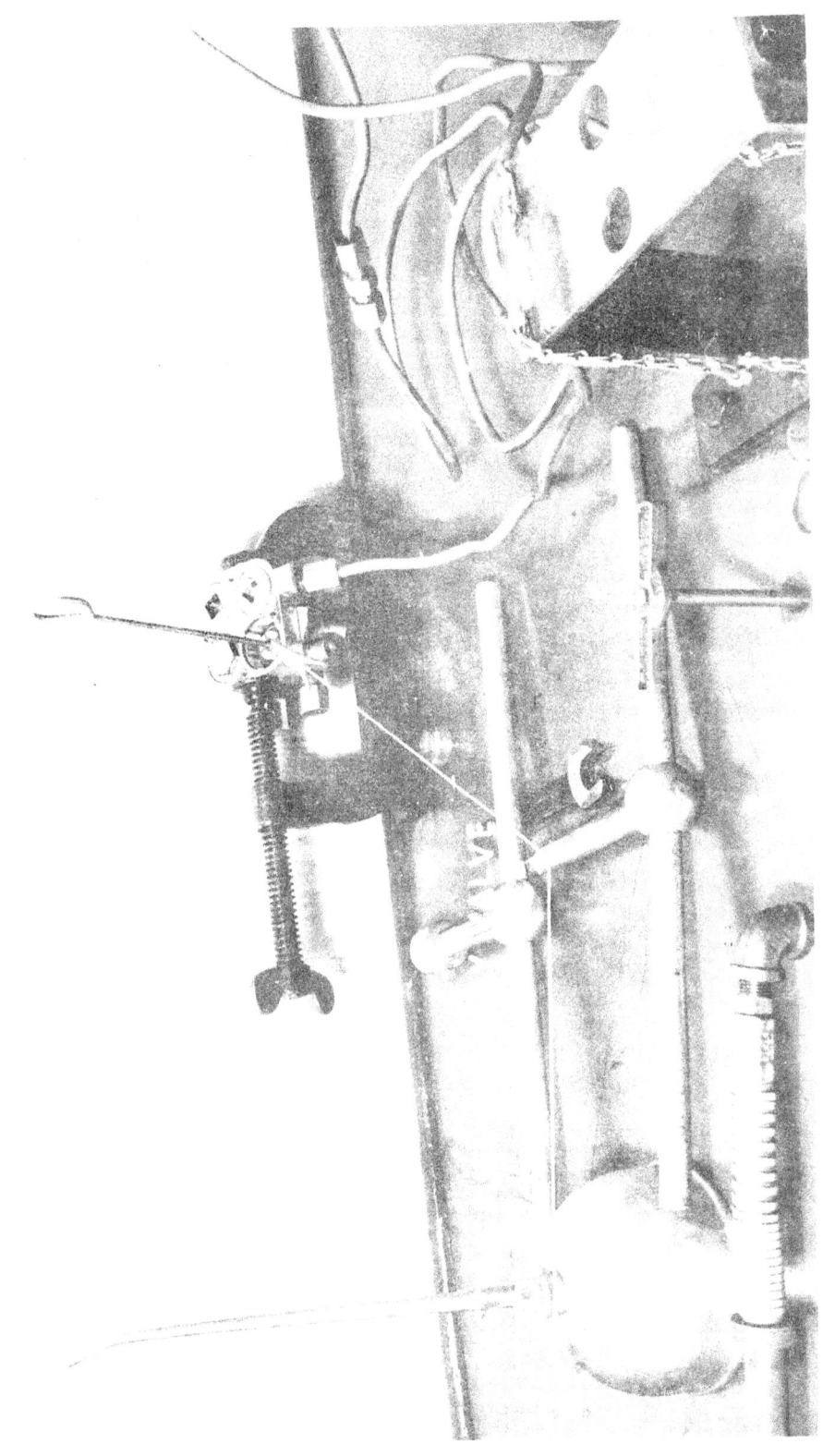

TOOL NO.

    (3) Use about 20 lbs. pressure in the tank.
    (4) Open outlet valve.
    (5) Note blow from pipe, should be vigorous and full size of the pipe. If not, inspect for dirty mesh.

5064. Wipe dry and re-oil (C) check valves and strain- 74,405
ers after test.................................... 406,407
Install on air flask and secure with holding
screws.. 41
Connect vent pipe................................. 141A

K. OVERHAUL AND TEST AIR CHECK VALVES.

5065.
    (1) Remove plugs and washers for check valve body.. 405
    (2) Remove check valve plug and guide with center- 12
        ing bushing, check valve and spring............. 75
  *(3) Reseat seats for valves, use air check valve
        plug with centering bushing removed for guiding
        tool when cutting................................ WE108
  *(4) Reseat seat for air check valve plug washer.... WE113
  *(5) Reseat seats on pipe nipples and remove burrs
        on threads....................................... WE83
  *(6) Reseat outer seat of valve on plug............. WE111
  *(7) Reseat outer seat on valve..................... WE111
  *(8) Lap hole in air check valve plug and sleeve on WE107
        check valve to fit (a tool is provided for hold- WE110
        ing valve for operations (5),(6) and (7) above) WE112
    (9) Clean, inspect, oil (C) and assemble air check
        valves, springs and plugs in air check valve
        body, holding valves in plugs with screw rods
        until threads are engaged, then remove screw 74
        rod and set up on plugs.......................... 12
   (10) Replace plugs and washers for check valve body. 405

TEST OF AIR CHECK VALVES.

5066. Test of inboard seats.
    (1) Connect inlet nipple of check valve body to outlet of test tank.
    (2) Turn on air, build up pressure slowly. Valve may show a slight leak at 3 to 5 lbs., which is of no consequence. A large leak must be rectified.
    (3) Build up pressure to 10 to 12 lbs., at which time valve should fully lift. (Spring calibration).
    (4) Turn off air and disconnect.

5067. Test of outboard seats.
    (1) Blank off outlet nipples.
    (2) Remove outboard plugs.
    (3) Connect inlet of check valve body to outlet of test tank.

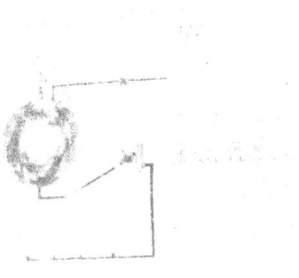

TEST PANEL FOR CHECK VALVES

TEST OF AIR CHECK VALVES, INNER SEATS MAY SHOW A VERY SLIGHT LEAK AT 3 TO 5 LBS. VALVES MUST LIFT AT 10 TO 12 LBS.

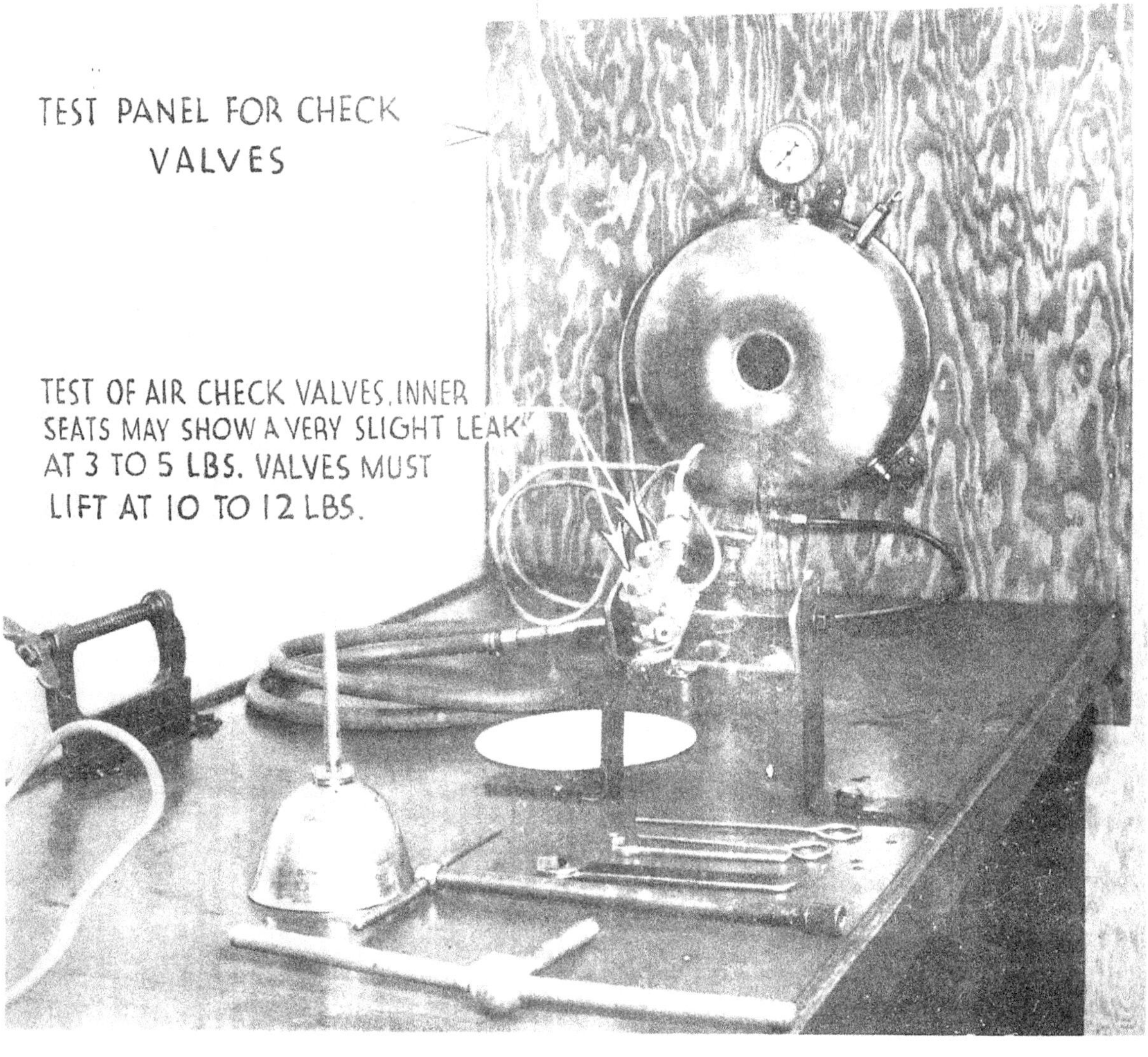

TS-5   V-20A

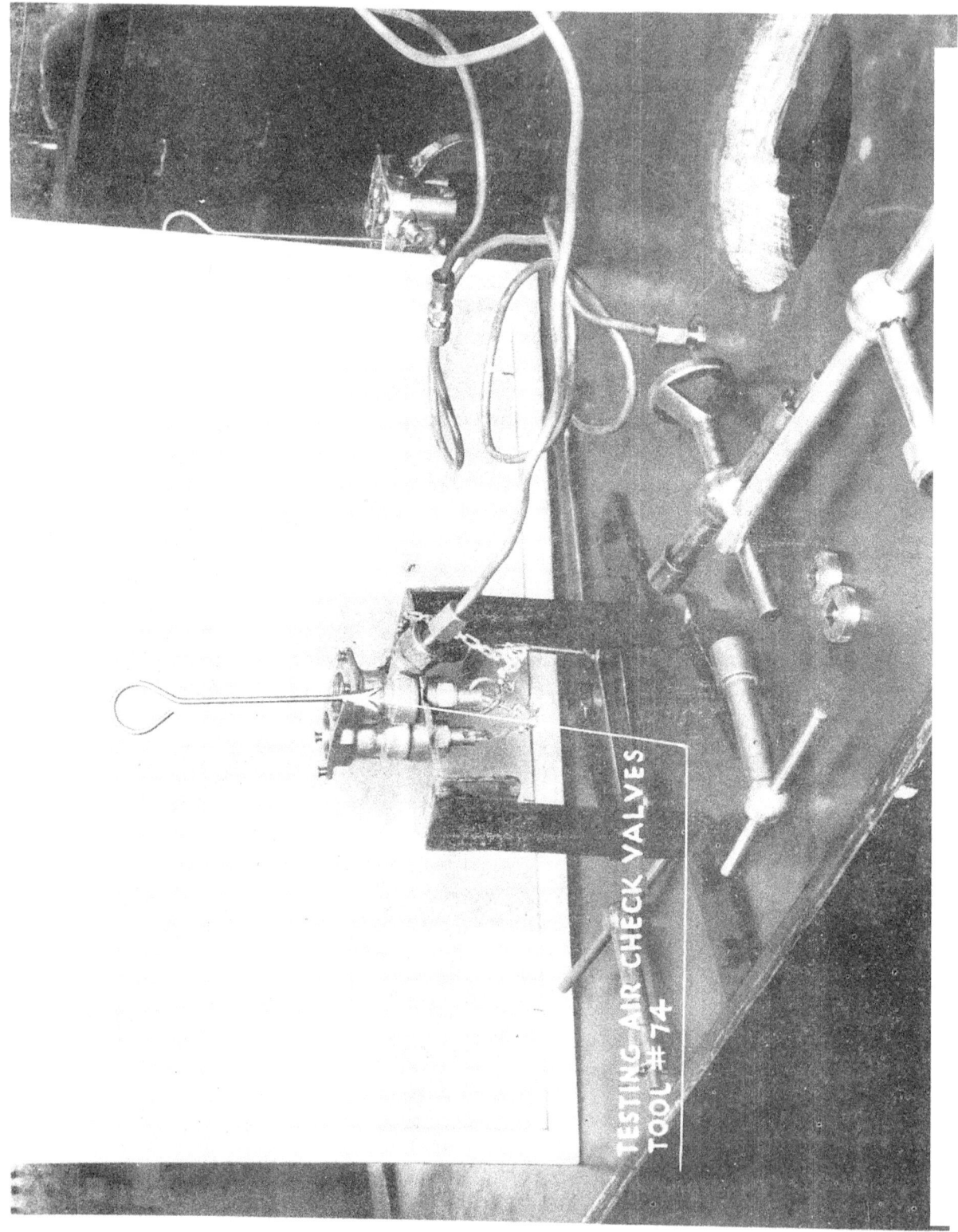

(4) Turn on 350 lbs. air.
(5) Squirt oil into opening in plugs. Leaks may be detected by bubbles.
(6) Turn off air and disconnect, remove blanks, and replace outboard plugs.

    NOTE: A small threaded hole on the inside of the valve for receiving a screw rod, is used to grind in and operate the valve. These valves should be well oiled and worked freely before the torpedo is fired.

5068. Wipe dry and re-oil (C) air check valve after test, install on air flask with holding screws and connect pipes....................................   TOOL NO. 12,74 41,229,141A

L. CHARGE AIR FLASK.

5069.
(1) Blank off outlet nipple on stop and charging valve body
(2) Remove water and fuel plugs.
(3) Close air flask stop and blow valve.
(4) Remove charging valve, plug and leather washer.
(5) Open stop valve wide and turn back $\frac{1}{4}$ turn.
(6) Install wing nut in charging valve plug bushing, secure safety strap in place.
(7) Open bleeder valve in separator and drain moisture from separator.
(8) Open inlet valve to charging line.
(9) Crack main inlet valve from charging source and charge torpedo slowly to 800 lbs.
(10) While charging, bleed moisture from separator frequently.
(11) When flask is charged to 800 lbs., close main inlet valve from charging source.
(12) Open bleeder on inlet valve to charging line and bleed air from line.
(13) Close stop valve.
(14) Remove safety strap and wing nut.

M. SAFETY PRECAUTIONS.

5070. The precautions in handling, charging and blowing down air flasks must be observed:

(1) Use safety straps when charging.
(2) Never charge above working pressure stamped on flask.
(3) Do not transfer air flask while charged.
(4) Use no artificial means to cool.
(5) Do not charge, or leave a charged flask stand in the direct rays of the sun.
(6) Blow down flask by means of flask stop and blow valve, and not main air valve group.
(7) In case flask has been given a heavy jar, it will not be charged until hydraulically tested.

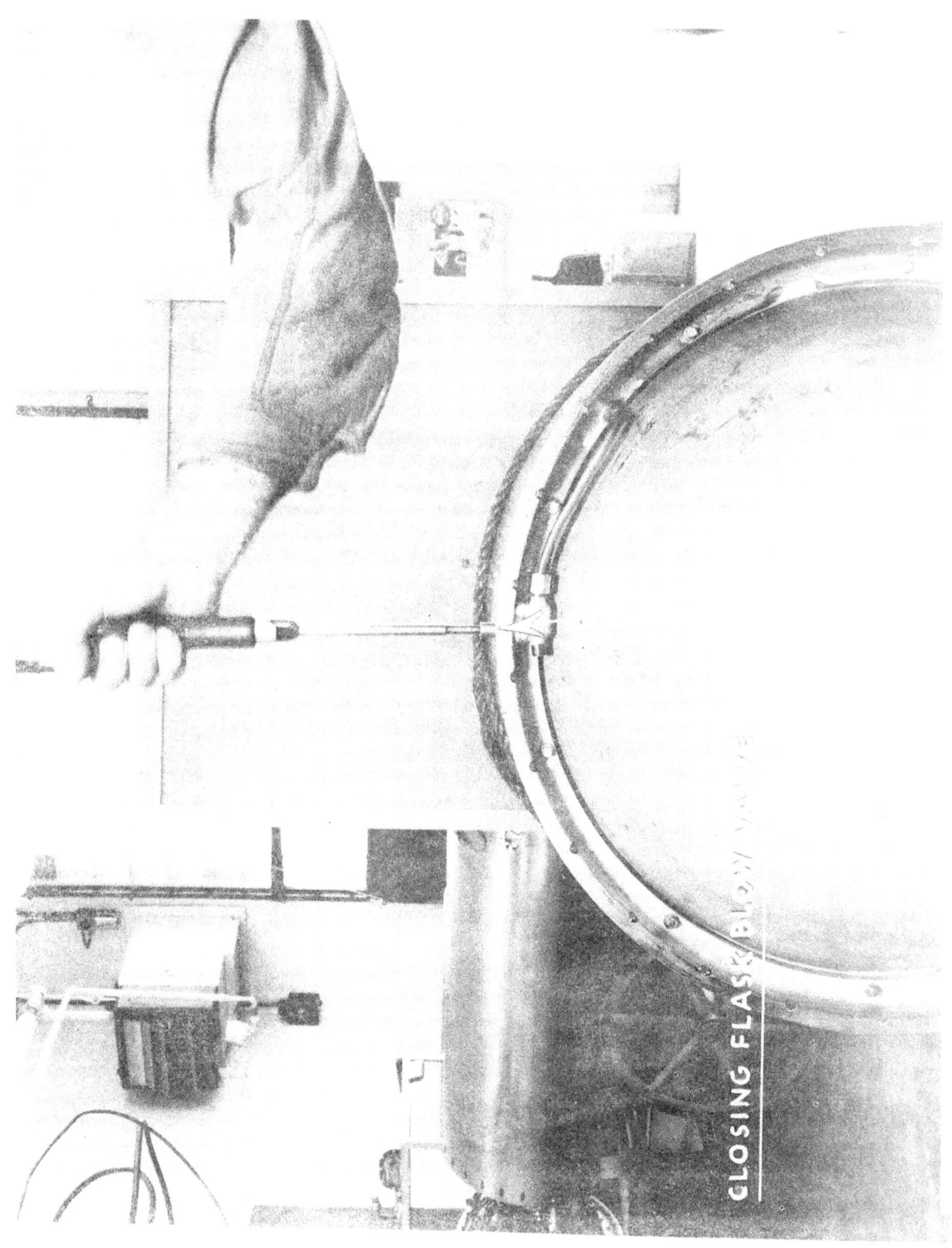

CLOSING FLASK BLOW VALVE

TS-5 V-21A

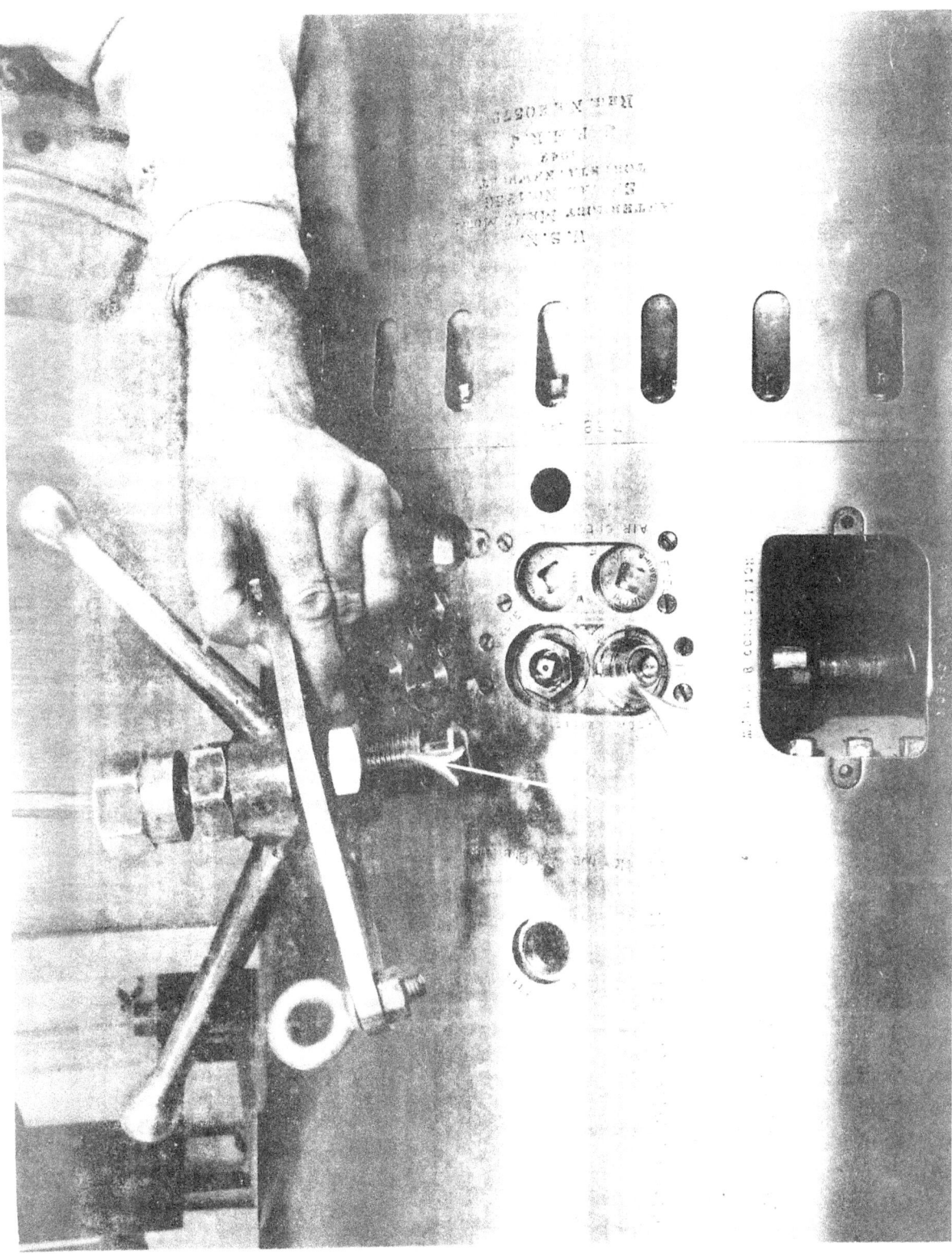

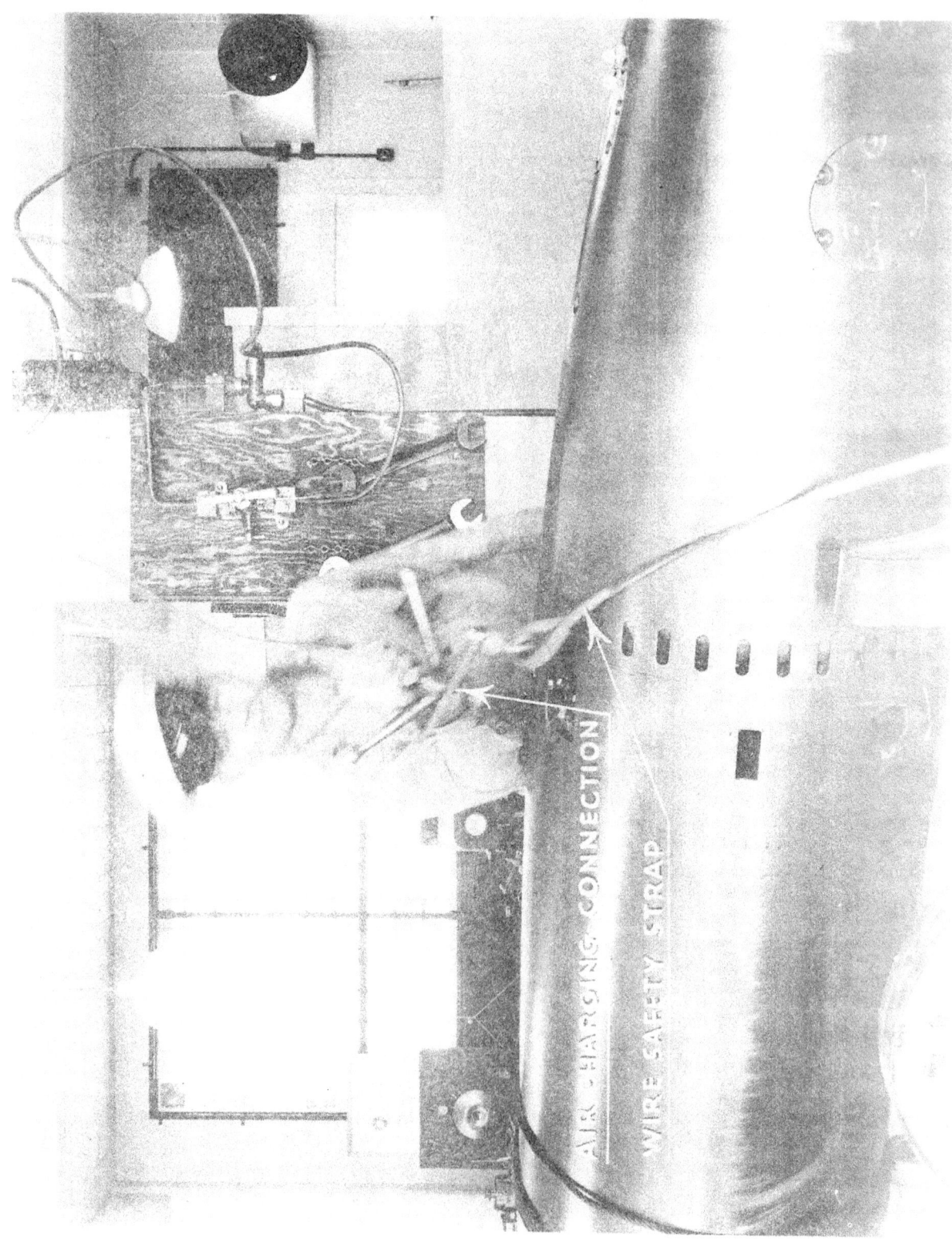

TS-5 V-21D

OPENING MAIN AIR INLET VALVE

TS-5   V-21E

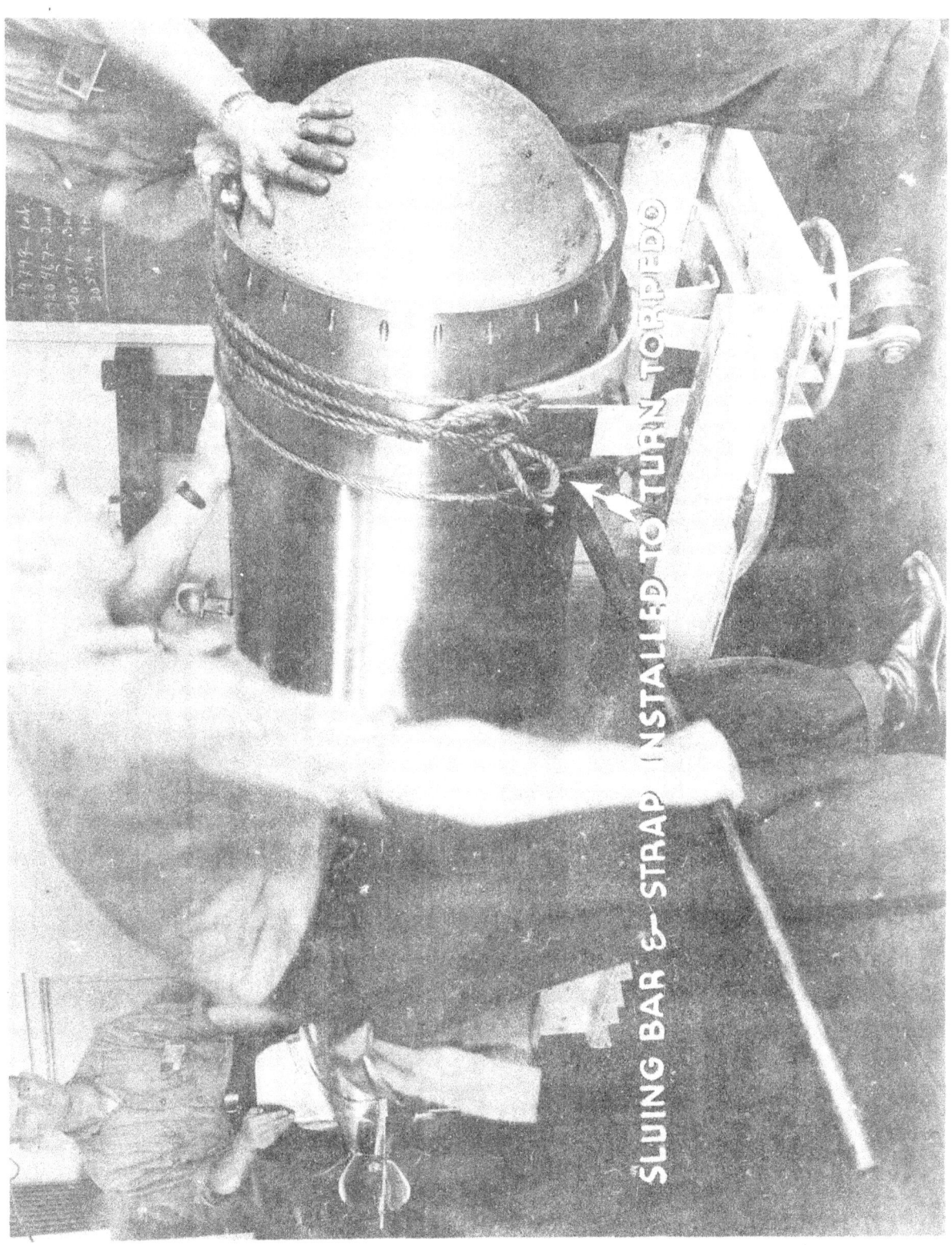

SLUING BAR & STRAP INSTALLED TO TURN TORPEDO

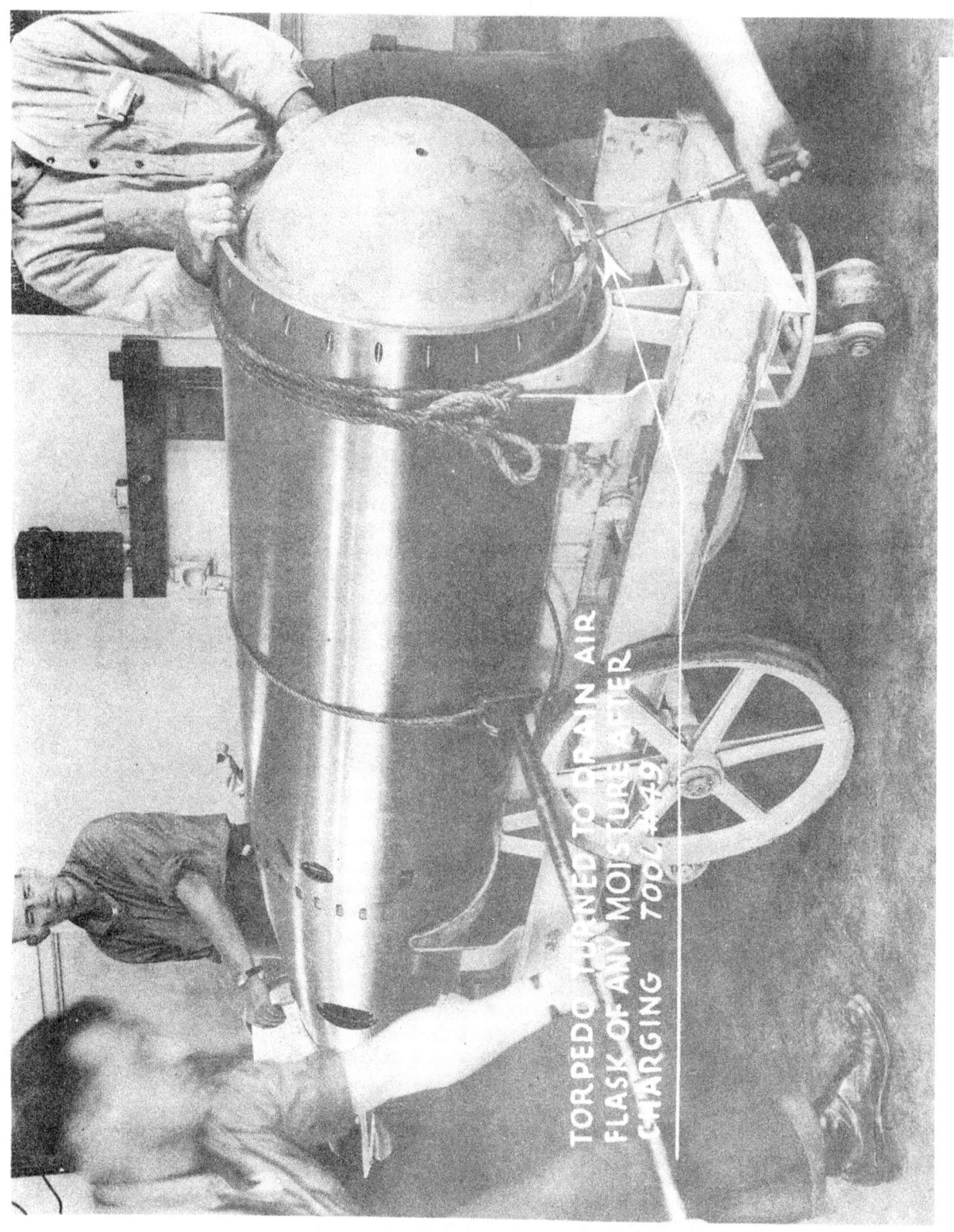

TORPEDO TURNED TO DRAIN AIR FLASK OF ANY MOISTURE AFTER CHARGING TOOL #49

(8) Explosions have occurred in main air lines, due to rapid opening of stop valves and sudden compression of oil vapor held in the line between air flask and starting valve. Bureau of Ordnance Manual, Chapter XV, paragraph 15B2(14) gives precautions which must be observed to prevent oil explosions in main air lines. These are brief:
    (a) When the starting piston is on its seat, open the stop valve gradually.
    (b) In making tests of torpedoes, control the air at all times from the stop valve. Keep the starting piston off its seat when testing torpedoes.
    (c) Remove all traces of gasoline or kerosene from valve before re-assembling after overhaul. Do not leave excessive oil in the valve group after assembly.
(9) Air flasks must be charged through separators.
(10) Gauges used in charging lines and test gauges should be frequently calibrated by the tender.
(11) After inspection of the air flask, the screw holes for forward bulkhead holding screws should be cleaned out with a hand tap.

N. TEST OF AIR FLASK FOR TIGHTNESS.

5071. When the air flask is charged to 800 lbs., test for leaks by use of oil (C) around the forward bulkhead flange and holding screws. A lighted taper may be used as an additional means for indicating leaks. The main air connection passing through the after bulkhead may be tested for a leak by placing a cigarette paper over the water compartment filling hole. The air flask should be allowed to cool to normal before testing main air connection as above outlined.

O. TEST FLASK STOP AND BLOW VALVE.

5072. With flask charged and valve seated, test around outlet pipe on valve body with oil. Blank off outlet pipe, open valve fully and test between copper washer and shoulder on valve stem with oil. If leaks are found between the copper washer and the shoulder on the valve stem, the washer must be annealed or renewed. If the valve leaks, it will have to be renewed, since it is of the compression type and cannot be readily ground in.

P. TEST OF WATER COMPARTMENT.

	TOOL NO.
5073.	
(1) Remove access plug to water compartment.......	11
(2) Remove fuel filling plug.....................	74 thru 217

                                                                TOOL NO.
 (3) Blank off main air connection from stop
   and charging valve.......................... 134
 (4) Charge air flask to 300 lbs..
 (5) Close stop valve................................ 227
 (6) Replace water compartment plug (the fuel
   filling plug should be left out so as not to
   damage the fuel flask during test)............ 11
 (7) Blank off the following nipples:
   (a) Fuel delivery lead........................ 141A
   (b) Water delivery lead....................... 144
   Vent fitting is not blanked off and will show
   leaks of the upper seats of the check valves.
 (8) Install low pressure test set in charging
   valve, connecting pipes so that air will pass
   through the reducing valve attached to the
   test set gauge.
 (9) Connect pipe from test set to inlet air nipple
   on the air check valve body.
 (10) Open stop valve and charge water compartment
    to pressure delivered through reducing valve
    (400-450 lbs.)
 (11) Close stop valve and test with oil for leaks
    around water compartment head joints, main air
    connection joint, plugs and nipples. Hold
    pressure for at least five minutes. A lighted
    taper may be used as an additional means for
    indicating leaks. Upon completion of test,
    release pressure by opening bleed valve on test
    set.

CAUTION: It is important in this test that the pressure
     in the air flask be kept within the above limits
     to avoid the possibility of excessive pressure
     in the water compartment, which is required
     to withstand a hydraulic test pressure of 1500
     lbs. Do not stand or permit others to stand
     in front of blow-out plug during test. Blow-
     out plug operates at 1200 lbs.

Q. INSPECTION AND CARE OF THE AIR FLASK.

5074.

 (a) Inspection of air flasks is required semi-annually in vessels afloat, when torpedo has not been fired. The spirit of the instructions is that as soon after firing as possible the interior of the flask shall be inspected and placed in a ready condition.

 (b) If excessive pitting or corrosion is detected, report to the Bureau of Ordnance, giving details as to location, extent and depth of corrosion.

 (c) An accurate and complete record regarding all charg-**ings of** each air flask must be kept and entered in the record book.

(d) A hydraulic test of the air flask is required:
   (1) After 50 or equivalent charges to full pressure. Equivalent charges are: 2 below full pressure and above 1500 lbs; six to, or below 1500 lbs.
   (2) If held under working pressure fully ready for four months.
   (3) If held between 1500 and 1800 lbs. for 6 months.
   (4) In case flask has been given a heavy jar.

(e) While inspecting flasks the following precautions should be observed:
   (1) Have all men entering flask remove shoes and any abrasive parts of clothing.
   (2) Be careful not to scratch bulkhead seats.

(f) Use no acids, lye or gasoline to clean flasks. Hot fresh water only shall be used. Metal brushes of any kind shall not be used.

(g) To preserve interior of flask, slush down with a mixture composed of one part hot running torpedo oil and one part 600-W, or its equivalent. About one quart should be left free in the bottom of the flask.

(h) As soon after firing as possible and when air flasks are examined, clean out water compartment, remove corrosion and re-oil.

CHAPTER SIX

		ARTICLES
A.	Reducing and Superheating System, General Description	6001
B.	Starting Gear	6002-6013
C.	Starting and Reducing Valve	6014-6026
D.	Combustion Flask	6027-6029
E.	Sprays	6030-6031
F.	Igniter	6032-6042
G.	The Nozzles	6043-6044
H.	Remove Parts on Turbine Bulkhead and Starting Gear	6045-6046
I.	Disassembly, Assembly, and test of the Starting Gear	6047-6059
J.	Disassembly, Assembly, and test of the Starting Valve	6060-6063
K.	Disassembly, Assembly, and test of the Reducing Valve	6064-6068
L.	Overhaul and test Combustion Flask and Nozzle Assembly	6069-6072
M.	Disassembly, Assembly, and test of the Sprays.	6073-6076

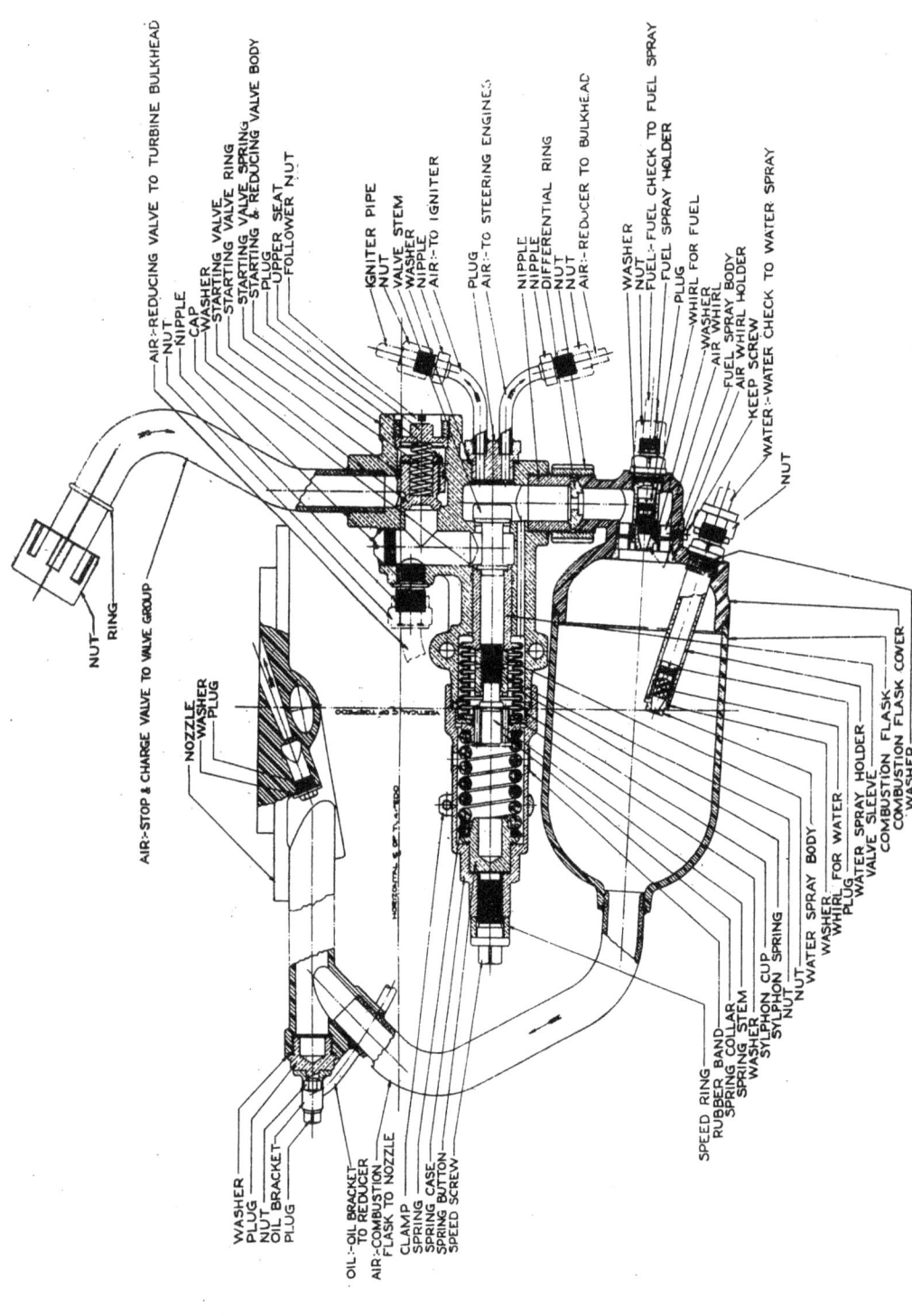

VALVE GROUP & SUPERHEATER – SECTIONAL VIEW

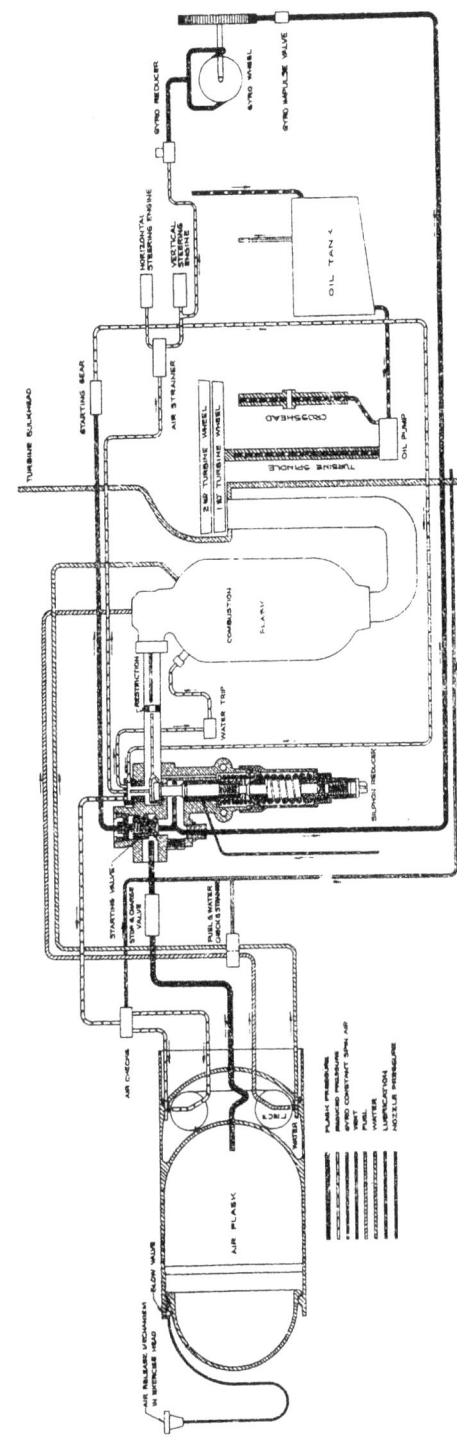

A. REDUCING AND SUPERHEATING SYSTEM - GENERAL DESCRIPTION

6001. The basic source of the power of the torpedo is the alcohol, the combustion of which is supported by the oxygen in the air contained in the air flask. The air is conveyed through suitable piping and controlling valves to the regulator where it is reduced to the working pressure and thence flows to the combustion flask. Due to the loss of pressure in the passage through the restriction ring, there is a lower pressure in the combustion flask than at the reducing valve. This difference of pressure is the "differential pressure". Reduced pressure from the reducing valve is led to the fuel and water compartments through suitable piping, and fuel and water, impelled by the differential pressure, flows to the combustion pot. An igniter is provided for ignition. The water and the nitrogen in the air reduce the temperatures of the gases. The products of combustion are delivered to the nozzle.

B. STARTING GEAR.

6002. The starting gear controls the following: (a) Starting of the torpedo; (b) Cut off by manual operation of starting spindle. The starting gear used in the Mark 13 torpedo is different from that used in previous types of aircraft torpedoes.

6003. Located at the top centerline of the afterbody, abaft the gyro mechanism, is an irregular bronze casting (starting gear body) secured to a bronze ring riveted and sweated to the shell of the afterbody and made watertight by a sheet packing gasket and containing the starting mechanisms.

6004. The starting gear is composed of the following essential parts: Starting gear body, starting spindle arm, starting spindle and head, connecting lever and catch, unlocking lever, index spindle and trip cam, starting piston and spring, and starting valve body.

6005. There are three openings through the starting gear body into the afterbody which require sealing against leakage of sea water, viz., the index spindle, starting spindle, and starting piston body.

6006. The index spindle is made tight by a leather washer held in place by a brass packing ring secured by three screws to the starting gear body. The starting spindle is made tight by a lapped fit in the starting gear body. The starting piston body is made tight on the starting gear body by a leather washer and nut.

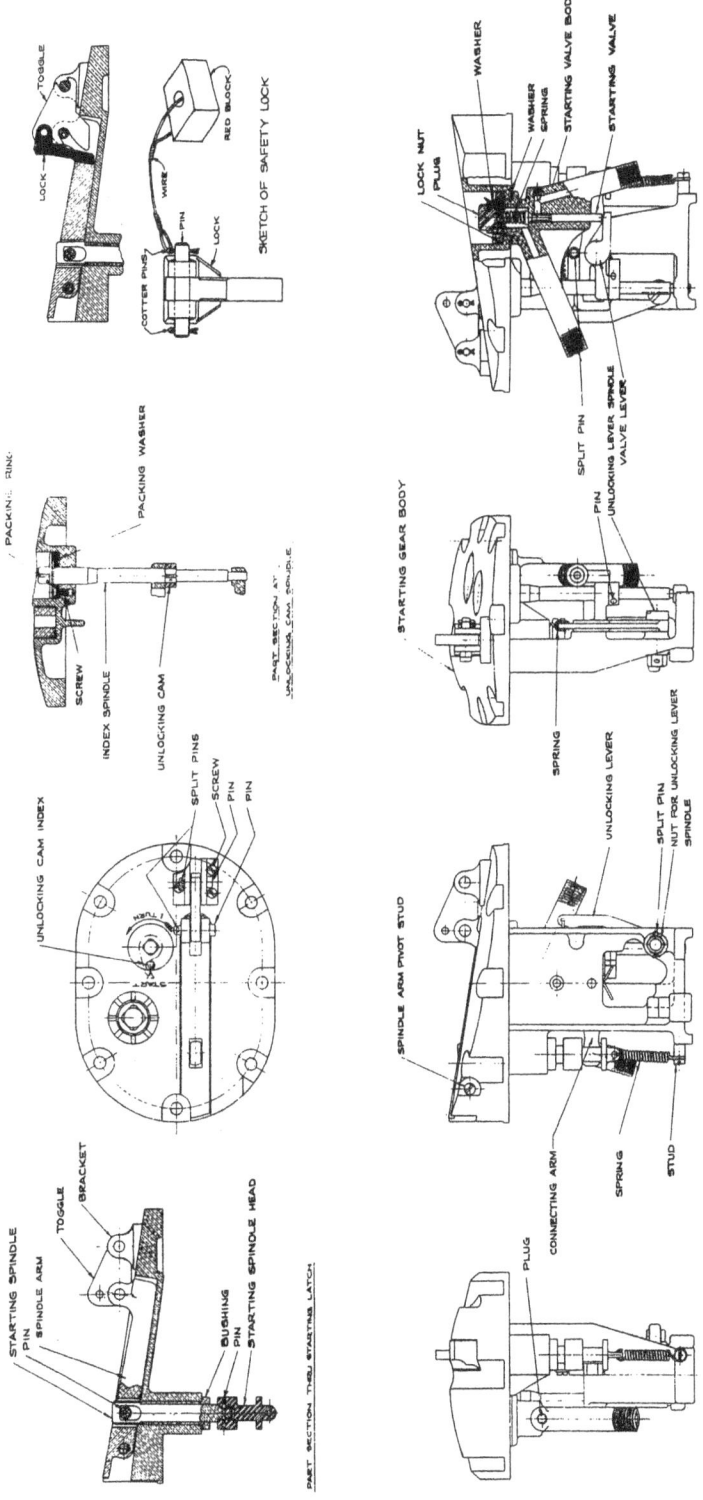

GENERAL ARRANGEMENT STARTING GEAR

6007. The starting spindle arm is contained in a longitudinal recess machined on the outer face of the starting gear body, the forward end being pivoted in this recess. A slotted eye is machined on the after end of the starting spindle arm in which is fitted a cam toggle between it and the forked bracket on the after end of the starting gear body. This cam toggle is permanently secured to the firing latch lanyard and remains attached to this lanyard upon release of a torpedo from the plane.

A forked bracket is secured on the starting gear on the after side of and in line with the starting spindle arm, and the after end of the starting spindle arm is machined with a similar forked projection, pins for cam slots in toggle inserted in the forked bracket and forked projection on the spindle arm.

The toggle is of suitable dimensions for insertion in forked bracket and forked projection on the after end of the spindle arms, cam slots being machined on the cam toggle to fit pins in the bracket and spindle arm. These cam slots are contoured to permit the insertion of toggle only when the spindle is in its extreme out position, after which by turning the starting index spindle until the starting piston is seated, the starting spindle arm will lock in its down (flush) position and thus, due to relative position of cam slots on pins of spindle arm and bracket, lock the toggle in position for firing.

Upon release of torpedo, the forward end of the toggle being secured to the firing latch lanyard, will, through engagement with the starting spindle arm, pull its after end out in a manner similar to that of the breaking link thereby causing the starting gear to operate (see paragraph 6011), after which the cam slots in toggle disengage the pins in the starting gear, separating toggle from starting gear.

The cam toggle attachment requires at least **18-20 lbs.** pull to lift the starting spindle arm with 1500 lbs. of air in the flask.

<u>CAUTION</u>: The toggle must be installed with the stop valve closed prior to final adjustment of torpedo for firing. Toggles should be used only in torpedoes to which they are fitted.

6008. The upper end of the starting spindle is suitably machined for passing through a slot in the starting spindle arm, to which it is joined by a pin, the hole in the starting spindle being elongated to permit the starting spindle arm to drop back into its recess after the torpedo is launched. The starting spindle passes through a lapped vertical bearing in the starting gear body and carries on its lower end a starting spindle head machined with shoulders for engaging the end of the connecting lever and catch. A spring, connecting between an eye on the lower end of this head and spring post attached to the frame, tends to keep the connecting lever catch against a notch in the unlocking lever. The connecting lever and catch is pinned to a small shaft passing through a transverse bearing in the starting gear body, the valve lever being attached to the other end of this shaft and so located that its end is directly under the starting piston stem.

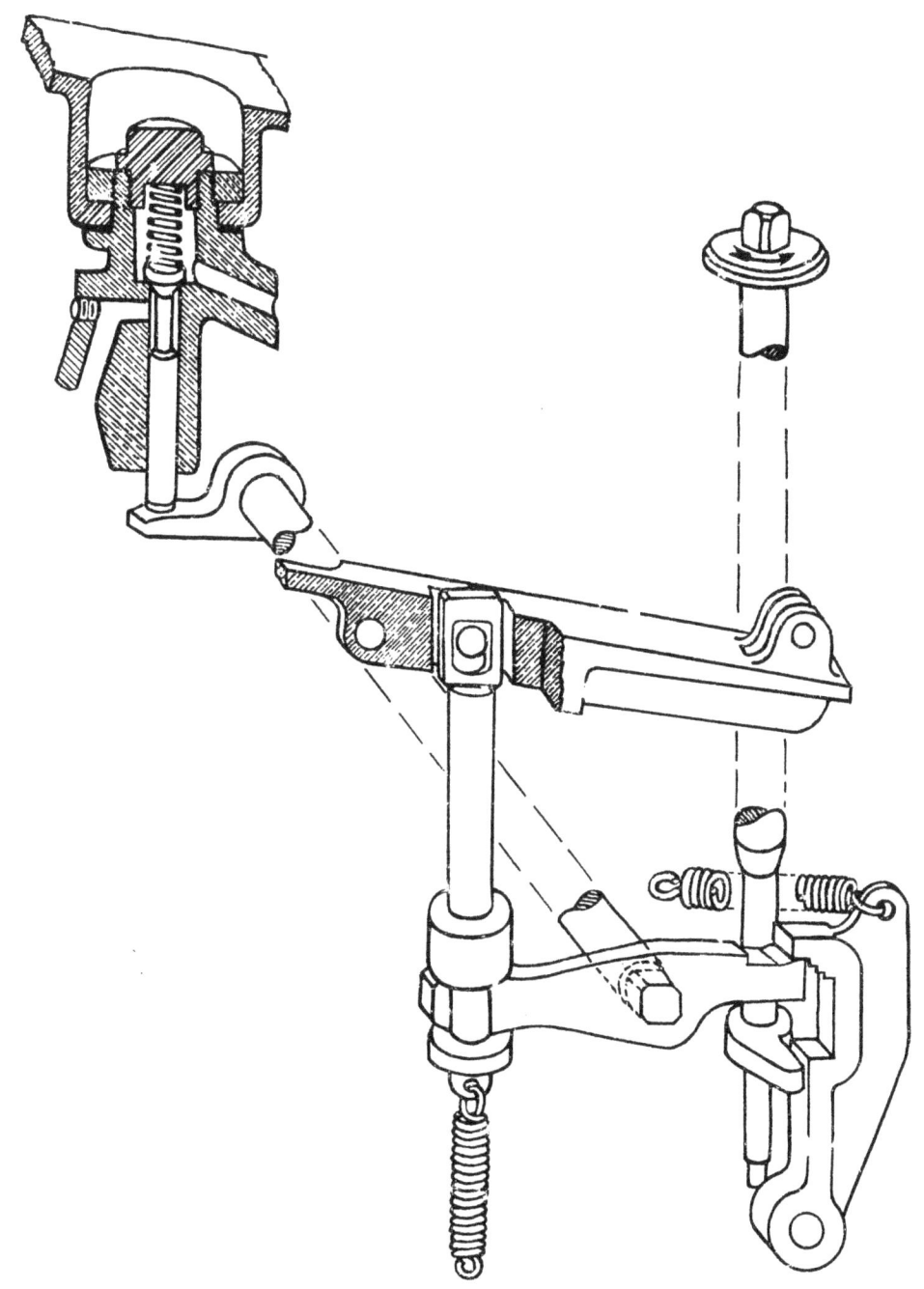

DIAGRAM OF STARTING GEAR

6009. The unlocking lever is pivoted to the starting gear body, its upper end being notched and resting against the rear end of the connecting lever catch. A tension spring tends to draw the parts into engagement.

6010. There are two pipes connecting to the starting gear. One of these is a lead from the top of the starting valve to the top of the starting piston and the other is a lead from the lower side of the starting piston valve seat to the low pressure side of the reducing valve. The starting piston is thus in effect a valve between the top of the starting valve and the low pressure side of the reducing valve which permits of complete control of the starting valve and thus the starting and stopping of the torpedo, by suitable manipulation of the starting piston.

6011. The operation of the starting gear is as follows: The starting spindle is normally held in its lower position by the starting spindle head spring. In this position the valve lift is at its lowest point, the starting piston is closed, and the escape of air from above the main starting valve is prevented. When the torpedo is released, the cam toggle attachment lifts the after end of the starting spindle arm out of its recess, moving the starting spindle and head upward, thus raising the forward end of the connecting lever. This causes the valve lever to raise the starting piston off its seat and allows the air to exhaust from above the main starting valve, permitting it to open and start the torpedo. At the same time, the catch on the end of the connecting lever engages in the notch on the unlocking lever and locks it in position, thus holding the starting piston open.

6012. To provide a means of stopping the torpedo on deck and for resetting the starting gear after a run, an unlocking cam is mounted on the index spindle in such a position that, if the index spindle is rotated, the unlocking cam will engage the unlocking lever and force it back. This allows the connecting lever and catch, and consequently the starting spindle and head, to resume their previous positions and allows the starting valve to seat under the action of the air pressure aided by the spring, or by the spring alone. The upper end of the index spindle is provided with a flange stamped with an arrow showing the direction of rotation and a small head squared to receive a setting tool, by means of which it may be rotated.

GOVERNOR GEAR.

6013. The governor which was used on the Mark 13 torpedoes as a safety device to stop the torpedo when the engine exceeded its maximum r.p.m. has been discontinued in the Mark 13 modification torpedoes, therefore no description of the gear is necessary.

C. STARTING AND REDUCING VALVE.

6014. The assembled mechanism composed of the starting valve and the reducing valve is contained in a forged naval brass body generally referred to as the valve group which is secured by means of two screws to the turbine bulkhead.

6015. The starting valve of the torpedo executes a function similar to that of a throttle valve of a marine steam engine, its opening being effected automatically at the desired time. Its essential function is to isolate the stored energy (compressed air) of the torpedo until the proper time for its release and then to open automatically.

6016. The starting valve situated and moving in a cylinder chamber of the valve group body is of a form as shown on Page VI-6A.

6017. It consists of a 1-3/16" diameter piston with two lapped rings fitted in the lap bore of the valve body. The valve is held against its seat in the valve body by means of a spring. A small hole or by-pass is drilled through the valve affording communication between the space above the valve and the high pressure cavity in the valve body. Upon opening the stop valve high pressure air passes through the by-pass equalizing the pressure above and below the valve, thus holding it on its seat.

6018. The starting valve chamber is closed by a plug with lapped seat secured by means of a threaded ring in the tapped hole of the valve body. A hole is drilled through the body to the space above the valve and a pipe connects it to the starting piston of the starting gear, which will be described under the latter heading. The plug is provided, on its inner face, with a tapered seat to fit a corresponding tapered seat at the upper end of the starting valve, serving as a stop to check the opening movement of the valve piston without danger of burring the edge.

6019. The area of the cylindrical portion of the valve is slightly larger than the area over the valve seat so that when the air is exhausted from above the valve by the action of the starting piston in the starting gear, the pressure on the difference of areas opens the valve which permits the flow of high pressure air to the reducing valve.

REDUCING VALVE.

6020. The reducing valve, interposed in the valve group between the starting valve and the restriction **ring** is for the purpose of automatically delivering air through the restriction **ring** to the combustion flask at a constant pressure and constant and economical rate throughout the torpedo's run. For purposes of demonstration, it may be treated diagrammatically as on Page VI-7A.

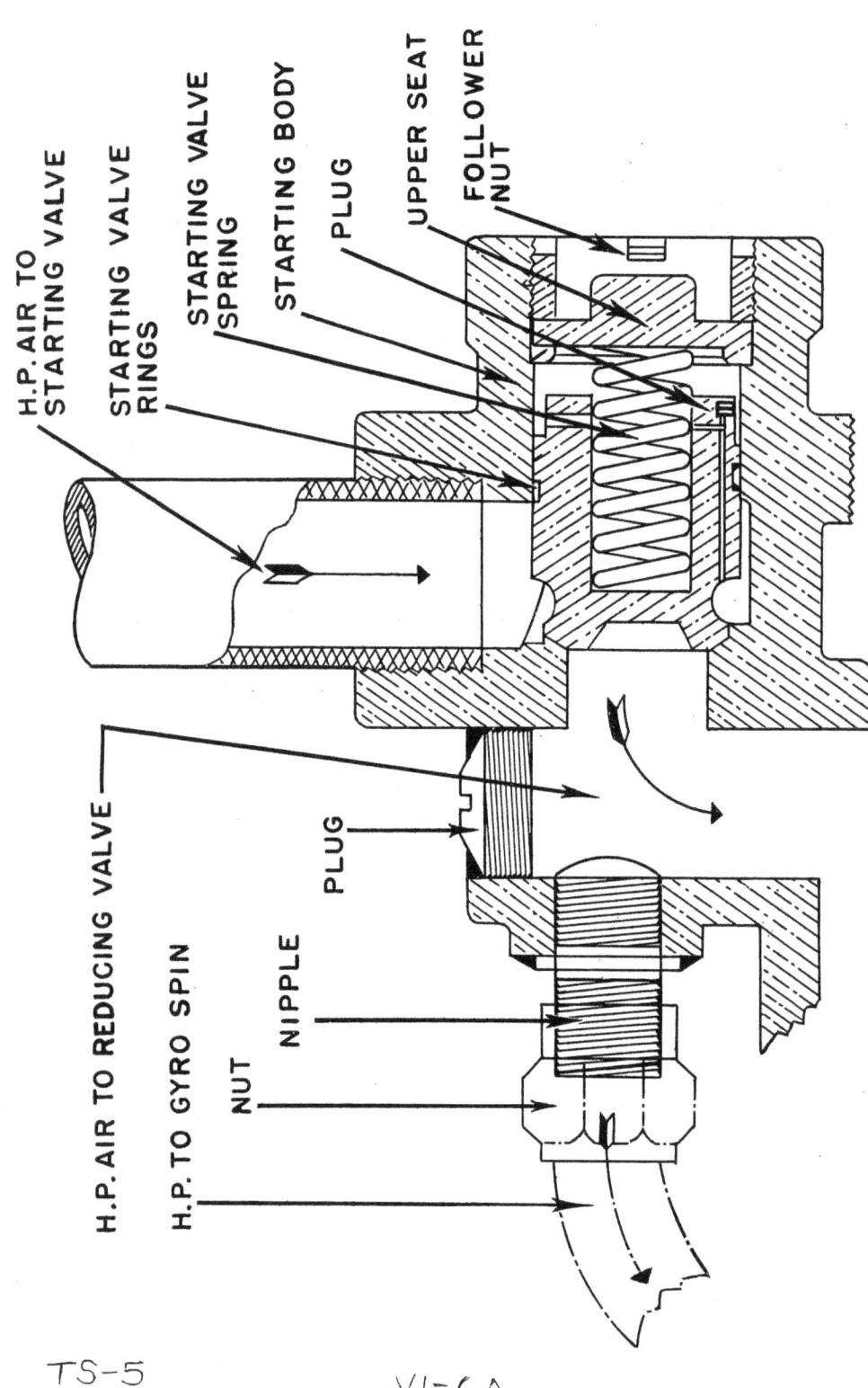

STARTING VALVE

TS-5 VI-6A

6021. For a clear understanding of the function performed by the reducing valve, attention is invited to the forces tending to open and close the valve, when control valve pressure is not applied. The spring pressure alone is adjustable and it is by virtue of this adjustable spring pressure that the air pressure delivered from the valve may be regulated and set to deliver the designated pressure before a run.

6022. The essential operation of the valve as shown on Page VI-7A is as follows: The natural and constant tendency of the valve, if free from spring pressure is to close, as the areas (A - B) + D subject to downward pressure exceed areas (C - B) + valve face, subject to upward pressure. Therefore, if spring pressure (E) should be applied on the end of the valve stem in excess of this natural force tending to close the valve when free, the valve would, on entrance of air, open fully and furthermore, providing the escape was unrestricted (as for instance direct escape into atmosphere through an exit equal in size to the inlet), the valve would remain thus open.

6023. The escaping air is not unrestricted, however, but on the contrary is confronted with considerable resistance as it must pass through a restriction prior to its entry into th combustion flask to support the combustion of the alcohol, and thence through the turbine nozzle before escaping to the atmosphere. Under these conditions the air between the reducing valve and restriction will, if the reducing valve remains open, soon increase in pressure sufficiently to cause the forces tending to close the valve to exceed those tending to open, with the resultant partial closure of the valve. When such closure occurs the air under the valve will be rapidly exhausted, with accompanying fall of pressure. As soon as the forces tending to open again exceed those tending to close, the valve will open more. From the above it will be seen that the air pressure prevailing between the reducing valve and the restriction (reduced pressure) will not be allowed to exceed or to fall below a certain point without accompanying closure or opening, respectively, of the reducing valve and further that the pressure of the air delivered from this valve permits regulation (within certain limits) by varying the spring pressure. At the start of a run, the reducing valve will quickly take up an amount of opening necessary for the required reduced pressure, delivering the pressure with a slightly rising characteristic during the first half of the run with steady pressure from there to the beginning of the tail off. The maximum rise of this pressure is about 15 lbs.

6024. The sleeve in this reducer is a finely lapped fit within the body, with a small definite clearance (".003 difference in diameter) between the surfaces. Great care is taken to secure a mirror-like finish on these surfaces. In order to conserve the leakage of high pressure air through the clearance between

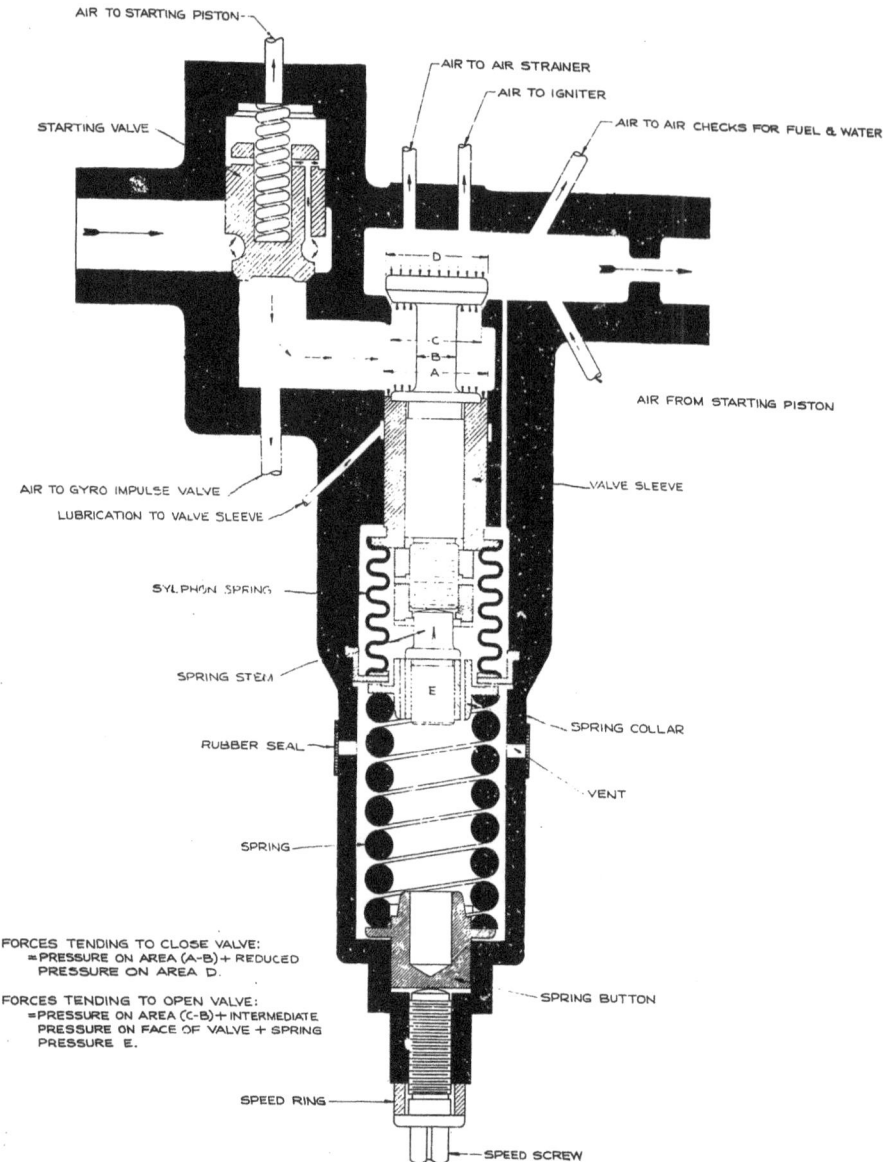

DIAGRAM OF SYLPHON REDUCER

TS-5   VI-7A

the sleeve and the body, the sylphon diaphragm is employed, which separates the reduced pressure side of the reducer from the interior of the spring case and yet provides sufficient flexibility for proper reducer adjustment and action. Communication between the sylphon chamber and the reduced pressure side of the regulator is made by a drilled hole through the body. This hole or port is much larger than the clearance space around the sleeve, and any leakage of high pressure air readily flows away and does not cause undue pressure in the external sylphon chamber.

6025. The sylphon diaphragm is formed of laminated tubing into a succession of concentric convolutions; their depth and form and laminated walls make a relatively flexible bellows, which sustains its working pressure fairly well and at the same time permits the valve to seek its position of equilibrium, so necessary for good pressure regulation.

6026. The regulator spring is contained in a chamber open to external pressure, and which includes the interior of the sylphon. It is formed by the spring case, and has a speed screw and a distance washer to adjust and fix the spring compression. A vent hole, closed by a rubber band, vents any air leakage that might get past or through the sylphon walls. The spring compression fixes the regulator pressure. The speed ring in these torpedoes should never be changed. The size of the speed ring is adjusted to a speed of 29.5 + .5 knots for the Mark 13 torpedo and to speeds of 33.5 + .5 knots and 40. for Mark 13-1 and 2 torpedoes.

D. **COMBUSTION FLASK**.

6027. The function of the combustion flask is to provide a pressure tight chamber of such proportions and arrangement that the flowing air and fuel may be thoroughly mixed and ignited, and that the resulting mixture of gases and steam be raised in temperature for delivery to the nozzles.

6028. After passing through the restriction the reduced air enters into the combustion flask through an annular concentric air passage in the combustion flask cover called the premixer top. In the premixer top is located a bronze ring machined with right hand helical guides to give the flowing air a whirling motion as it passes through. The fuel spray holder is located in the center of this passage and, passing the helical guide ring, locates the fuel spray in a position most efficient for vaporizing and igniting the fuel for combustion.

6029. The combustion flask is a cylindrical steel forging, one (exit) end of which is hemispherical in form and the other end closed by a dome shaped cover. The exit of the flask is extended to form a boss which is bored and tapped for the insertion of a 1" x 20 thread copper pipe connecting the combus-

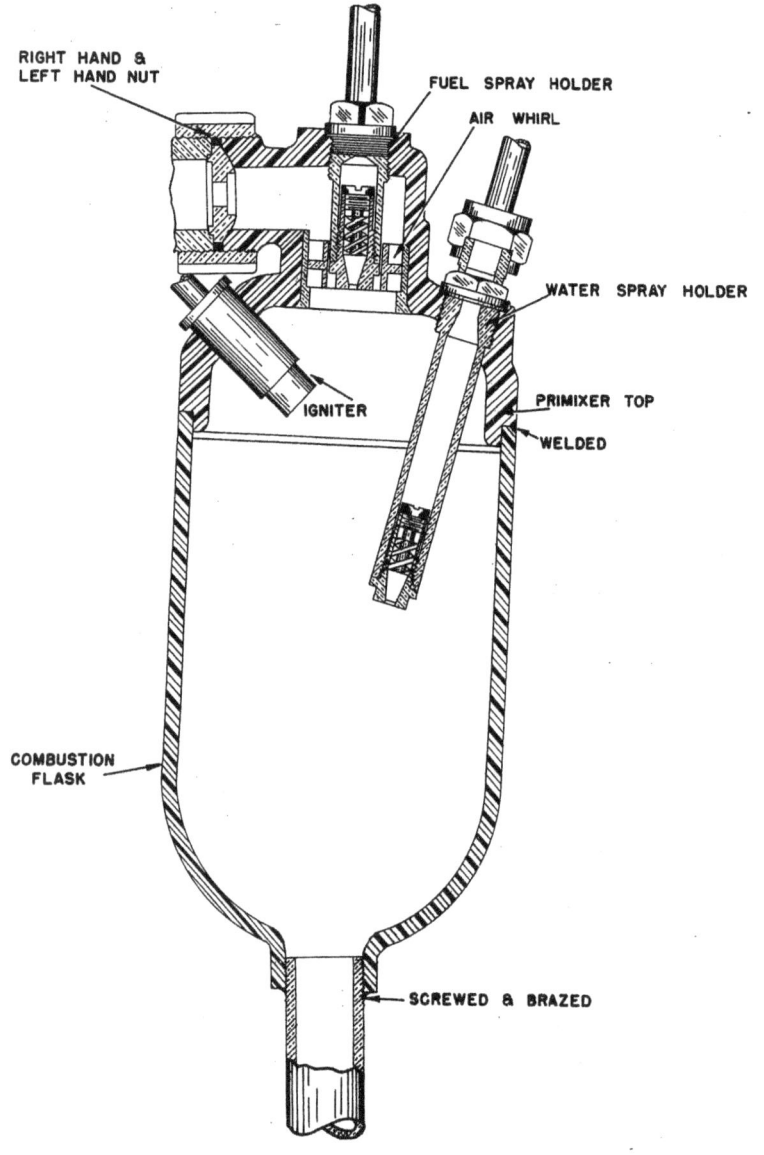

COMBUSTION FLASK

tion flask with the turbine nozzle. After insertion and alignment this pipe is brazed in position and thus becomes fixed in the exit end of the flask. As previously stated, the interior central portion of the combustion flask cover is machined in the form of an annular concentric air passage into which is fitted the bronze helical guide ring after fitting. A nipple machined on a boss on the cover at right angles with the concentric air passage affords a means for connecting the combustion flask to a nipple on the reducing valve by means of a right and left hand threaded nut. A hole is drilled and tapped in the center of the cover for the insertion of the fuel spray, a similar hole being drilled and tapped on the side of the cover at a suitable angle for directing the water spray to the center of the combustion pot at a point most desirable for conversion to steam. A boss extending above the cover adjacent to its central projection is drilled and tapped for the insertion of the igniter. The joint of the cover with combustion flask is formed by a circular lip machined on the cover, fitting snugly into the interior diameter of the flask, after which the outside of the joint is welded.

E. SPRAYS.

6030. The function of the sprays is to regulate the rate of flow and vaporize the fuel and water entering the combustion flask and thus, supported by air from the reducing valve, introduce the elements of combustion in their proper ratios.

6031 The fuel and water sprays each consist of the holder, spray whirl and spray body. The holders are cylindrical in form and of sufficient length to properly locate the sprays in the combustion flask. The outer diameters of the holders are extended to form seats for washers, the portion over and under the extended flanges being threaded for insertion in the combustion flask and for pipe connections from the check valves The interiors of the spray holders are suitably machined and tapped for the insertion of the spray bodies. The spray holders each screw into the combustion flask cover against copper washers. The spray whirls are made in the form of double thread worms snugly fitted against the inner walls of the spray body, the fuel spray whirl having a fine pitch and the water spraywhirl a coarse pitch. The spray bodies are bored out to fit the spray whirls, the holes tapering toward the tip end to near the exit holes, thus forming cone shaped atomizing chambers. The holes in the inner ends of the spray bodies are closed by screw plugs. Four holes are drilled through these plugs to permit the passage of fuel and water into the atomizing chamber. The spray holes are sized to regulate the spray delivery, but due to minor variations in the dimension of the passages in the atomizing chamber, the sizes of the spray holes may vary slightly in different torpedoes, as the sprays are not calibrated for size of opening but for rate of flow. The Mark 13 and Modification 1 and 2 torpedoes carry one fuel spray and one water spray. NOTE: For instructions relative to spray calibrations see NTS Newport O.D #750.

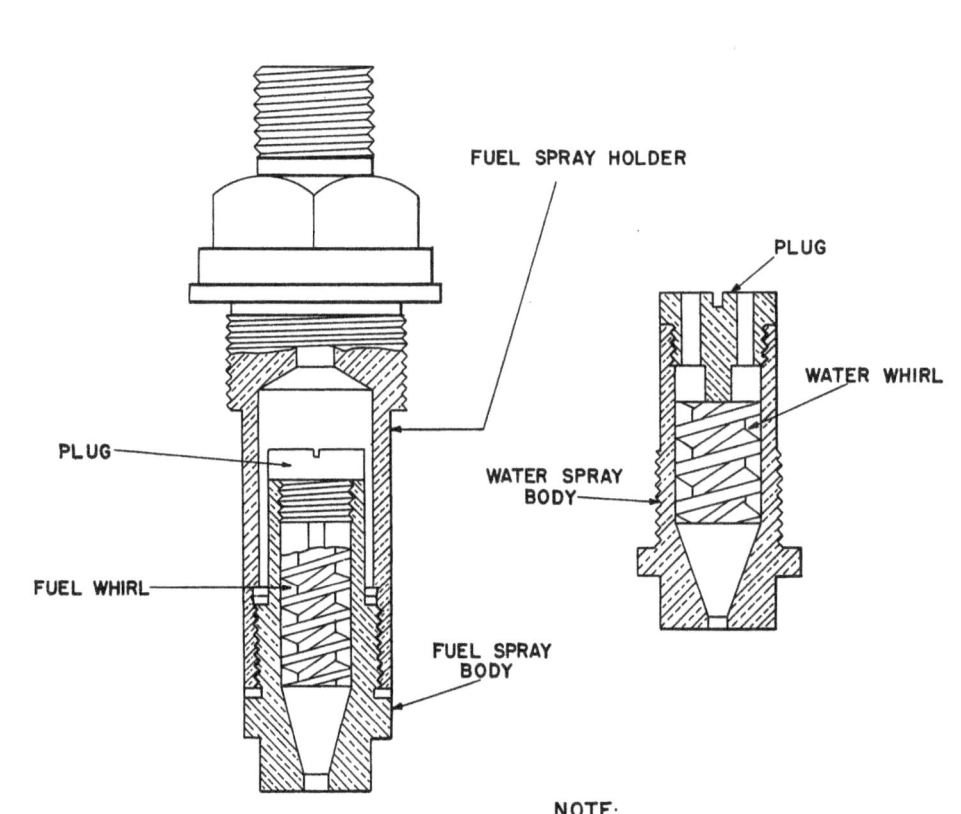

FUEL & WATER SPRAY

F. **IGNITER.**

6032. The combustible mixture of air and alcohol entering the flask at the instant of firing is ignited by a combined pistol and double fuse, called the igniter. The igniter is fired by air from the reduced pressure side of the reducing valve almost as soon as the torpedo is fired. It consists of a body containing a piston assembly (composed of piston, spring, and plunger as one unit) which is sealed by a diaphragm on its upper or pressure side and which normally rests on two firing pins. The firing pins rest on the ignition tube. This member is screwed in the body from the top against a copper washer, its lower section is flat and divides the inside of the body into two half sections, each side of which is sealed against each other by packing and into which the double charge is pressed, the lower end being sealed by a lead disc. The ignition tube is drilled from its upper end to take the primer caps, with holes extending nearly to the bottom and so perforated that the flame from each cap is directed to both ignition charges. The piston assembly, which rests upon the two firing pins, is held in the firing position by shear nibs on the firing pins, which take up on the shear plate and the sides of the holes in which the caps are seated. The shear nibs hold the piston and firing pins away from the caps until air pressure of about 250 lbs., per square inch is built up above the diaphragm, when by suddenly yielding, the piston assembly and firing pins are forced down on the caps by the firing air pressure plus the energy stored in the plunger spring during the time the pressure was building up to the firing pressure. The firing pin nibs rest on a small steel shear plate with holes of lesser diameter than the holes in the igniter tube; this is to insure that the sheared areas are not dragged along on their way down to the firing caps. The igniter is installed in the combustion flask against a copper washer.

6033. The burning igniter projects its flame into and across the combustion flask and burns for about ten seconds. The two primer caps and the double loading reduce failures far below the misfire record of single cap or a single loading.

6034. Composition of the burning charge - Mark 6-2 Igniter.
   Main load 2000 lbs. applied pressure.
   153 grains of celluloid.
   17 grains of black powder with 10% gum arabic.
   16 grains of parafined magnesium.
   Ignition load 1600 lbs. applied pressure.
   10 grains of 4 and 1 mixture.
   (8 grains of black powder with 10% gum arabic.
    2 grains of pyro cellulose).
   $\frac{1}{2}$ grain of gun cotton.
   40 grains of black powder with 10% gum arabic.

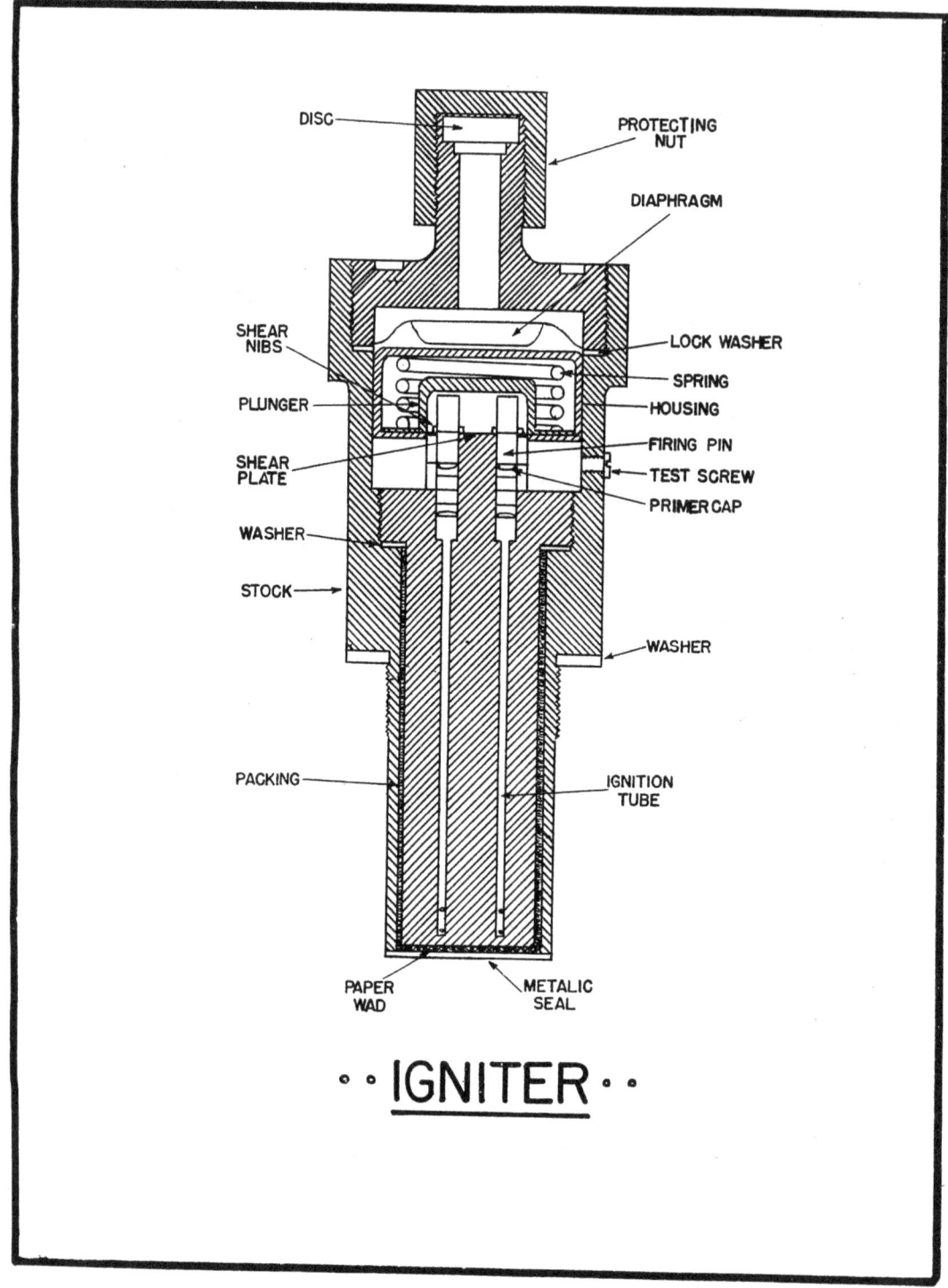

6035. __Precautions:__

(1) Do not disassemble igniter.
(2) Do not use igniters that have unusually swollen end seals as dampness has probably reached the powder and igniter would not fire.
(3) Some igniters **have** the metallic seal coated outside with shellac. If such is the case, do not use one that has remained in a combustion flask long enough to allow alcohol fumes to eat away the shellac. Newer igniters have seal coated with lacquer which is not soluble in alcohol.
(4) Do not use an igniter that has the plug nipple burred or cross-threaded.
(5) Be sure that the lead protective disc is removed when removing the protective nut before firing.

6036. Igniters are issued to the service in rectangular sheet copper boxes, each containing eight igniters, and weighing, with its contents, about 10 lbs. All seams of the boxes are soldered to the edges of both the box and cover, making an air tight joint. This sealing strip is provided with a loop, by means of which it may be torn free when the box is to be opened. Each igniter is carefully wrapped in a sheet of paraffin paper and an outer wrapper on which are printed complete instructions, for care installation and use of the igniter. A copy of these instructions is also pasted on the inside of the box cover, and additional instructions relative to storage and care are printed on labels pasted on the outside of the box and cover. These instructions should be carefully followed.

6037. __Care of Igniter after firing:__

(1) Immediately upon recovery of torpedo, remove igniter and screw on protecting cap.
(2) Do not disassemble igniter.
(3) Clean out all burned explosive from stock and wash thoroughly with hot water and soap, and dry.
(4) Coat with vaseline, wrap in paper; see that washer and protecting cap are in place.
(5) When all igniters in box have been used, immediately return box as noted below:
   (a) Vessels operating with tenders will, at first available opportunity after holding torpedo practice, transfer to tenders all empty igniters used on that practice, first carrying out above instructions. After all igniters in box have been used, the empty box shall be transferred to tender.
   (b) Tenders will forward igniters in lots of eight by express or parcel post, to Newport, if on the East coast or Mediterranean, and to Keyport, if on the Pacific Coast or Asiatic Station, using regular containers if available, or else suitable packing boxes.

(c) Vessels operating away from tenders will, after each torpedo practice, pack and ship expended igniters by parcel post or express to stations as indicated above.

6038. All igniters which fail to fire or are considered defective for any reason will be forwarded promptly, via government transportation or express, to the nearest torpedo station (Newport or Keyport) for examination and test. Destroyers will turn their defective igniters in to their tender for forwarding.

6039. When forwarding a defective igniter, a report giving the following data will be forwarded direct to the torpedo station concerned, with copies to the Bureau of Ordnance, the other torpedo station and the tender:

(a) Mark, modification and number of igniter.
(b) Station at which loaded.
(c) Date loaded.
(d) Date received aboard (and date received by tender).
(e) Mark, modification and register number of torpedo in which fired.
(f) Date fired.
(g) Brief statement of performance.
(h) If not fired, state defects noted.

6040. Each igniter will be tagged by the vessel forwarding it. The tag will bear the name of the vessel and the data under (a) to (f) inclusive, of Article 6039.

6041. In order that tests of defective igniters may be accomplished as soon after the igniter failure as practicable, shipment by express, when government transportation is not available, has been authorized by the Bureau of Supplies and Accounts. They may <u>not</u> be forwarded by mail. Igniters are classified as "Percussion fuzes" in the Interstate Commerce Commission Regulations. When shipped by express, igniters must be packed in strong, tight, wooden or metal containers with each igniter firmly secured. Each container must be plainly marked "Percussion Fuzes" not Torpedo Igniters.

6042. Defective igniters will be examined and tested by the torpedo station as soon as practicable after receipt, and the results will be reported to the vessels concerned via force commanders. Copies of this report will be forwarded to the Bureau of Ordnance, the other torpedo station and the tender or base.

G. <u>THE NOZZLES</u>.

6043. The nozzle is a single piece steel forging, suitably machined to close the opening on the horizontal projection of the upper portion of the turbine bulkhead and to convey and distribute the hot gases from the combustion pot connection to the turbines.

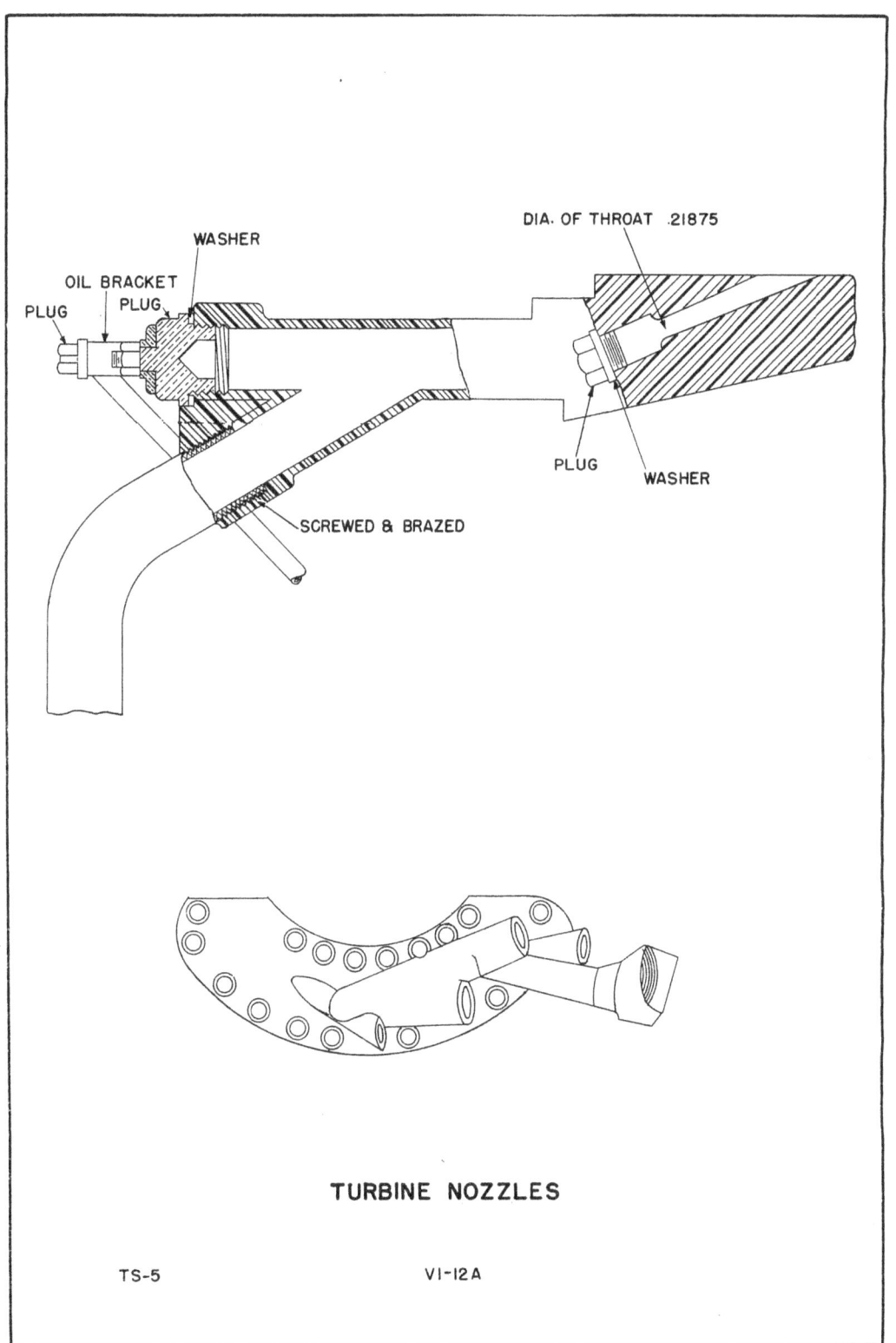

TURBINE NOZZLES

6044. The three nozzles' jets are of the so-called conical type. The least diameter is at the throat, the nozzle entrance being slightly rounded at its entering edge to give easy approach for the gases. The diameter of the throat fixes the rate of flow of the gases at any given pressure and is accordingly determined by the power requirements of the torpedo. From the throat, the nozzle flares in a conical form, the amount of taper being fixed by the ratio of expansion desired. The nozzles are placed so as to properly direct the gases to the turbine buckets. In passing through the conical portion of the nozzles, the gases expand from nozzle pressure to approximately afterbody pressure, and emerge from the mouth at high velocity, (about 4000 f.s. in modern torpedoes).

NOTE: Steps preceded by an asterisk (*) are necessary only when repairing or replacing damaged parts.

H. REMOVE PARTS ON TURBINE BULKHEAD.

6045.

		TOOL NO.
(1)	Remove air pipe to igniter	141A
(2)	Remove air pipe to air check valve	229
(3)	Remove oil pipe to reducing valve	48,141A
(4)	Remove gyro spin pipe	229
(5)	Remove fuel and water pipes to spray	144,141A
(6)	Remove air pipe from starting valve to bulkhead.	141A
(7)	Remove air pipe from bulkhead to reducing valve (starting gear return)	141A
(8)	Remove air pipe from reducing valve to air strainer	141A
(9)	Remove fuel spray and washer	388
(10)	Remove water spray and washer	388
(11)	Remove dummy igniter and washer	391A
(12)	Disconnect main air connection between valve group and combustion flask. (Loosen coupling nut, then remove valve group holding screws before disconnecting coupling).	134A 227A
(13)	Remove vent pipe	141A
(14)	Remove nuts for nozzle stud and remove combustion flask with nozzle and gasket	227A

6046. Remove starting gear.
   (1) Disconnect air pipe from starting valve to starting gear.................................. 141A
   (2) Disconnect air pipe from starting gear to reducing valve............................... 141A
   (3) Remove holding screws, starting gear and gasket. 41

I. STARTING GEAR DISASSEMBLE.

6047. Remove starting piston body.
   (1) Remove plug for starting piston and washer..... 12
   (2) Remove starting piston and spring.
   (3) Remove holding nut, starting piston body, and washer.. 432

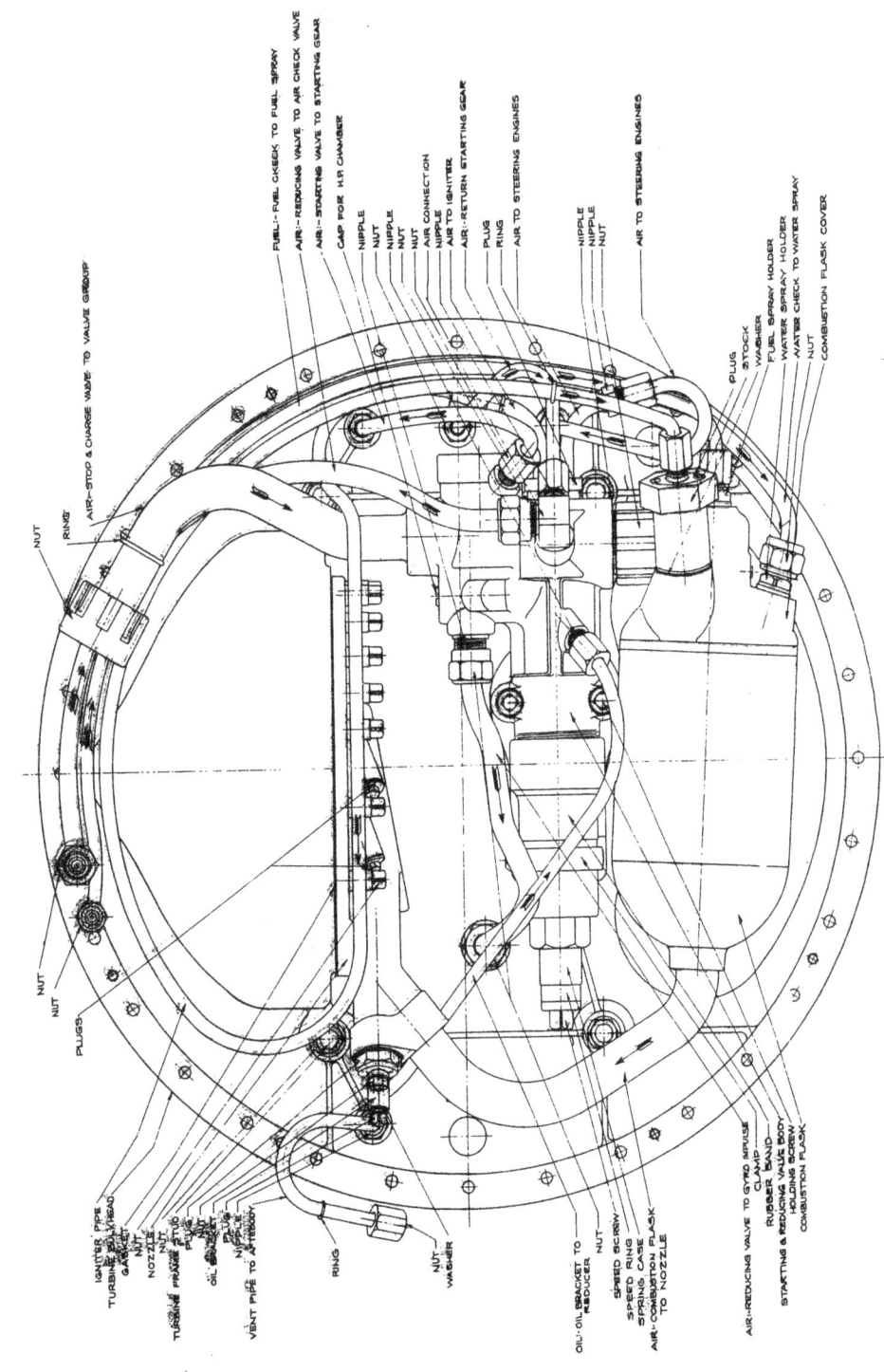

VALVE GROUP & SUPERHEATER — END VIEW

TS-5 VI-13A

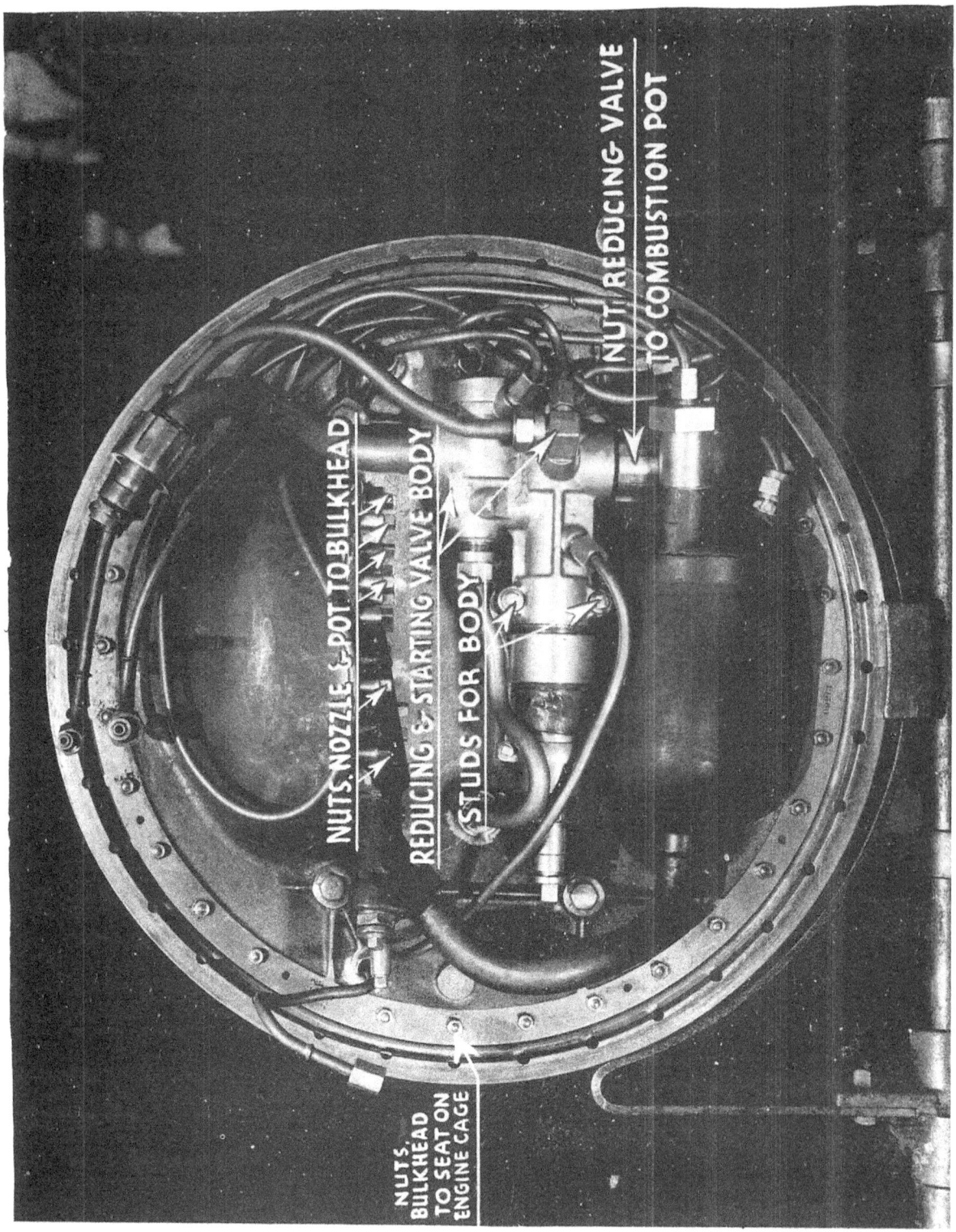

TS-5

                                                                TOOL NO.
6048.   Remove index spindle.
    (1) Remove taper pin for unlocking cam.
    (2) Withdraw index spindle through unlocking cam
        and starting gear frame.
    (3) Remove holding screws for packing ring.......... 41
    (4) Remove cup leather packing and packing ring.

6049.   Remove unlocking lever.
    (1) Remove unlocking lever spring.................. 72
    (2) Remove cotter pin and nut for unlocking lever
        pivot...................................... 72,48,40
    (3) Remove pivot pin and unlocking lever.

6050.   Remove connecting lever and valve lift.
    (1) Remove cotter pin for connecting lever on valve
        lift shaft..................................... 72
    (2) Push valve lift and shaft clear of connecting
        lever.
    (3) Remove connecting lever.

6051.   Remove starting spindle and arm.
    (1) Disconnect spring on starting spindle head...... 72
    (2) Remove pin holding starting spindle head on
        starting spindle.
    (3) Unscrew starting spindle head from starting
        spindle.
    (4) Remove pivot screw for starting spindle arm..... 41
    (5) Remove starting spindle, spindle arm, and pin...

OVERHAUL AND ASSEMBLE.

6052.   Replace starting spindle and arm.
    (1) Insert starting spindle in its bearing.
    (2) Replace starting spindle arm and pin for spindle
        arm.
    (3) Replace pivot screw for starting spindle arm.... 41
    (4) Replace starting spindle head on spindle, line
        up holes for pin, and insert.
    (5) Replace spring for starting spindle head........ 72,41

6053.   Replace connecting lever and valve lift.
    (1) Insert connecting lever in its slot.
    (2) Insert valve, lift until holes in connecting
        lever and square end of spindle line up.
    (3) Insert cotter pin through lever and spindle..... 72

6054.   Replace the unlocking lever.
    (1) Replace unlocking lever.
    (2) Replace pivot pin through unlocking lever and
        frame.

		TOOL NO.

(3) Secure pivot pin with nut and cotter pin.... 40,48,72
(4) Replace spring for unlocking lever............ 72

6055. Replace index spindle and unlocking cam.
   (1) Renew packing washer.
   (2) Replace packing ring and three screws......... 41
   (3) Insert index spindle through starting gear frame and unlocking cam.
   (4) Replace taper pin in unlocking cam and spindle.

6056. Replace starting piston body.
   (1) Reseat valve seat in body and lap valve to seat. 41,WE75
      NOTE: If necessary to lap in a new valve, lap hole in body with lap tool WE56 and new valve with lap tool WE57.
   (2) When finished lapping, wash parts thoroughly in spirits.
   (3) Replace starting piston body and washer in starting gear frame. Replace lock nut......... 432
   (4) Oil (A) and replace starting piston and spring.
   (5) Replace valve plug and washer................. 12

ADJUSTMENTS AND TESTS OF STARTING GEAR.

6057. With thickness gauge, obtain the clearance between bottom end of starting piston and valve lever. This should be ".010; if not, remove stock from end of starting piston until this clearance is obtained. Do not exceed this clearance as too much may result in insufficient lift of the starting piston when torpedo is fired.

6058. Install starting gear in test fixture and connect air intake nipple to a high pressure air line, seat starting piston and turn on air. Test for leaks at: (1) outlet nipple from starting piston; (2) around plug for starting piston.

6059. Blank off starting piston discharge nipple and pull up on starting spindle arm, thus unseating starting piston. Test for leaks around bottom end of starting piston. The starting piston is a lapped fit in the starting piston body, and should leaks appear to be excessive, it will be necessary to lap in a new piston.

EXCERPT FROM BUORD CIRCULAR LETTER NO. T 6-42.

1. It has recently been determined that some previously unexplained failures of subject torpedoes to start at the beginning of the run was due to vibration or shock disengaging the connection lever from the steps of the unlocking lever, thus permitting the starting piston to re-seat and thus stop the torpedo with consequent loss.

STARTING MECHANISM ON TEST STAND

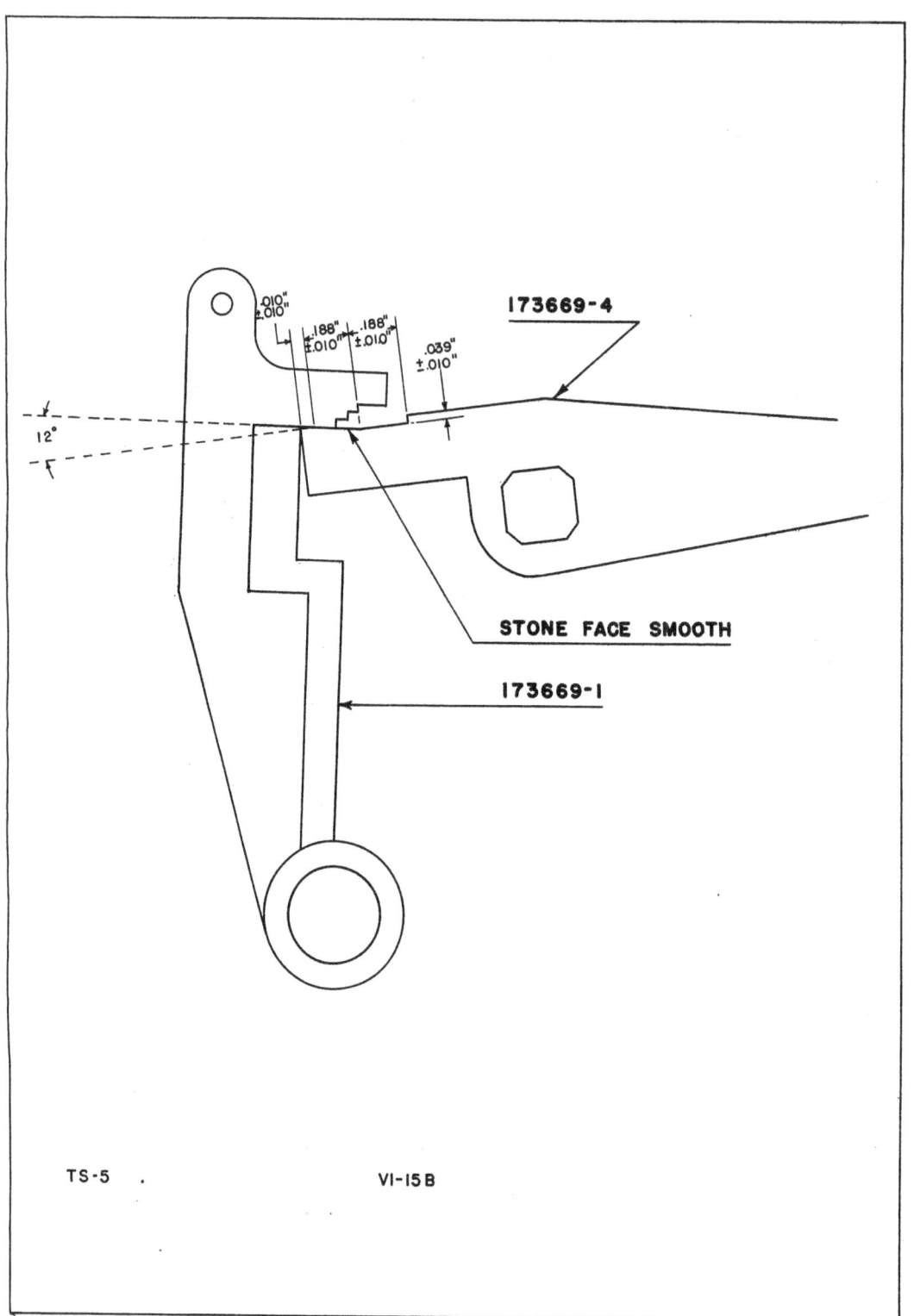

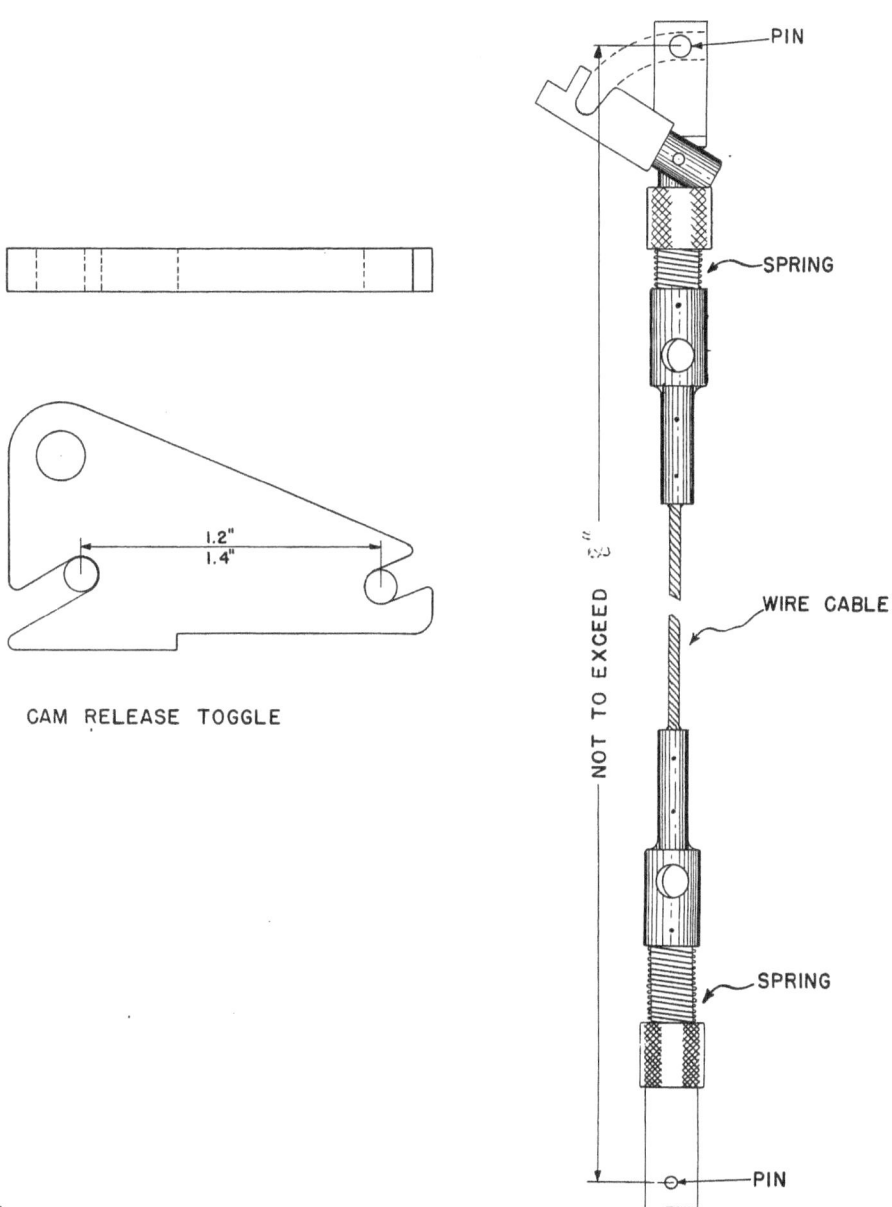

TRIPPING LANYARD FOR T.B.F.

2. In order to correct the condition which permits this casualty the following changes in parts concerned should be made:
   (a) Modify the connection lever (drawing No. 173669-4) by grinding the upper surface of its after end so this surface will be parallel to the last (bottom) step of the unlocking lever when engaged. This modification not only produces <u>surface</u> contact between these two parts but also eliminates the tendency toward disengagement referred to above.

3. Tests of starting mechanisms fitted with parts altered as above indicate that no amount of shock or vibration normally to be expected will produce disengagement from the last step.

4. The following features of starting gears of Mark 13 type torpedoes should likewise be given careful attention:
   (a) If necessary, the unlocking cam (drawing No. 79238-2) should be reduced in length to prevent excessive extension of the unlocking lever spring (drawing No. 79236-7) during unlocking of the starting gear. It is thought that over extension of this spring may have introduced permanent set which contributed to sensitiveness to shock of the connection lever.
   (b) Renew all unlocking lever springs which fail to pull the unlocking lever forward smartly when the after end of the connection **lever** is depressed below the last step.

5. The following tests should be given starting gears of Mark 13 torpedoes to insure that they are in proper operating condition:
   (1) With starting gears in "ready" condition, apply air pressure of 2800 pounds per square inch.
   (2) Insure that starting piston is properly seated by observing for air leakage and by inspecting for clearance between bottom of piston and the valve lever (about 0".01). By any convenient method (such as a pencil mark on the exposed stem) establish the initial starting piston position relative to its body.
   (3) By means of lanyard, withdraw starting toggle smartly in a vertical direction. Examine mechanisms to insure that: (a) connection lever is well engaged on last step of unlocking lever and (b) that starting piston has been moved upward at least 0".06. Repeat this test at least 5 times.

6. It is directed that the design of the present starting gear be modified in accordance with the above and that torpedoes be not issued by Torpedo Stations until this modification has been made.

J. DISASSEMBLY, OVERHAUL AND ASSEMBLY OF STARTING VALVE.

6060. Disassembly: TOOL NO.
   (1) Remove starting valve follower and upper seat... 151
   (2) Remove spring and starting valve.

6061. Overhaul:
   The following workshop equipment tools are supplied for effecting overhaul of starting valve:
   (1) Reseat seat in body for starting valve.......... WE50,24
   (2) Lap hole in body for starting valve piston...... WE53
   (3) Grind starting valve to its seat................ WE54,212
   (4) Lap starting valve piston rings................. WE51&52

6062. Assembly:
   (1) Inspect starting valve seat in body, starting valve, hole for starting valve in valve body, piston rings, upper seat, and threads of follower nut.
   (2) Insert starting valve in guide sleeve with chamfered end of guide sleeve out................... 212
   (3) Oil (A) and push starting valve through guide sleeve down on its seat (screw rod may be inserted through gyro spin nipple under starting valve and screwed into valve to pull down on its seat).. 74
   (4) Oil (A) and insert starting valve spring.
   (5) Replace upper seat and secure with follower nut. 151

TEST OF STARTING VALVE.

6063.
   (1) Blank off connection to starting piston.
   (2) Connect air lead to inlet pipe of starting valve.
   (3) Turn on high pressure air.
   (4) Submerge assembly in water; look for leaks.
   (5) Turn off air. Disconnect inlet lead. Remove blank.

K. DISASSEMBLY, OVERHAUL AND ASSEMBLY OF REDUCING VALVE.

6064. Disassembly.
   (1) Remove plug with pipes over top of differential chamber.. 191
   (2) Remove speed screw and speed ring............... 12
   (3) Remove clamp from spring case................... 49
   (4) Remove rubber band from spring case.
   (5) Remove spring case.............................. 377
   (6) Remove spring and spring button from spring case.
   (7) Remove spring collar and stem from sylphon cup..
   (8) Remove lock nut and nut from valve stem......... 252,180

                                                                            TOOL NO.

(9) Lift out sylphon, remove valve and stem from sleeve.

(10) Remove washer from annular recess in end of sleeve.

6065. **Overhaul.**

The following workshop equipment tools are supplied for effecting overhaul and renewal of parts for reducing valve:

(1) Lap (male) hole in valve body for reducing valve sleeve... WE59
(2) Lap (female) reducing valve sleeve.............. WE60
(3) Lap (male) reducing valve sleeve to stem........ WE61
(4) Lap (female) reducing valve stem to sleeve...... WE62

As a rule, the use of the above tools is necessary only for the renewal of parts. Before attempting renewal of parts, the inside and outside diameters of mating parts must be determined with micrometers. The differences should not exceed that given on the drawings. Clean and examine reducing valve for scratches and grooves. The beveled diameters of reducing valve seat are made to a tolerance of "001 and may be reproduced only by the application of the utmost skill. Clean and examine sylphon bellows for hardening, distortion, or surface defects. Examine reducing valve sleeve and hole in reducing valve body for scratches and a mirror finish. Examine washer between end of reducing valve sleeve and flange on valve stem for hardness; anneal or renew if necessary.

6066. **Assembly.**

(1) Oil (C) and replace sylphon diaphragm assembly through hole in valve body.
(2) With copper washer in place, oil (C) and replace reducing valve with stem through hole in valve sleeve.
(3) Replace reducing valve stem nut and lock nut.... 252,180
(4) Screw sylphon tester wing nut on to valve body, thus clamping sylphon cup rigidly against end of valve body.
(5) Oil (C) and replace plug with washer over reducing valve 191
(6) Blank off all openings of valve body, except igniter nipple, and proceed to test sylphon diaphragm for leaks.
*(7) Connect low pressure air line to igniter nipple.
(8) With regulator placed sylphon end up, fill core of sylphon with oil (C).
(9) Turn on air and look for bubbles.
    Bubbles may show between:
    (a) Valve stem and sleeve (copper washer).
    (b) Sweated ends of bellows to sleeve or sylphon cup.
    (c) Convolutions of bellows.
    (d) Bottom of cup and end of valve body.

* - Use 450 lbs. air pressure for testing.

SYLPHON DIAPHRAGM TEST

SYLPHON TESTER WING NUT ON
REDUCING VALVE BODY OVER SYLPHON CUP

AIR TO (GN) TER NIPPLE ON
REDUCING VALVE BODY

TOOL NO.

(10) When satisfactory test is completed, turn off air.
(11) Disconnect all blanks and air lead.
(12) Remove sylphon tester wing nut.
(13) Blow out excessive oil used in testing with air.
(14) Oil (C) and replace spring stem and collar.
(15) Oil (C) and replace spring button in spring case.
(16) Replace reducing valve spring in spring case.
(17) Screw spring case to valve body and set up hard... 377
(18) Replace speed ring and speed screw............... 12
(19) Replace new rubber band over vent hole in spring case.
(20) Replace clamp with clamp screw for spring case.... 49

TEST OF REDUCING VALVE.

6067. Mount assembled reducer on testing block and make the following connections:

(1) Air from banks through heater to starting valve... 134
(2) Air from regulator to combustion pot on test block. 134
(3) Combustion pot to atmosphere. In this line is placed a "212 restriction, designed to pass air at same rate as in hot run of torpedo. Lines from location of restriction out of air ports must be of sufficient diameter to prevent back pressure from building up.
(4) Air check valve nipple on valve group to reduced pressure recording gauge...................... 229
(5) Starting valve to firing valve...................... 141
(6) Oil regulator, Oil (D) and blank off nipple........ 141
(7) Blank off the following nipples:
  (a) Air to gyro spin............................. 229
  (b) Return air from starting gear................ 141
  (c) Air to igniter............................... 141
  (d) Air to steering engines...................... 141
(8) Turn steam on heater and bring water temperature to 205° F.
(9) Have speed ring in place, but speed screw backed off about one turn.
(10) Charge banks to 1500 lbs.
(11) Open firing valve, and note pressure obtained, valve operation, and air tightness. If operation and pressure are satisfactory, set up on speed screw. This is a necessary preliminary step to check set up and to prevent damage to sylphon by over pressure.
(12) Charge banks to pressure designated for air flask of torpedo. Fire and note the pressure obtained, observing closely for fluctuations and variations in pressure.

6068. **Performance and instruction data for testing reducing valves.**

The relative pressure, which is correct for the indivi-

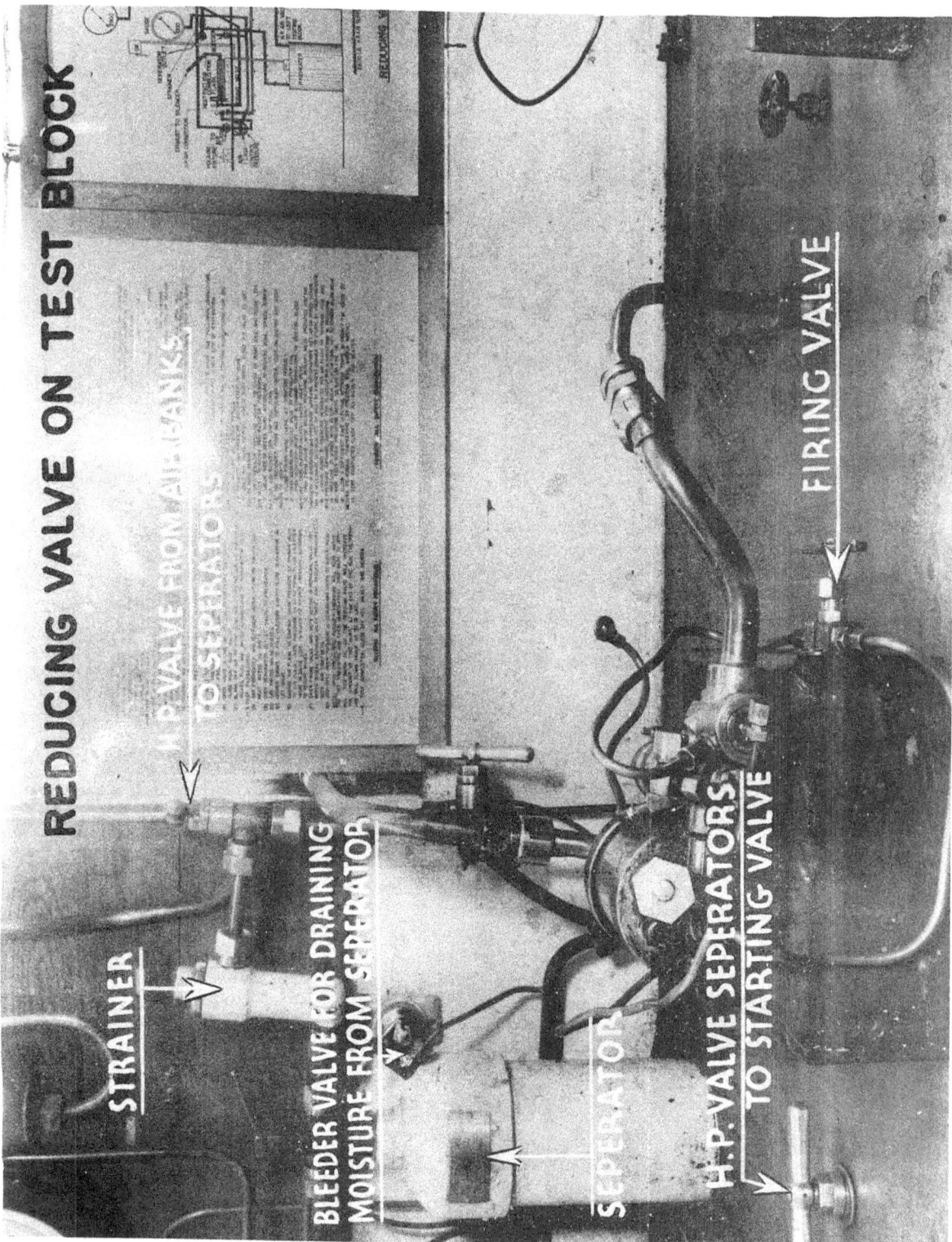

dual torpedo, will be shown on the torpedo work sheet inserted in the record book at the time torpedo was last run in the dynamometer tank or proof ranged.

L. OVERHAUL AND TEST COMBUSTION FLASK AND NOZZLE ASSEMBLY.

6069.   TOOL NO.

   (1) Inspect and clean interior of combustion flask.
   (2) Reseat seat and clean threads on nipples on combustion flask..................................WE89,WE85
   (3) Run nozzle clearance tool through nozzles....... WE37B

BLANKING OFF TOOLS FOR COMBUSTION FLASK AND NOZZLE.

6070. Plates with gaskets for blanking off nozzle openings of the various types of torpedoes when testing combustion flasks for leaks, are now being furnished with the workshop equipment. These plates and gaskets are shown on Bu.Ord. Dwg.Nos. 226063, 226065.

6071. Blanks for blanking off igniter hole, main air nipple, or triangular joint to combustion flask, blanks for water and fuel spray holders, are also being furnished with the workshop equipment in order that facilities may be available to perform the test outlined below.

COMBUSTION FLASK TEST.

6072. To test combustion flask for leaks, proceed as follows:

   (1) Clamp combustion flask in clamp (referred to under "Cover Wrench and Clamp" above) and secure in vise with nozzle plate uppermost and at an angle to bring test plug to highest point.
   (2) Blank off all holes in combustion flask except nipple for connection to air strainer.
   (3) Blank off nozzle plate. Remove test plug in nozzle plate.
   (4) Connect $\frac{1}{2}$" test pipe from hydraulic test pump to nipple for air strainer pipe on combustion flask top.
   (5) Ascertain that gauge furnished with hydraulic test pump is accurate.
   (6) Pump pipe line and flask full of water until it begins to come out through test plug; replace test plug and washer.
   (7) Continue to pump until obtaining a pressure of 1,500 pounds per square inch. If flask is tight this pressure will hold.
   (8) Inspect carefully for leaks around joints, with particular reference to joint of combustion flask cover. Should any leaks be found, it will be necessary to remedy prior to use.

M. **DISASSEMBLY, OVERHAUL AND ASSEMBLY OF SPRAYS.**

6073. **Disassemble Sprays.**
                                                                             TOOL NO.
- (1) Remove spray body from holder.................... 12,51A
- (2) Remove plug from body........................... 12,40
- (3) Push whirl from body with small brass rod.

6074. **Overhaul and Assemble Spray.**
- (1) Inspect whirl, push it into body with letter toward the hole in body so it will be visible when assembled.
- (2) Replace plug for body noting 4 holes are clear.. 12,40
- (3) Replace body in spray holder, noting washer in place... 12,51A

6075. **Assemble Sprays in Premixer Top.**
- (1) Place annealed copper washer on spray holder and screw into premixer top......................183,208,51A
- (2) Do not interchange spray holders; short holder is for fuel.

TEST OF SPRAYS.

6076.
- (1) Connect spray at end of outlet line of test set.
- (2) Fill bottle of test set with water.
- (3) Turn on air.
- (4) Make trial test and set reducer to 35 lbs. pressure.
- (5) Turn off air.
- (6) Refill bottle.
- (7) Turn on air; start stop watch.
- (8) Check time until bottle is empty. Stop watch.
- (9) Fuel spray should deliver 6 pints of water in 115 seconds; water spray in 46 seconds.

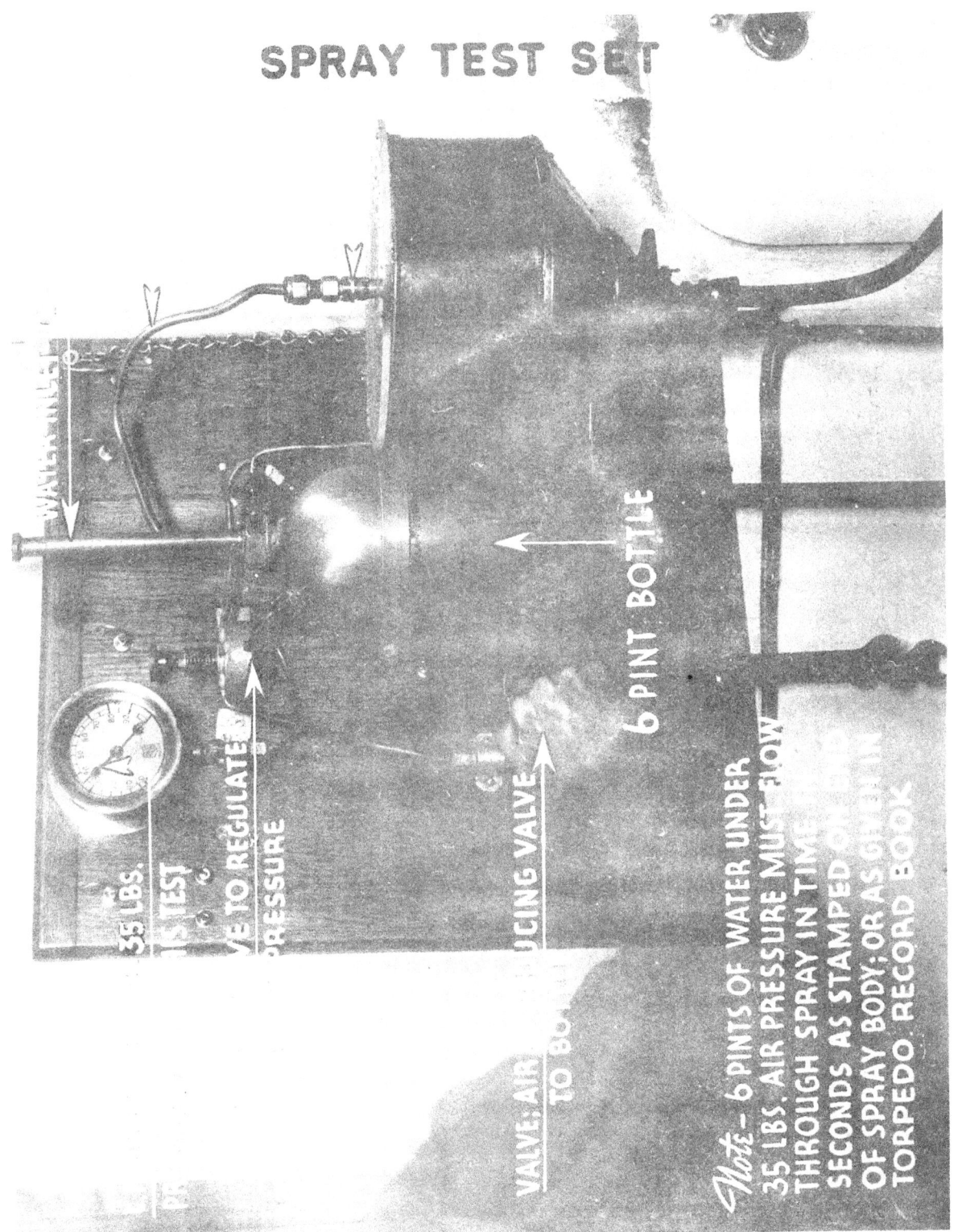

CHAPTER SEVEN

		ARTICLES
A.	Remove Gyro and Immersion Mechanism	7001
B.	Remove the Tail	7002
C.	Remove the Main Engine	7003
D.	Description Turbine Bulkhead	7004-7005
E.	Description Main Engine	7006-7012
F.	Propeller shafts, sleeves, hubs and propellers	7013-7020
G.	Exhaust System	7021-7024
H.	Smoke Prevention	7025-7027
I.	Lubrication System	7028-7034
J.	Disassembly, Overhaul, Assembly and Test of Oil Pump	7035-7038
K.	Disassembly, Overhaul, Assembly and Test of Main Engine	7039-7064
L.	Description of Afterbody and Tail	7065-7077
M.	Overhaul, Test and Assemble parts assembled with Afterbody Shell Group	7078-7088
N.	Install Starting Gear	7089
O.	Install Main Engine assembly in Afterbody	7090
P.	Build up of Turbine Bulkhead	7091

NOTE: Steps preceded by an asterisk (*) are necessary only when repairing or replacing damaged parts.

A. REMOVE GYRO AND IMMERSION MECHANISM FROM AFTERBODY.

7001.
                                                                                  TOOL NO.

(1) Remove hand hole plates (with afterbody bottom up).. 200,48
(2) Disconnect air leads from horizontal and vertical steering engines........................... 141A
(3) Remove pipe from steering engine to gyro reducing valve.. 141A
(4) Remove valve connection clamp screws for both engines... 246A
(5) Remove engine holding screws, detach engines from gyro pot and lay aside in afterbody leaving engines connected to rudder rods.... 72,49,49A
(6) Disconnect gyro spin lead...................... 229
(7) Remove transportation pin, and holding screws for gyro and immersion mechanism base......... 49,459
(8) Remove tension on depth spring................ 135A
(9) Put lifting screws in gyro and immersion mechanism base and remove mechanism from afterbody. 200
(10) Remove rudder rod pins and take horizontal and vertical steering engines out of afterbody.... 449

B. REMOVE TAIL (MARK 13 MODIFICATION TORPEDOES).

7002.
(1) Remove holding nut, after propeller and sleeve with four (4) bushings....................... 183
(2) Remove two (2) lock screws from lock nut for forward propeller and remove lock nut....... 185C,185D
(3) Remove forward propeller and hub from forward propeller sleeve (use lead maul if necessary).
(4) Remove keys from sleeve, use screw driver in milled slot in key............................ 41
(5) Remove four (4) drain plugs, disconnect and remove rudder rod connecting pins............ 13-14
(6) Remove sixteen (16) tail joint screws and remove tail....................................... 184
(7) Remove wire and four (4) locking clip screws.. 72,41
(8) Turn forward propeller sleeve until locking clips are in alignment with milled slot in after bulkhead and remove four (4) locking clips; pry out with a screw driver............ 41
(9) Remove forward propeller sleeve...

C. REMOVE GEAR TRAIN ASSEMBLY FROM AFTERBODY.

7003.
(1) Disconnect gyro spin pipe at nipple in strengthening ring.. 229

REMOVING GYRO AND IMMERSION MECHANISM

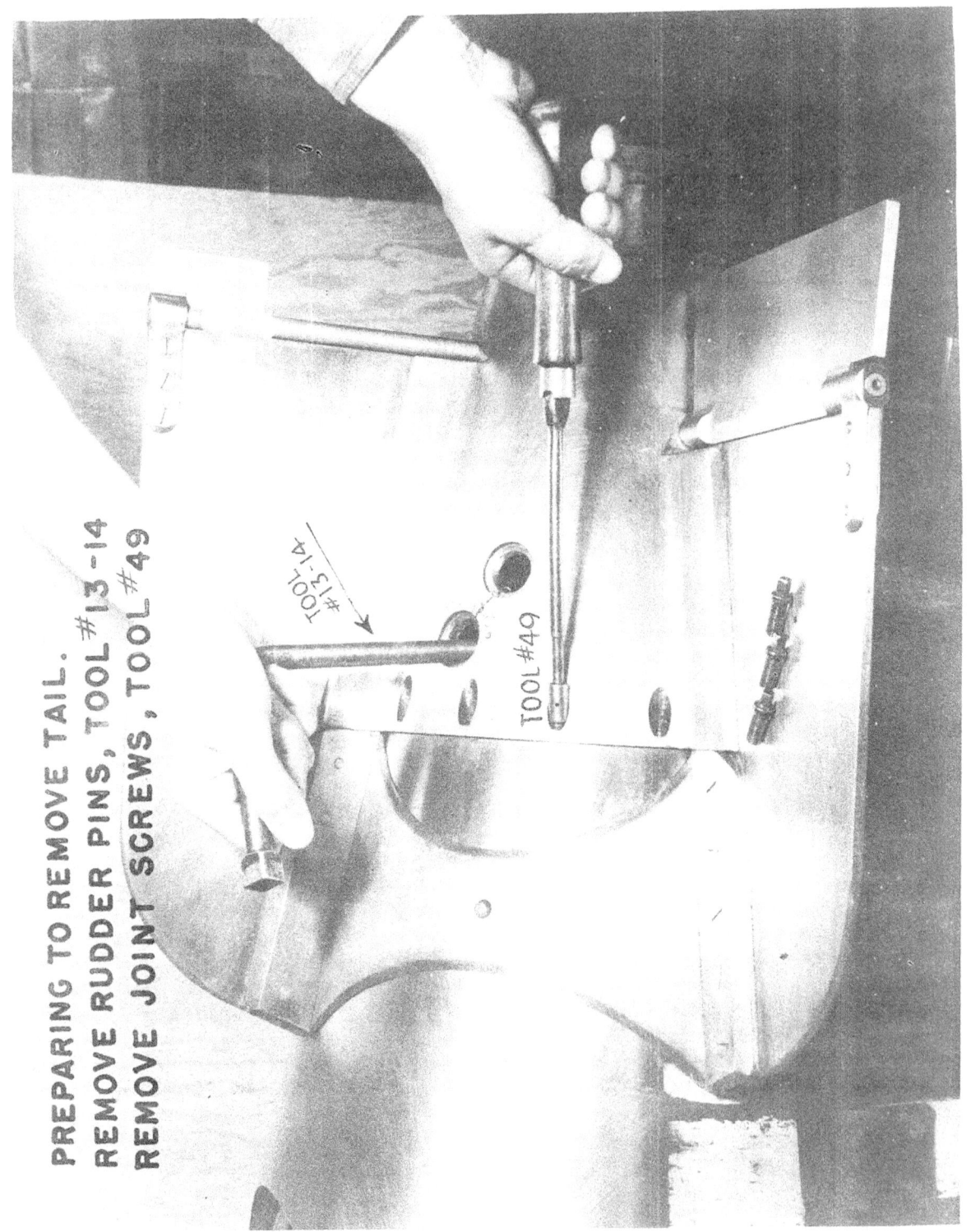

PREPARING TO REMOVE TAIL.
REMOVE RUDDER PINS, TOOL #13-14
REMOVE JOINT SCREWS, TOOL #49

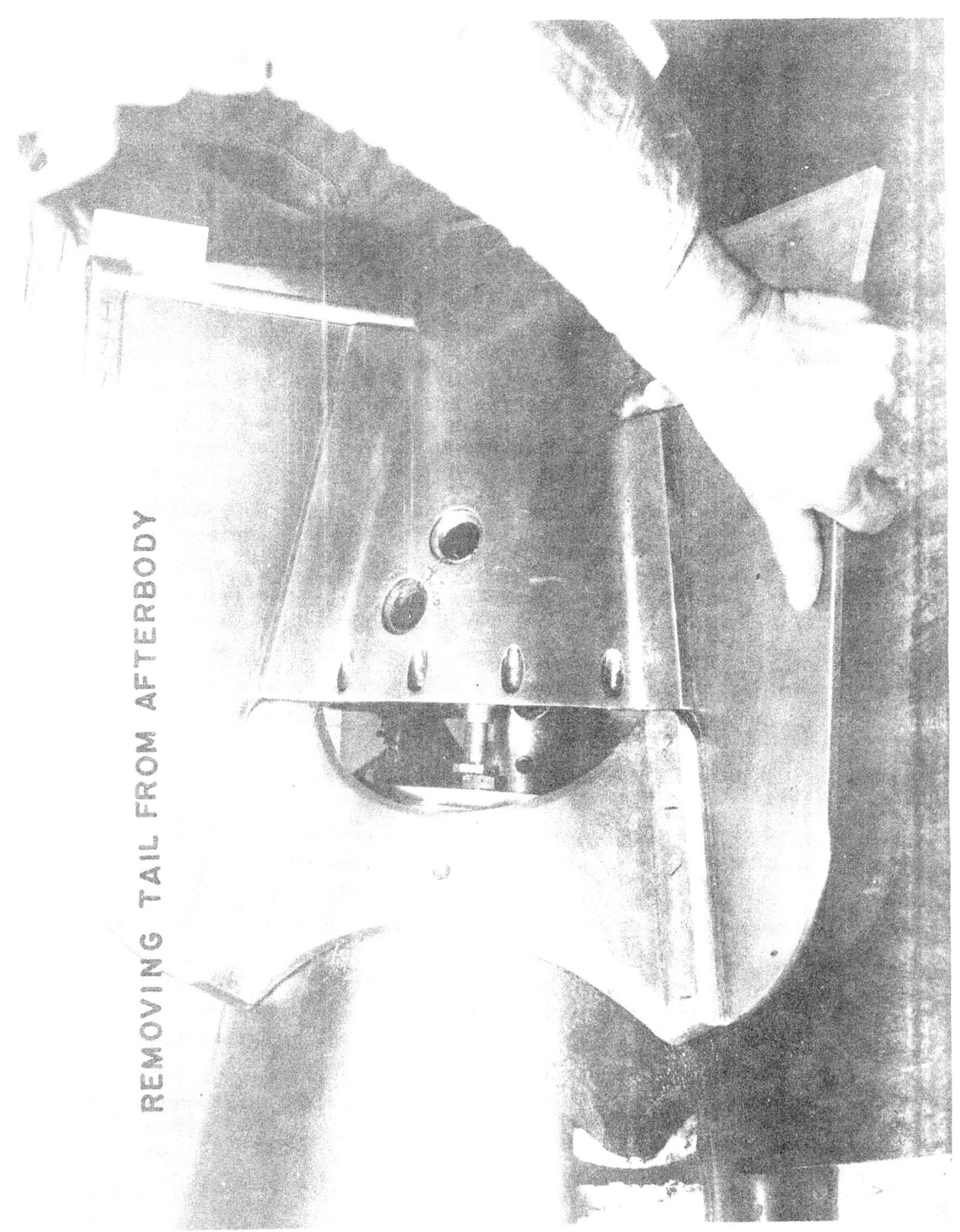

REMOVING TAIL FROM AFTERBODY

	TOOL NO.

    (2) Disconnect oil pipe, from pump to oil tank at oil tank.................................. 229
    (3) Disconnect air pipe from bulkhead to air strainer at manifold on strengthening ring.
    (4) Disconnect air pipe from starting valve to starting gear at manifold on strengthening ring... 141A
    (5) Disconnect air pipe from starting gear to reducing valve at manifold on strengthening ring. 141A
    (6) Remove nuts for engine bulkhead studs......... 48
    (7) Install bearing guide on after propeller shaft. WE34
    (8) Install lifting handles on bulkhead and remove gear train assembly from afterbody........... 446A, 446B

D. <u>TURBINE BULKHEAD</u>.

7004. The forward end of the afterbody is closed by a large bronze casting known as the turbine bulkhead. It is secured to the engine cage by means of screws sweated in the latter, and nuts. A gasket of high pressure packing is interposed between the surfaces to secure pressure tightness.

7005. The upper portion of the bulkhead projects forward and has on its lower horizontal surface an opening and tapped holes with studs for attaching the nozzle. The engine frames are bolted to this bulkhead on its after face, and thus it supports the entire engine assembly; also at its forward side it supports the nozzle, combustion flask and starting and reducing valve bodies. Suitable nipples are mounted in the bulkhead for connecting the air pipe lines that are required to pierce the bulkhead.

E. <u>THE MAIN ENGINE</u>.

7006. The main engine is of the turbine driven, gear reduction type. It is frequently spoken of as a balanced turbine engine, due to the fact that the turbines and gearing are in pairs revolving in opposite directions, thus balancing the gyroscopic effect on the torpedo. It consists essentially of two turbine rotors each mounted on a suitable spindle, one within the other, and a system of gearing for conveying the drive to the propeller shafts.

7007. The turbine spindles are mounted in a spindle casing which in turn is bolted to the upper and lower engine frames. The crosshead supports the main driving gears and bevel pinions, and is also bolted to the engine frames. The after ends of the engine frames are tied together by a strut, which also serves as housing for a ball bearing. This bearing serves as a radial bearing for the outer propeller shaft and as a thrust bearing to take the rearward thrust of the after bevel gear. The forward ends of the engine frames are bolted to the engine bulkhead.

REMOVING MAIN ENGINE FROM AFTERBODY

REMOVING MAIN ENGINE FROM AFTERBODY

7008. The ball bearings for the turbine spindles are three in number, the outer races being threaded and screwed into threaded seats in the spindle casing. The forward side of the spindle casing is slotted, and at each bearing a bolt is provided for drawing down on the casing and thus clamping the ball races. The edge of each outer race is notched, and locking levers mounted in the slot, one for each bearing, engage with the notches. By this means a degree of adjustment is provided whereby the desired clearance between the face of the nozzle and the first turbine wheel, and clearance between the two turbine wheels may be obtained. The first turbine wheel spindle is carried in two, single row, combined radial and thrust bearings. The middle bearing takes the downward thrust, and the upper bearing takes the upward thrust. These bearings are adjusted so that the first turbine wheel will have a maximum of "060 clearance over the face of the nozzle, and a clearance of "018 between the two bearings. The lower end of the second turbine wheel spindle is carried in the bottom bearing which is a double row, combined radial and thrust bearing, and takes thrust in either direction. The second turbine wheel has a maximum clearance of "060 between it and the first turbine wheel. The bottom bearing has a vertical clearance of "015 between the thrust collars. The bearing between the turbine spindles consists of four bushings of bearing bronze, each provided with oil holes and grooves. The fit at this point is relatively loose, the difference between each two bearing diameters being "007, resulting in a total play between the inner and outer turbine spindles of "014.

7009. Due to the high rate of rotation of the turbines and the close proximity of the bearings to the hot exhaust gases, it is important that these bearings receive a sufficient supply of oil during the run of the torpedo.

7010. As previously stated, the main driving gears and bevel gear train are mounted on the cross head. The main driving gears carry the bevel pinions which mesh with the bevel gears, one on the forward end of each propeller shaft. This arrangement is such that the unequal turning effort of the turbines is converted to equal turning effort on the propeller shafts. Further, at the point of greatest tooth pressure viz., at the bevel pinions, the bearing pressure is partially balanced and the driving torque is carried by a greater number of teeth. The bearings for the main driving gears and bevel pinions are plain bushings of bearing bronze, provided with oil holes and grooves.

7011. The turbines are secured to the turbine spindles by means of a tapered fit and two keys, diametrically opposite, and are pulled home by means of a locking nut. A locking screw prevents the nut from accidentally unscrewing. A high degree of perfection in the taper fit is required. This is

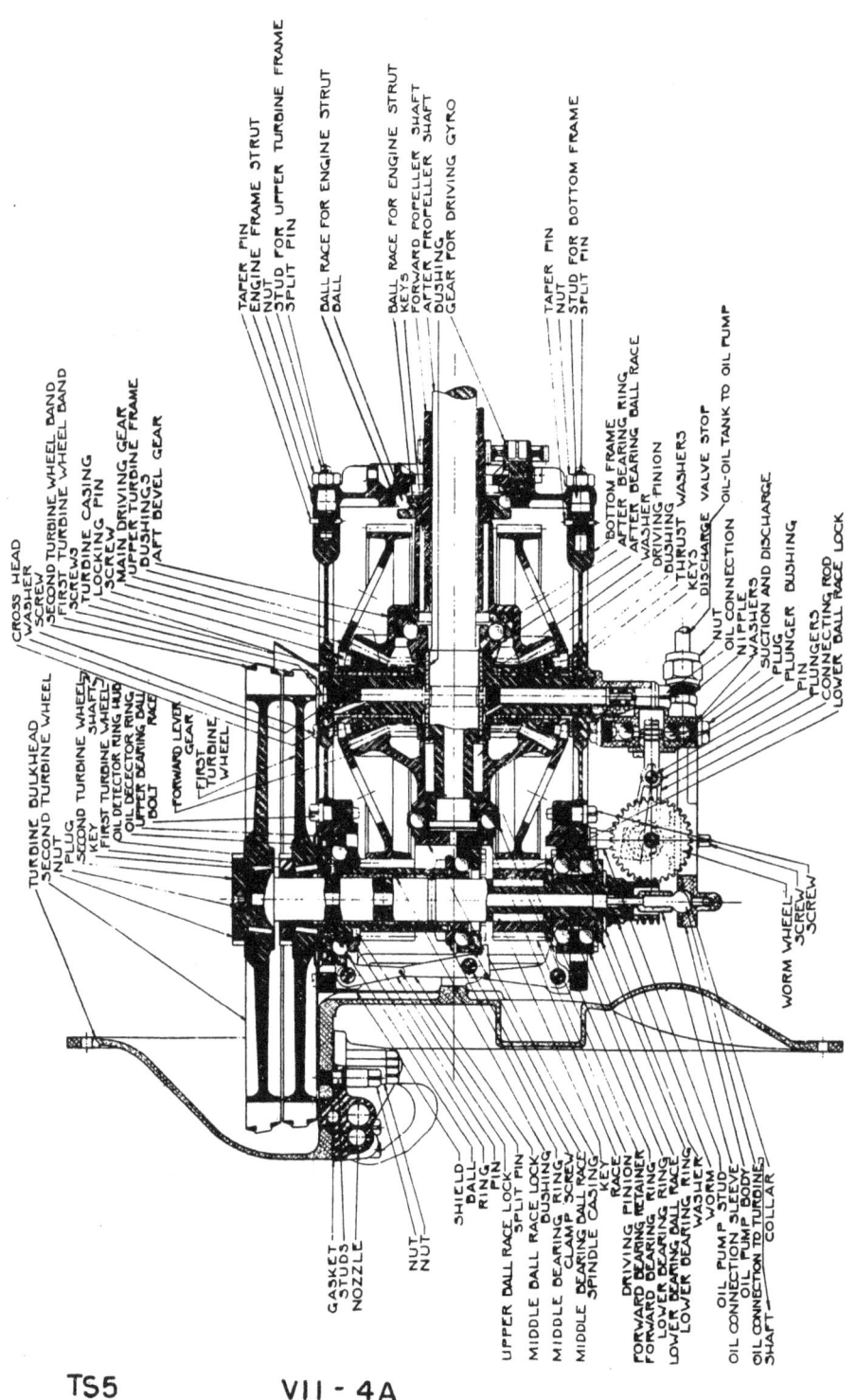

TURBINE & GEAR TRAIN – SECTION

TS5   VII - 4A

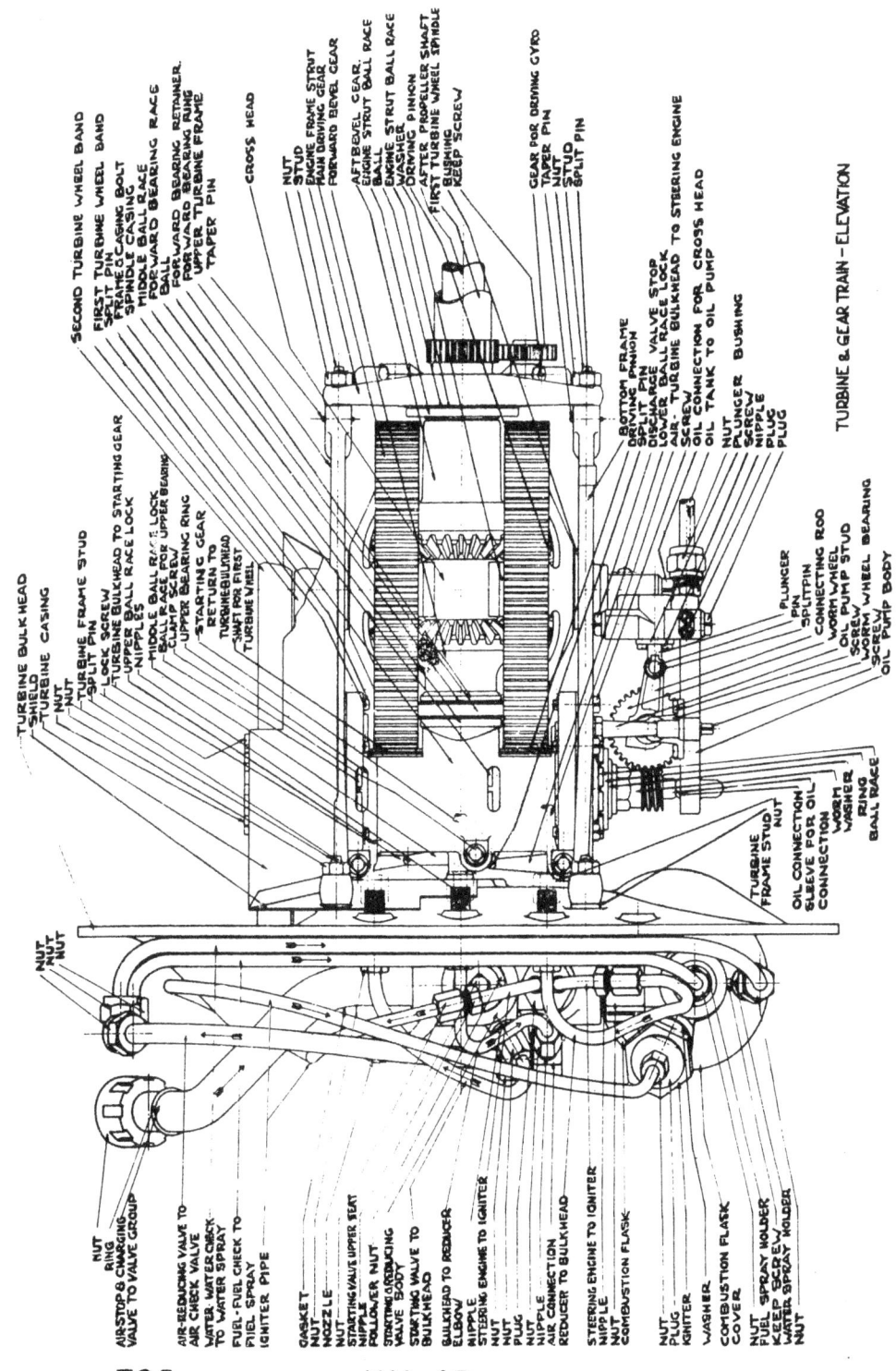

MAIN ENGINE ON
OVERHAUL STAND,
SIDEVIEW
SHOWING
SHAFT, SLEEVES,
AND
PROPELLER
ASSEMBLY

MAIN ENGINE, TOPVIEW

MAIN ENGINE, BOTTOM VIEW

accomplished by first grinding both surfaces and then lightly lapping them together. At assembly the turbines are drawn strongly to the spindles. This is essential, as in operation, the wheels enlarge due to a combination of centrifugal stresses and temperature changes; therefore an equivalent initial stress must be set up at assembly in order to prevent loosening of the fit during a run.

7012. The turbine buckets are cut from the solid metal of the wheel. A band at the periphery of the turbine serves to shroud and complete the form of the bucket. The band is secured by means of rectangular projections, one at each bucket, which fit corresponding holes in the bands. These projections are riveted over after assembly of the band.

F. PROPELLER SHAFTS, SLEEVES, HUBS AND PROPELLERS.

7013. The drive from the engine to the propellers is conveyed through the propeller shafts, of which there are two. The forward propeller shaft is tubular in form, passes through the engine strut and is keyed at its forward end to the after bevel gear. A ball thrust bearing is interposed between this gear and the cross head. A collar is turned on the shaft just forward of the strut and serves as a bearing for the inner race of the ball bearing in the strut. This bearing is designed to carry both radial and thrust loads. At its after end this shaft is supported in a bearing in the after bulkhead. Its extreme after end is provided with keyways for attachment to the forward propeller sleeve. Both the driving torque and the thrust of the forward propeller are carried through this shaft, the latter being partially resisted by the tendency of the bevel gear and pinion to separate under the driving load. The thrust of the propeller is somewhat the greater, and this difference in the forces is carried by the ball bearing to the cross head, and by it to the engine frames and thence to the turbine bulkhead. The shaft has, just aft the engine strut, a spur pinion integral with it, for driving the pallet mechanism.

7014. The after propeller shaft is solid and is mounted within the forward propeller shaft. At the cross head it runs in a floating bronze bushing. At its forward end it is keyed to the forward bevel gear. Interposed between the forward bevel gear and a collar on the shaft at this point, and the turbine spindle casing is a ball thrust bearing. The after end of the after propeller shaft is supported in a bronze bushing between it and the forward propeller shaft. Its extreme after end is provided with keyways for attaching to the after propeller sleeve, and a nut for securing the sleeve. The driving torque and thrust of the after propeller are carried through this shaft. This thrust, in addition to the tendency to separate between the forward bevel gear and the pinions is taken by the ball thrust bearing and carried by it to the spindle casing. Thus the thrust of the after propeller is taken by the spindle casing and carried by it to the turbine bulkhead, through the engine A frames.

7015. To prevent the leakage of sea water into the afterbody at the end of a run and prior to recovery of the torpedo, the forward propeller shaft is packed with a flanged felt washer, set up by a follower nut screwed into the after bulkhead. This nut has a left hand thread in order that the direction of turning of the shaft will not tend to unscrew it. The after propeller shaft has a hole drilled from its after end, and joined by four radial holes communicating with the bushing between the two shafts. This bushing is provided with helical grooves and radial holes communicating with radial holes in the outer **propeller** shaft. Thus, when grease is forced into the end of the after shaft, it thoroughly grease packs the two shafts and after bearing, excluding sea water. A beveled grease packing ring installed in the forward propeller sleeve entraps the grease that tends to come out between the propeller shafts at the after end. This ring must be installed with the beveled edge forward. The hole in the after end of after propeller shaft is closed by a threaded plug.

7016. The propeller sleeves are short shafts connecting to the propeller shafts and carrying the propellers. They are mounted in the tail cone, and are separated from the torpedo when the tail is removed.

7017. The forward propeller sleeve is tubular in form and connects at its forward end with the forward propeller shaft by means of keys fitting the keyways in the latter. A neat slip fit is provided. The sleeve is pinned to the propeller shaft to prevent slipping. Its after end is supported in a bronze bearing secured to the after end of the tail cone. On its extreme after end is mounted the forward propeller. This sleeve has a shoulder forward of the tail cone bearing. There is no thrust contact between this shoulder and the tail cone bearing, a considerable degree of longitudinal play (".125 minimum) being provided for this sleeve in order that the expansion of the forward propeller shaft, due to temperature in the afterbody, may not bind the bearings. The towing drag is taken up by the ball thrust bearing in the engine strut. The forward propeller sleeve is fixed on the forward propeller shaft by 4 pins engaging in holes in the shaft. These pins are riveted on extension clips which in turn screw to the sleeve by set screws.

7018. The after propeller sleeve is also tubular in form and its forward end is secured to the after propeller shaft by means of keys (the fit being similar to that of the forward sleeve) and a nut on the threaded end of the latter. Near its after end it carries a bronze bearing in the form of four separate bushings interposed between it and the forward propeller sleeve. At its extreme after end it carries the **after** propeller.

7019. A considerable space is provided between these sleeves within the tail cone, and in addition, each sleeve has a series of large holes. This construction is for the purpose of permitting the escape of exhaust gases from the interior of the tail cone to the interior of the after propeller sleeve, and thus to the sea.

7020. The propellers are of forged steel, and four bladed. Each propeller has a steep taper hole in the central body which fits a corresponding taper on the propeller hub to which it is also keyed. The propeller hub fits neatly over the rear end of the propeller sleeve and is keyed thereto, with one key. The hub is slotted through one side. The propeller hub and propeller sleeve are locked together by means of a nut screwed to the propeller sleeve and forcing the propeller on the taper of the hub. By virtue of the slot in the hub, the latter is forced tightly to the propeller sleeve. The nut is locked from unturning by means of a keep screw partially in the nut and partially in the body of the propeller.

G. EXHAUST SYSTEM.

7021. The exhaust system is devised to carry off the turbine gases with the least back pressure and with least circulation through the afterbody. The system consists of a semi-closed chamber above the turbines, two exhaust tubes leading aft, into the after bulkhead, two exhaust valves seating on the after bulkhead, and hollow propeller sleeves with escape holes in them.

7022. The turbine exhaust chamber is formed within the afterbody shell by a vertical and horizontal bulkhead, which, supplemented by the splash pan below the turbines, and the turbine bulkhead, forms a semi-tight chamber.

7023. The exhaust tubes are telescoped over short sleeves, with flanged collars, attached to the vertical bulkhead with screws. The after end of each tube is slipped into suitable openings in the after bulkhead and held in place by rolling a beaded projection into a recess cut around opening. By removing the flanged sleeve and prying under the beaded ends, the tubes may be withdrawn from the afterbody.

7024. The two exhaust valves are identical and are carried in a bracket secured to the after bulkhead by studs. The valves are of monel metal and have beveled seats lapped into corresponding seats on the after bulkhead. The valve stems slide in reamed holes in the bracket and are held up on their seats by springs of heat treated stainless steel. The springs hold the valves on their seats until opened by exhaust gas pressure, which at the valve is about 1 lb. 4 ozs. above the depth head.

SMOKE PREVENTION.

7025. Unless means are taken to prevent it, all hot running torpedoes in addition to the wake of the bubbles have a smoke wake. This is due to engine oil entering the exhaust and being partically burned by the heat of the gases. It is desirable for a war shot to render the wake of a torpedo as nearly invisible as possible, and to this end it is necessary that engine oil be excluded from the exhaust.

7026. Underneath the turbine and attached to the top engine A frame is a sheet metal pan flared at its periphery, and this, with the horizontal bulkhead serve to prevent the splash of oil from the gearing into the exhaust chamber. It also serves to check the circulation of the hot exhaust gases in the afterbody. An oil deflector ring is installed over #1 turbine spindle, above the top bearing which serves to prevent the entry of oil from this bearing to the exhaust chamber.

7027. With the splash pan, bulkhead and oil deflector ring installed, the wake of the torpedo is practically smokeless. It should be noted, however, that if the torpedo is rolled over in preparation for run when there is much loose oil in the afterbody, so that the oil enters the exhaust space, the torpedo will smoke heavily until the oil is burned up, usually in three to eight minutes.

I. LUBRICATION SYSTEM.

7028. The oil pump is a worm driven plunger pump secured to the lower "A" frame of main engine by two screws and the oil connection. The pump is a flat framework with a vertical integral extension at its after end, housing the suction and discharge valves. In the forward face of this housing are formed the suction chambers, each connecting with vertical holes drilled from the bottom of housing, into which the suction plugs are screwed. The upper ends of the suction plugs are machined cup-shaped to form a seat for the ball suction valves. When the plugs are in place in pump body, an annular groove with two radial holes are aligned with horizontal channels drilled in the pump body. The oil nipple, connected by piping to the oil tank is silver soldered in place with its forward end terminating at the annular groove of the starboard suction plug.

7029. The suction chambers are connected with two vertical holes drilled from the top of the housing which carry the discharge ball valves and discharge plugs. The discharge valves seat on a restriction formed in lower part of vertical holes, just over the suction chambers, and operate vertically between their seats and tips machined on the lower end of discharge plugs. Midway between the suction chambers is drilled a large vertical hole which snugly houses the oil connection. The oil connection has an annular groove machined on its outer surface which connects with a channel from the space above the port discharge valve seat. Two radial holes are drilled in the annular groove of oil connection to the hollow interior of it. The upper end of oil connection screws into the hollow lower spindle of the crosshead hub. The oil pump is thus forced securely against the crosshead spindle and the lower "A" frame. The starboard discharge chamber is connected by suitable channels in the pump frame to a recess in the forward upper end of pump body. Into this recess is fitted the ball end of a hollow oil connection. A tubular sleeve fitting over the end of oil connection extends up into the hollow lower end of #2 turbine shaft, with a shoulder on the sleeve bearing against a shoulder on the shaft, thus making an oil tight joint.

7030. The pump plungers operate in flanged bushings secured by two screws and soldered to the forward end of the horizontal suction chambers. The after ends of these plungers extend into the suction chambers between suction and discharge valves. Four annular grooves are cut in the plungers to form an oil seal between the plungers and bushings. The after ends of the connection rods are pinned to the forward ends of the plungers, and their forward ends are fitted over eccentric surfaces machined on the port and starboard sides of the worm wheel. The worm wheel meshes with the worm nut on the bottom of #2 turbine shaft.

7031. The operation of the oil pump is as follows: The worm nut on the end of #2 turbine shaft rotates with that shaft causing the worm wheel to rotate. Eccentrics on the wheel impart a reciprocating motion to the connecting rods which in turn transfers that motion to the plungers. A suction is created in the chambers on the forward stroke of the plungers, tending to draw the discharge valves down against their seats and the suction valves up off their seats. Oil from the oil tank passes through the piping and pump nipple to the space around the annular groove of the starboard suction plug, and through the cross connecting channel to the space around the annular groove of the port suction plug. Oil passes to the hollow interior of these plugs via the radial holes and as the suction valves are off their seats, passes around these ball valves filling the suction chamber around the plunger.

7032. On the after stroke of the plunger the suction valve is forced down by gravity on its seat and the discharge valve off its seat by oil pressure from the bottom. The oil in the suction chamber now passea up around the discharge valves and into the space between the stops and the restriction. From the port discharge chamber the oil passes through the channel in the body to the annular space around the oil connection, and thence to the crosshead. From the

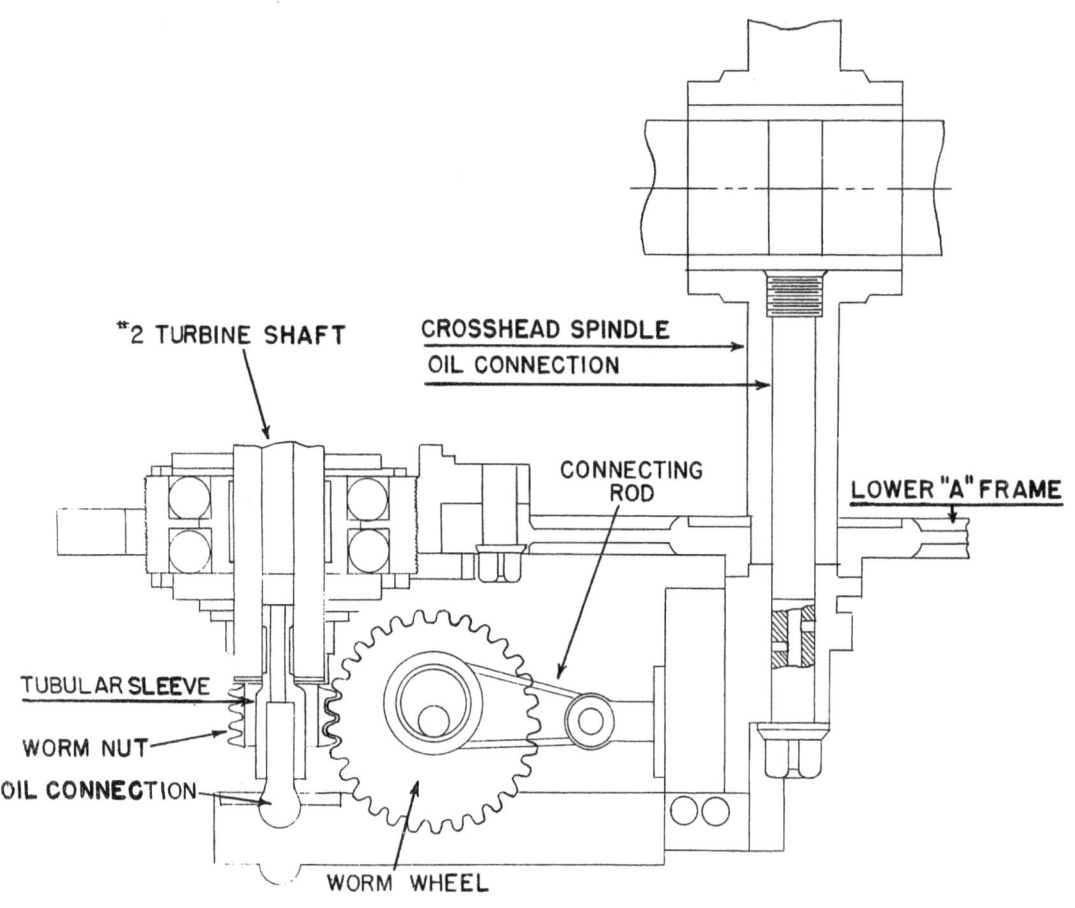

OIL PUMP

TS-5   VII-9A

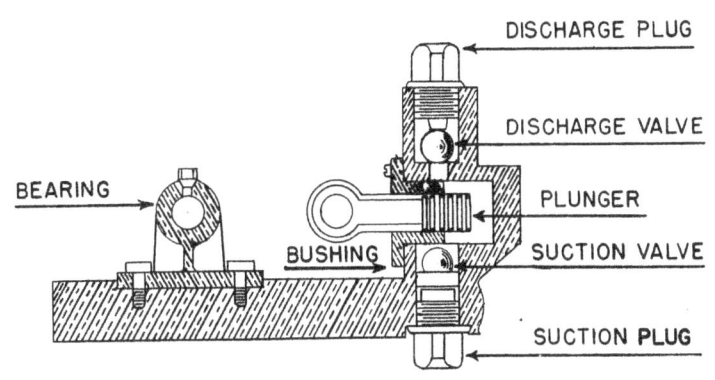

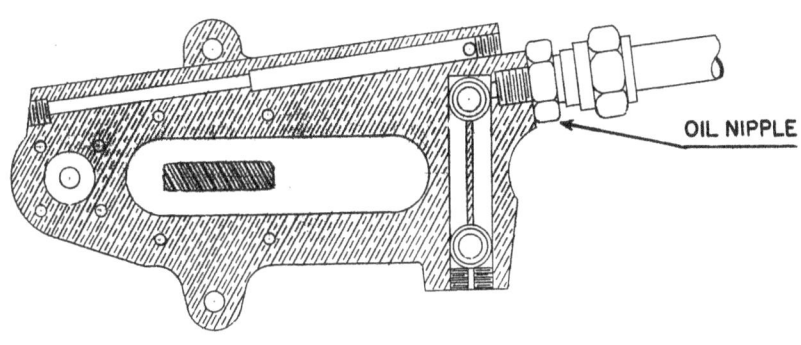

OIL PUMP

starboard discharge chamber, oil passes through a channel down the body, thence through the horizontal channel in the frame work and finally to the lower end of the oil connection and sleeve to #2 turbine shaft.

7033. Oil delivered from port discharge chamber of oil pump is distributed to crosshead and bearings as follows: The oil connection is machined with a collar at the lower end which takes up against the under side of the oil pump body, thus making an oil tight joint. Opposite the discharge from the port side of the oil pump, the oil connection is machined with an annular space which is drilled with two radial holes extending into the hollow oil connection. Above this groove a collar is machined which makes an oil tight connection with the oil pump body, and just above this collar the diameter of the oil connection is reduced to form a space between it and lower spindle of crosshead. The oil connection is screwed into the crosshead hub, passing through the interior of the lower spindle. All oil for the crosshead is delivered through the hollow oil connection to the wide annular groove in the inside of the crosshead hub, encircling the crosshead bushing, and rising into the hollow upper spindle of crosshead. From a point near the top of this crosshead spindle, oil is forced out through two radial holes to the upper end of the bushing for crosshead and bevel pinion. Through radial holes and helical grooves in this bushing, the oil is distributed inside and outside the bushing. Oil is also delivered through a diagonal hole in the spindle to the upper main drive gear bushing and via grooves and channels in this bushing to the "A" frame thrust washer. Oil passes through radial holes in the crosshead bushing to an annular groove machined on the inner propeller shaft, and from here it is distributed by helical grooves in the inner surface of the bushing over the outer surface of the inner shaft. From the wide annular groove in the inside of the crosshead, oil is lead by helical grooves in the outside of crosshead bushing to the forward narrow annular groove of crosshead. A diagonal hole is drilled from this narrow annular groove to the hollow lower spindle of crosshead. Through this diagonal channel, oil is delivered to the lower spindle of the crosshead outside the oil connection. The lower chamber of crosshead spindle is closed by a collar machined on the oil connection as previously explained. A reservoir of oil is thus built up, and passes out through two radial holes near the top of lower crosshead spindle to the bushing for bevel pinion and crosshead. Through radial holes and helical grooves in this bushing, oil is distributed inside and outside, and to the lower bushing for main drive gear and "A" frame thrust washer. From the after end of the crosshead bushing, oil is carried to the after thrust bearing.

7034. Oil delivered from starboard discharge chamber of oil pump is distributed to turbine spindle bearings as follows:

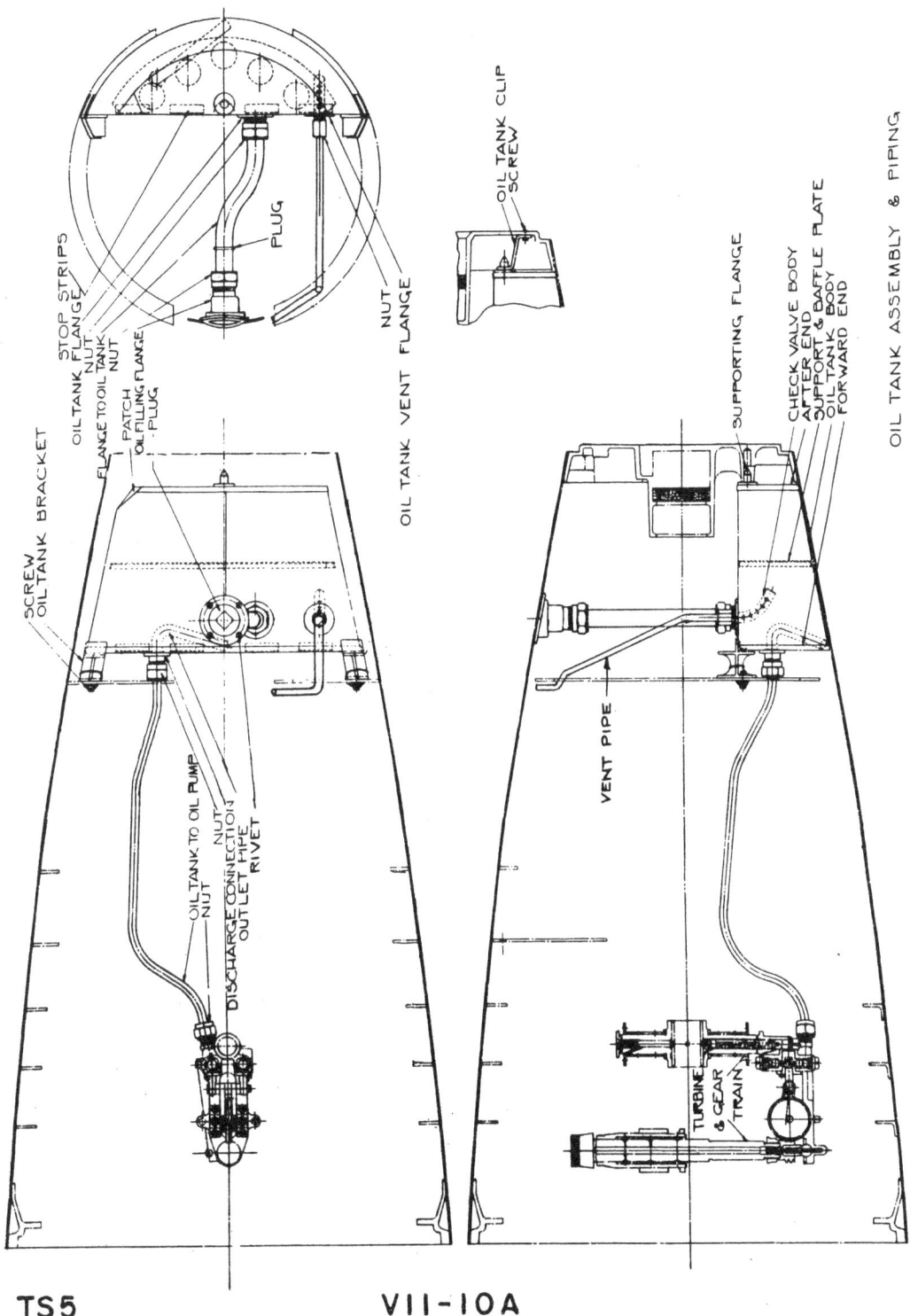

From the oil connection and sleeve in #2 turbine shaft, up through the oil channel in shaft. From this channel, oil is forced out through the radial hole in the mid-annular groove of shaft and further to four radial holes in top annular groove. Over the shaft at these points is located the four part bushing. Through helical grooves in the bushings and the three annular grooves in #2 shaft, oil is distributed evenly over the shaft inside the bushings. Through four radial holes in each of the bushings, oil is carried to the outside of the bushings to the bearing space inside #1 shaft, and is circulated by annular grooves in the outer sides of the bushings. Opposite the radial holes in the upper four part bushing is a single radial hole in #1 turbine shaft. This hole terminates at a point immediately above the ring for the upper bearing. Oil thrown off radially from the four part bushings passes through hole and lubricates the upper bearing. The middle bearing is lubricated by overflow from the upper bearing and the oil from the lower end of the four part bushing. In the after side of the spindle casing at a height just above the middle bearing ball race is a single radial hole terminating in the forward thrust bearing. Through this hole the thrust bearing receives oil from the overflow of the upper end of the middle bearing. The bottom bearing is lubricated by overflow from the top and middle bearings.

J. DISASSEMBLY, OVERHAUL, ASSEMBLY AND TEST OF OIL PUMP.

7035. Disassembly:

TOOL NO.

(1) Remove oil connection sleeve, oil connection for turbine shaft, four screws and oil connection collar.............................. 41
(2) Remove discharge valve stops (tap on large screw driver with a hammer to loosen threads for removal) and remove discharge valves..... 165A
(3) Remove suction plugs and valves............. 12
(4) Remove (2 each) holding screws and two worm wheel bearings.............................. 41
(5) Remove worm wheel, shaft, connecting rods with pin and plungers.
(6) Remove cotter pin, washer and connecting rod pin with 3 distance collars (note position of collars before disassembly so that the collars may be replaced in their proper places.

7036. Overhaul.

During overhaul of plunger pumps the following conditions must exist:

(a) Suction plugs - Ball seat free from burrs and reasonably oil tight. The plug must fit the oil pump body so that, on the discharge stroke, there is no flow of oil from the pump cylinder back through the suction line. Such a condition makes it necessary not only for the ball seat to be tight but also for the end of the plug to be a tight push fit with the oil pump body so that oil can not pass between the plug and body. This is accomplished by spreading the end of the suction plug in W.E.T.40, using W.E.T.40A. When removing plug form W.E.T.40, use block W.E.T.40B, to hold W.E.T.40. W.E.T.40C is a lap for the ball seats in oil pump body and the suction plug.

(b) Discharge plugs - These plugs are designed to give the ball a clearance of ".024 plus or minus ".009 when on its seat. Too much clearance allows the ball to obstruct the oil discharge hole, too little clearance restricts the oil passing through the discharge valve. This clearance may be varied by varying the thickness or number of discharge plug washers. Actual clearance may be measured by use of micrometers W.E.T.9, and W.E.T.6. Vary washer thickness to get correct clearance, or file tit of plug.

(c) Plungers and plunger bushings - Must be within drawing tolerances. Comparatively little fault will be found here.

(d) Pump stroke - Must be within drawing tolerances. Comparatively little fault will be found here.

(e) Pump casting - Must be free from imperfections that allow oil to leak past the valves. Some castings have been found to have blow holes at these points. When discovered the pump must be replaced.

(f) Oil channels - Must not be restricted. There have been a few cases where oil channels have been plugged with solder and where the channels, having been drilled from both ends, have failed to line up, causing restrictions. Inspections of channels can only be made by drilling out plugs and should only be made as a last resort.

7037. Assembly.

(1) Clean, inspect, blow air through and clear out all oil holes and passages. Note if suction nipple is soldered in place with silver solder or is loose. If not soldered in place, use silver solder in replacing it, being careful to use only sufficient heat to melt the solder.

(2) Assemble worm wheel and shaft in bearings, connecting rod, distance collars, plungers, connecting rod pin washers and split pin for pin, assembling the whole on pump body. Secure bearings with 4 holding screws. Note that bench marks coincide when assembling.

(3) Assemble discharge valves and plugs.
(4) Assemble suction valves and plugs.
(5) Assemble oil connection for turbine shaft on oil pump. Secure collar with 4 screws.

7038. **Test of Oil Pump.**

(1) Install pump on stand, make suction and discharge connections. During the run, the oil discharge line to the crosshead must be held firmly over its male slip connection, to prevent excessive leakage and consequent false readings.

(2) See ample oil in lower or supply tank.

(3) Open supply valve and run pump slowly until pump is discharging oil.

(4) Close supply valve and speed up pump to 1725 R.P.M.'s.

(5) Read vacuum gauge. If the suction joints and lines are airtight, pumps should show a vacuum of 22". A vacuum of less than 22" is a definite indication of (a) leaks in the suction line, (b) faulty suction valve, (c) undersize pistons or oversize bushings.

(6) Open supply valve. See valve at bottom of discharge oil tank tightly closed. Open valves to pressure gauges.

(7) As oil passes a graduation in gauge glass, start stop watch.

(8) Read pressure gauges. Static head should be as follows:

    (a) Spindle pressure -4.2 lbs.
    (b) Crosshead pressure - 5 lbs.

These are the minimum requirements. Overhaul pump if less than required minimum. Continue test if static head is satisfactory.

(9) Continue run for five minutes, taking care to keep the speed at correct R.P.M.'s. The total amount of oil pumped for a five minute run is 2.8 pints (minimum). If below standard, overhaul pump and continue test until satisfactory.

K. **MAIN ENGINE DISASSEMBLY.**

7039. Remove turbine bulkhead.   TOOL NO.

    (1) Hold assembly in special jaws in vise........ 300
    (2) Disconnect gyro spin pipe on bulkhead........ 229
    (3) Disconnect air pipe to air strainer on bulkhead.. 141A
    (4) Disconnect air pipe to starting gear on bulkhead.. 141A
    (5) Disconnect air pipe from starting gear on bulkhead.. 141A

    NOTE: It is good practice to tag pipes in steps 3, 4, and 5 above prior to removal. These pipes are very similar and can easily be replaced on the wrong nipple if not tagged.

    (6) Take out cotter pins, and remove nuts for turbine A frame studs........................ 72, 141A
    (7) Remove bulkhead.

OIL PUMP ON TEST STAND

                                                                    TOOL NO.
7040. Remove oil pump.
   (1) Remove wire and two holding screws for oil
       pump... 72,141A
   (2) Remove oil connection for crosshead.......... 12
   (3) Remove oil pump, being careful not to burr
       oil connection or sleeve.

7041. Remove second turbine, shaft and bearing.
   (1) Remove worm for driving oil pump (holding   229
       turbine and shaft from dropping out with
       clip)... 454
   (2) Remove washer under worm.
   (3) Remove ring for lower bearing end.
   (4) Remove 12 balls in lower bearing end.
   (5) Remove cotter pin, clamp screw and nut for
       lower bearing................................. 13-14,72
   (6) Remove unlocking lever for lower bearing.... 72
   (7) Remove cotter pins, and loosen up on bolts
       for A frame and spindle casing................ 72,227A
   (8) Unscrew bottom bearing race and remove....231,54,228
   (9) Remove 12 balls, center race and upper ring
       of lower bearing.
  (10) Remove clip for holding turbines together,
       and push second turbine and shaft out through
       first turbine spindle, removing driving pin-
       ion from spindle casing.

7042. Remove first turbine wheel, spindle and bearings.
   (1) Remove cotter pins, lock screws and turbine
       wheel nut..................................... 72,41,179
   (2) Remove first turbine (it may be necessary to
       start turbine by tapping with a lead hammer).
   (3) Remove 2 keys in shaft for turbine wheel
       (pry off with screw driver). Note if keys
       are marked; if not, do so..................... 41
   (4) Remove wire and 4 holding screws and turbine
       casing with shield............................ 72,41
   (5) Remove cotter pins, clamp screws with nuts,
       upper and middle ball race locks.............. 13-14,72
   (6) Remove oil deflector ring and hub........... 41
   (7) Remove cotter pins and loosen bolts for
       frame and casing.............................. 72,13-14
   (8) Unscrew upper bearing race (it may be neces-
       sary to use wedge between splits in casing
       to ease removal).............................. 231,54,228
   (9) Remove 16 balls from upper bearing, lift out
       first turbine wheel shaft, inner race for
       middle bearing and 4 part bushing, and re-
       move 13 balls from middle bearing.
  (10) Do not remove outer race for middle bearing
       unless necessary.

		TOOL NO.
7043. Remove engine frame strut.
  (1) Remove cotter pins and nuts for turbine frame studs.................................. 72,141A
  (2) Slip engine frame strut back clear of engine frame studs.
  (3) Remove 22 balls in strut bearing.

7044. Remove upper and lower turbine A frames.
  (1) Place clip over main driving gears.
  (2) Remove bolts for upper A frame and spindle casing.. 227A
  (3) Remove screw and washer for crosshead (upper A frame) and remove upper A frame....... 39
  (4) Remove thrust washer in frame.
  (5) Remove 6 bolts for lower A frame and spindle casing.. 227A
  (6) Remove lower A frame and spindle casing with after propeller shaft thrust bearing and ring.
  (7) Remove thrust washer in lower A frame.
  (8) Remove thrust bearing for after propeller shaft (ring from casing, retainer and race from shaft).

7045. Remove main driving gears and crosshead.
  (1) Remove clip tool holding main driving gears together.
  (2) Remove main driving gears and 2 washers for crosshead and pinions, crosshead bushings and washers for crosshead.
  (3) Slide after propeller shaft out through forward shaft about eight inches, catching 15 balls from after bearing.
  (4) Push after propeller shaft in and remove propeller shaft bushing.
  (5) Remove after propeller shaft, crosshead, bushing and inner race for after bearing.
  (6) Remove forward propeller shaft from vise jaws and slide engine strut and bearing race over end of shaft.
  (7) Remove outer race from bevel gear on outer shaft.

7046. Remove gears from forward and after propeller shafts.
  (1) Due to infrequent necessity for removal of bevel gears and strut bearing races, tools for this operation are not furnished.
  NOTE: Appropriate sleeves may be made from brass piping of a size to slip over end of shafting, and by the use of arbor press on tender no difficulty should be experienced in the removal or replacement of bevel gears on their shafts.

NOTE: Steps preceded by an asterisk (*) are necessary only when repairing or replacing damaged parts.

The above steps complete disassembly of main engine.

OVERHAUL, ASSEMBLY AND TEST.

7047. Propeller shafts.   TOOL NO.
   (1) Clean, inspect and stone out rough spots on bearings where necessary.
   (2) Place shaft between centers in shaft straightening press or a lathe, or lay shaft in roller brackets placed on a surface plate.
   (3) Place a dial indicator on surface plate, and turning shaft slowly in press, lathe or on roller brackets, note the amount shaft runs out of true. If out of true over 0."002 at any point, straighten in shaft straightening press (furnished with W.E. Equipment) before assembling.......................... WE10
*(4) Inspect and replace bearing race for strut thrust bearing, on forward propeller shaft..
*(5) Replace bevel gears on propeller shafts.

7048. Engine Strut.
*(1) Inspect and replace bearing race in engine strut.
   (2) Clean, oil (B) and inspect engine strut.
   (3) Slip strut over forward propeller shaft and place assembly in vise.................. 300

7049. Crosshead.
   (1) Clean, inspect, remove burrs where necessary and oil (B) crosshead with parts.
   (2) Replace ball race for after bearing on the after side of crosshead (after side is identified as being the side without oil grooves in bearing hub or oil hole in top after end of crosshead spindle).
   (3) Oil (B) and replace crosshead bushing.
   (4) Slip crosshead with bushing over after propeller shaft and insert after propeller shaft through forward propeller shaft.
   (5) Clean, inspect, stone out burrs where necessary, oil (B) and replace propeller shaft bushing.
   (6) Install shaft guide over after end of propeller shaft catching a few threads......... WE34
   (7) Remove assembly from vise and place with shaft guide resting on floor.
   (8) Hold crosshead up against forward bevel gear for clearance, and drop 15 inspected bearing balls into race.
   (9) Hold bevel gears together, replace shafts in vise and remove shaft guide................ WE34
  (10) Install aligning bushing over propeller shaft ends, so that key in bushing will line up the propeller shaft. Secure aligning bushing with nut for after propeller shaft.. WE35A

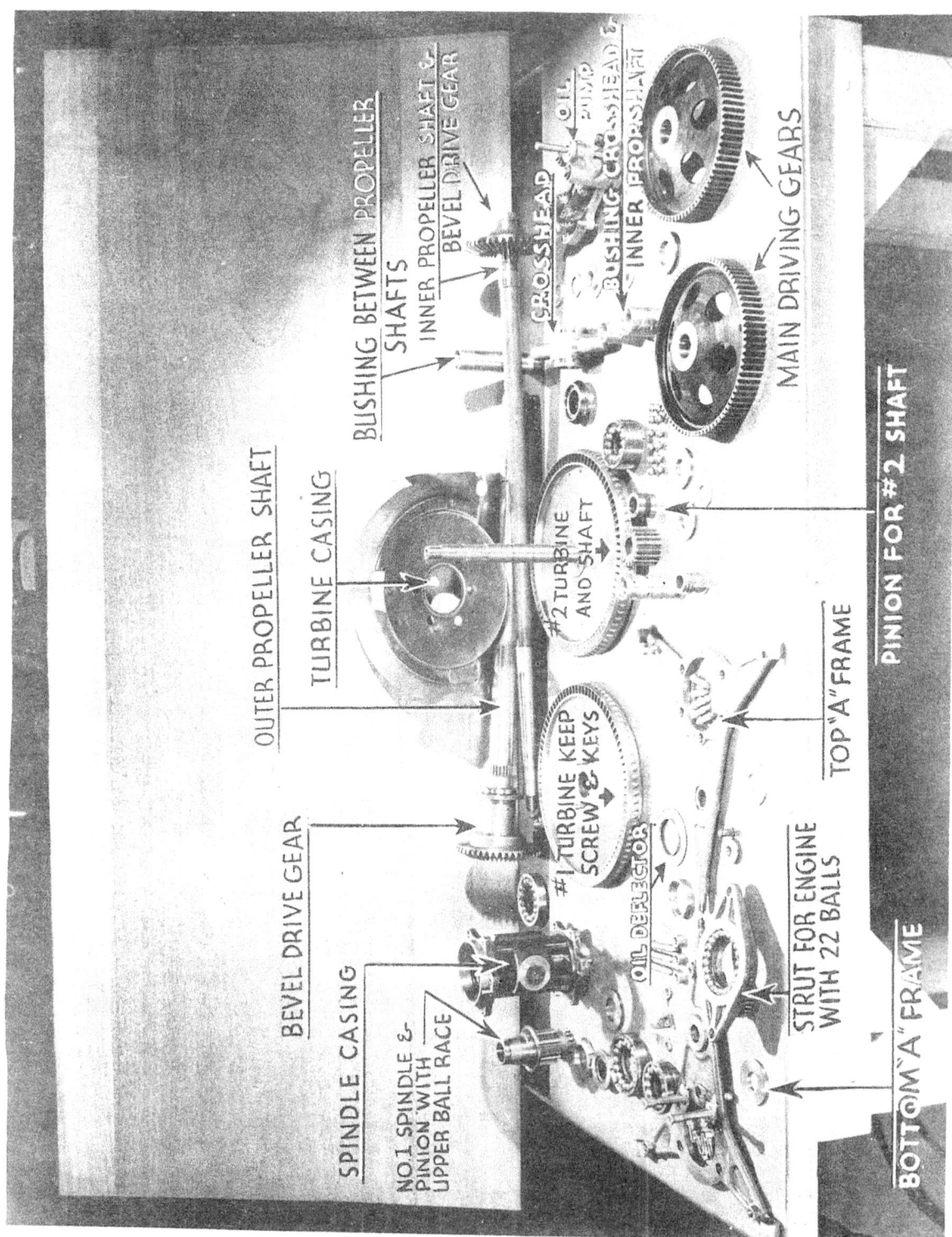

TS-5 VII-16 A

ALIGNING PROPELLER SHAFTS WITH MAIN DRIVING GEARS

- LEVEL ON FLAT OF W.E.T. #35A
- ALIGNING BUSHING W.E.T. #35A ON PROPELLER SHAFTS, SECURED WITH NUT FOR AFTER SHAFT
- DRIVING GEARS
- LEVEL ON MAIN
- BALL RETAINER & BALLS FOR FORWARD BEARING
- STRUT & 22 BALLS FOR ENGINE FRAME

TS-5

                                                                TOOL NO.
   (11) Level shaft in a transverse direction
        (placing level on flat of tool WE35A)........ WE3
   (12) Clean, inspect, remove burrs, oil (B) and
        replace washers and bushings for crosshead
        and driving pinions.

7050. Main driving gears.
   (1) Clean, inspect with particular attention to
       loose rivets, remove burrs where necessary,
       oil (B) and install main driving gears. Place
       a level across teeth of gears and bring to
       mesh with bevel gears when main driving gears
       are level across the teeth. Put clip tool
       across main driving gears to keep gears to-
       gether. To level, one bevel gear may have to
       be shifted on shaft 90°..................... WE3
   (2) Oil (B) and replace crosshead outer washers..

7051. Forward bearing.
   (1) Clean, inspect, oil (B) and replace ring
       (bevel aft) and retainer with balls for for-
       ward bearing.

7052. Turbine frames and spindle casing.
   (1) Clean, inspect and oil (B) spindle casing.
   (2) Clean, inspect (if necessary insert in special
       chuck furnished with W.E. bench lathe and
       polish out rough spots or rust) oil (B) and
       insert middle bearing race, obtaining approx-
       imate position by the use of depth gauge..... 424A
   (3) Clean, inspect, oil (B) and replace race for
       forward bearing on spindle casing.
   (4) Clean with particular attention to oil holes,
       inspect, remove burrs from teeth of shaft.
   (5) Clean, inspect for roughness and size, oil
       (B), and install in first turbine shaft, the
       4 turbine bushings........................... WE6
   (6) Clean, inspect and if necessary (hold in
       special chuck furnished with Workship Equip-
       ment bench lathe) polish out rough spots and
       rust, oil (B) and replace ring for upper
       bearing and ring for middle bearing.
   (7) Clean, inspect, oil (B) and replace 13 balls
       in middle bearing (insert handle end of
       large screw driver through bottom of spindle
       to keep balls from falling out).
   (8) Insert first turbine shaft with middle and
       upper bearing rings in turbine spindle cas-
       ing, removing screw driver.
   (9) Clean, inspect and replace 16 balls in upper
       bearing.
  (10) Replace top bearing outer race:
       (a) Clean, inspect and if necessary polish out
           rough spots and rust (hold in special chuck
           furnished with Workshop Equipment bench lathe).

DIAL INDICATOR SHOWING ".018 END PLAY FOR #1 SPINDLE BETWEEN UPPER AND MIDDLE BEARINGS, IN SPINDLE CASING

TOOL NO.

    (b) Oil (B) and replace upper bearing race in spindle casing; spread spindle casing with wedge, if necessary, to prevent binding of threads..............................231,228,54
    (c) Screw race until all the way down. Note if notch on race lines up with split in spindle casing for upper ball race lock. If not, back off bearing to nearest notch.
    (d) Back bearing out three (3) notches.
        NOTE: This will give the required end play for first turbine spindle in upper and middle bearing, and must be maintained during adjustment of these bearings.
(11) Clean, inspect and replace oil deflector ring and hub.
(12) Replace turbine keys (having been previously fitted with bench marks corresponding).
(13) Place spindle casing on surface plate and with dial indicator measure end play and trueness of #1 turbine spindle and middle bearing.

7053.
(1) Clean, inspect for true, note if oil pump studs are loose, oil (B) and replace thrust washer in lower turbine A frame.
(2) Clean, inspect for true, oil (B) and replace thrust washer in upper turbine A frame.
(3) Assemble A frames, spindle casing and engine strut:
    (a) Enter lower end of spindle casing in lower turbine A frame about half way.
    (b) Line up hole in lower A frame with lower crosshead spindle, slide in place and hold.
    (c) Slip upper turbine A frame over upper crosshead spindle and spindle casing.
    (d) Insert bolts for A frames and turbine spindle casing; leave bolts loose........ 227A
    (e) Insert screw and washer for upper turbine A frame and crosshead; leave loose.. 39
    (f) Replace oil connection bolt with a spacer for lower turbine A frame and crosshead; leave loose........................ 12
    (g) Replace 22 inspected bearing balls in engine strut ball race and slide strut over bolts in ends of A frames.
    (h) Replace nuts for engine strut; leave loose................................ 141A
    (i) Remove clip holding main driving gears together.
(4) Place assembly on a surface plate and shake down until the four turbine frame bearings to

	TOOL NO.
bulkhead are in contact with surface plate.	141A
Set up on bolts for A frame and spindle	227A
casing, screw for washer and crosshead and	39
nuts for turbine frame studs to strut.	

NOTE: If a bearing does not contact surface plate by this method, it is bent out of true and should be tapped in line with a lead hammer.

(5) Place dial indicator on surface gauge base (both furnished in Workshop Equipment) and with base for indicator on surface plate, regulate height to contact inner and outer shafts; lift each shaft and obtain reading of end plays, which should be 0."025 to 0."030 for forward propeller shaft and 0."025 to 0."030 for after propeller shaft.......... WE10

NOTE: To obtain this clearance in after propeller shaft thrust bearing, it may be necessary to place a shim between spindle casing and forward race for bearing.

7054. First turbine wheel.

(1) Clean, inspect, oil (B) and install turbine on spindle with numbers or bench marks on key ways corresponding with keys.

NOTE: If band has parted so as to show a clearance of more than ."004 (state clearance) with bucket, turbine should be discarded and a new one installed.

(2) Clean, inspect and replace first turbine wheel nut, setting up for full due, but do not replace lock screw............................ 179

(3) Install dial indicator on one of the upper turbine frame bearings to bulkhead, using "C" clamp upright furnished with dial indicator. Place dial indicator button on top outer edge of wheel and, turning wheel, obtain reading of amount out of true. If more than ."006, it will be necessary to find cause and remedy........................... WE10

7055. Adjust clearance of first turbine and nozzle.

(1) Replace turbine bulkhead temporarily........ 141A
(2) With depth micrometers measure height from nozzle flange on bulkhead to first turbine wheel.. WE9
(3) Measure height of nozzle lip from flange to opening end.
(4) With micrometer caliper measure thickness of copper gasket................................ WE5

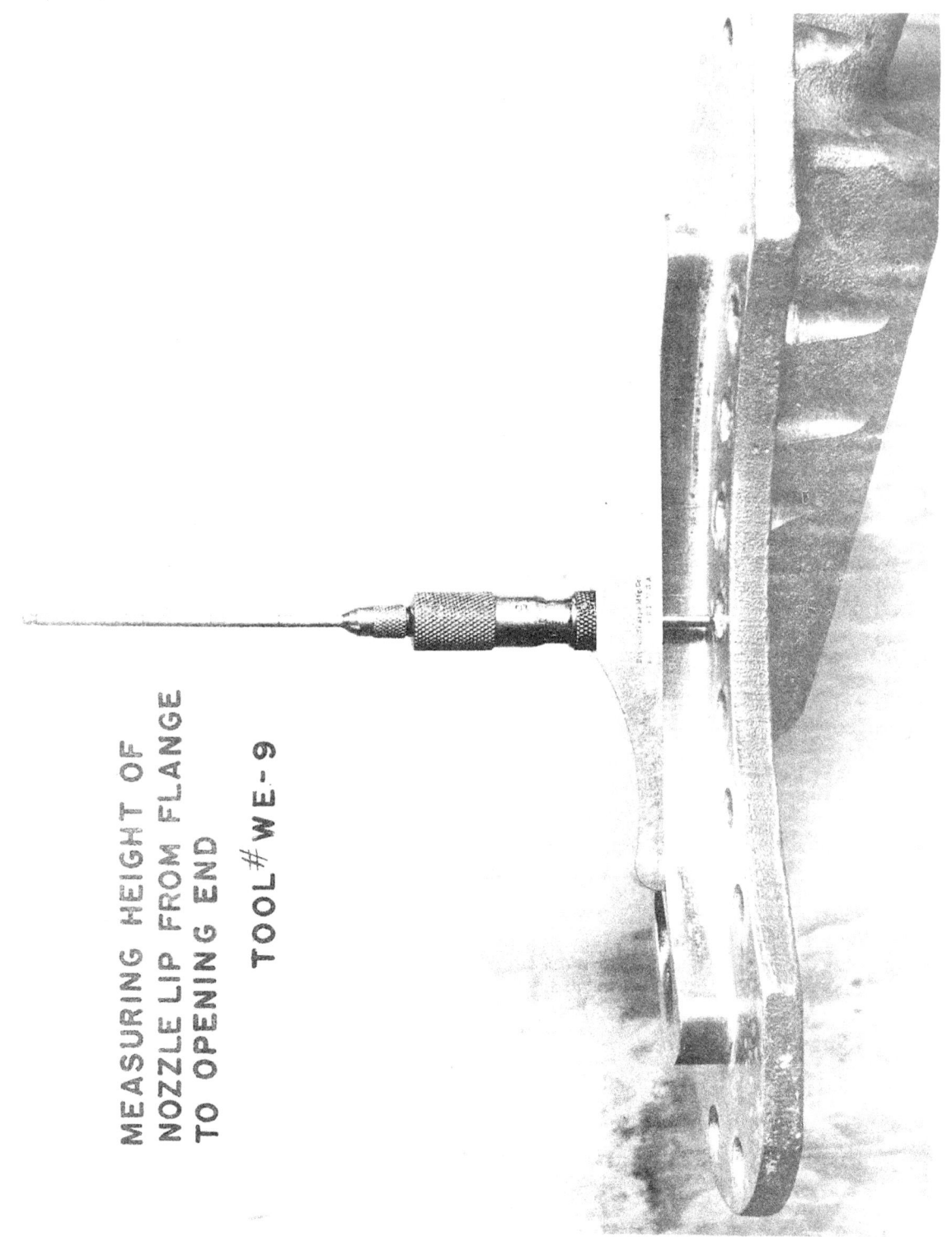

MEASURING HEIGHT OF NOZZLE LIP FROM FLANGE TO OPENING END
TOOL #WE-9

DIAL INDICATOR SHOWING ".025 TO ".030 END PLAY OF OUTER PROPELLER SHAFT

TOOL NO.

   (5) Assuming that measurement taken
        in step (3) is ".400
        in step (4) is ".025
        in step (2) is ".435
   subtract step (4) from step (3) and subtract
   sum obtained from step (2). This will give
   sum of clearance between nozzle and first
   turbine.
   Thus:
      ".400 - ".025 = ".375 and
      ".435 - ".375 = ".060, which is the pro
   per clearance of first turbine with nozzle.

   (6) **With thickness gauge measure clearance** be-   WE2
   tween **turbine and turbine A frame, this**
   should be not less than 0".015.

   (7) For alignment of turbine buckets with nozzle,
   see O.D. #750, Page 91-92........W.E.T.26, 27, 28

   (8) Remove turbine bulkhead, first turbine nut
   and turbine............................. 179,141

7056. Secure adjustments of first turbine.
   (1) Clean, inspect, oil (B) and replace upper
   and middle ball race locks, clamp screws
   and nuts for casing and cotter pins in clamp 72
   screws. After engaging locks in slots in
   upper and middle ball races, secure ends
   together with cotter pins. Set up on upper
   and middle clamp screws.................... 227A
   (2) Set up on 5 bolts for upper turbine frame   227A
   and casing, and replace cotter pins through
   holes in ends.............................. 72

7057. Turbine casing.
   (1) Clean, inspect, straighten if necessary and
   replace turbine casing. Secure with 4
   screws, and wire screw heads............... 72,41

7058. Replace first turbine wheel.
   (1) Replace first turbine wheel (note that bench
   marks correspond on keys and keyways) and
   turbine wheel nuts, setting up for full due
   (see that bench marks on nut line up with
   mark on end of spindle).................... 179
   (2) Inspect lock screws for turbine wheel nut
   for burrs on head and replace; insert cotter pin through head of screw.............. 41,72

7059. Second turbine wheel and shaft.
   (1) Clean, inspect, if necessary stone smooth
   bearing surface on shaft, note that oil
   holes are clean, that turbine keys are properly fitted to hub, and that driving pinion has a sliding fit without binding on
   keys in shaft.

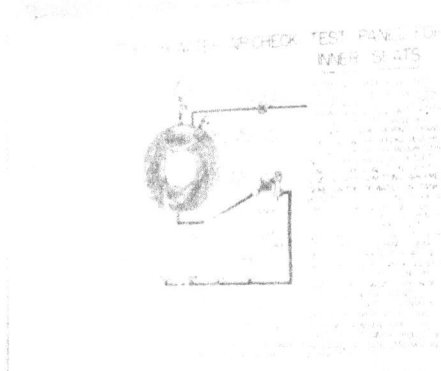

TEST PANEL FOR CHECK VALVES

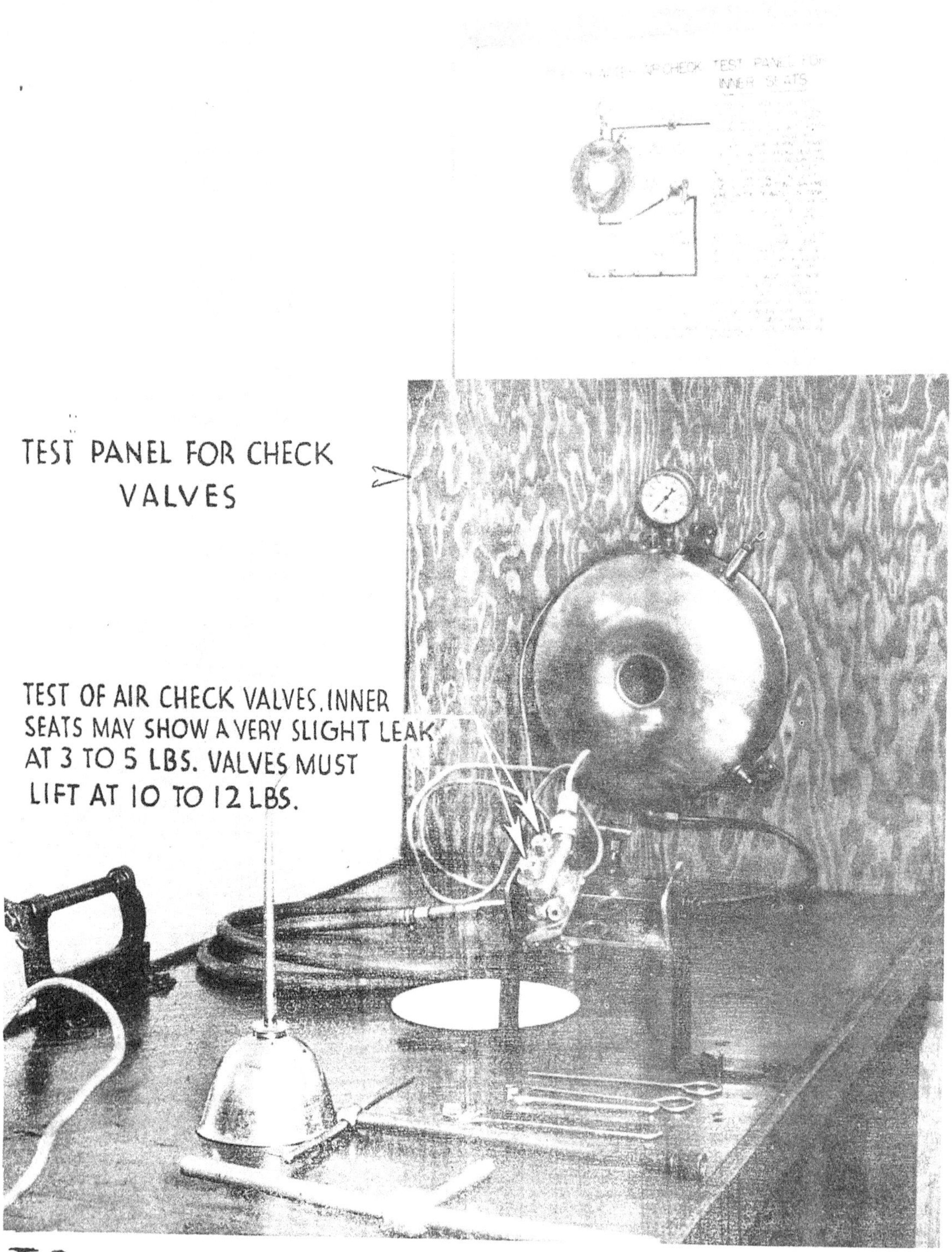

TEST OF AIR CHECK VALVES. INNER SEATS MAY SHOW A VERY SLIGHT LEAK AT 3 TO 5 LBS. VALVES MUST LIFT AT 10 TO 12 LBS.

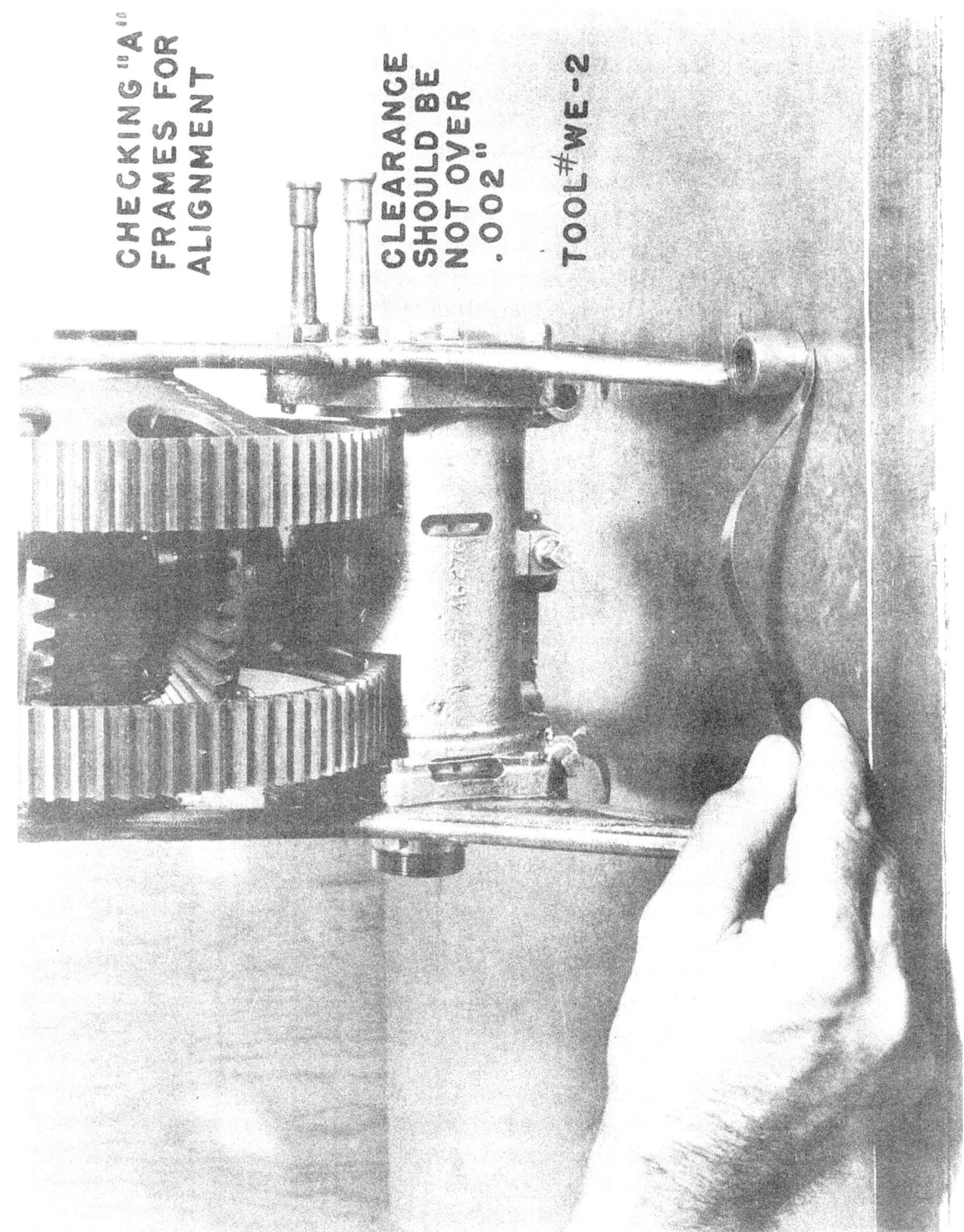

CHECKING "A" FRAMES FOR ALIGNMENT

CLEARANCE SHOULD BE NOT OVER .002"

TOOL #WE-2

CHECKING CLEARANCE BETWEEN "A" FRAME AND #1 TURBINE
CLEARANCE SHOULD BE .018" TOOL # WE-2

TOOL NO.

(2) Clean, inspect with particular attention to lifting of turbine band and smoothness of keyways.
NOTE: Turbine wheels found to have bands lifted must be discarded.
(3) Install turbine wheel on second turbine wheel shaft, with marks on keys and keyways matching.
(4) Secure with nut and lock screws, place cotter pins in lock screws, install in aligning fixture............................. 176,41,72
Check #2 turbine for warp using spindle aligning fixture, plate 56, O.D. #750 ".006 maximum shop allowance.
(5) Insert second turbine shaft through first turbine shaft and bushings, and install clip to hold turbines together................... 454
(6) Clean, inspect, oil (B) and replace driving pinion, lining up with punch mark on end of shaft.
(7) Clean, inspect, stone out rough spots where necessary, oil (B) and replace upper washer for lower bearing race.
(8) Clean, inspect, stone out rough spots, oil (B) and replace inner race for lower bearing.
(9) Oil (B) and replace 12 inspected balls...... 231
(10) Clean, inspect, stone out rough spots where necessary, oil (B) and replace outer race for lower bearing....................... 54,228
(11) Oil (B) and replace 12 inspected balls.
(12) Clean, inspect, stone out rough spots as necessary, oil (B) and replace bottom washer for lower bearing.
(13) Clean, inspect, oil (B) and replace worms for driving oil pump with washer. Remove   229
clip holding turbines together.............. 454

7060. Adjust clearance between the turbines.
(1) Regulate clearance between the turbines to 0".060, using two thickness gauges provided in tool box. If, when proper clearance is obtained, notch on bearing race does not line up properly with slot in spindle casing, 228
move nearest notch to line up............... 424A

7061. Secure lower bearing adjustment.
(1) Clean, inspect, oil (B) and replace lower ball race lock, clamp screw for casing and nut for clamp screw.
(2) See that end of lock engages in a notch on lower bearing race, and secure clamp screw for casing, placing cotter pin on end of   72
screw. Secure end of lower bearing lock to middle bearing lock with a cotter pin....... 227

TOOL NO.

(3) Set up on 5 bolts for A frame and spindle
casing and thread a copper wire through
ends of bolts to secure...................... 72
227

7062. Oil pump.
(1) Insert oil connection sleeve in turbine shaft.
(2) Replace oil pump, guiding oil connection to
turbine shaft into sleeve and joint on pump
for crosshead into recessed seat in lower A
frame.
(3) Clean, inspect, oil (B) and replace oil connection for crosshead...................... 12
(4) Replace holding screws for oil pump and
thread wire through holes in heads of screws
to secure.................................. 49,72

7063. Turbine bulkhead.
(1) Clean, inspect, chase burrs off threads and
reseat seats on nipples where necessary,
check up alignment of bulkhead.
(2) Install bulkhead and secure with 4 nuts for
turbine frame studs and cotter pins.......... 141A,72
NOTE: Do not re-use cotter pins as erratic
runs may result.

Test of assembled gear train.
7064.
(a) This test is made for the purpose of testing assembled
gear train for smoothness of running and for testing
the oil pump assembled with the engine.

(b) Test procedure:
(1) Insert special "0625 restriction in the air inlet
connection to the combustion flask.
(2) Attach special test nozzle to the turbine bulkhead
using gasket as necessary to maintain approximately
"060 nozzle clearance.
(3) Place bearing sleeve on after end of forward propeller
shaft.
(4) Place engine assembly in the test stand and secure.
(5) Connect oil pump to oil reservoir, fill oil reservoir,
open oil valve.
(6) Make connections to air gauges and air connections
to test nozzle.
(7) Heat water around air coil to 205° F.
(8) Crack air valve slowly and run engine until oil
shows on gear train assembly.
(9) Shut off air and install gear train guard.
(10) Open valve and build nozzle pressure to 50 lbs.
(11) With tachometer on end of inner shaft the shaft
R.P.M. should be 700 in not more than two minutes.
(12) After satisfactory run has been completed, and
immediately upon shutting off the air, snap a stop

MAIN ENGINE IN TEST STAND
LEAD TO OIL RESERVOIR
OIL PUMP

TS-5 VII-22A

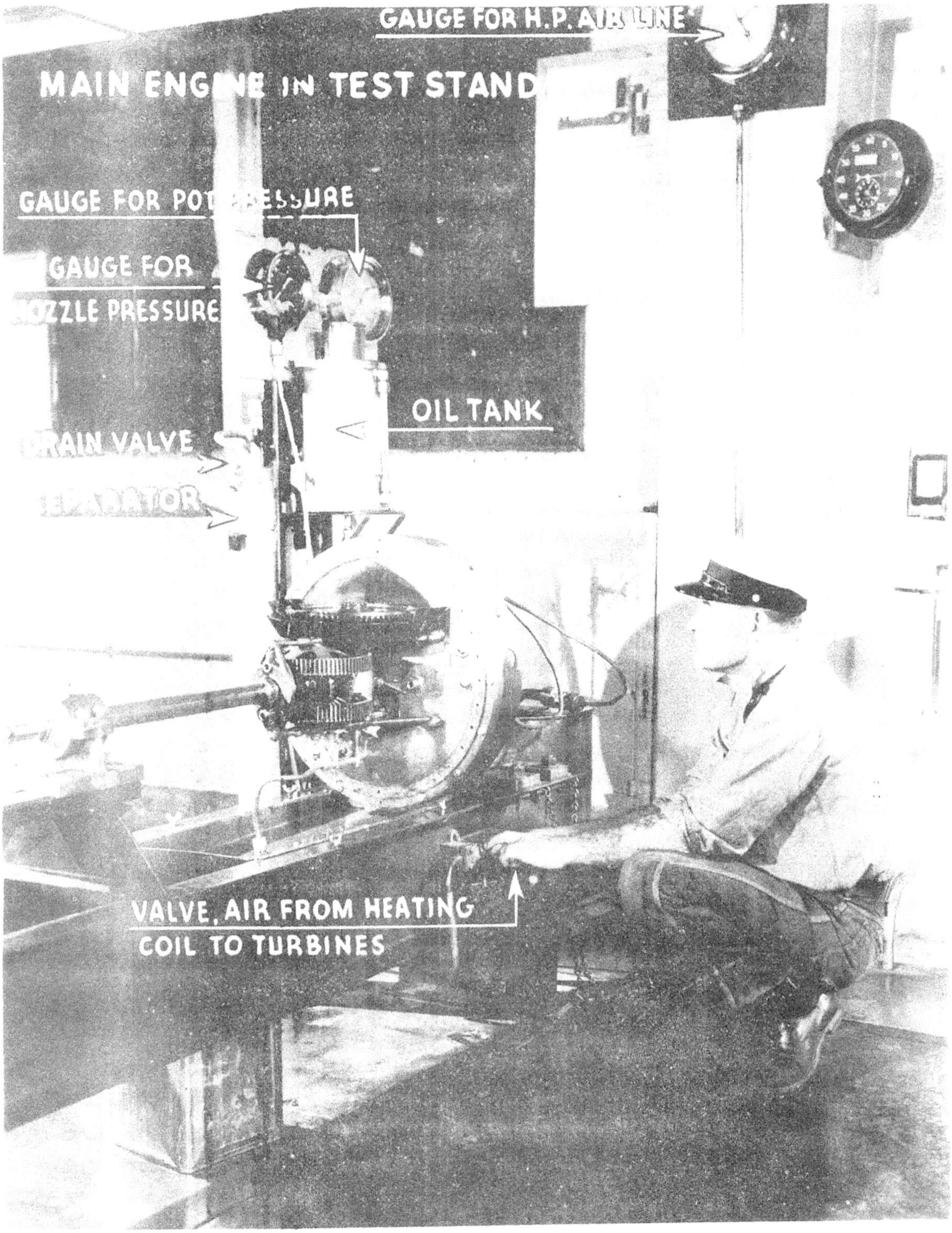

MAIN ENGINE IN TEST STAND

GAUGE FOR H.P. AIR LINE

GAUGE FOR POT PRESSURE

GAUGE FOR NOZZLE PRESSURE

OIL TANK

DRAIN VALVE

SEPARATOR

VALVE, AIR FROM HEATING COIL TO TURBINES

VII-22B

watch and note time required for shafts to come to a complete stop. With a properly running engine, the time should be less than 75 seconds.

(13) Should the engine fail to make the desired number of revolutions in two minutes after the admittance of air pressure the test should be stopped. A second failure should be cause for checking the assembly for excessive friction, faulty assembly or improper adjustments.

(14) Whether or not the oil pump is functioning properly can also be determined during this test by noting drop in oil level in oil reservoir

(15) The above test is not intended for use in determining delivery of oil through the oil pump of engine being tested but intended to test functioning of oil pump in the engine to which assembled.

L. **THE AFTERBODY.**

7065. The afterbody shell is made of ".072 thick sheet steel in the form of a truncated cone, the sides of which are slightly paraboloidal. The interior of the shell is reinforced by six strengthening rings, three strengthening angles and three strengthening tees which are riveted and soldered at points along its interior.

7066. To its forward end is riveted and soldered the engine cage, a ring of forged steel, which, in addition to reinforcing the end, forms a means for centering and attaching to the air flask, and a seat for the engine bulkhead. The seat for the joint between air flask and afterbody is slightly taper in form to facilitate assembly, insure a tight fit and proper alignment. The joint is made by 36 - 3/8" steel joint screws.

7067. The engine bulkhead is secured to the engine cage by studs and nuts, a gasket being interposed to obtain a watertight joint.

7068. The after end of the afterbody shell is closed by the after bulkhead, a bronze casting, which is held to the shell by screws and by soldering, making the after end watertight when all parts are assembled. This bulkhead provides an after bearing for the propeller shafts and seats for the exhaust valves. Both rudder connections pass through it, stuffing boxes being provided to insure watertightness. The propeller shaft bearing is water cooled during a run, an annular cavity being provided connected by pipes to small scoops on the exterior of the shell, the upper scoop pointing forward and the lower scoop pointing aft. Thus a forced circulation of water results from the motion of the torpedo (see note). Guides are sweated to the after end of the shell on the vertical and horizontal centerline into which the horizontal and vertical tail vanes are fitted when tail is assembled on the afterbody.

VIEW FROM THE REAR, SHOWING AFTERBODY AND TAIL ASSEMBLY

TS 5   VII 23A

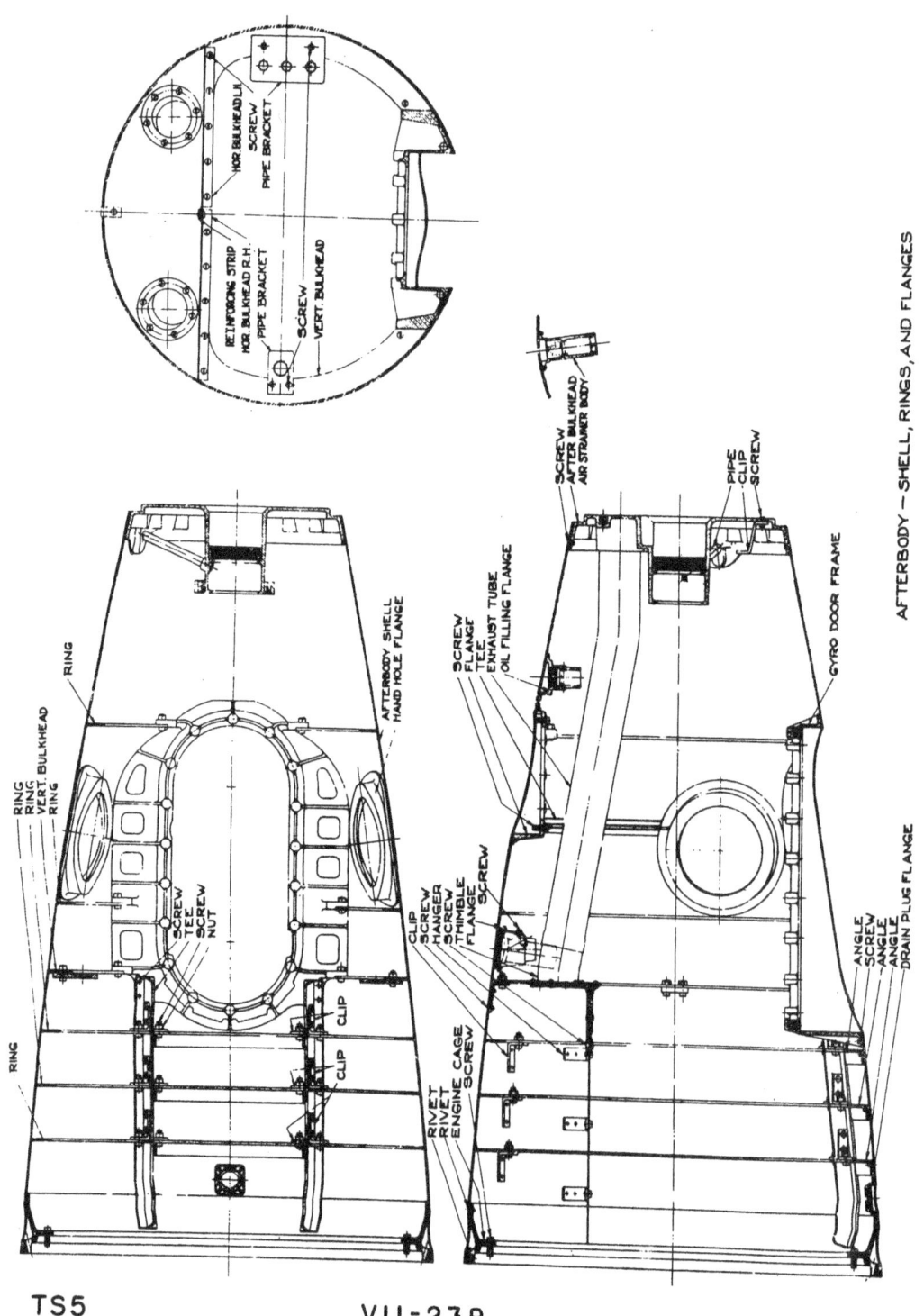

VIEW OF INTERIOR OF AFTERBODY FROM FORWARD END

AFTERBODY AND AFTER BULKHEAD

TS-5 VII-23D

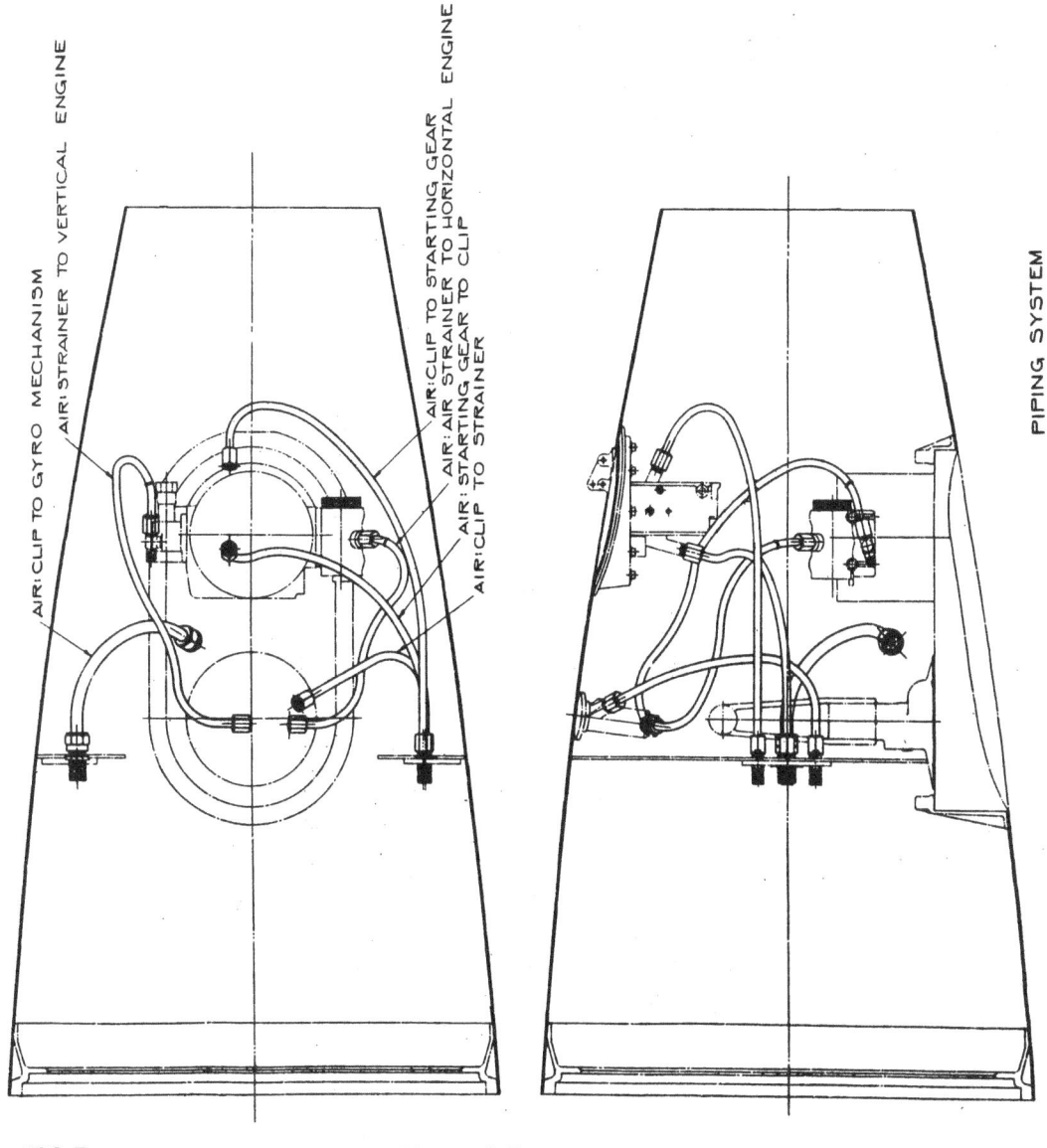

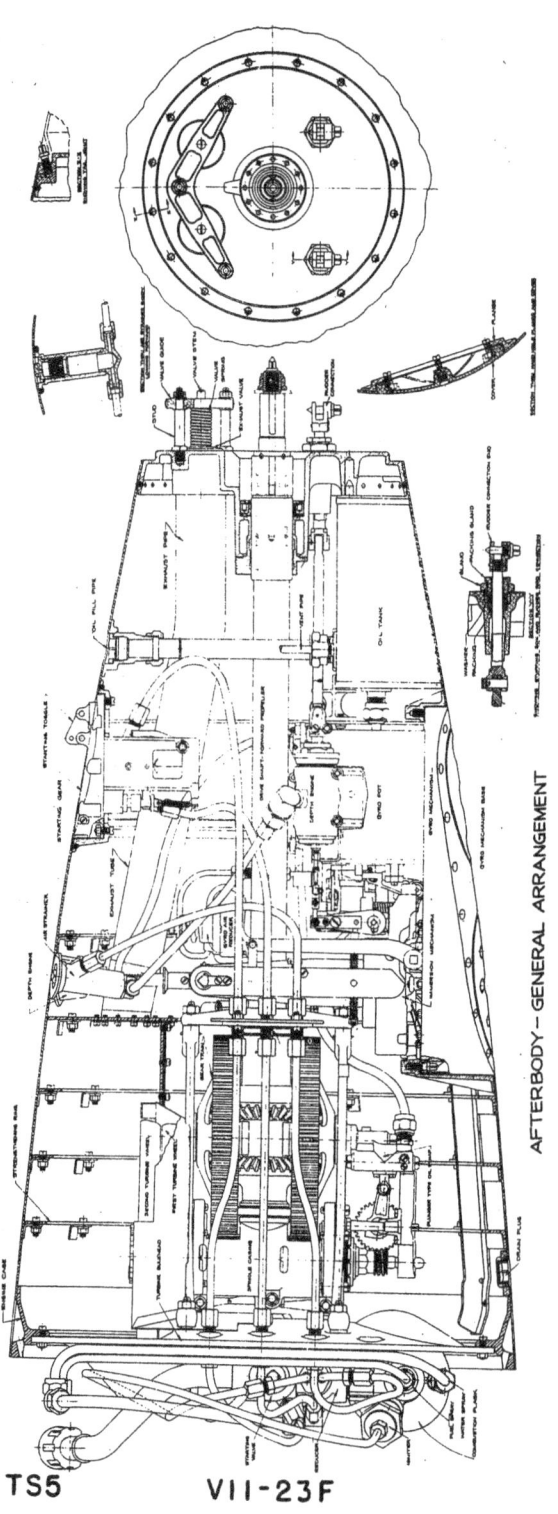

AFTERBODY – GENERAL ARRANGEMENT

TS5 VII-23F

NOTE: Water cooling of propeller bearing will be discontinued after Mark 13 torpedo No. 13482. Guides will be discontinued after torpedo No. 13482 and a spider substituted to brace the forward ends of the tail vanes. No water cooling or guides are used in the Mark 13-1 torpedo.

7069. The exhaust valves, two in number, are of monel metal with beveled seats, lapped to corresponding seats at the exhaust openings in the bulkhead. They are described in detail in connection with the exhaust system.

7070. Permanently riveted and sweated to the interior of the afterbody shell are receptacle frames for the various mechanisms and access opening covers; fittings and attachments for air strainer, setting devices and drain plug. The engine oil tank is located in the bottom of the afterbody directly in back of the gyro mechanism. This tank is connected to a filling flange located on the top in rear of starting gear. In the forward upper part is the turbine chamber, consisting of sheet steel horizontal and vertical bulkheads. Running from the vertical bulkhead to the after bulkhead are two exhaust tubes, the forward ends of which are held in place by exhaust tube thimbles, the after ends being rolled into expansion grooves in their seats in the after bulkhead. The entire afterbody and its closures are made as nearly watertight as possible and then fully assembled, must stand an external hydrostatic pressure of 50 lbs. per sq. in. for 30 minutes with no leakage in excess of 4 oz. of water, and must be capable of withstanding a 200 ft. drop with a plane speed of approximately 150 knots without showing any undue strain or sustaining any damage.

7071. The afterbody is a container for the propelling and controlling mechanisms of the torpedo, each unit of which is described under its own heading.

THE TAIL.

7072. In the Mark 13-1 torpedo, the depth and steering rudders are mounted forward of the propellers instead of on the afterside as in the Mark 13 tail. This arrangement eliminates interference caused by the tail rails, rudder linkage, rudders and their supports in the propeller stream, thereby increasing the speed and range of the torpedo.

7073. The tail cone is a hollow drop forging of alloy steel of the general form of a frustum of a cone. It has four slotted projections, two vertical and two horizontal at right angles to each other on its surface for securing the after portions of the vertical and horizontal tail blades. The forward end of the cone is flanged internally to permit angular holes for the joint screws by which the tail is held to the afterbody. A flange in the after end contains tapped holes for the holding screws with which a bronze bearing for the forward propeller sleeve is held in place. This flange is also machined with an annular grease cavity for lubricating the tail bearing. The cone is further reinforced by a two step band, the after portion of which is machined parallel to the axis on the inner surface of the tail cone. This band serves as a stiffening ring and provides support for the depth rudder bearings.

7074. In addition to the holes drilled in the after end of the cone for the inner bearings of the rudders, a grease plug hole is drilled and tapped in the after port side connecting with the grease cavity for the tail bearing and four access holes are cut on the under side of the cone to facilitate connecting and adjusting the rudder rods. Flanges are secured and soldered to the interior of the cone at these openings and provide, by being threaded, a means of closing the same with plugs and copper washers, when not in use as access openings. At the center of the port side of the tail cone are graduations by which the depth rudder readings are made. A zero graduation is stamped on the starboard side for the purpose of aligning the top surfaces of the horizontal rudders.

7075. The vertical and horizontal tail blades are secured in the slotted projections on the tail cone by screw rivets. The blades extend forward over the end of the afterbody 6"903 inches, the extended portions being interconnected by spider shaped straps, thus forming a truncated support for the forward end of the tail blades around the afterbody. These straps with their reinforcing pieces are secured to each tail blade by five holding screws. The after outer ends of the blades support the outer bearings for the depth and steering rudders.

7076. Enclosed in the tail cone are the depth and steering rudder yokes, the rear terminals of the rudder rod connections and fittings and the forward and after propeller sleeves. To the latter are attached the propeller hubs which in turn support the propellers. The propellers are secured to their hubs with nuts, thus completing the general outline of the torpedo in the rear of the tail cone.

7077. The depth and steering rudders are made of alloy steel forged in one piece with their spindles. The steering rudders are located on their spindles so that when assembled on the tail the effective rudder area will be outside of the diameter of the propellers, thus minimizing rudder interference with the water ahead of the propellers, with consequent rolling of the torpedo. The area of the lower rudder is smaller than the area of the upper rudder. The rudder spindles are machined square on their inner ends for engagement in the square holes in the rudder yokes. The portion of the spindles next to the square ends is sized to fit snugly with freedom to rotate in the inner bearings, the outer ends of the rudder spindles being similarly fitted to rotate in the bushed outer bearing.

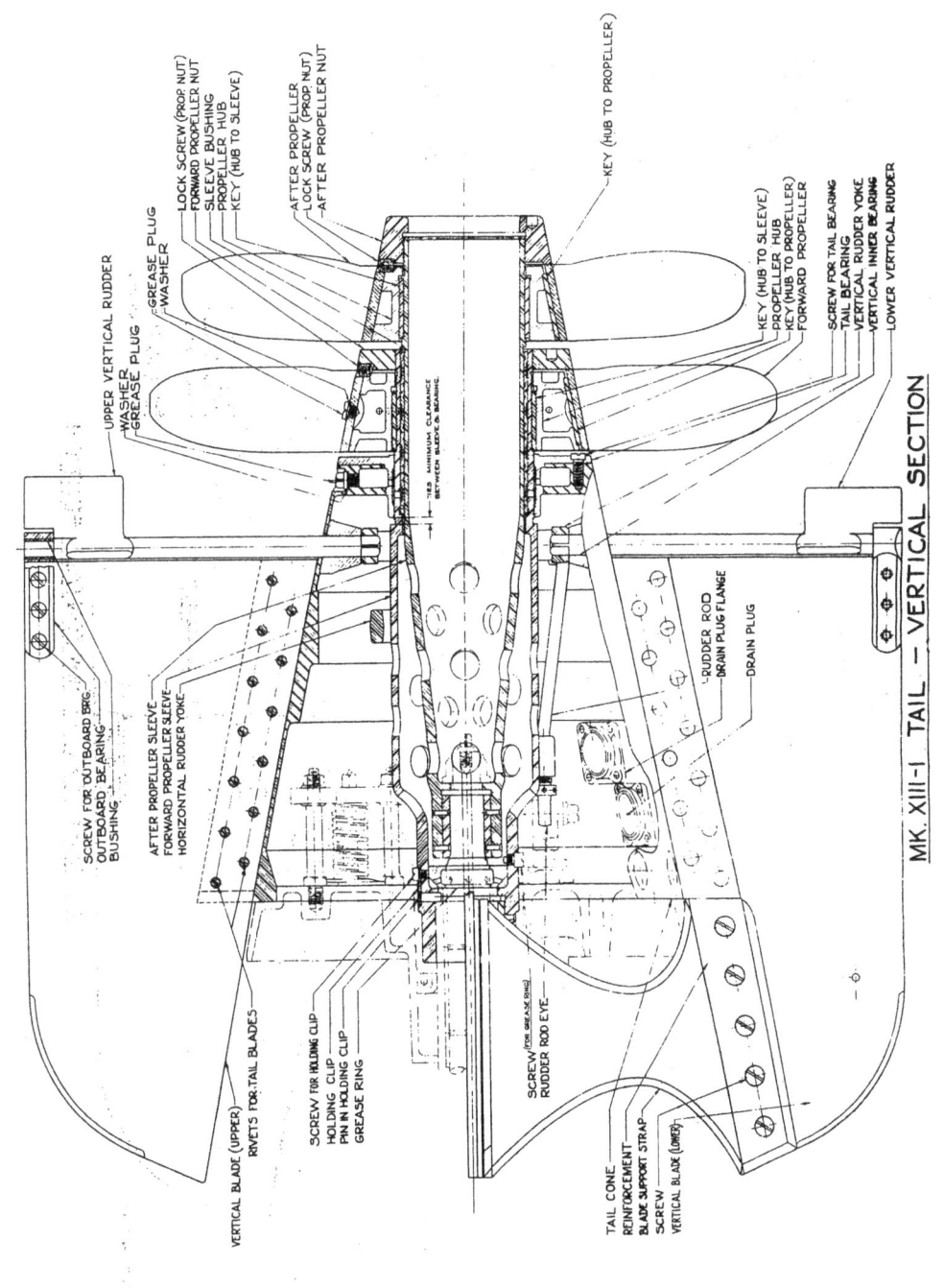

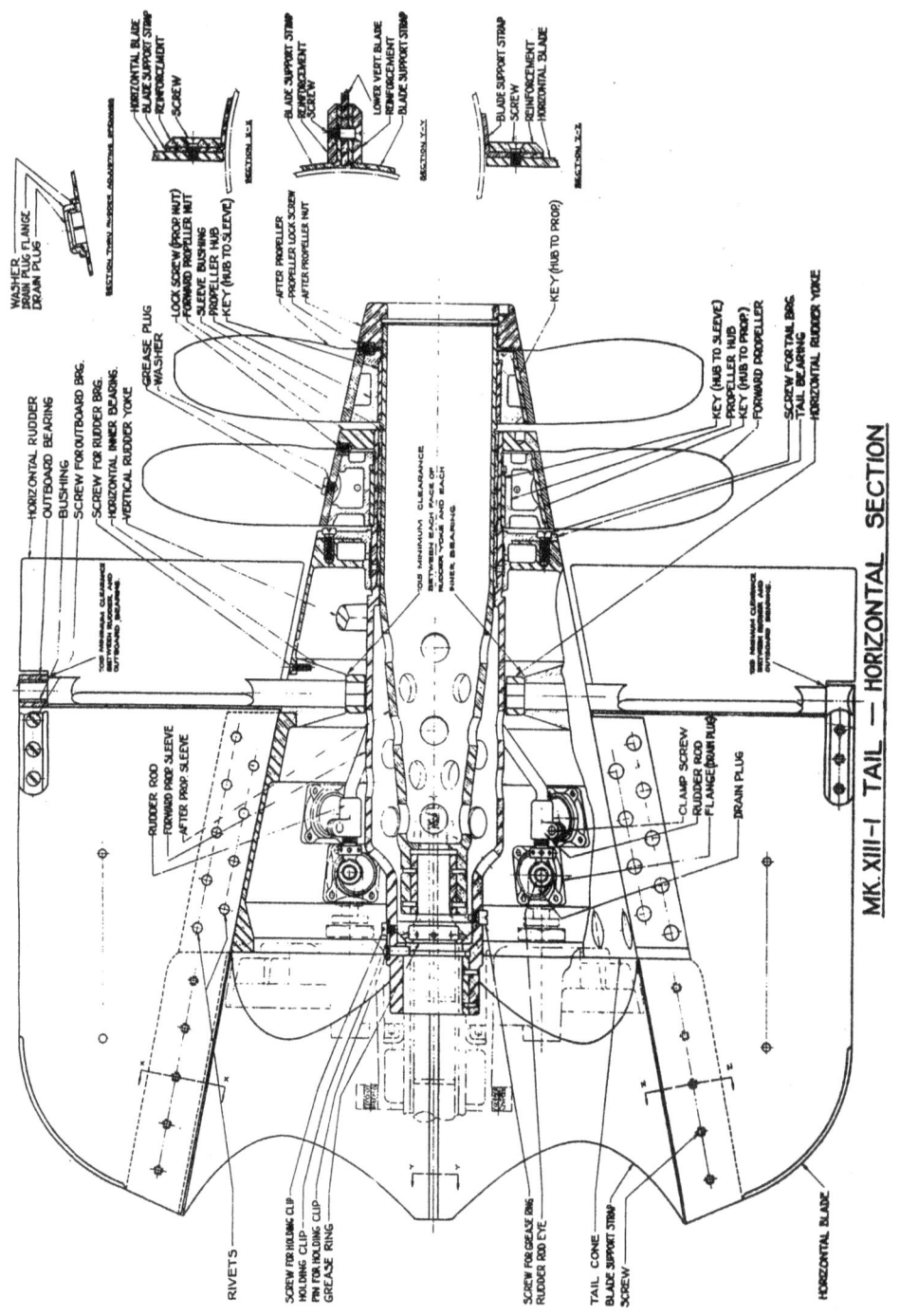

MK. XIII-I TAIL — HORIZONTAL SECTION

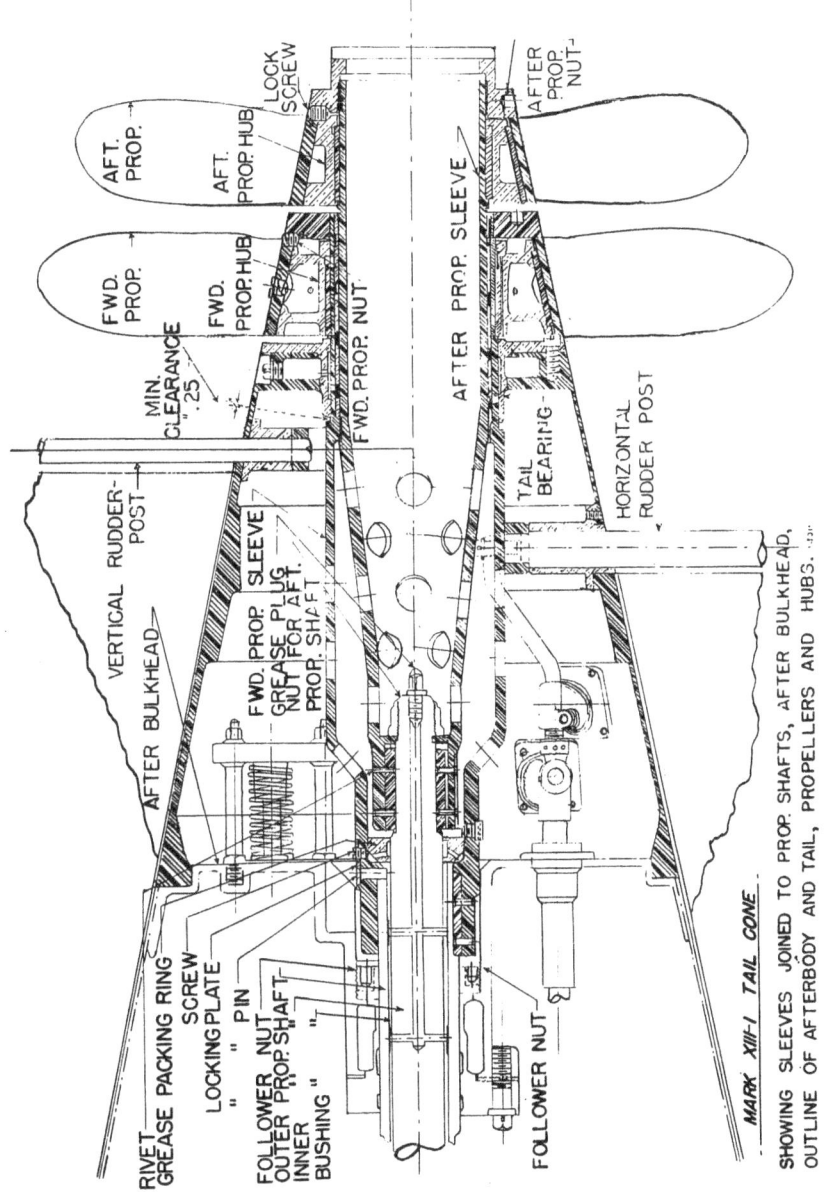

MARK XIII-I TAIL CONE

SHOWING SLEEVES JOINED TO PROP. SHAFTS, AFTER BULKHEAD, OUTLINE OF AFTERBODY AND TAIL, PROPELLERS AND HUBS.

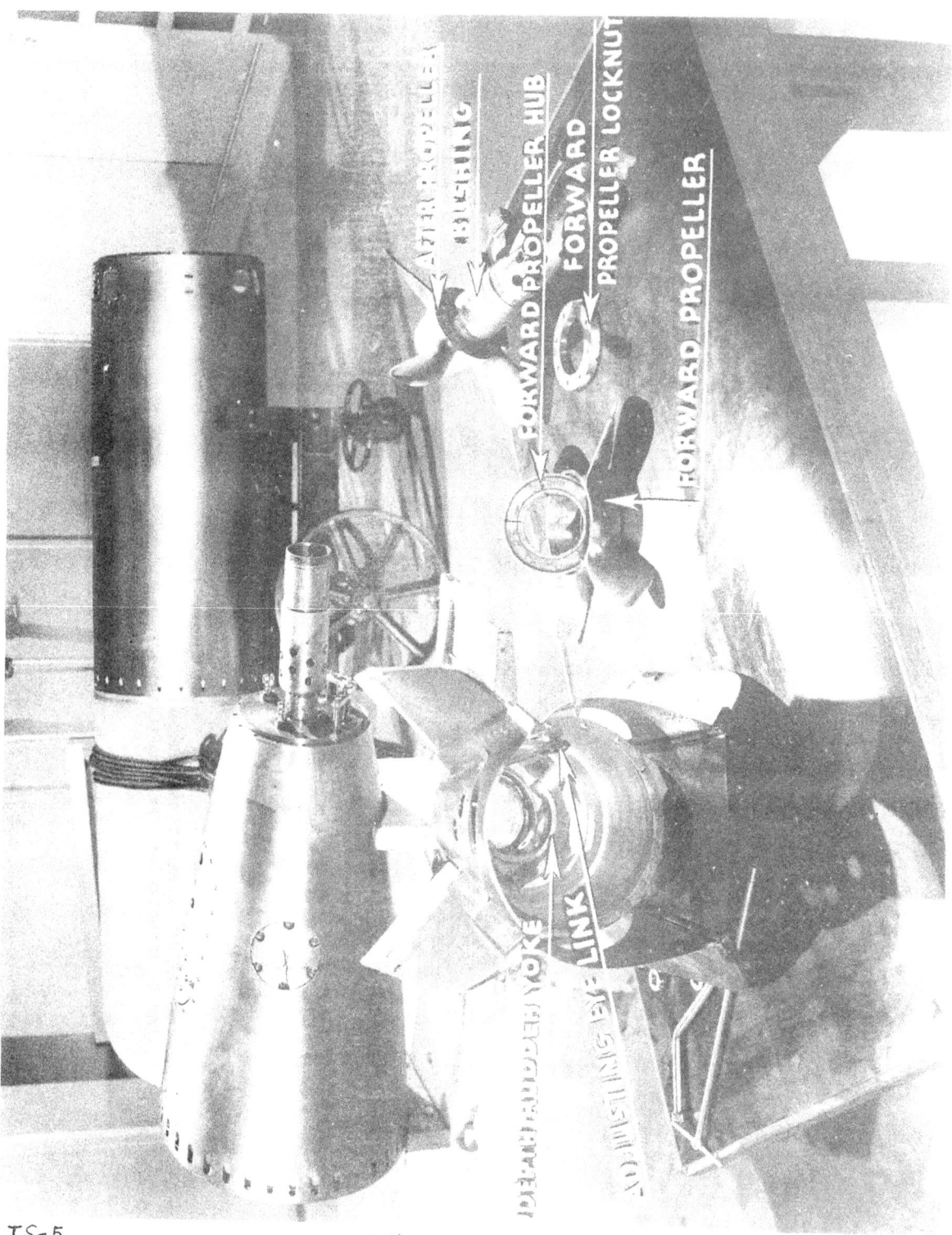

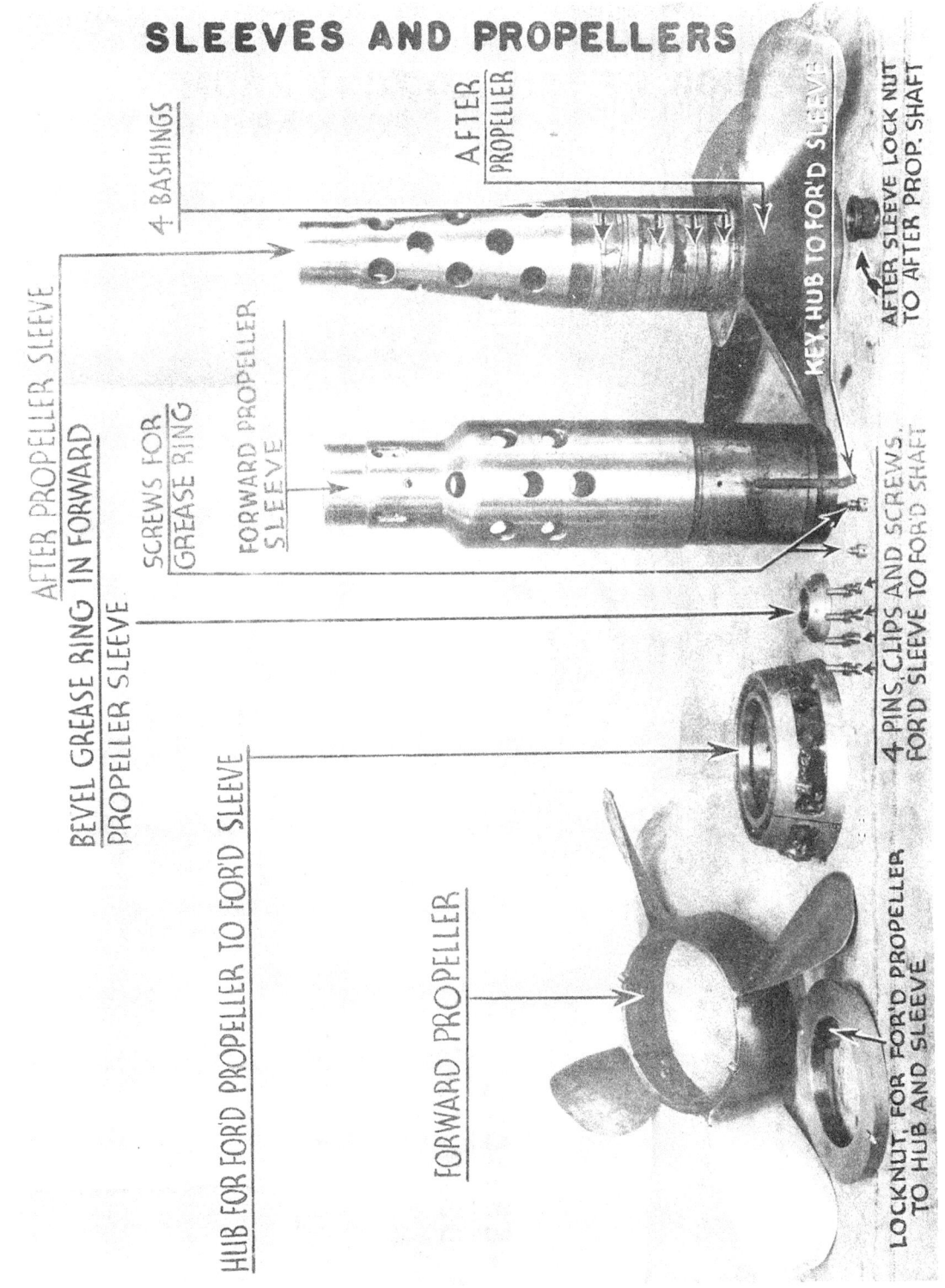

SLEEVES AND PROPELLERS

INTERIOR OF TAIL, LOOKING FROM FORWARD AFT

HORIZONTAL RUDDER YOKE

VERTICAL RUDDER YOKE

ADJUSTING EYES

TAIL, VIEW FROM BOTTOM

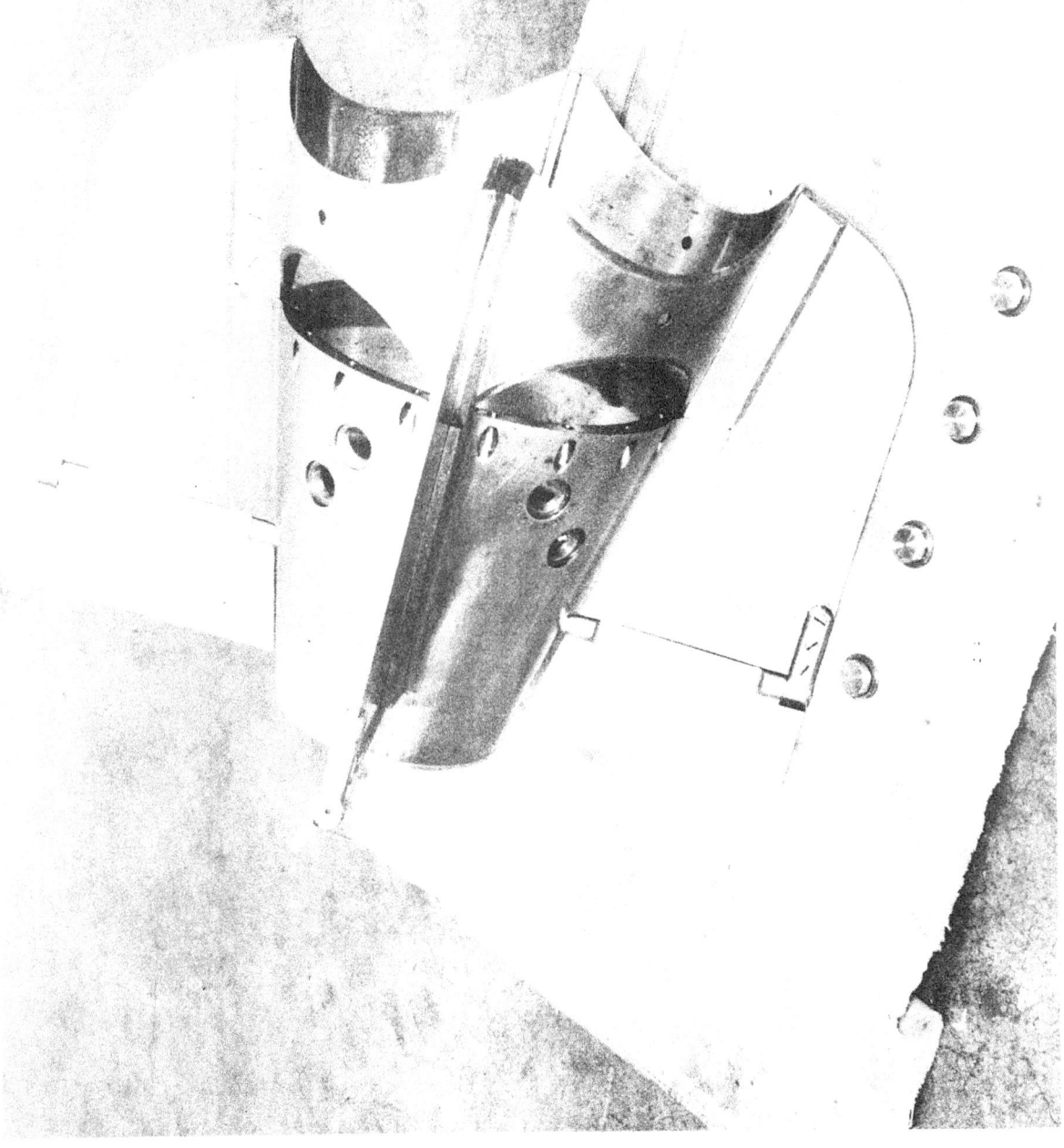

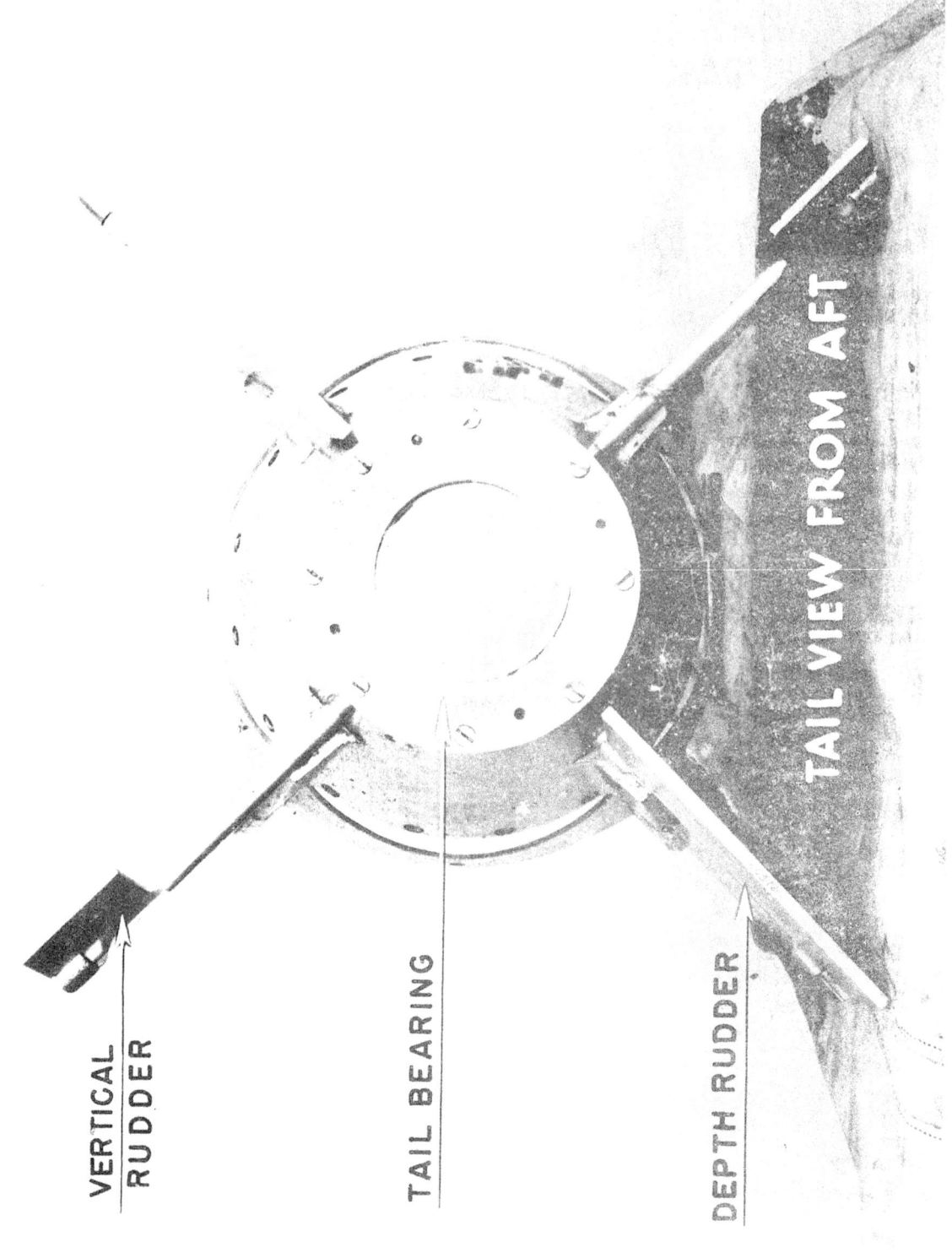

M. OVERHAUL, TEST AND ASSEMBLE PARTS ASSEMBLED WITH AFTER-BODY SHELL GROUP.

TOOL NO.

7078. Afterbody shell rings and flanges.
   (1) Clean and inspect afterbody for loose rivets, loose solder around flanges, bulkhead, engine cage, water circulating pipes, after bulkhead etc...

7079. Exhaust tubes.
   *(1) Reform after ends of exhaust tubes (if using tubes previously removed).   WE210
   *(2) Insert exhaust tubes through forward end, guide ends into holes in after bulkhead and push in place. Roll into groove with expander tool, line up forward ends with holes in strengthening ring (vertical bulkhead).   WE209
   *(3) Insert thimbles for exhaust tubes and secure with (6 each) screws, wiring screws together.   72
   41

7080. After bulkhead bearing.
   (1) Clean, inspect and dress burrs off where necessary on after bearing.
   (2) Replace a new gasket and after bearing securing with 2 holding screws................   386

7081. Oil tank.
   (1) Clean and wash interior of tank with solution of tri-sodium phosphate.
   *(2) Small dents may be removed by the application of hydraulic pressure to the interior; about 5 lbs. should be sufficient.
   *(3) Dress burrs found on seats or threads of nipples.
   (4) Replace oil tank in afterbody pointing after holding stud into slotted clip on after bulkhead.
   (5) Replace securing brackets and secure brackets to strengthening rings with screws. Wire screws...   72,184

7082. Rudder rods and connections.
   (1) Clean, inspect and remove burrs where necessary.
   (2) Remove old packing. Clean and inspect stuffing boxes for rudder connections.
   (3) Replace rudder rods and connections through stuffing boxes, with guides on connections in alignment with guide slots.
   (4) Repack rudder rod stuffing boxes using 2 pieces 7" long of 3/16" diameter asbestos

	TOOL NO.

packing. Set up on packing gland for stuffing box, gradually working rudder rods back and forth; use oil (D) on packing when working into place. It may be necessary to add more packing, in which case split 3/16" packing to the amount required. When properly packed, rods should move with a push balance test of not more than 9 lbs. with the gland nut set up flush.. 229 / 98

NOTE: It is important that gland nut be set up flush as otherwise the rudder rod movement may be restricted.

  (5) Replace rudder connection ends and pin in place with cotter pins........................ 72

7083. Exhaust valves.

  (1) Clean and inspect exhaust valves and their seats.

  (2) Assemble valves in exhaust valve bracket without springs. (Note that numbers on valves correspond with numbers on seats).

  (3) Replace bracket with exhaust valves over studs and secure with nuts................... 457

  (4) Check valve alignment to seat by inserting strips of cigarette paper dipped in oil (C) about 90° apart, across seats of valves. Hold valves against seats and try to remove paper strips. If this can be done without binding, valve is out of alignment with seat.

*(5) Insert valve in a collet on bench lathe and determine if seat on valve runs true by placing indicator in tool post and against valve. If seat runs out more than ".002, it will be necessary to turn the seat true in bench lathe.................................. WE10

*(6) If valve seat is found out of alignment, install reseating tool in valve bracket in line with valve seat to be refaced and secure bracket on studs. Place forked end of expander clamp over stem between cutter and bearing boss on bracket, set up on screw in center of expander clamp and turn reseating tool until new seat is obtained............ WE214 / 457 / 40,227

  (7) Having obtained true seats, grind valves to seat using grinding compound sparingly...... 40

  (8) Remove valves and bracket, clean and oil (B) valves and seats.

  (9) Replace valves with spring on bracket, numbers corresponding with numbers on seat, inserting cotter pins in ends of valve stems to hold valves in bracket during assembly on after bulkhead............................. 72

                                                              TOOL NO.
   (10) Replace exhaust valve bracket with valves
        and springs on studs on after bulkhead, se-
        cure with nuts and cotter pins.............   72,457
   (11) Remove assembly cotter pins on exhaust valve
        stems.

7084. After bulkhead.
   (1) Clean tapped holes for joint screws for tail.

7085. Propeller shaft packing.
   (1) Unscrew follower for packing, remove old
       packing, clean and inspect..................   452
   (2) Replace follower, setting up by hand.

7086. Circulating water holes.
   (1) Clean and blow air through to see if holes
       are clogged up..............................   41

7087. Depth index.
   (1) Clean, inspect teeth in fixed wheel and re-
       move burrs where necessary.
  *(2) Remove fixed wheel from casing..............   41,72
  *(3) To replace fixed wheel, remove burrs if
       necessary, push in place, secure with lock
       screw and resolder..........................   41,72
   (4) Oil (B) and replace pinion, index wheel and
       depth index spindle.........................   135A
   (5) Replace felt packing ring and gland around
       depth index spindle and tighten up on gland.   18
   (6) Replace spring seat and secure with cotter
       pin...   72
   (7) Replace spring.
   (8) Replace depth index socket extension and
       spindle and secure with cotter pin..........   72

7088. Pipes assembled with afterbody.   141A
   (1) Clean, inspect, reseat seats on collars,   229,191
       chase burrs from threads and connect pipes.   WE85,WE83
                                                     WE94,WE95

   The above completes assembly of afterbody ready for
installation of major units and tail.

N. REPLACE STARTING GEAR.

7089.
   (1) Place a new gasket on starting valve flange
       and soak with oil (D).
   (2) Install starting gear.
   (3) Secure with eight holding screws............   40
   (4) Connect pipes - strengthening ring to
       starting gear...............................   141A
   (5) Connect pipe-strengthening ring to air
       strainer....................................   141A

TS-5   VII-28   Original Page.

INSTALLING STARTING GEAR IN AFTERBODY

O. INSTALL MAIN ENGINE ASSEMBLY IN AFTERBODY.

7090.   TOOL NO.

   (1) Replace gyro spin lead on bulkhead.
   (2) Replace lead to starting gear from reducer.
   (3) Replace lead from starting gear to reducer.
   (4) Replace lead to air strainer from reducer.
   (5) Attach propeller guiding tool................ WE34
   (6) Connect pipe, oil tank to oil pump, to nipple on oil pump.................................. 229
   (7) Swing rudder rods clear.
   (8) Install lifting handles.....................446A,446B
   (9) Place a new gasket on bulkhead seat and soak with oil (D).
  (10) Insert main engine assembly in afterbody guiding propeller shaft through after bearing, and push all the way in on bulkhead seat.

      NOTE: It requires two men to install main engine assembly in afterbody; one to guide shaft and one to carry forward end. It is important that forward end be carried high all the way in, as otherwise the propeller shafts may bind in after bearing.

  (11) Remove lifting handles.....................446A,446B
  (12) Replace nuts for turbine bulkhead screws.... 48
  (13) Connect the following pipes to manifold bracket on strengthening ring:
     (a) Bulkhead to strengthening ring, gyro spin.229
     (b) Bulkhead to strengthening ring, starting gear... 141A
     (c) Bulkhead to strengthening ring, air strainer...................................... 141A
     (d) Bulkhead to strengthening ring, starting gear return.................................. 141A
  (14) Connect pipe (oil pump to oil tank) to oil tank.. 229

P. BUILD UP OF TURBINE BULKHEAD.

7091.

   (1) Replace gasket and turbine nozzles with combustion flask; secure with nuts.............. 227A
   (2) Install valve group:
     (a) Hold valve group in place and attach to combustion flask with coupling nut....... 134A

      NOTE: Care must be exercised when securing coupling nut to see that threaded partf of the valve group and combustion flask enter the same distance into the coupling nut, to insure proper seating of the joint.

	TOOL NO.

    (b) Secure valve group to bulkhead with
        two (2) holding screws.................... 227A
    (c) Secure heads of holding screws with wire. 72
 (3) Connect the following pipes on bulkhead:
    (a) Valve group to turbine bulkhead(gyro spin) 229
    (b) Combustion flask to bulkhead(air strainer) 141A
    (c) Starting valve to bulkhead............... 141A
    (d) Valve group to air checks................ 229
    (e) Valve group to igniter................... 141A
    (f) Bulkhead to fuel sprays.................. 141A
    (g) Venting fitting to bulkhead.............. 141A
    (h) Bulkhead to water spray.................. 144
 (4) Pack after propeller shaft bearing.
    (a) Remove follower for packing............. 452
    (b) Oil (D) and replace felt packing in after
        propeller shaft bearing.
    NOTE: If packing appears to be too small to
    go over the shaft, do not attempt to force
    and tear the end; by heating the packing it
    will go over the shaft without being forced.
 (5) Remove propeller guide tool.................. WE34
 (6) Replace follower for packing and set up tight. 452,54

CHAPTER EIGHT

		ARTICLES
A.	General Description of Depth Control Mechanism............................	8001-8007
B.	Immersion Gear Base.......................	8008-8009
C.	Immersion Casing..........................	8010
D.	Pendulum..................................	8011-8014
E.	Hydrostatic Diaphragm.....................	8015-8017
F.	Depth Setting Mechanism...................	8018-8024
G.	Adjustments - Tests.......................	8025-8027
H.	Depth Engine..............................	8028-8037
I.	Depth Steering Line.......................	8038-8043
J.	Air Strainer..............................	8044
K.	Connecting Linkage........................	8045-8046
L.	Operation.................................	8047-8049
M.	Disassembly, Overhaul and Assembly, Adjustments and Tests.................	8050-8059
N.	Instructions for complete overhaul and replacement of parts in depth engine......	8060-8067

A. THE DEPTH CONTROL MECHANISM - GENERAL DESCRIPTION.

8001. The depth control (**immersion**) mechanism has for its sole function the maintaining of the torpedo at the desired depth below the surface of the water while running.

8002. In the Mark 13 modification torpedoes the horizontal rudders are mounted in bearings on the after end of the horizontal tail vanes, just forward of the propellers, with their inner ends secured to the rudder yoke in the tail.

8003. The rudder yoke in turn is connected to an adjustable extension; the forward part of this extension, called the rudder connection, passes through a stuffing box in the afterbody bulkhead, being made watertight by 3/16 round packing and a suitable gland. Inside the afterbody, connected at one end to the rudder connection and at the other end to the **dep**th engine piston fork is the steel rudder rod which completes the steering line.

8004. The piston of the horizontal steering engine is operated by air at reduced pressure, and is controlled by a very sensitive valve located inside the piston. This valve is connected by means of a system of levers to the pendulum of the immersion mechanism and the pendulum in turn is connected to the lower depth spring socket by inter-connected levers and links. As thus connected, the pendulum, depth spring socket, and horizontal steering engine valve move simultaneously. Motion of the valve may be produced by the pendulum when the torpedo inclines from the horizontal, or by the lower depth spring socket (hydro-diaphragm, and diaphragm lever) when the torpedo is off set depth, or by both in combination; in any case, the pendulum and lower depth spring socket (hydro-diaphragm) move together, each exercising a modifying effect upon the other. This latter is the Uhlan principle, which is now installed in all the later Marks and modifications of service torpedoes.

8005. The diaphragm, diaphragm lever, and lower depth spring socket, are acted upon by pressure of the sea water while the torpedo is running. The diaphragm is held against this sea water pressure by the depth spring, connected to the lower depth spring socket, plus atmospheric pressure in a pressure tight air chamber. Since the pressure in the air chamber is constant for all practical purposes, the depth spring loading controls the running depth of the torpedo. The pendulum is acted upon by gravity and serves to limit the angle of inclination at which the torpedo may seek set depth and at the same time tends to steady the torpedo when at set depth.

8006. The immersion mechanism is assembled on the same removable base as the gyro mechanism and is therefore removed from the afterbody with the gyro mechanism.

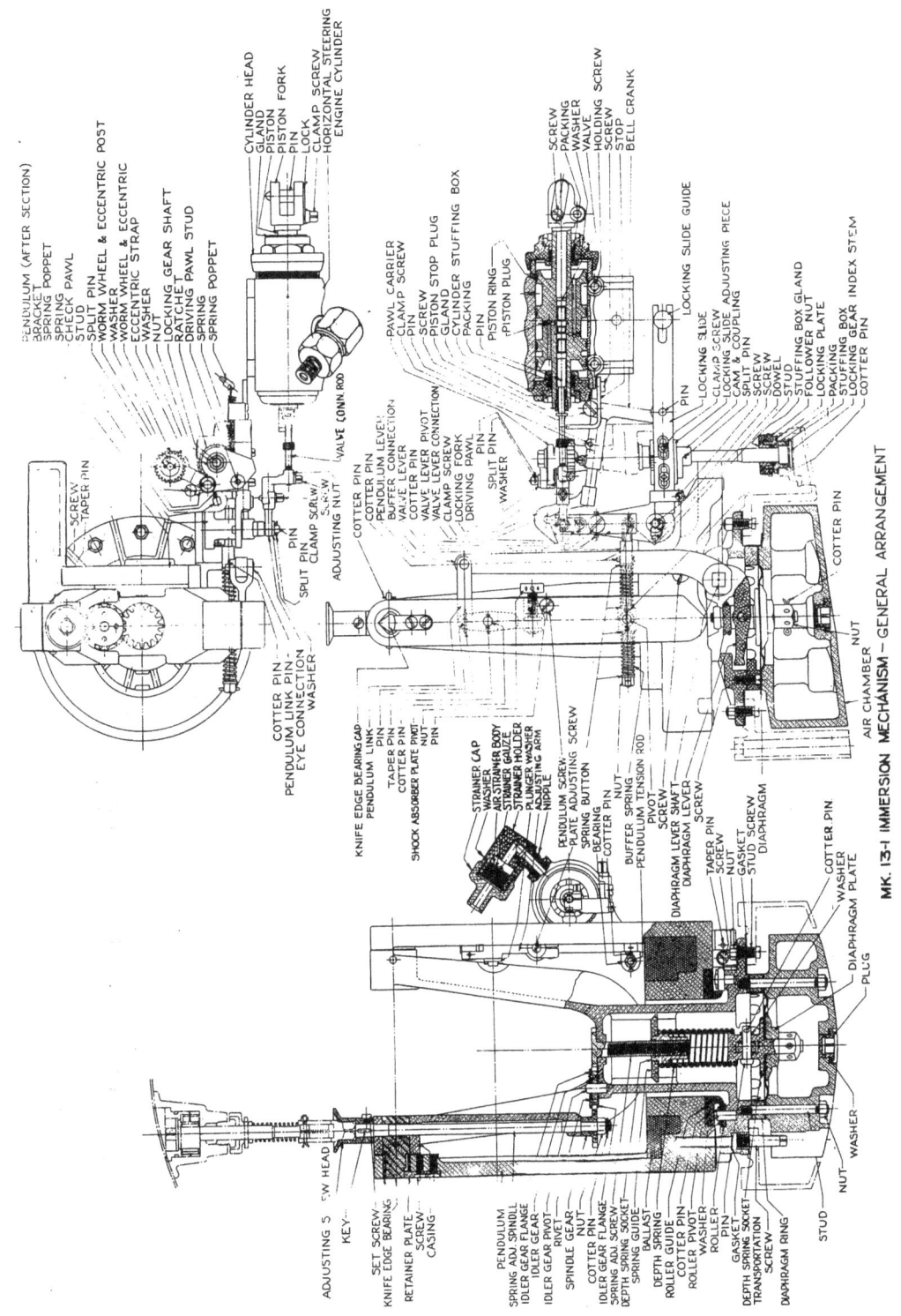

MK. 13-1 IMMERSION MECHANISM—GENERAL ARRANGEMENT

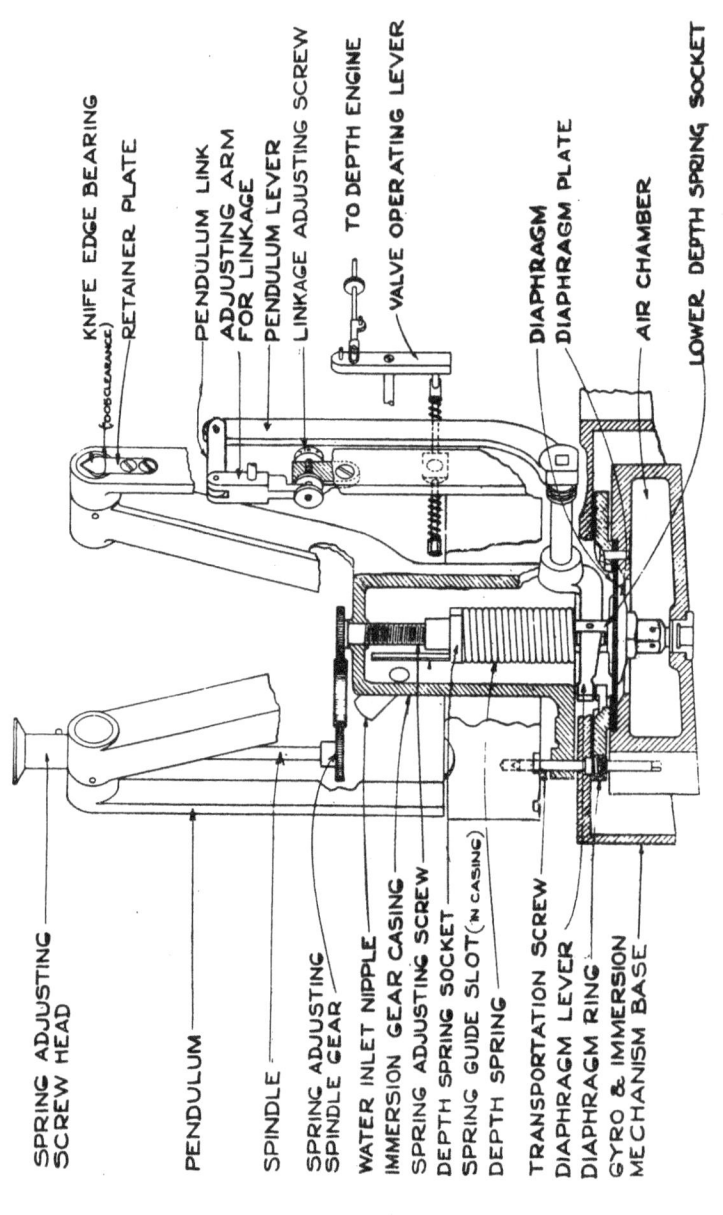

DIAGRAMMATIC VIEW OF IMMERSION MECHANISM

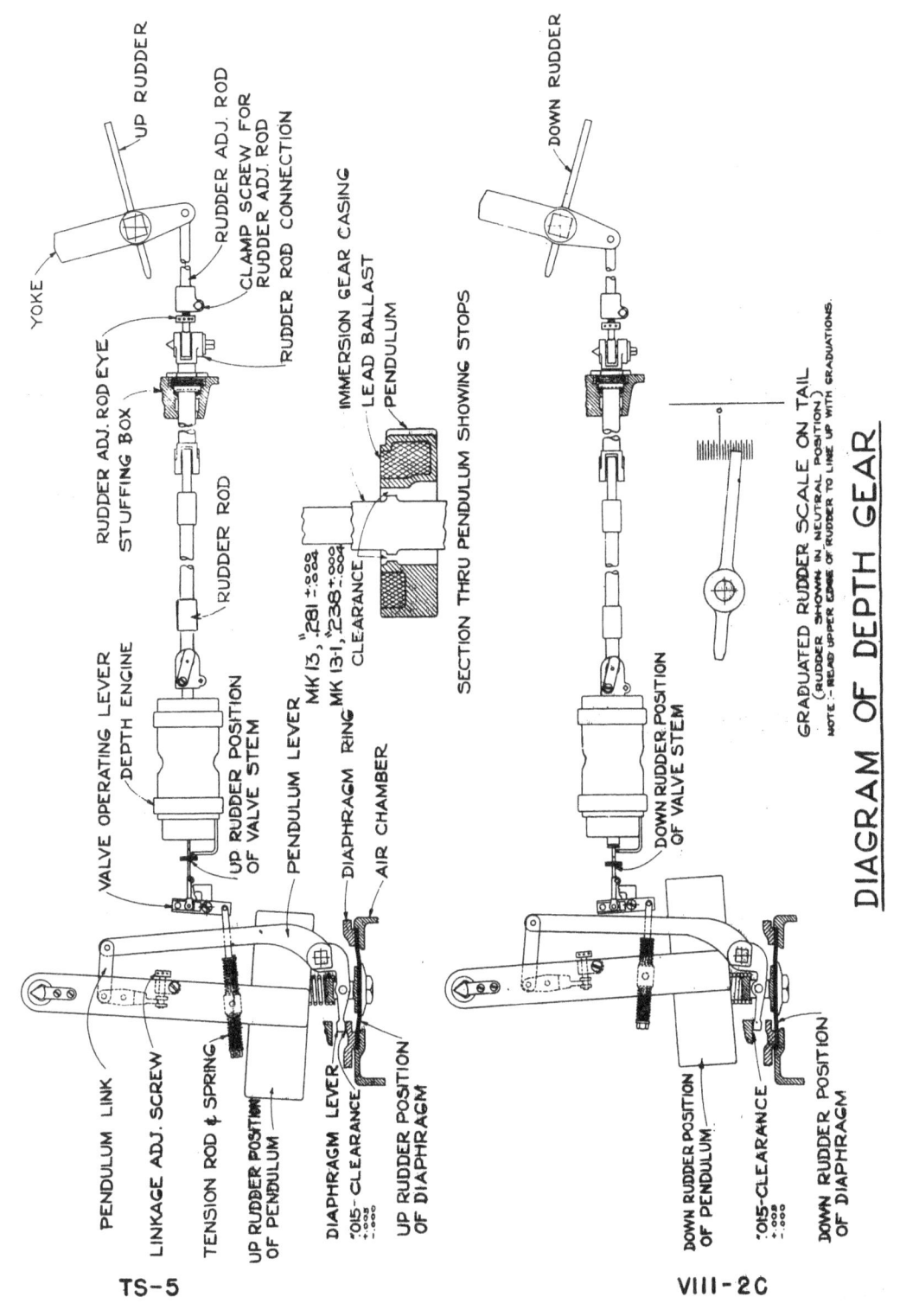

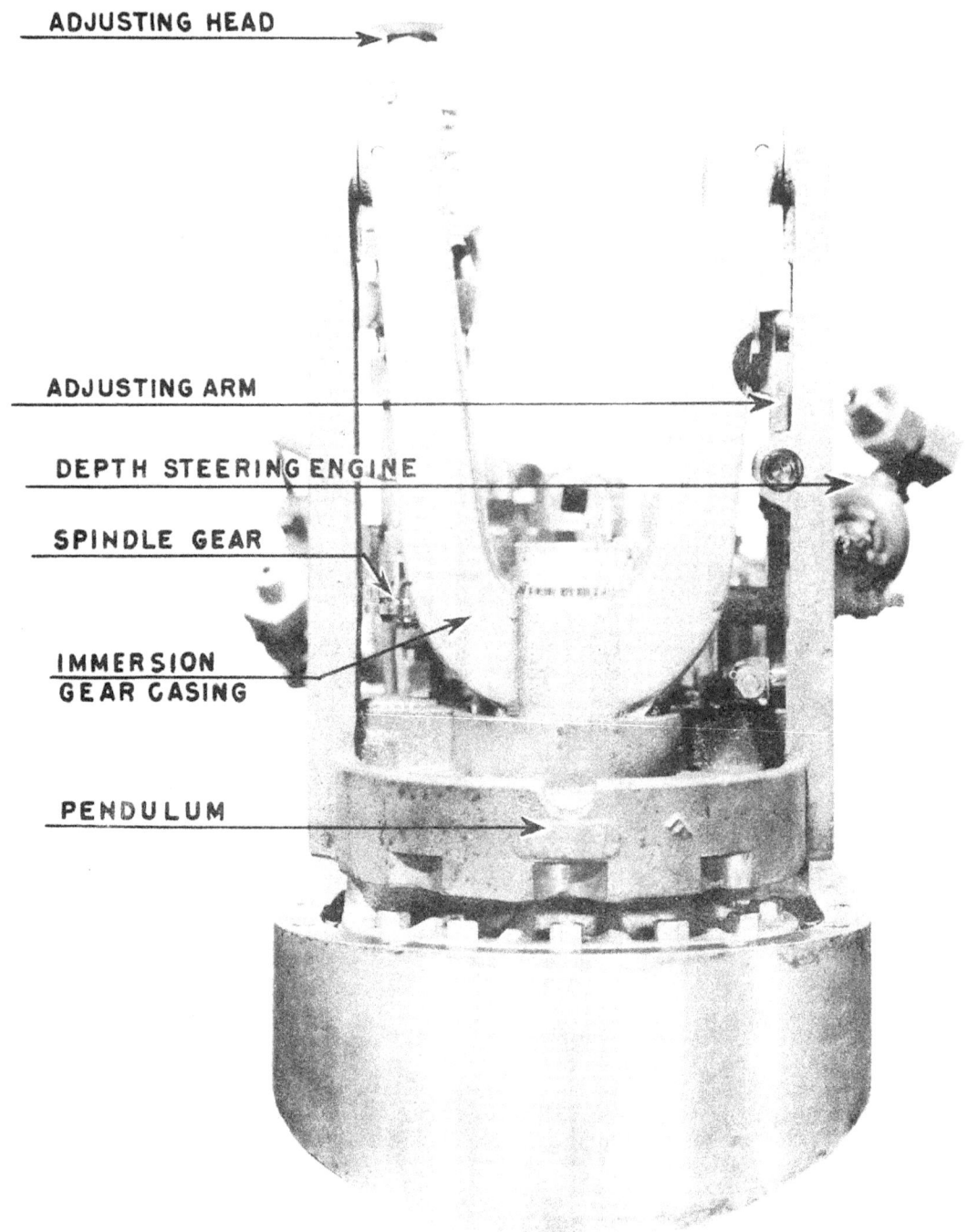

FRONT VIEW OF HOUSING

VIII-2E

8007. **The major units of which the depth control mechanism is composed are as follows:**

    Immersion gear base.
    Immersion casing.
    Pendulum.
    Hydrostatic diaphragm.
    Depth setting mechanism.
    Depth engine.
    Depth steering line.
    Rudders.

B. <u>THE IMMERSION GEAR BASE.</u>

8008. As above stated, the gyro mechanism and the immersion mechanisms are both contained on the same base, which is a large bronze casting fastened by holding screws to a frame riveted and sweated to the bottom of the afterbody shell, a gasket being fitted between the base and frame to make this joint watertight.

8009. The inner surface of the mechanism is tinned as necessary to enable it to withstand an external hydraulic pressure of 135 pounds per square inch without leakage. The immersion mechanism is installed on the forward end of the base and the gyro mechanism on the after end. The base is provided with necessary access openings for the various units of the gyro and immersion mechanisms.

C. <u>THE IMMERSION CASING.</u>

8010. This unit is a bronze casting secured to the forward end of the gyro and immersion mechanism base by suitable studs and nuts with a gasket between. Its lower section is hollow, it is open to sea water and contains the hydro-control elements. The upper section branches in two arms, the ends of which are fitted with knife edge bearings for the pendulum. The starboard arm is directly under the depth setting spindle and carries the depth setting spindle socket, spindle and gear. The depth setting gear on the lower end of the spindle meshes with an idler gear carried on a pin pressed into the immersion gear casing and in turn meshes with the gear that is the head on the adjusting screw. Two connections are made from the interior of the immersion casing to the interior of the afterbody, one where the spring adjusting screw comes through at the top of the casing and one where the diaphragm lever shaft passes through near the bottom of the casing. Both of these openings are lapped with the parts which pass through. There is no packing at these points, but due to the lap fits the joints are practically watertight.

D. <u>THE PENDULUM.</u>

8011. The pendulum is a heavy bronze casting suspended on knife

THE LATEST MK XIII MOD. I TORPEDO HAS TWO AFTER PENDULUM STOPS MACHINED ON THE CASING OF THE IMMERSION GEAR AS SHOWN ABOVE.   CLEARANCE IS .234"

edge bearings from the arms of the immersion casing. The base of the pendulum encircling the immersion casing is contoured to prevent interference with the spinning and unlocking mechanism and is ballasted with lead to obtain the necessary power and smoothness for effective control of the depth rudders. The extended after portions of the pendulum base are joined by a steel member secured with screws to the pendulum. The upper face of this member is milled out to give clearance around the impulse rotor. In the upper ends of the pendulum arms are bushings, also called knife edge bearings. These bushings, or knife edge bearings, are locked in place by set screws and have triangular openings which fit over the knife edge bearings which project from the immersion casing arms. In the Mark 13-1 and 2 torpedoes, the maximum swing is decreased to "238 + .000 - .004 in order to effect proper depth control. A retainer plate set in each pendulum arm makes near contact with the under side of the immersion casing knife edge bearing and prevents the pendulum from jumping or pounding excessively on its support. This clearance is "005.

8012. The transportation, or centering screw, is on the starboard side of the immersion gear casing; it is inserted from outside and extends through the mechanism base and into the pendulum. When in place, this screw securely locks the pendulum in mid-position. By locking the pendulum in its mid-position, there is provided a reference point for adjusting the valve to its mid-position and the diaphragm lever to its mid-position. Before firing, the transportation screw is removed in order to permit the pendulum to swing freely when controlling the torpedo during the run. The replacement screw for the transportation screw should be set up against its copper washer to insure watertightness of this fitting.

8013. The pendulum may be subjected to severe side forces as when the plane is "crabbing" at the time of release and in order to prevent damage to its supports, the under side of the pendulum has fore and aft guides which just clear rollers attached to the immersion casing. Normally these guides do not make contact with the rollers, but any heavy side forces which might otherwise cause the pendulum to lift on its suspensions will bring one of the guides against its mating roller which thus absorbs the shock. The clearance between guides and rollers is such that contact will not be made until the torpedo has been slowly heeled about 45°. These rollers can be inspected from the side of the mechanism and should be free to turn on their pins. They are not removable and are of case hardened material.

8014. A boss is cast on the pendulum on the forward inside surface where it encircles the immersion casing, and the steel member connecting the after ends of the pendulum is suitably machined for making contact with bosses on the immersion casing and thus act as stops for the pendulum. These bosses are so machined as to permit a movement of "238 + .000 - .004 of the pendulum either fore or aft from its mid-position in the Mark 13-1 and 2 torpedoes. These stops are therefore fixed permanently and should not be altered so long as the above permissible swing is obtained.

E. **THE HYDROSTATIC DIAPHRAGM.**

8015. Concentric with and beneath the immersion casing is the diaphragm ring which is secured to the immersion casing by screws. The diaphragm ring carries six studs which are used to fasten the air chamber to it. The air chamber is concentric with the diaphragm ring and among other purposes secures the diaphragm between it and the diaphragm ring. The air chamber is a circular hollow bronze casting tinned inside, and must be sealed air and watertight. Its lower surface conforms to the afterbody exterior; its upper face is surfaced for the diaphragm and with a central opening which is closed by the diaphragm and its connection to the depth spring. It has two openings, one for the diaphragm and one for the access plug for the depth spring testing weight. These openings must be kept sealed airtight during a run to maintain the atmospheric pressure therein. Clearance around the side of the air chamber and holes drilled through the diaphragm ring afford sea water access to the diaphragm and interior of the immersion casing, thereby establishing hydrostatic pressure within the immersion casing which acts upon the upper face of the diaphragm. The air chamber is in effect completely surrounded by sea water. The depth spring socket, and adjusting screw are thus all immersed in sea water inside the immersion casing and with the diaphragm and air chamber are kept isolated from the effects of afterbody temperature and pressure. This fact produces a marked superiority of this depth gear over previous types.

8016. The diaphragm is of oil resistant, rubberized fabric clamped on its outer edge between the diaphragm ring and the air chamber. Its center is clamped between the diaphragm plate in the air chamber and the lower depth spring socket flange in the immersion chamber. These parts on each side of the diaphragm hold it rigidly at its center, thereby providing the necessary connecting movements whereby the unbalanced forces on the two sides of the diaphragm may be transferred to the pendulum via the interconnecting linkage and thence to the depth engine valve. The diaphragm has an effective area of five square inches and a very limited movement which is so multiplied that the ratio of the valve movement to diaphragm movement is 18 to 1. The small amount of free diaphragm area between the outer and inner supported areas gives ample flexibility to the small movement required up to 50 feet depth setting.

8017. The lower depth spring socket, the depth spring, and the upper depth spring socket with the upper guide are assembled as a unit, all parts being screwed and lightly sweated together. They should not be disassembled. The upper depth spring socket is threaded internally for the adjusting screw and has guide arms sweated to it to prevent its turning (hence to prevent the depth spring turning) when the adjusting screw is turned. The lower depth spring socket is slotted vertically and has a pivot pin to position the diaphragm lever in the slot; below this slot is the flange against which the diaphragm is clamped. The lower end of the depth spring socket is threaded for the clamping nut which is for the purpose of holding the diaphragm plate against the diaphragm on its lower side. This clamp nut is locked by a cotter pin and its lower extremity is tapped to receive the hook rod for the testing weight which, as will be seen later, is for the purpose of obtaining the proper setting of the depth index wheel on the afterbody.

F. THE DEPTH SETTING MECHANISM.

8018. The depth setting mechanism is essentially a device for varying the tension of the depth spring to establish a desired balance for the hydrostatic pressure at running depth. The setting is indicated on a graduated dial called the index wheel, which surrounds the depth index spindle on the top of the afterbody.

8019. In the depth setting mechanism there are two major units; the depth index casing assembly, and the spring and socket assembly.

8020. The depth index casing contains all the parts secured to the afterbody shell and connections. The casing is riveted and sweated to the afterbody shell. In its center it carries the depth index spindle, square on its outer end for the setting tool. The depth index spindle has an eccentric machined on its stem upon which is mounted, freely, the annular index spindle pinion for driving the depth index. The depth index spindle passes through a stuffing box in the depth index casing and the lower end is squared to fit the depth setting socket extension. A spring seat is pinned to the spindle directly under the stuffing box gland.

8021. Attached loosely to the lower end of the depth index spindle is the socket extension and spring which form the flexible connection between the depth setting side gear spindle and the socket attached to the spring adjusting spindle on the immersion mechanism. This spring loaded socket extension is for the purpose of facilitating engagement of the depth setting spindles referred to above when the immersion mechanism is installed in the afterbody. In the event the spring socket extension does not fully engage the socket attached to the top of the spring adjusting spindle on the immersion mechanism, the index spindle at the top of the afterbody should be turned sufficiently to make such engagement possible. (This should not be in excess of 1/8 turn).

8022. The depth index spindle and spring socket assembly transfer the movement of setting the depth spring from the outside of the afterbody to the immersion mechanism.

8023. The depth index wheel is a graduated dial which takes on a reduced motion from the depth index spindle through a system of planetary gears.

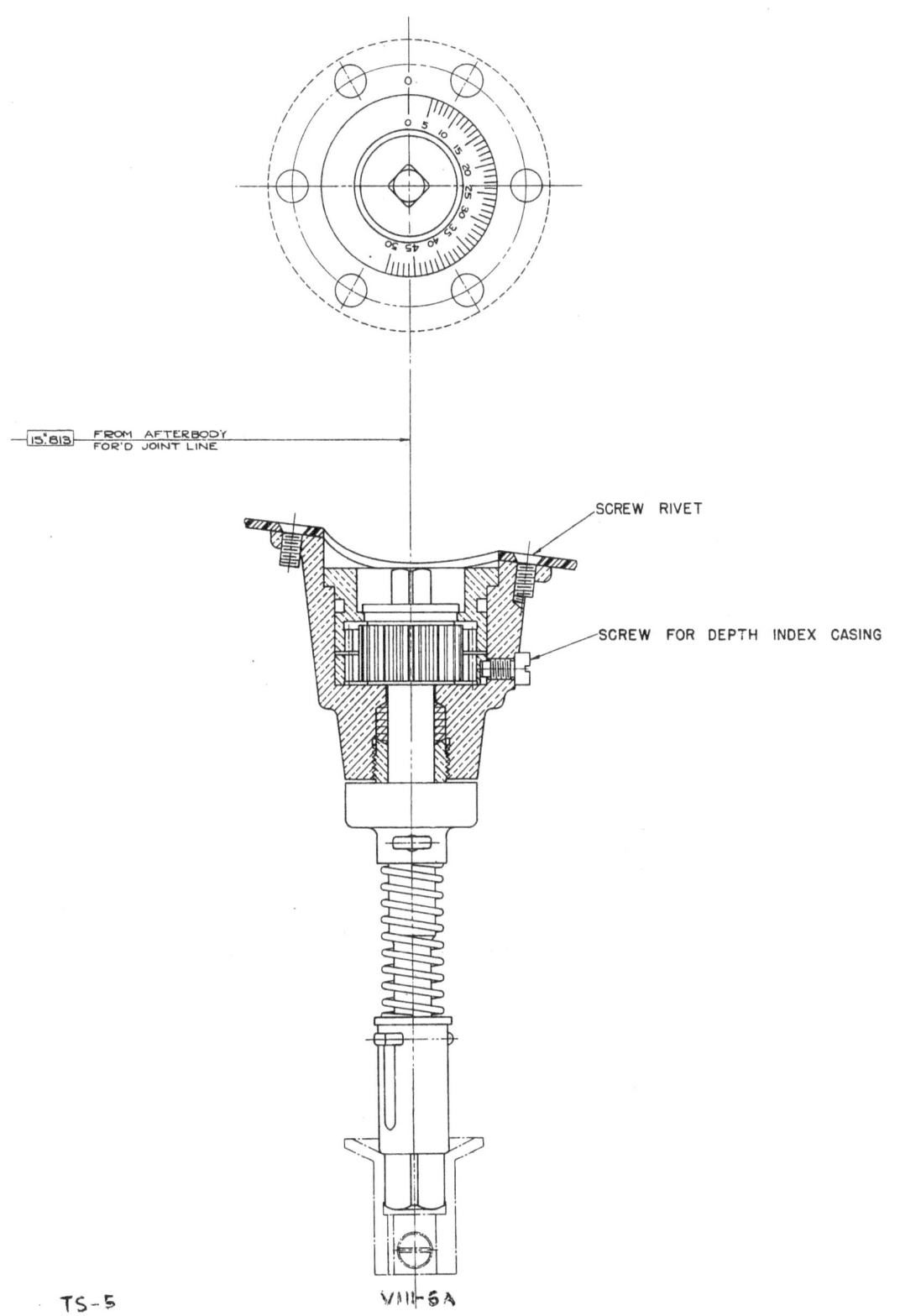

TS-5

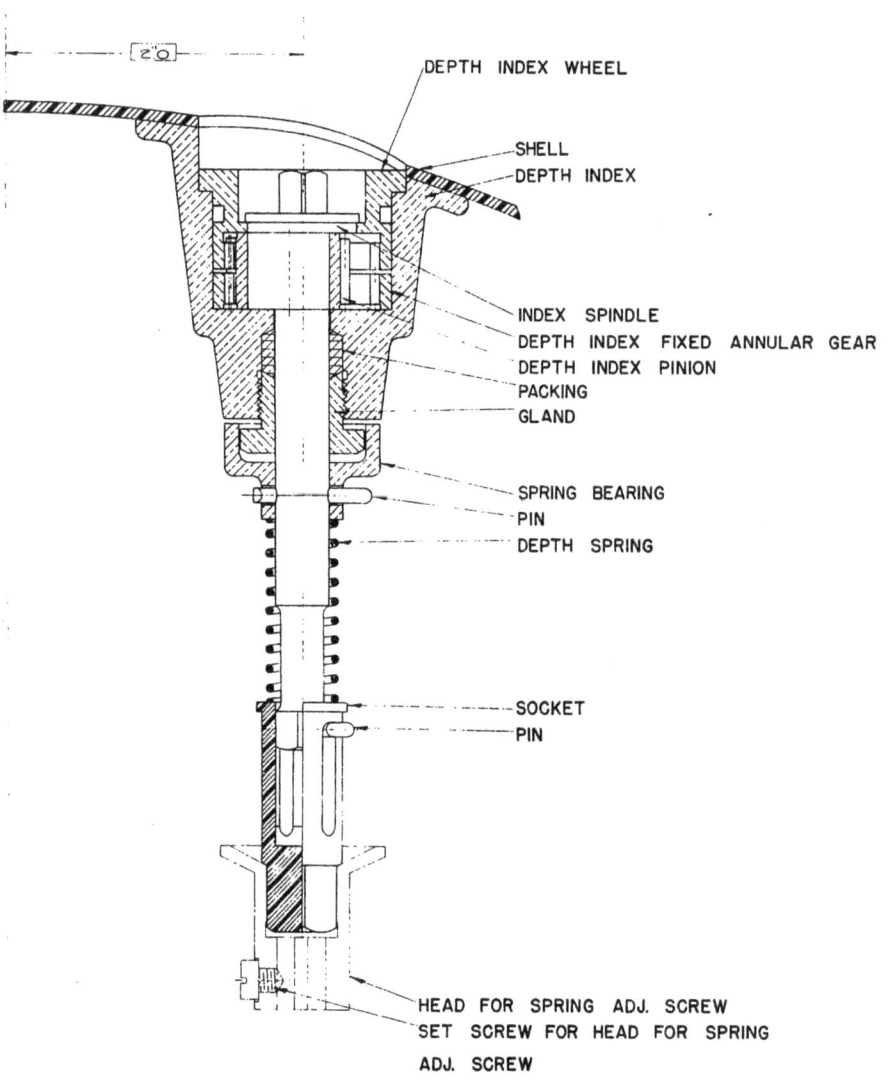

8024. In the Mark 13 and modification torpedoes the index wheel is graduated to a scale which is based on the fact that a 16 lb. downward pressure on the diaphragm would be obtained were the torpedo at a depth of 10 feet. NOTE: BuOrd Circular letter T-145 of October 1940 authorizes the use of a 16 lb. weight for all Mk. 13 and modification torpedoes.

IMPORTANT NOTES: The depth setting mechanism should never be set below the zero mark on the depth index. Frequently in service the depth index has been set below the zero mark causing the adjusting screw to jam or be improperly positioned because of damage to the lower washer on the idler gear, probably resulting in a tension being placed on the depth spring which would cause the torpedo to run at a depth varying from that indicated. If the depth index is turned below the zero mark the adjusting screw tends to rise, lifting the idler gear assembly with it. The depth setting gear is fixed in position, therefore when the idler gear rises the lower thin washer, riveted to it, hits the depth setting gear and is bent causing the trouble described above. To eliminate this condition the depth setting gear on the lower end of the depth setting spindle has been redesigned with a collar to fit over the upper washer on the idler gear, located in such a manner that when the idler gear rises the upper washer hits this collar. This washer is much thicker than the lower washer and will therefore not be bent by the upward pressure.

G. ADJUSTMENTS - TESTS.

8025. First, there is the adjustment to insure that the diaphragm lever is in mid-position when the pendulum is likewise in mid-position as determined by insertion of the centering screw. Secondly, there is the depth engine valve centering test which is made with the transportation screw in place and with air pressure on the engine. The valve is adjusted to its mid-position by means of the adjustment provided on the valve stem previously mentioned. Thirdly, there is the setting of the depth index wheel so that it will truly represent the depth at which the torpedo will run. These adjustments, together with the locking adjustments, are fully described in later chapters.

8026. Testing of the immersion gear is done at stations and on tenders by means of a special testing fixture in which the mechanism base can be installed and water pressure actually applied to the diaphragm by means of a hand pump, the water line being provided with a pressure gauge graduated in "feet of salt water". By means of this testing fixture the sensitivity of the depth engine valve and inter-connecting linkages to the hydro-diaphragm can be determined. This test should be conducted on tenders during annual overhaul. Sluggishness of the depth engine valve when disconnected from the valve operating lever can also be determined by means of Tool 222 which is a small leaf spring whose deflection is indicated on an attached sector when the tip of the spring is used to move the depth engine valve while air at working pressure is on the engine.

8027. Tenders and bases likewise are provided with a testing stand for measuring air leakage past the valve. This test is for the purpose of replacing depth engines which are abnormally wasteful in air consumption. The outfit merely consists of a device for measuring the drop in pressure in an air line, the drop being a direct measure of the leakage through the engine.

H. DEPTH ENGINE.

8028. The depth steering engine is mounted in a cut out and flattened section of the gyro pot and is for the purpose of moving the horizontal rudders. It consists essentially of a piston moving in a cylinder and controlled by a valve concentric with the moving longitudinally inside the piston rod. The assembled engine is so built that only a very slight force is required to move the engine valve and as the result, movement of the piston follows the movement of the valve but with an enormous increase in force.

8029. The depth engine valve is so delicate in its operation that it must be kept scrupulously clean and free from any foreign particles which might increase the resistance to valve movement. Obviously, this high sensitivity of the valve is necessary by reason of the extremely small forces available through the pendulum and hydrodiaphragm as the torpedo varies slightly from set depth. Smoothness of depth performance is therefore dependent upon smoothness of depth engine control valve performance.

8030. As previously stated, both the horizontal and vertical steering engines are air operated with air which is probably only slightly warmer than room temperature. For maximum efficiency, it is necessary that air leakage be kept as small as possible. This requires that valves and pistons be lap fitted to their working surfaces.

8031. In assembling the various parts of an engine, interchangeability is lost to a considerable extent as the degree of fit of the parts must be such as to permit the final lapping of mating parts. When the control valve is finally lapped in place it is required to operate the engine for a full stroke with a push or pull on the valve of $\frac{1}{4}$ ounce with 400 pounds air pressure on the engine. At this pressure, slight leakage is allowed around the valve and through the exhaust ports but should disappear when the pressure has been reduced to 100 pounds per square inch. As previously stated, a suitable testing outfit is issued tenders and bases for determining whether or not the air leakage through the engine is abnormal.

8032. On the piston of the engine there are three grooves or air chambers machined at right angles to the axis, the central one being a supply chamber and the two end ones are exhaust chambers.

8033. The operation of the depth steering engine is as follows: Starting with the piston in mid-position in the cylinder and the

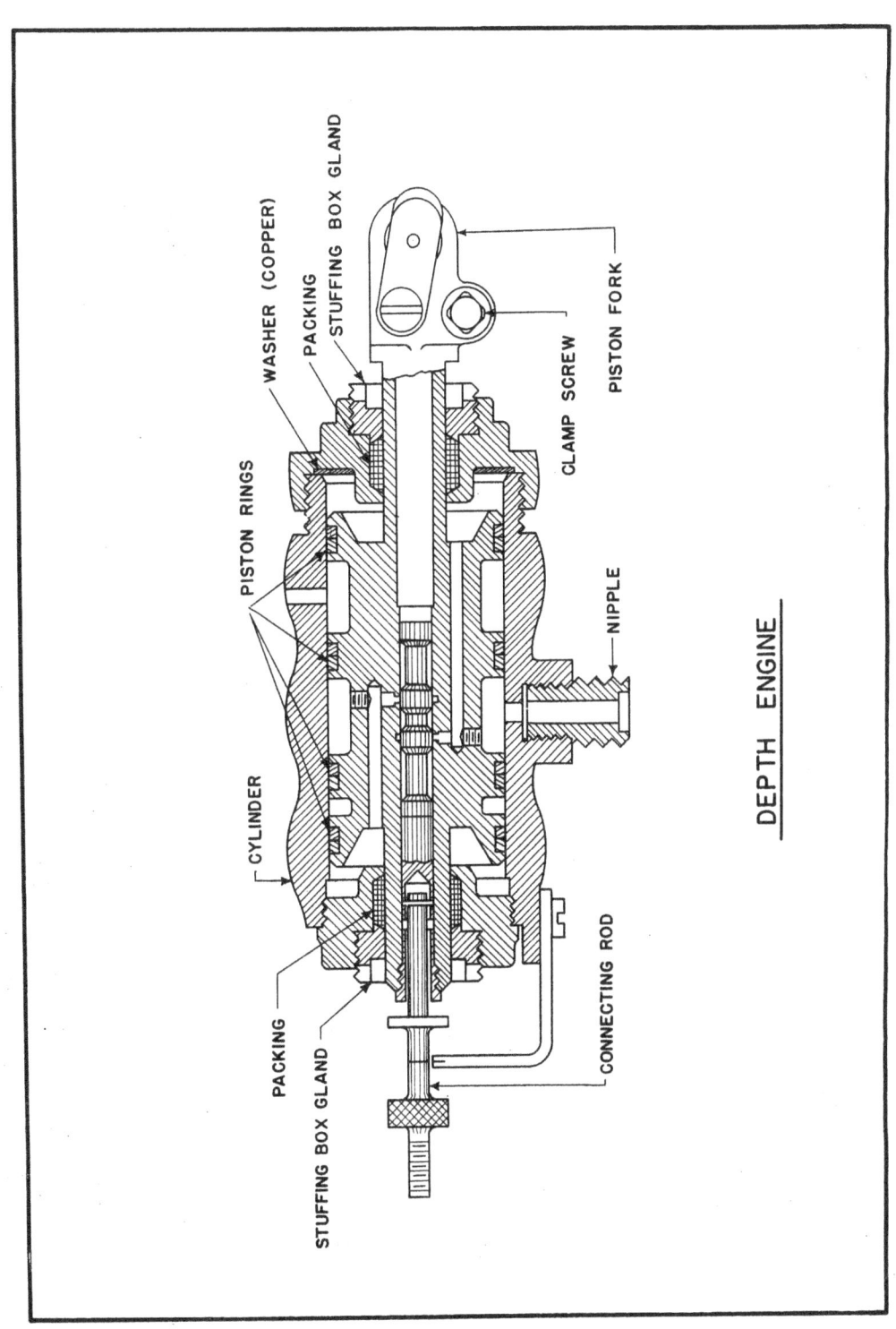

control valve in mid-position in the piston, air from the supply line fills the central chamber of the piston and through the supply ports also fills the chamber in the center of the valve. Disregarding leakage, air is confined to these spaces so long as the control valve remains stationary and the piston does not move.

8034. Moving the valve forward (or to the left) the forward supply port is opened and air passes through it to the after end of the piston where sufficient pressure is built up to move it forward following closely the movement of the valve. If the valve stops, the movement of the piston also stops as soon as the forward supply port is again covered. The movement aft of the control valve accomplishes an after movement of the piston in the same manner.

8035. The movement of the piston is normally so close behind that of the valve that the air at the exhaust end of the cylinder is forced out by leakage over the end piston rings into the two exhaust chambers machined on the outer surface of the piston. This action results in a slowing and cushioning of the piston stroke through which extremely smooth action of the engine is obtained. It can be readily seen that if the exhaust ports opened simultaneously with the supply ports the hammering action of the piston would soon wear out the engine.

8036. Four pairs of rings are fitted in grooves in the periphery of the steering engine piston. These rings are for the purpose of providing suitable bearing surfaces which at the same time reduce air leakage past the contact surface to a minimum.

IMPORTANT NOTES.

8037. To insure satisfactory operation of the torpedo in depth, the depth engine valve must move freely in the piston and must be kept well lubricated with the approved gyro oil. With working pressure on the depth engine, the piston should move when a pressure of $\frac{1}{4}$ to $\frac{1}{3}$ ounce is applied on the valve stem, and if the piston fails to move when a pressure of $\frac{1}{2}$ ounce is applied, the valve is not considered sufficiently sensitive. The flat spring gauge, tool No. 222, previously referred to in paragraph 8026 of this chapter is furnished for this test; this tool is graduated to each $\frac{1}{2}$ ounce up to four ounces. Many faulty runs have been caused by dirt or grit collecting in the valve chamber, also by dirt or foreign matter lodged under or between the packing rings of the piston. Air to this engine is required to pass through a 200 mesh strainer in the air strainer body prior to admission to the engine, but obviously, with such finely lapped surfaces particles of dirt which can pass through this strainer might easily cause trouble. The packing gland in the piston rod should be kept properly packed and this packing should be renewed if necessary upon annual overhaul and tested for tightness. When the depth engine valve is properly adjusted, it should be possible to push the piston about 1/64 of an inch further by hand on each end of the stroke than it will go by air when the valve is against its stops.

I. **THE DEPTH STEERING LINE.**

8038. The depth steering line is that portion of the horizontal rudder operating mechanism between the depth engine and the rudders.

8039 The motion of the depth engine piston is transferred to the horizontal rudders through the following parts: Piston fork, depth rudder rod, rudder connection (through stuffing box in afterbody bulkhead), rudder adjusting rod (in tail cone), rudder yoke, rudder post, lever arms, rudder arms and the rudders.

8040. The eye on the forward end of the rudder rod fits in the piston fork and is held in place by a pin locked by a flat brass retaining spring. This rod passes along the afterbody and is pinned to the connection which passes through the stuffing box in the after bulkhead. This rod, and hence the entire steering line is prevented from turning by a square section in the stuffing box.

8041 On the after end of the connection which passes through the stuffing box is fitted a fork, and this fork is pinned to the forward eye of the adjusting rod in the tail cone. This connection provides a means for lengthening or shortening the depth steering line for the purpose of obtaining proper rudder throws. This is accomplished when the tail cone is assembled to the afterbody through two plugged access openings in the tail cone beneath the forward end of the adjusting rod. The eye which screws into the forward end of the adjusting rod can be rotated when disengaged from the connection through the afterbody bulkhead by turning it after the clamp screw on the forward end of the adjusting rod has been released. When the adjustment has been completed, as determined by reading the rudder throws at the graduations on the after end of the tail cone, the clamp screw is carefully set up and the adjusting rod is again engaged with the fork on the steering connection, and the inter-connecting pin carefully screwed home to maintain this connection.

8042. The after end of the adjusting rod in the Mark 13 Modification torpedoes connects to a projecting pin on the rudder yoke. The rudder yoke has square holes broached in its ends which fit over the square shank at the inboard end of each rudder. Where the shank of each rudder passes through the tail cone, bronze bearings are riveted and sweated in place on the tail cone. The shanks must be close fits in these bushings to prevent excessive gas leakage from getting into the water stream forward of the propellers. The rudder yoke is so designed as to inter-connect the two horizontal rudders on each side of the tail cone and at the same time pass around the outer propeller sleeve. Obviously, there must be no interference of the rudder yoke with any of the tail cone parts during a run.

1043. **IMPORTANT** NOTES.
The parts of the steering line are subject to considerable heat during a run, especially the parts in the tail cone. As a result

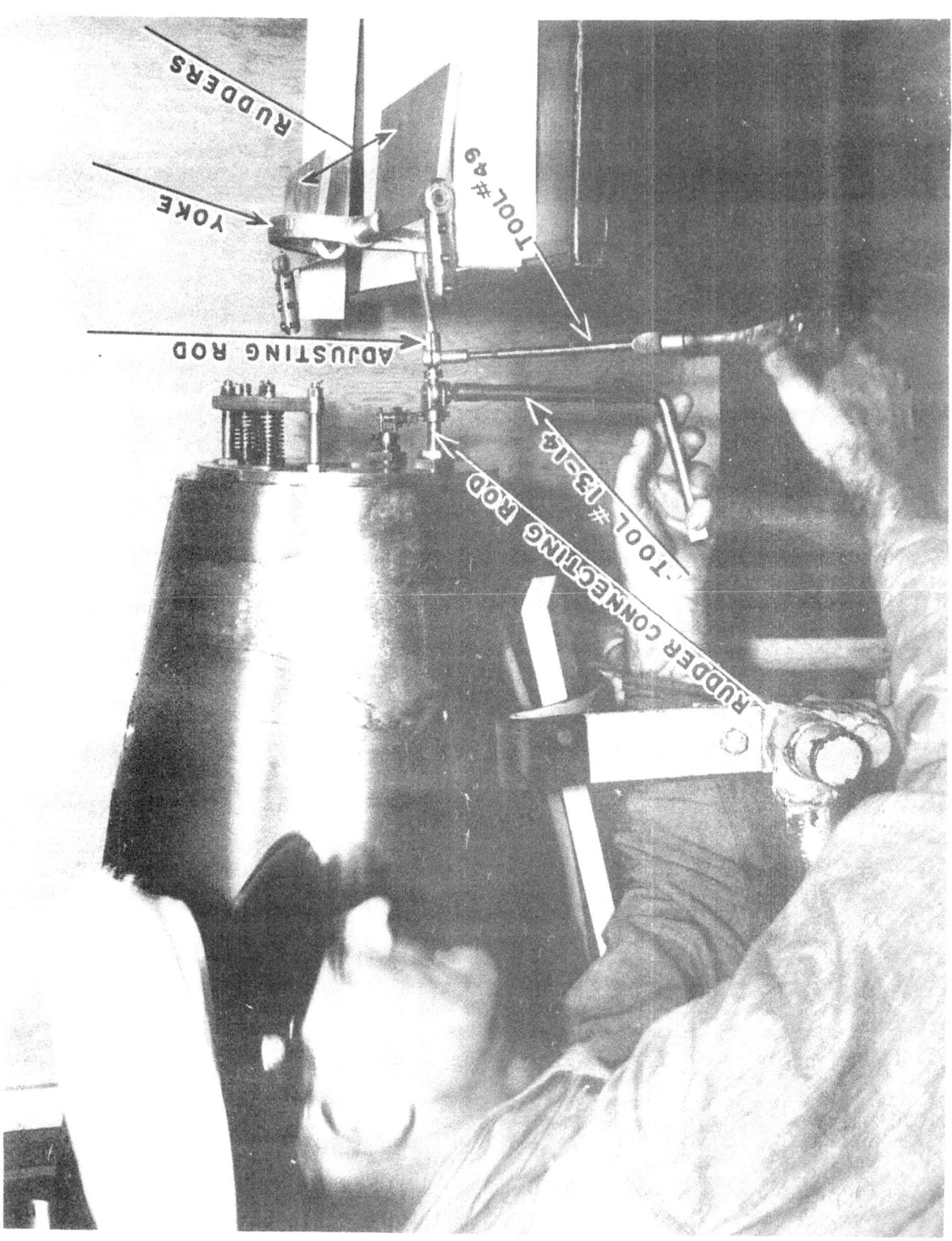

of these temperatures, expansion of the steering line takes place which requires a margin of clearance at the end of the stroke both at the rudder yoke and at the steering connection which passes through the stuffing box in the afterbody bulkhead. The several parts of the steering line are so designed that no interference should be encountered at these points, but, obviously if the adjusting rod in the tail cone becomes bent or if the stuffing box gland in the afterbody bulkhead be allowed to project too far into the tail cone, interference can be expected which will cause erratic performance. These features should be carefully inspected during annual overhaul.

J. AIR STRAINER.

8044. The air strainer is contained in a flanged body, the flange of which is riveted and sweated under an opening in the afterbody shell forward of the starting gear. The strainer is 100 mesh and screws into the body. The opening is closed by a plug seating against a leather washer. The air strainer body has one inlet and two outlet nipples as follows:

Inlet
1. Source of supply from top opening on reduced side of reducing valve. This opening is closed by a plug with two short leads sweated in place; one of which connects to the air strainer.

Outlets
1. Depth engine.
2. Two way nipple on gyro steering engine where the air divides, one lead going to gyro engine and the other to the gyro reducing valve.

NOTE: 200 mesh airstrainers are now contained in the depth and steering engine inlet nipples; therefore, the 100 mesh strainer has been removed from the air strainer body.

K. CONNECTING LINKAGE BETWEEN VALVE, PENDULUM AND HYDROSTATIC DIAPHRAGM.

8045. The forward end of the valve connecting rod is connected to upper end of valve operating lever. This lever is pivoted centrally by a screw to a bracket on the front plate of the spinning and unlocking mechanism. Secured to the lower end of the valve operating lever is the after end of the tension rod, which in turn is secured centrally to a boss on the lower left pendulum arm. Two springs each fitted with buttons at their extremities fit over the tension rod, one interposed between the central bearing and after fork, and the other between the forward nut and the central bearing. The assembly permits slight fore and aft motion of the tension rod, resisted in each case by the tension springs. In the upper end of pendulum arm is pivoted the forward end of the short horizontal pendulum link. In the forked after end of this link is pivoted the upper end of the pendulum lever. The lower end of pendulum lever fits over the squared end of the diaphragm shaft. This shaft passes through and operates in lapped bearings in immersion casing. The squared inner end of this shaft fits into a squared hole in the curved after end of the diaphragm lever. This diaphragm lever is pivoted centrally in the lower spring socket and housed in a small recess in the casing flange. Its forward end is free and operates vertically between its upper and lower stops on the lower side of the immersion casing and the upper side of the diaphragm ring respectively.

8046. With no tension on the depth spring and the pendulum hard up against its stops, there must be a minimum of ".015 clearance between the free end of diaphragm lever and its stops. This clearance is equalized by the adjusting arm pivoted at its center to the inboard face of the left pendulum arm. In the upper forked end of the arm is pivoted the forward end of the short horizontal pendulum link, and in the forked after end of this link is pivoted the upper end of the pendulum lever as described above. The lower end of the adjusting arm fits into an annular groove formed by two collars machined on the forward end of the horizontal adjusting screw. The adjusting screw passes through a threaded boss on the inner face of the left arm. Over the after end of the screw and against the boss is screwed a lock nut. A clamp screw which passes through the pendulum and split portion of the boss clamps the adjusting screw firmly in the boss once an adjustment is made.

L. OPERATION.

8047. When depth is set from outside the afterbody, the operation is as follows: Rotation of the index spindle rotates the side gear through the center gear. Rotation of the side gear will rotate the socket spindle, socket, depth spring adjusting spindle and gear. As this lower gear through an idler gear is in mesh with the gear on the upper end of the adjusting screw, this screw will rotate within the upper spring socket. This socket is prevented from turning by the guides and guide slots, so the upper spring socket will rise vertically, putting a tension on the depth spring and causing the lower spring socket to rise slightly. As the diaphragm lever is pivoted centrally to the lower spring socket, the free end of the lever will be pulled upward, rotating the diaphragm shaft. The pendulum lever secured rigidly at its lower end to the outer end of the diaphragm shaft will also rotate, its upper end moving aft. This end being connected to the pendulum through the pendulum link, adjusting arm and screw will pull the pendulum aft against its forward stop. With the pendulum against this stop, the toe of the diaphragm lever is up but not contacting its upper stop. Any further turning of the index spindle beyond this point will only cause a rise of the upper spring socket and a greater tension on the spring. Movement of the pendulum aft will move the lower end of the valve operating lever aft through the buffer spring and tension rod. As this lever is pivoted centrally to a bracket, its upper end will move forward, moving the depth engine control valve forward. This initial setting will not change the position of the rudders so long as no air is on the steering engine, which is the case at this time. As a rule before firing, the horizontal rudders are given down rudder by hand. Never put on a depth setting with the transportation screw in as too great a strain will be placed on the linkage.

8048. Operation when torpedo is fired is as follows: Air from the air strainer in the afterbody enters the steering engine and as the control valve is forward, immediately gives down rudder. This occurs before torpedo takes the water. After the torpedo has submerged the rudders start to take effect and nose the torpedo down towards its set depth. The torpedo being down by the nose the tendency of the pendulum is to swing forward against the resisting spring tension, removing some of the down rudder effect. Water enters through the holes in the diaphragm ring, filling the immersion casing above the diaphragm. Water pressure acting on top of the diaphragm forces the diaphragm downward. As the diaphragm is secured to the lower end of the lower spring socket, the spring socket is also moved downward, extending the depth spring. Downward motion of this socket by the same system of linkages described above causes the pendulum to move forward and ultimately move the control valve aft, removing some of the down rudder. As the torpedo reaches its depth the water pressure on the diaphragm is equal to the spring tension and the diaphragm and lower spring is in the mid-position. This places the control valve in mid-position and the rudder in neutral. As the torpedo travels downward the down rudder diminishes, the torpedo approaches and even keel and the resisting pendulum effect is reduced, thus causing the torpedo to seek its initial depth gradually. During the run the pendulum effect on the rudder always resists the diaphragm effect and gives smooth rather than jerky operation to the depth engine.

8049. Motion of the valve may be produced by the pendulum when the torpedo inclines from the horizontal, or by the lower depth spring socket (hydro-diaphragm) when the torpedo is off set depth, or by both in combination. In any case, the pendulum and lower depth spring socket (hydro-diaphragm) move together, each exercising a modifying effect upon the other. The service performance of this immersion mechanism built on the Uhlan principle is a vast improvement over that of previous types and especially in the rapidity with which the torpedo is brought to set depth.

M. DISASSEMBLY, OVERHAUL AND ASSEMBLY, ADJUSTMENTS AND TESTS.

8050. Disassemble immersion mechanism.   TOOL NO.

(a) Remove gyro reducing valve and pipes.
   (1) Disconnect air pipe to reducing valve........... 24
   (2) Disconnect air pipe from reducing valve at nipple on mechanism........................... 24
   (3) Remove 2 holding screws, and remove reducing valve with "T" nipple and pipes assembled....... 246A

                                                                TOOL NO.
(b) Remove air chamber.
    (1) Remove transportation screw.....................   49
    (2) Remove access plug for weight..................   11
    (3) Remove holding nuts for air chamber............   48
    (4) Install lifting tool and remove air chamber....   409

(c) Remove diaphragm.
    (1) Remove cotter pin through lower depth spring
        socket nut.....................................   72
    (2) Remove nut on lower depth spring socket........   407,461
    (3) Remove diaphragm plate (pry off with screw
        drivers).
    (4) Remove diaphragm.
    (5) Remove 3 holding screws and the diaphragm ring..   40

(d) Remove pendulum lever and diaphragm lever shaft.
    (1) Remove cotter pin and pin for pendulum link to
        pendulum lever.................................   72
    (2) Remove keep screw for diaphragm lever shaft....   41
    (3) Remove pendulum lever and diaphragm lever shaft
        assembled.
        NOTE: Insert a soft brass rod in back of
              pendulum lever close to its hub and tap
              out the shaft. Great care must be exer-
              cised in removal of this shaft not to bend
              the pendulum lever.

(e) Remove depth spring and sockets assembled and
    diaphragm lever.
    (1) Turn depth spring adjusting spindle until
        thread on adjusting screw disengages threads
        on upper depth spring socket...................   180
    (2) Remove depth spring with sockets and diaphragm
        lever assembled.
    (3) Remove cotter pin, washer and pin for diaphragm
        lever in depth spring socket...................   72

(f) Remove linkage from pendulum to valve.
    (1) Remove cotter pin for tension rod bearing on
        pin..   72
    (2) Slip pendulum tension rod assembly clear of its
        pivot on pendulum.
    (3) Remove wire and holding screw for valve lever
        bracket..   72,49
    (4) Remove valve lever bracket with valve lever and
        pendulum tension rod assembly.

(g) Remove after section of pendulum.
    (1) Lock gyro spinning mechanism...................   208
    (2) Remove 2 screws holding after section of
        pendulum.......................................   39
    (3) Remove after section of pendulum.

TOOL NO.

(h) Remove depth setting spindle and socket.
   (1) Remove cotter pin and nut........................180,72,155
   (2) Lift out depth setting spindle and remove pinion gear.

(i) Remove pendulum.
   (1) Remove cotter pins and taper pins for knife edges.. 72,166
   (2) Remove knife edges, pushing through from the outside.
   (3) Lift out pendulum.
   (4) Remove depth spring adjusting screw and idler gear.
     NOTE: The removal of parts preceded by an asterisk are necessary only to make room for removal of pendulum. They are not otherwise associated with "disassembly of immersion mechanism".

*(j) Remove steering engine valve rockshaft linkage.
   (1) Remove 2 screws for bearing cap............... 41
   (2) Remove rockshaft assembly from bearing.
   (3) Remove spring buttons and springs under rockshaft bearings.

*(k) Remove pallet mechanism driving gear shaft and bearing.
   (1) Remove 6 holding screws....................... 41
   (2) Remove driving gear bearing cap.
   (3) Remove pallet driving gear and shaft.

*(l) Remove driving gear bracket.
   (1) Remove hexagonal nut.......................... 48
   (2) Remove screw with hexagonal head.............. 48
   (3) Remove holding screw.......................... 41
   (4) Remove driving gear bracket.

*(m) Remove clamp for gyro spin pipe nipple.
   (1) Remove 2 holding screws....................... 40
   (2) Remove clamp.

*(n) Remove gyro spinning rotor.
   (1) Hold spinning shaft gear with a special spanner, pins of spanner meshing across teeth of spinning gear...(126 special)
   (2) Insert pins of tool #25 in holes in rotor and unscrew rotor from spinning shaft............... 25
   (3) Remove bronze spacing washer.
   (4) Unlock gyro spinning mechanism.

*(o) Remove gyro spinning mechanism frame and front plate assembled.
   (1) Remove 6 holding screws for gyro spinning mechanism....................................... 41,233
   (2) Remove gyro spinning mechanism.

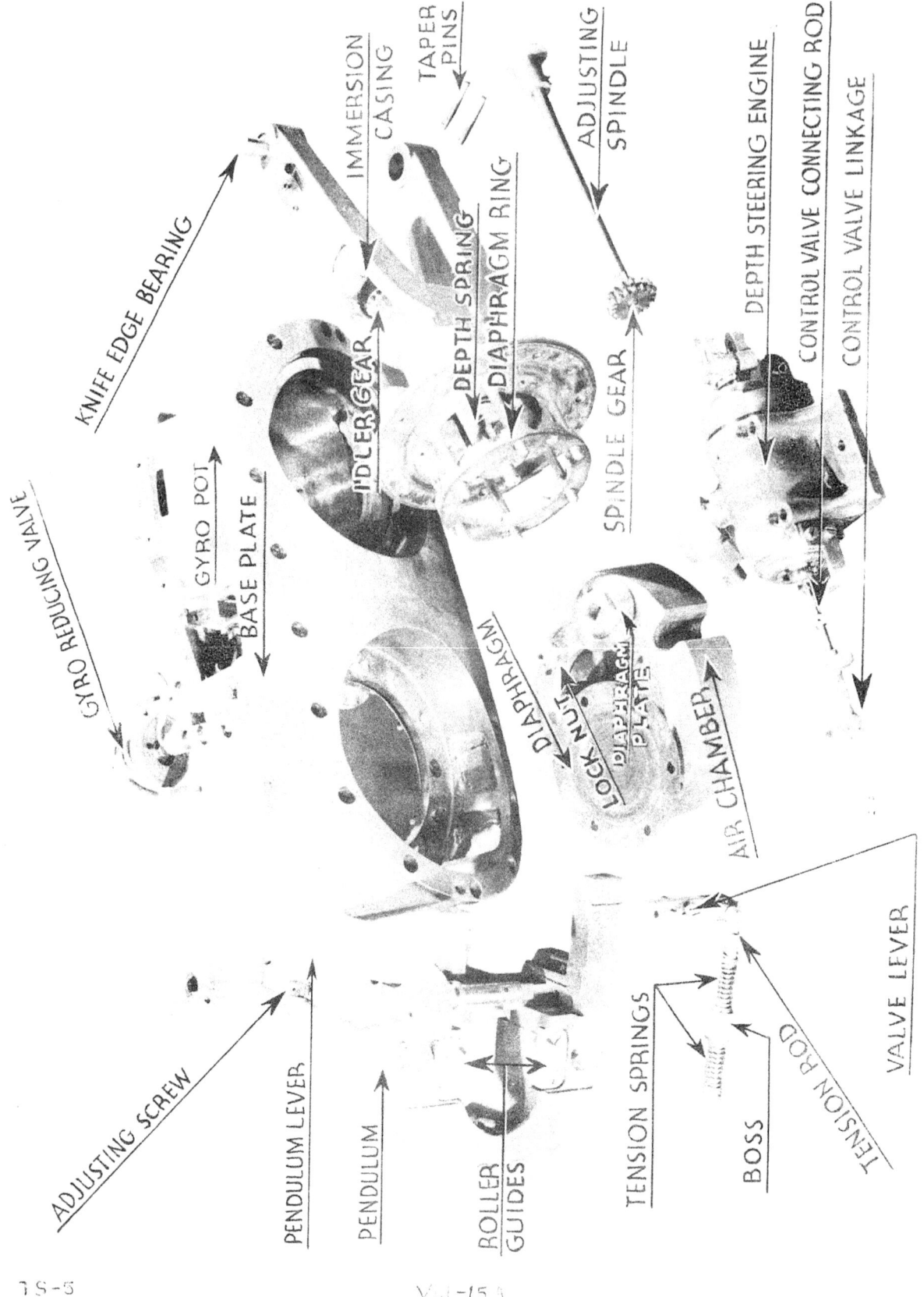

TOOL NO.

(p) Remove immersion gear casing.
    (1) Remove 10 holding nuts.......................... 48
    (2) Remove immersion gear casing.
    (3) Remove gasket for immersion gear base.

8051. *Overhaul and Assembly of Immersion mechanism.*

(a) Check alignment of knife edges on immersion casing.
    (1) Temporarily install knife edges on arms of immersion casing. Secure with taper and cotter pins.
    (2) Extend a line through the center of immersion casing. A line through center of hole for transportation screw and pivot for roller on opposite side is in the center. Scribe marks on edge of casing on this line.

(b) Replace immersion gear casing.
    (1) Inspect rollers on immersion casing. Note for freedom of movement. Remove all dirt and oil lightly.
    (2) Clean old gasket and rough spots from seat for gasket.
    (3) Replace a good gasket over studs on seat.
    (4) Replace immersion casing and secure with 10 holding nuts................................... 48
    (5) Replace adjusting screw and idler gear in immersion casing. Note that thin side of idler gear is facing against casing.
    (6) Level base.
    (7) Place hand pointer gauges on each knife edge and note if pointer comes to rest on scribe mark. If not, knife edges are out of line and realignment will be necessary........................ WE186

(c) Install gyro spinning mechanism.
    (1) Replace spinning mechanism in unlocked position and secure with 6 screws....................... 41,233
    (2) Replace bronze spacer washer on spinning shaft.
    (3) Replace rotor.
    (4) Lock mechanism.................................. 205
    (5) Insert pins of tool #25 in holes in rotor and hold.
    (6) Insert tool #126(special) in teeth of spinning gear and tighten............................... 126
    NOTE: Make sure bronze washer is in place as rotor cannot be drawn up flush.

(d) Replace pendulum.
    (1) Inspect roller guides and remove any rough spots or burrs.
    (2) Clean and inspect pendulum arms for alignment.

	TOOL NO.

Check alignment of bushings in pendulum arms; insert aligning bar through these bushings. The bar should slide freely when passed through bushings. If not, it will be necessary to bend pendulum arms back in line until freedom of movement is obtained.......................... WE184

    (3) Note that lead weight is secure and place pendulum in position around immersion casing.
        NOTE: For assembly of steps preceded by an asterisk, see NOTE in Article 8050, step (i).

(e) Replace steering engine valve rockshaft linkage and holding clamp for air impulse nipple.
    (1) Replace springs and spring buttons under rockshaft bearings.
    (2) Replace rockshaft assembly on bearing.
    (3) Replace 2 screws for bearing cap............... 41
    (4) Replace holding clamp for impulse nipple.
    (5) Replace 2 screws............................... 41

(f) Replace after section of pendulum.
    (1) Clean, inspect and replace after section of pendulum.
    (2) Secure after section of pendulum with 2 holding screws... 41

(g) Replace knife edges in bearings.
    (1) See that knife edges are free from burrs. Insert in bearings, pushing through from inside immersion casing arms. Secure with taper and cotter pins.

(h) Replace retainer plates.
    (1) Inspect retainer plates and smooth down any burrs that may be found. Clean up that section of pendulum arms on which these plates seat.
    (2) These retainer plates are not interchangeable. Replace in proper order and set up evenly on holding screws................................. 41
    (3) With feelers measure clearance between knife edges and retainer plates. This clearance should be "006 when pendulum is swung to any position.

(i) Check oscillations of pendulum.
    (1) See that pendulum has small lateral clearance.
    (2) Level housing on work bench and measure clearance between rollers and guides. This clearance should be "018.
    (3) Bring pendulum against one of the stops of

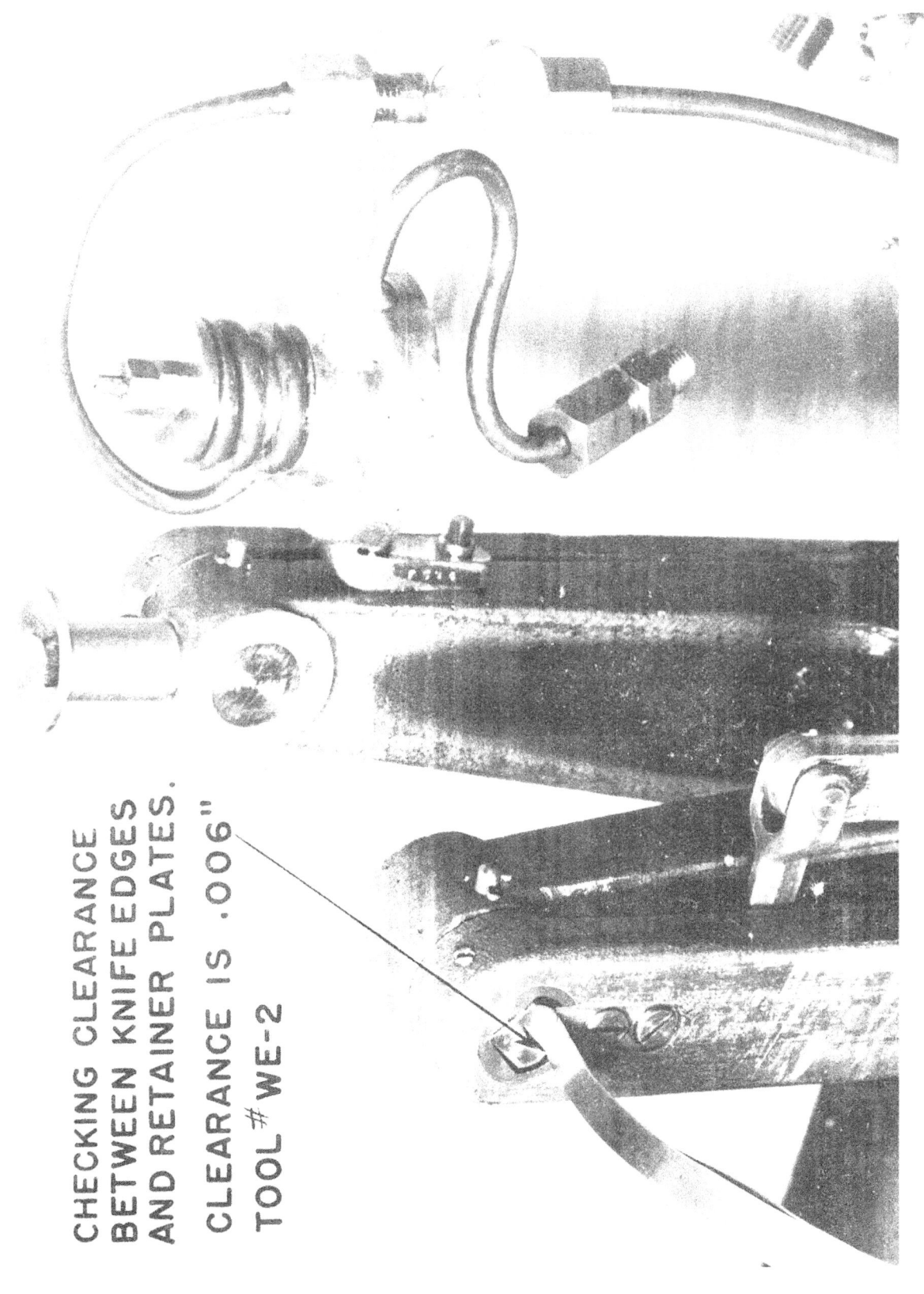

CHECKING CLEARANCE BETWEEN KNIFE EDGES AND RETAINER PLATES. CLEARANCE IS .006" TOOL #WE-2

WITH HOUSING LEVEL — CHECK CLEARANCE BETWEEN ROLLERS AND GUIDE
CLEARANCE SHOULD BE .018"
TOOL #WE-2

TOOL NO.

immersion casing. Release pendulum; it should make at least 16 half oscillations before coming to rest. Failure to pass this test means:
  (a) Arms of immersion casing out of alignment.
  (b) Improper clearance between knife edges and retainer plates.
  (c) Loose pins in knife edge bearings, permitting them to rotate slightly.
  (d) Dirt or excessive oil around or under pendulum.
  (e) Knife edge bearings or bushings burred.

(j) Measure clearance between pendulum and its stops.
  (1) With transportation screw in, take clearance between pendulum and stops on immersion casing (both sides). This clearance should be "234 for a Mark 13-1 torpedo.

(k) Replace depth setting spindle and socket.
  (1) Insert depth setting spindle and replace pinion gear.
  (2) Replace nut and cotter pin..................180,72,155

(l) Replace depth spring, diaphragm lever and sockets assembled.
  (1) Remove and inspect diaphragm lever for true. Free end of diaphragm lever should fit snugly in pins of inspection fixture. Replace diaphragm lever in slot of lower spring socket. Secure with cotter pin...................... WE175
  (2) Inspect depth spring for signs of corrosion, flaking or damage of plating. Note that sockets and spring are firmly sweated. Turn adjusting spindle when inserting, thus screwing the adjusting screw into upper socket until square hole in diaphragm lever lines up with bearing hole for diaphragm shaft.

(m) Replace pendulum lever and diaphragm lever shaft.
  (1) Remove pendulum lever from diaphragm lever shaft and check for true on inspection fixture... WE175
  (2) Replace lever on shaft and secure with taper pin.
  (3) Note that diaphragm lever shaft is not burred or scratched. Insert shaft in bearing hole, guiding through square hole in after end of diaphragm lever. Set up on keep screw that keeps shaft from backing out. Note that this screw does not bind on shaft.

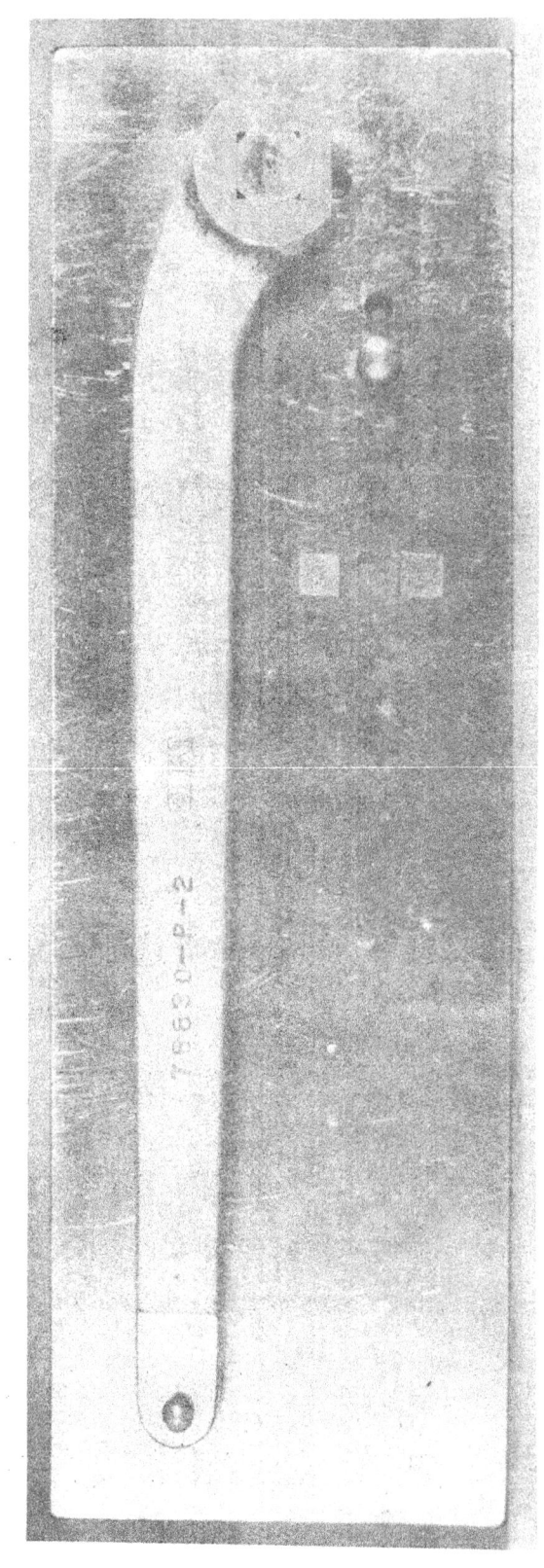

TS-5   VIII-18A

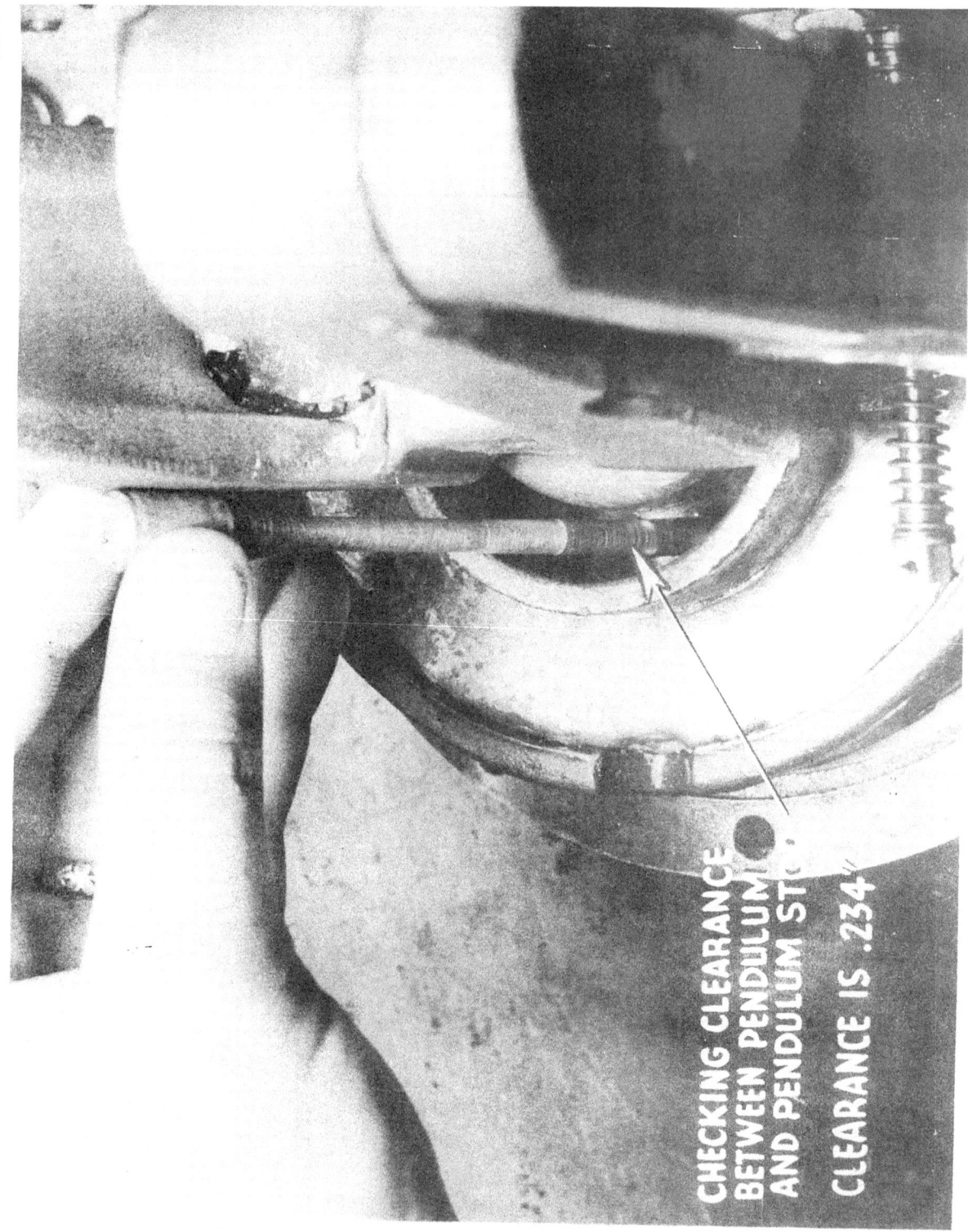

CHECKING CLEARANCE BETWEEN PENDULUM AND PENDULUM STOP. CLEARANCE IS .234"

TOOL NO.

    (4) Connect horizontal pendulum link from left pendulum arm to top of pendulum lever. This link should not hang or bind on end of pendulum lever but should drop over the end freely. Secure connection with cotter pin.

(n) Replace linkage from pendulum arm to valve lever.
    (1) Slip pendulum tension rod over its pivot on pendulum arm.
    (2) Replace valve lever bracket with valve lever.
    (3) Replace wire holding screw for valve lever bracket.
    (4) Replace cotter pin for tension rod on pivot bearing.

(o) Replace diaphragm ring.
    (1) Install diaphragm ring on its seat with projecting lip in line with free end of diaphragm lever. Secure with 3 holding screws.................... 41
    (2) Take all tension off depth spring, and swing pendulum aft, positioning free end of diaphragm lever towards its upper stop. Measure clearance between lever and its upper stop. Clearance should be not less than ".015.
    (3) Swing pendulum forward, positioning free end of diaphragm lever toward its lower stop. Clearance between end of lever and its lower stop should be not less than ".015.
    (4) Clearance between the end of diaphragm lever and its stops depends on the thickness of the lever at its free end, which should be ".320. This clearance can be equalized by the adjusting arm provided on the left pendulum arm.

(p) Replace diaphragm.
    (1) Install diaphragm with bulged side facing out.
    (2) Clean, inspect and replace diaphragm plate.
    (3) Replace holding nut and cotter pin........461,407,72

(q) Replace atmospheric chamber.
    (1) Replace chamber over diaphragm.
    (2) Secure with holding nuts..................... 48
    (3) Test atmospheric chamber for leaks.

8052. Depth Engine Leakage Test.
Adjust stand as follows:
    (1) Connect test restriction, ".018, to end of air line on test stand.
    (2) Open needle valve until #2 gauge registers 400 lbs. Note reading on #1 gauge, which should be 800 lbs. If this reading is not obtained, disassemble restriction in test stand and clean thoroughly, as water or oil frequently obstructs its proper functioning. It is important that restriction be kept clean and free from water and oil.

SWING PENDULUM FORWARD POSITIONING FREE END OF DIAPHRAGM LEVER TOWARD ITS LOWER SIDE. CLEARANCE BETWEEN END OF LEVER AND ITS LOWER STOP SHOULD BE NOT LESS THAN .015" TOOL #VE-2.

PENDULUM AFT. DIAPHRAGM LEVER TOWARDS ITS UPPER STOP. CLEARANCE BETWEEN END OF LEVER & UPPER STOP OF NOT LESS THAN .015" TOOL #VE-2.

TEST OF AIR CHAMBER
AIR CHAMBER MUST HOLD
15 POUNDS PRESSURE
FOR 5 MINUTES

TS-5 VIII-19B

(3) Having ascertained that stand is in proper working order, remove test restriction, install depth engine in stand and connect air load to nipple.

(4) Open needle valve until #2 gauge maintains a steady pressure of 400 lbs. at any position of piston. Note pressure on #1 gauge. If this pressure exceeds 1200 lbs. depth engine should be considered as unfit for further use until overhauled and worn parts replaced.

(5) The following table gives the readings on gauges of leakage test stand for fair, standard, and excellent engines:

	#1 Gauge	#2 Gauge
Fair engine............	1200	400
Standard engine........	1000	400
Excellent engine.......	800	400

8053. Test for tightness of packing.
(1) Turn off air.
(2) Place push balance on end of piston fork. Note on scale at what pressure the piston moves in the cylinder. This should not be over 18 lbs.

8054. Test for sensitivity of depth engine.
(1) Turn on air.
(2) Place feather spring (tool #222) against knurled nut on valve connecting rod and move control valve. Note at what reading on scale valve moves. With 400 lbs. air on line, this should be from $\frac{1}{4}$ to $\frac{1}{2}$ ounce. Good engines move at $\frac{1}{4}$ ounce.

8055. Center depth engine piston.
(1) Clean and oil, place in holding fixture.
(2) Measure thickness of control valve stop. This should be ".094.
(3) Connect nipple on engine to low pressure air. Turn on air.
(4) Test with oil around cylinder head, packing glands and around where valve connecting rod passes through stop plug.
(5) Pull valve connecting rod out against control valve stop. Shut off air supply to engine and allow all air to bleed from engine. Push piston in against stuffing box, noting amount of piston travel. This clearance should be about 1/64" which insures that piston will not stick at the end of its stroke.
(6) Turn on air to engine. Move valve connecting rod in against control valve stop. Shut off air supply

DEPTH ENGINE IN TEST STAND

TS-5   VIII-20 A

TOOL NO.

to engine and allow all air to bleed from engine. Pull piston out against cylinder head, noting amount of travel. This should be about 1/64".

(7) If the piston clearances are equal and approximately 1/64" on either end, the piston is centered in the cylinder. Otherwise the control valve stop should be adjusted in such direction as to make these clearances equal.

8056. Center depth engine valve with connecting linkage.
(1) Install depth engine on gyro pot and connect linkage of valve lever to valve connecting rod....49,246
(2) Install transportation screw........................ 49
(3) Place housing on workbench and connect low pressure air line to depth engine nipple.............. 141
(4) Turn on air and note if scribe mark on valve connecting rod is aligned with scribe on control valve stop. If not, loosen clamp screw in eye connection and turn knurled nut of valve connecting rod until scribes are aligned. After making adjustment, leave the hole in knurled nut in a horizontal position. Set up on clamp screw.................. 246

8057. Calibrate depth spring.
(1) Starting with housing leveled on workbench and air turned on depth engine, scribe marks of control valve stop and valve connecting rod should be aligned.
(2) Install screw hook and 16 lb. weight in lower spring socket. The weight moves pendulum forward and control valve aft.
(3) Set tension on depth spring, thus moving pendulum aft and control valve forward until scribe marks are again aligned. Oscillate pendulum and note that it comes to rest with scribe marks aligned. This calibrates depth spring for a 10 ft. setting.
(4) With depth spring thus calibrated and weight in place, a tilting test should be given as a further check on smoothness of operation of the depth engine valve and its connecting linkage. The same results should be expected as obtained in Article 8058.

8058. Make tilting test.
(1) Remove transportation screw..................... 49
(2) Verify all tension off depth spring.............. 180
(3) Level housing on workbench.
(4) Turn air on depth engine. The scribe marks of control valve stop and valve connecting rod should now be in line. Swing pendulum against one of its stops and let go. Pendulum should make at least 5 to 6 half swings before coming to rest and should stop with scribe marks aligned.
(5) Offset bevel protractor 2 degrees and place on base of housing. Pendulum and control valve should move at $\frac{1}{2}°$ tilt in either direction.

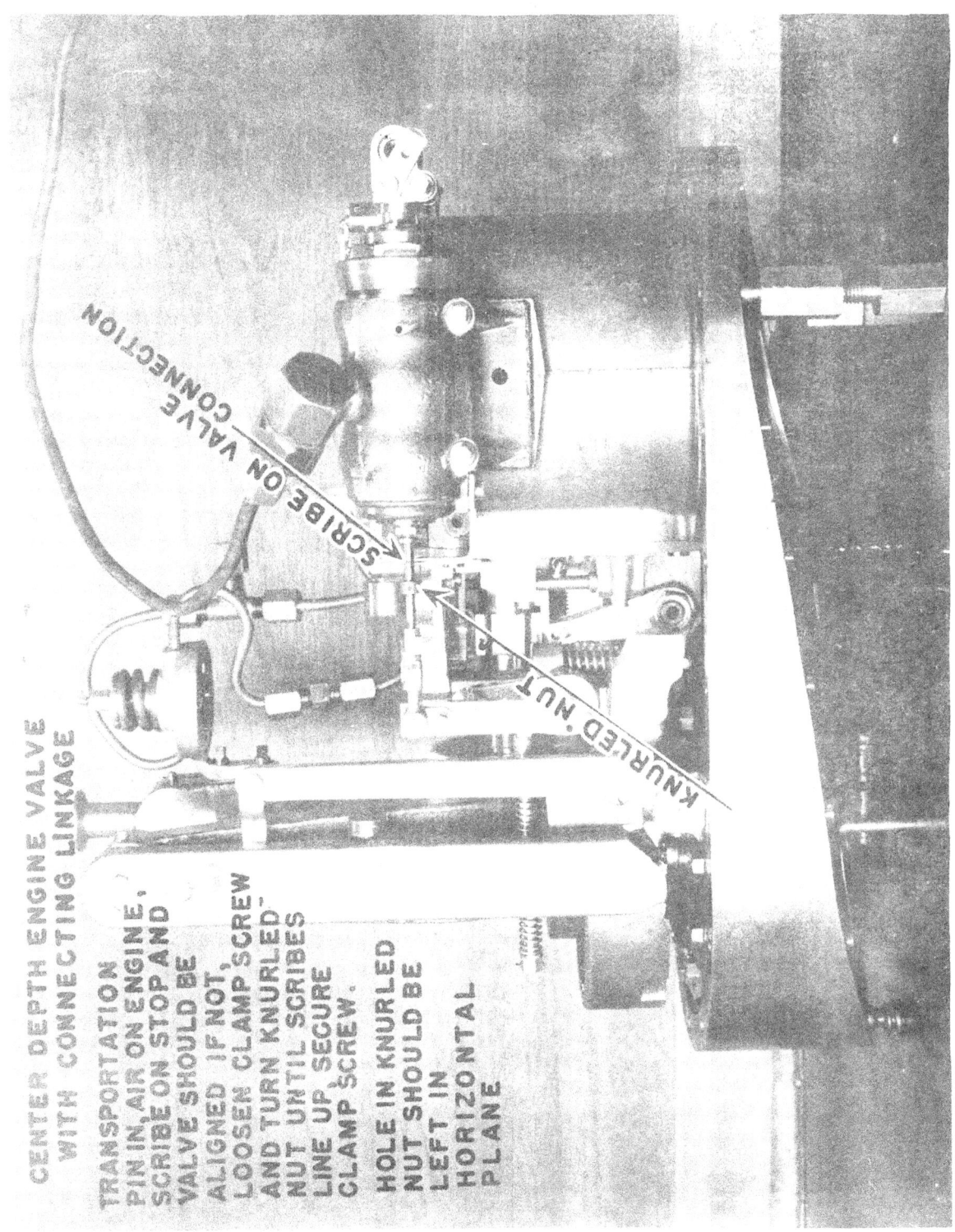

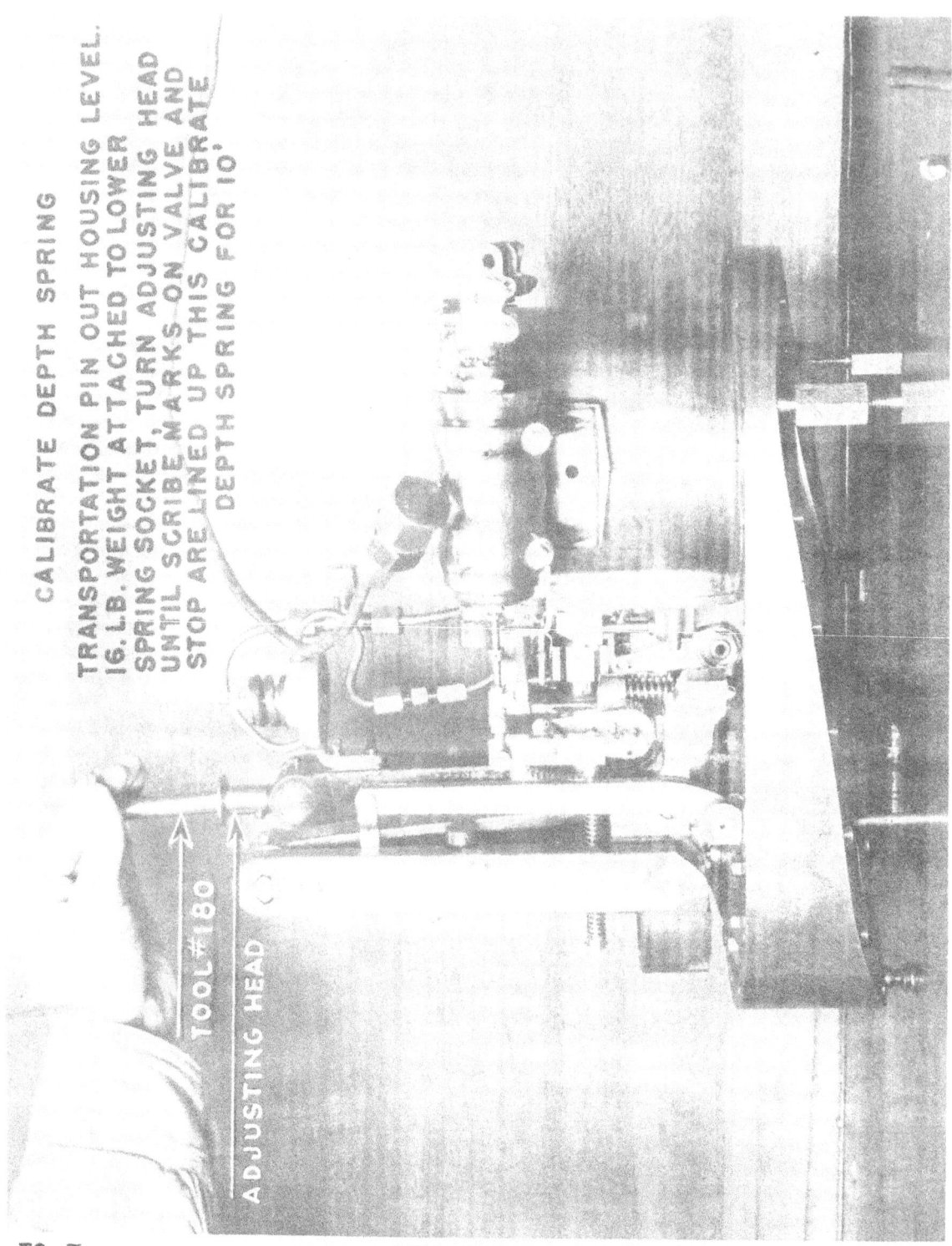

CALIBRATE DEPTH SPRING

TRANSPORTATION PIN OUT HOUSING LEVEL. 16 LB. WEIGHT ATTACHED TO LOWER SPRING SOCKET, TURN ADJUSTING HEAD UNTIL SCRIBE MARKS ON VALVE AND STOP ARE LINED UP THIS CALIBRATE DEPTH SPRING FOR 10'

TOOL #180

ADJUSTING HEAD

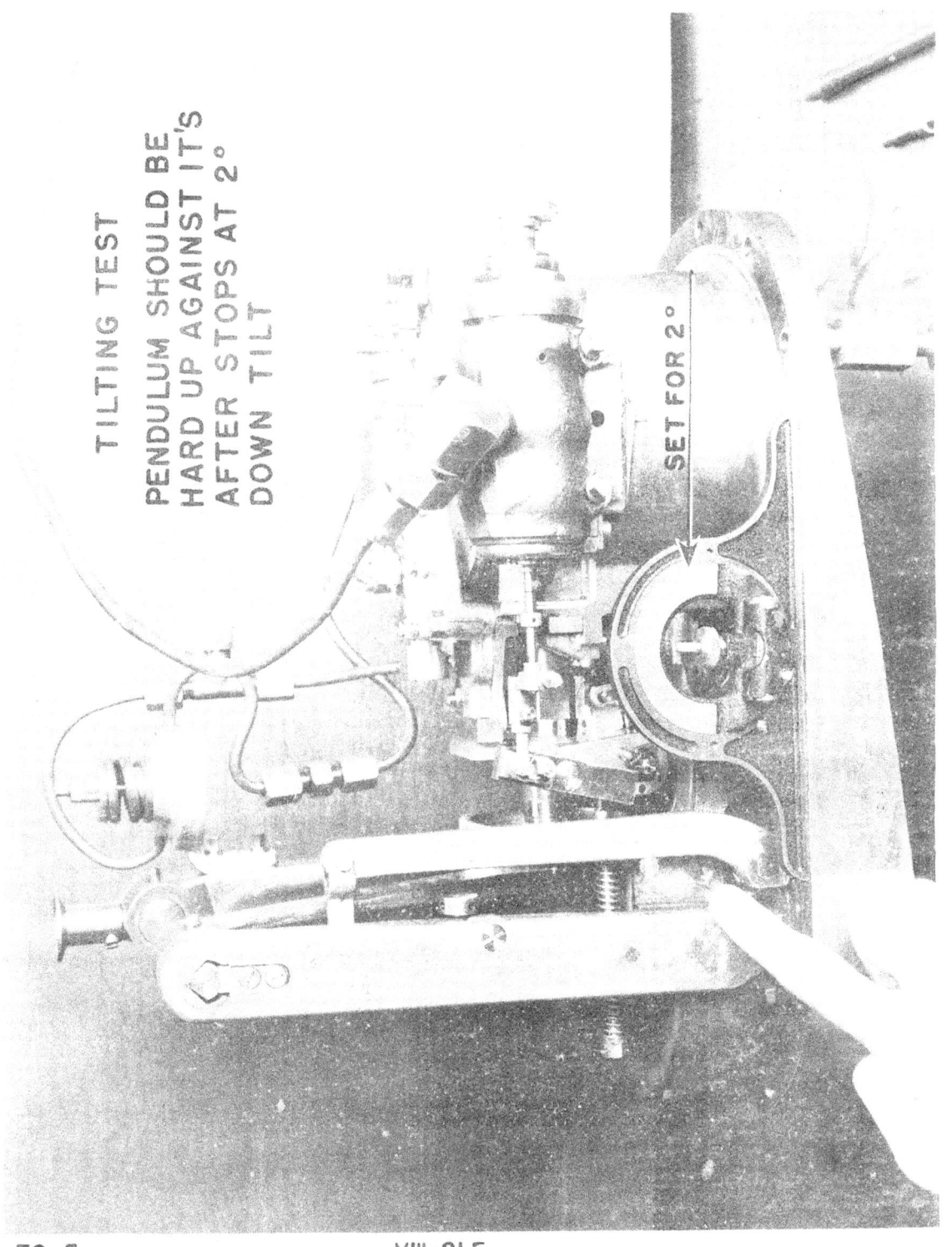

TILTING TEST

PENDULUM SHOULD BE HARD UP AGAINST IT'S AFTER STOPS AT 2° DOWN TILT

SET FOR 2°

TILTING TILT

PENDULUM SHOULD BE HARD UP AGAINST IT'S FORWARD STOP AT 2° UP TILT

TS-5 VIII-21 F

(6) Depress housing until bubble of bevel protractor is showing in center. Note movement of pendulum and valve for smoothness. Pendulum should be all the way forward and firmly contacting its after stop on immersion casing.

(7) Bring housing back to a level position and reverse bevel protractor.

(8) Elevate housing 2 degrees. Note movement of pendulum and valve for smoothness. Pendulum should be all the way aft, firmly contacting its forward stop on immersion casing.

8059. Hydraulic test of depth mechanism.

The hydraulic test is for the purpose of testing the hydrostatic control system for sensitivity. It accomplishes this purpose by permitting hydrostatic pressure to be applied on the diaphragm in the same manner as it would be applied with the torpedo running in the water and noting the corresponding movement of the depth engine valve. It will therefore be apparent that this test will likewise be a test for the sensitivity of the rubber diaphragm and of the various joints for tightness as well as of the general operation of the intermediate control rods and levers up to and including the steering engine valve itself.

(1) With depth and gyro engines removed from gyro pot, mount housing in test stand over a good gasket.
(2) Secure housing to stand with holding screws.
(3) Verify that gyro clamp plate is set up tight.
(4) Install replacement screw.
(5) Secure test stand cover in place over a gasket.
(6) Rotate stand to opposite side, thus placing housing in the upright position.
(7) Install depth engine on gyro pot and connect linkage to valve lever.
(8) Remove central plug in test stand cover.
(9) Remove atmospheric chamber plug.
(10) Install screw hook and 16 lb. weight in lower spring socket.
(11) Verify that stand and housing are level.
(12) Connect air line (400 lbs.) to nipple of depth engine.
(13) Turn on air.
(14) Set tension on depth spring until scribe on valve connecting rod comes in line with scribe on control valve stop.
(15) Remove screw hook and 16 lb. weight.
(16) Replace atmospheric chamber plug and central plug in test stand cover. Note that washers are in place under both these plugs.
(17) Connect water pipe from test tank to central nipple on test stand cover.
(18) Open drain cock in water line and pump water into testing stand until it emerges from drain cock, thus expelling all air from test stand.
(19) Close drain cock.

THE DEPTH ENGINE CONTROL VALVE HAS STARTED TO MOVE AT 3 FEET PRESSURE AS SHOWN ON GAUGE

CONTROL VALVE SCRIBE MARK AND SCRIBE MARK ON CONTROL VALVE STOP SHOULD BE IN ALIGNMENT WHEN PRESSURE REGISTERS 10 TO 11 FEET ON GAUGE

WITH PRESSURE BUILT UP TO 25 FEET AND PUMP HANDLE RELEASED, NOTE IF PRESSURE ON GAUGE HOLDS. IF NOT, LOOK FOR LEAKS.

(20) Start pumping water into test stand. As soon as pressure comes on the one, one observer should watch the hydrostatic gauge and another the depth engine valve. With depth setting of 10 feet already set on the mechanism, the depth engine valve should start to move when the hydraulic pressure registers for 2 to 3 feet. As the pressure is gradually increased, the valve should travel smoothly and continuously until it reaches its central position at about 10 to 11 feet gauge. Still continuing to build up the pressure, the valve should reach the position of full throw when the gauge reads from 18 to 20 feet, the travel meanwhile being continuous and steady, not in jerks.

(21) The pressure is now built up to 25 feet and the pump handle released. Note if pressure on gauge holds; if not look for leaks at the following places:
  (a) Depth spring adjusting screw.
  (b) Diaphragm shaft.
  (c) Around roller pivots.
  (d) Between immersion casing base and mechanism base.
  (e) Around replacement screw.
  (f) Around clamp plate of gyro door and gasket.
  (g) Around locking index spindle.

(22) The test is now continued, but with a falling pressure. The pressure is permitted to fall by turning the two-way valve on the test stand so that the pressure will decrease very slowly. With decreasing pressure, the valve should leave its full throw position at from 18 to 20 feet gauge, reach its central position at from 10 to 11 feet gauge, and cease its movement when the gauge reads from 2 to 3 feet.

Three successive tests of the mechanism should be made. Slight leaks which may not be apparent during a single test are sometimes indicated by the necessity for increased pressures on succeeding tests.

When the test has been completed, the water is drained from the testing stand housing by means of the two-way valve. The lines are disconnected and housing rotated vertically to the other side. The mechanism is now ready to be removed from the test stand.

N. **INSTRUCTIONS FOR COMPLETE OVERHAUL AND REPLACEMENT OF PARTS IN DEPTH ENGINE.**

                                                                                        TOOL NO.

8060. To disassemble depth engine:
  (a) Place depth engine in vise jaws and clamp in bench vise.. 248
  (b) Remove valve stop.. 41
  (c) Unscrew valve stop plug and remove valve from piston... 159
  (d) Unscrew depth engine packing glands about a turn... 169

TOOL NO.

    (e) Remove forked connection on after end of piston rod by slacking up on clamp screw and wedging screw driver in clamping slot until threads on connection turn freely.............. 41,49
    (f) Unscrew cylinder head and remove............... 377
    (g) Push piston out of cylinder and place where lapped surface will be protected against marring.

8061. Inspection of parts for wear, etc.
    (a) Wash cylinder with spirits and note if any scratches or small scores appear on inside walls.
    (b) Examine piston rods for burrs or scratches. Examine piston rings for wear; piston rings should show even wear throughout their convex surface. There should be no tendency for rings to bind or stick in grooves, butt ends should be free with no noticeable clearance.
    (c) Observe butt clearance by inserting piston with rings in cylinder; notice rings as they enter cylinder; if butt clearance is noticeable by eye the rings are worn too much and should be replaced.
    (d) Examine valve hole for burrs or scratches, particularly around annular air grooves for valve.
    (e) Examine valve for scratches and burrs, also for alignment.
    (f) Check up on condition of packing and washers.
    (g) Check up on condition of air inlet nipple.

8062. To renew depth engine valve:
    (a) Adjust male lap to size of hole in piston....... WE147
    NOTE: In making this adjustment, great care must be taken not to mar the end of lap. Using a piece of copper stock having a 3/16" hole, line end of lap up in this hole and tap lap handle lightly with small bronze hammer, trying lap in hole for valve until final adjustment is obtained.
    (b) If any burrs or scratches are found in hole of piston, use a small quantity of fine grinding compound, applying evenly over surface of lap; in lapping, move lap with reciprocating motion over total length of hole to obtain a uniform diameter throughout its length. Do not lap more than necessary to remove burrs and scratches. Carefully wash lapping material off lap and out of hole with spirits, blow out carefully with low pressure air jet and finish lapping with oil until hole shows a well polished surface.

	TOOL NO.

(c) With micrometer calipers, measure diameter of lap and pick out a spare valve about "0025 larger than this diameter.................... WE5
NOTE: As valves are furnished to the service assembled in groups, it will be necessary to disassemble and remove stop plug to facilitate lapping.

(d) Disassemble valve and connection by placing valve in resting block, with pin lined up with hole in same and driving pin out with 1/16" drift.

(e) Remove stop plug and replace valve on connection.

(f) Apply lapping compound sparingly on surfaces to be lapped and insert valve in female lap; regulate adjusting screws in lap for proper friction and proceed to lap, moving valve in lap with a turning and reciprocating motion throughout its length........................ WE148

(g) Remove, examine and measure diameter occasionally during lapping, change to finest grade of lapping compound when valve measures "0005 large and lap down to size.

(h) Wash compound from lap and valve with spirits, and finish lapping with oil; try valve in piston occasionally when finished lapping. The valve must be as perfect a fit in piston as is possible to get without any appreciable friction when finished.

(i) Remove pin from valve and connection and replace stop plug.

(j) Replace valve on connection.

8063. To renew depth engine cylinder:
(a) Lap out any scratches or burrs found in depth engine cylinder with male lap, using lapping compound only if necessary, and finish lapping with oil. Do not lap any more than is necessary to remove scratches or burrs............. WE140

8064. To renew depth engine piston:
(a) Lap out any scratches or burrs found on piston rods, using female lap. Use lapping compound sparingly as necessary........................ WE139

8065. To renew depth engine piston rings:
(a) Adjust male lap to fit depth engine cylinder, after which measure diameter of this lap and readjust to "005 over diameter obtained for cylinder................................... WE140,WE5
(b) Insert male lap in female lap and adjust female lap to that size..................... WE141

TOOL NO.

(c) See that butt ends of piston ring are free from burrs, fit rings in female lap, selecting as far as possible rings that will fit without dressing down to the butt ends. When necessary to remove metal on butt ends to fit rings in lap, use a small flat file, and finish with lapping compound between butt ends, lapping both together.

(d) Measure width of rings (the average ring before fitting will measure about ".100)....... WE 5

(e) Slip ring over lap holder and rub down to a width of ".096 against a piece of No. 0 emery cloth laid on a smooth and level surface...... WE142

(f) Rub down to ".095 on the smooth side of combination oil stone, thus leaving all rings ".001 oversize in width.

(g) Place all rings in female lap................ WE141

(h) Remove clamp bolt, plate and bushing on lap plug holder................................... WE138

(i) Insert lap plug holder through piston rings in female lap, slip bushing and plate back over end of lap and secure tightly with clamp bolt.

(j) Remove lap holder with rings and examine all rings to see that butt ends are closed together.

(k) Proceed to lap rings, using the medium fine grade of compound at first and gradually tightening up on a lap until within ".001 of the size of cylinder, then change to fine grade of lapping compound and finish lapping with clear oil.

(l) Remove occasionally during lapping and note if total convex surface of each ring contacts with lap; should there by any doubt about any rings not lapping out, when lapped to size, it is best to replace such ring before proceeding any further as the whole set must be finish-lapped together.

(m) After having lapped the piston rings to size of cylinder, wash clean with spirits, coat with oil and try plug holder with rings in cylinder; if too snug a fit, do not attempt to force into cylinder, but replace in female lap and continue to lap and try again until fit is obtained.

(n) Remove rings from lapping plug holder and proceed to fit in grooves in piston. Rings having been previously lapped to ".001 oversize in width, slip ring over lap holder tool and lap against smooth side of combination oil stone, trying rings in groove until a snug fit without binding is obtained.................................... WE142

(o) Wash pistons and rings thoroughly in spirits and blow out with air and coat lightly with oil.

8066. To assemble depth engine.   TOOL NO.

   (a) Wash all parts thoroughly in spirits, blow
       out with air and give a light coating of oil.
   (b) Remove old packing and clean stuffing boxes.
   (c) Place butts of piston rings opposite each other
       and insert piston in cylinder, being careful
       not to mar in entering rings in cylinder.
   (d) Place cylinder in vise jaws and place in bench
       vise.. 248

       NOTE: In securing depth engine cylinder in vise
       jaws and bench vise, extreme care should be
       exercised to see that inside edges of slot cut
       in wake of depth engine nipple on vise jaws do
       not take up on boss for nipple on cylinder.
       If this slot requires removal of metal to obtain
       necessary clearance, a round file will be found
       most convenient to use. Make sure that enough
       metal is being removed to obtain sufficient
       clearance to prevent crushing of the cylinder
       when securing in vise.

   (e) Inspect condition of cylinder head gasket and
       renew same if necessary.
   (f) Before clamping cylinder in vise, rotate in clamp
       in the direction it is intended to use the wrench
       until all play is out. This will allow the boss
       for nipple on cylinder to take the strain of
       turning and prevent cylinder from turning in
       the clamp with possible damage to the clamping
       groove.
   (g) Replace cylinder head......................... 377
   (h) Repack piston rod stuffing boxes being careful
       not to mar piston rods by using any sharp tools
       for pushing packing into place.
       NOTE: 3/32" round packing should be cut in 11"
       lengths and kept submerged in a jar containing
       600-W, until used.
   (i) Replace forked connection on piston rod, screwing
       same all the way in and clamp.
   (j) Tighten up on packing glands gradually, using
       spanner wrench; turn and push piston back and
       forth by hand to work packing in around rods... 169
   (k) Insert depth engine valve in piston and set up
       on stop plug.................................... 159
   (l) Center piston in cylinder..................... WE 8
   (m) Center valve in piston.
   (n) Install and adjust valve stop so that centerline
       on stop will line up with centerline on valve
       connection.
   (o) Center engine as described in Article 8055.
   (p) Test for sensitivity as described in Article
       8054.

8067. Depth engine valves are furnished in sizes from #252 to #258. Pistons having a valve hole larger than #258 should be discarded and the engine surveyed for overhaul at a torpedo station.

CHAPTER NINE

		ARTICLES
A.	General Description, Gyro Mechanism	9001-9010
B.	Spinning and Unlocking Mechanism	9011-9027
C.	Disassembly, Overhaul and Assembly of Spinning and Unlocking Mechanism	9028-9032
D.	Pallet Mechanism	9033-9052
E.	Top Plate Centering Device	9053-9055
F.	Steering Engine	9056-9062
G.	Steering Line	9063-9065
H.	Gyro Reducing Valve	9066-9069
I.	Test, Disassembly and Assembly of Steering Engine	9070-9072
J.	Disassembly and Assembly of Pallet Mechanism	9073-9077
K.	Test of Gyro Reducing Valve	9078

A. **GENERAL DESCRIPTION, GYRO MECHANISM.**

9001. The gyro mechanism is the housing in which the gyroscope is contained and to which the impulse mechanism, together with the mechanism necessary for relaying gyrostatic control of the torpedo, is assembled.

9002. In all present service torpedoes, the mechanism together with the immersion and depth control mechanism, is assembled on one large base, which may be removed or replaced in the afterbodies of torpedoes without affecting the adjustments of the individual mechanisms assembled thereon. The depth and vertical steering engines must be removed, however.

9003. The gyro pot is a large cylinder, the lower end of which is seated in a circular recess in the after end of the base, and opening from the outside for the insertion of the gyro. The upper end of this cylinder is closed by a bronze top plate thus forming a housing for the gyro when installed in a torpedo.

9004. In the gyro mechanism used in this torpedo, the pot is a fixed part of the base, being rigidly secured to same by screws and solder and suitably machined with bosses and bearing surfaces on its outer cylindrical wall for the installation of the gyro impulse mechanism on the forward side, and the depth and steering engines on the port and starboard sides. The top plate on which the pallet mechanism is assembled, is held against a lapped surface on the top of the gyro pot with retainer plates and is capable of a small movement ($24°$) in azimuth for the purpose of centering the axis of the gyro gear with the axis of the torpedo. No provisions are made for angle firing in this torpedo.

GYRO MECHANISM BASE.

9005. The gyro mechanism base is a bronze casting of suitable dimensions for mounting the depth and gyro mechanisms; the outer side of the base is machined to conform with the contours of the afterbody, the inner side being machined with a seat to fit against a gasket in the gyro door flange on the afterbody shell to which it is secured with screws. A circular opening in the forward end is machined with a flange for mounting the depth mechanism, the gyro pot being mounted in a circular recess in the after end; the base is bored out to fit the outside diameter of the pot, so that the pot may be inserted and sweated in place. Two holes are drilled and tapped on the exterior of the base, between holes for the holding screws, for the insertion of lifting screw tools when removing or assembling in afterbody.

GYRO POT.

9006. The gyro pot is a large cylindrical bronze drum of appropriate dimensions for housing the complete gyroscope assembly

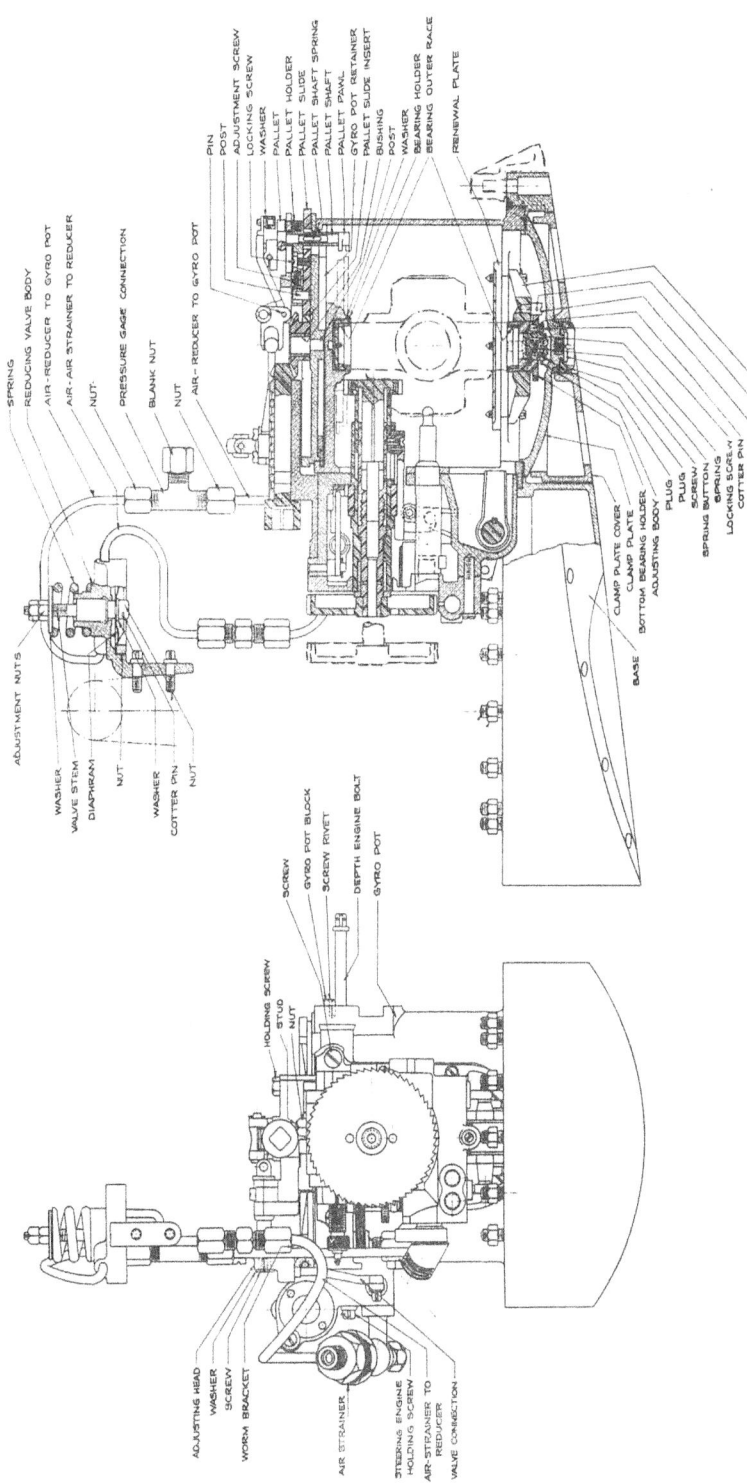

GYRO MECHANISM—SECTION & END VIEW

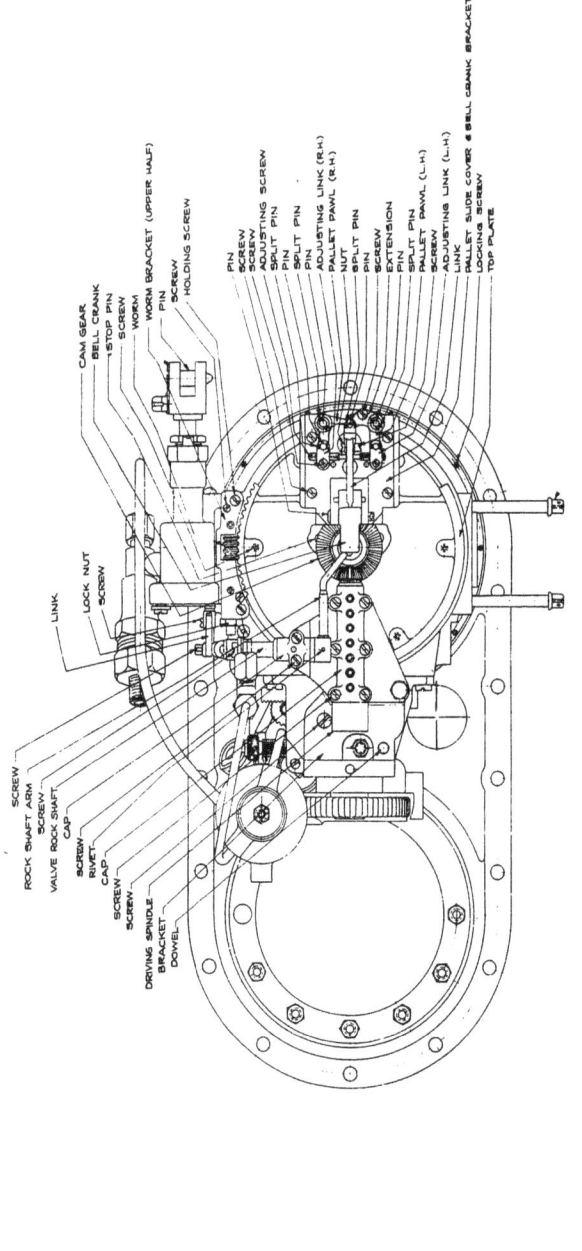

GYRO MECHANISM – PLAN VIEW

TS-5

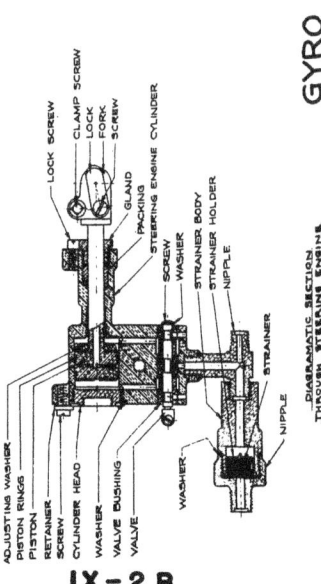

DIAGRAMATIC SECTION THROUGH STEERING ENGINE

IX-2B

with freedom of same to turn 360° on its vertical axis inside the pot without interference. A seat is machined on the lower inside of the pot for the clamp plate cover, an interrupted flange or rib being machined directly under this seat to provide a means for holding the clamp plate. Four segments of this flange are cut away, leaving 4 ribs an equal distance apart, each 45° of arc to permit the insertion of a similarly cut flange on the clamp plate. Thus, after insertion of the clamp plate and cover, the clamp plate is rotated until the ribs are in line, after which the cover may be clamped against its seat by setting up on the plug. The top of the pot is machined with a tight joint permitting a slight turning of the top plate. The seat must be accurate and is scraped and lapped during original assembly. A shelf extending from the top of the pot at the forward end and extending toward its center holds the top bearing for the gyro outer gimbal ring, suitable means being provided for the introduction of air into this bearing to sustain the speed of the gyro wheel. Two segments are extended from the lower end of the inner wall of the pot for the support of the bottom head to which it secures with six holding screws. The complete gyroscope is supported through its vertical axis between a bearing in this bottom head and a bearing mounted in the shelf referred to above. A seat is machined on the forward side of the pot for mounting the spinning and unlocking mechanism and means are provided for mounting the depth steering engine on the port side. A boss projecting from the forward side of the flange for the vertical steering engine on the starboard side of the pot, is drilled and tapped for the introduction of air to the top bearing for constant spin of the gyro. A bracket containing bearings for a small worm is located on the upper side of the gyro pot, just inside the vertical steering engine. This worm, meshing into gear teeth cut over a small portion of the periphery of the top plate, forms the means for adjusting the relation of the pallet on the top plate with the pallet cam on the gyro.

GYRO CLAMP PLATE AND COVER.

9007. The gyro clamp plate and cover close the lower opening of the gyro pot. The clamp plate is a composition forging in the form of a circular dished plate, the outer circumference of which is machined with a flanged seat to fit a similar seat in the bottom of the gyro pot, a gasket being interposed to make this seat water-tight. The reenforced center of the clamp plate is drilled and tapped for the screw holding the clamping plug. The clamp plate cover is machined with four interrupted ridges on its circumference as previously described, the center of the cover being drilled and tapped for the clamping plug, which is similarly threaded. This plug passes through the threaded center of the cover and is secured to the plate with a screw in such a manner as to permit the plug to turn freely about the securing screw, thus holding the cover and plate together and yet permitting rotation of either one.

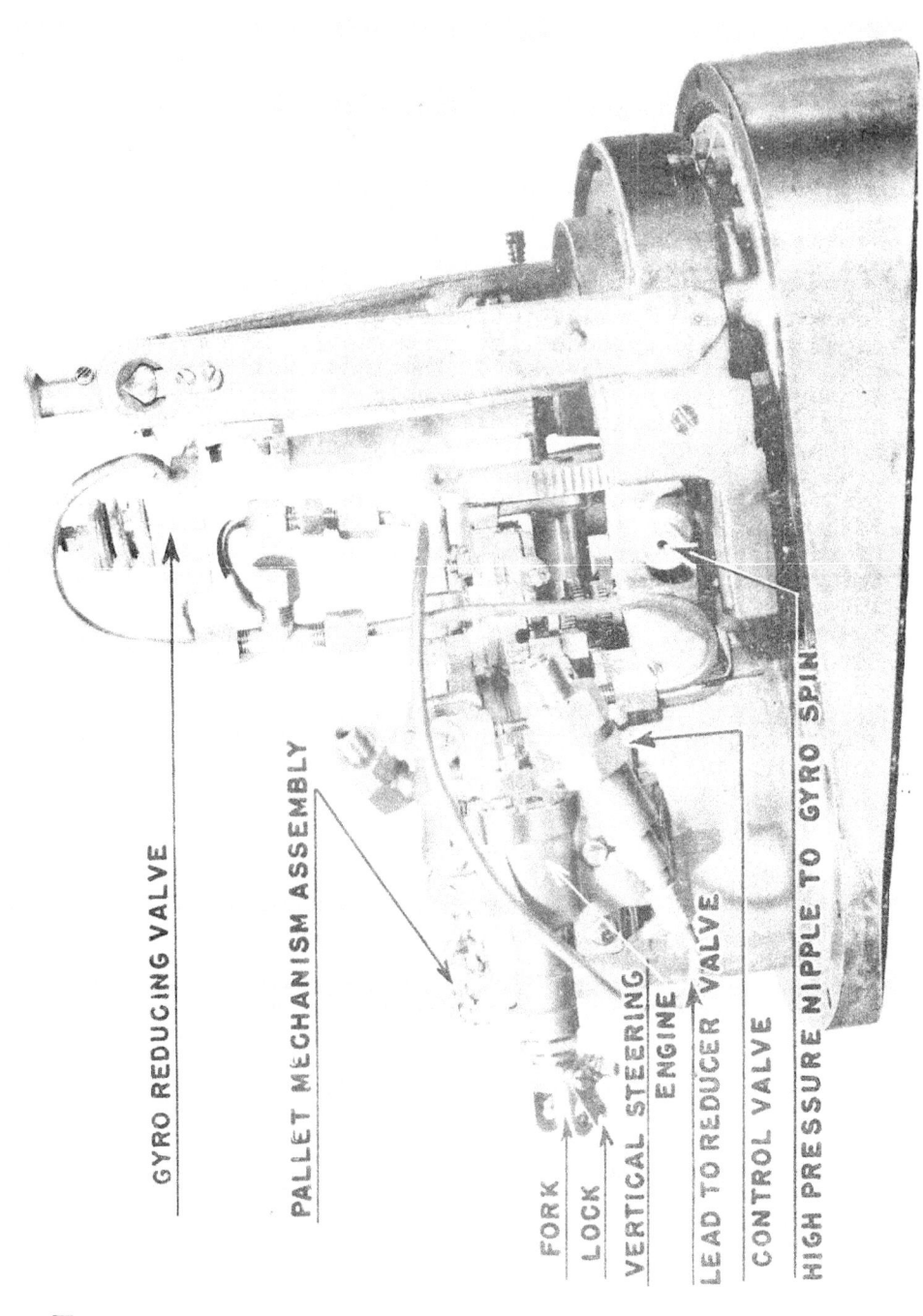

SIDE VIEW

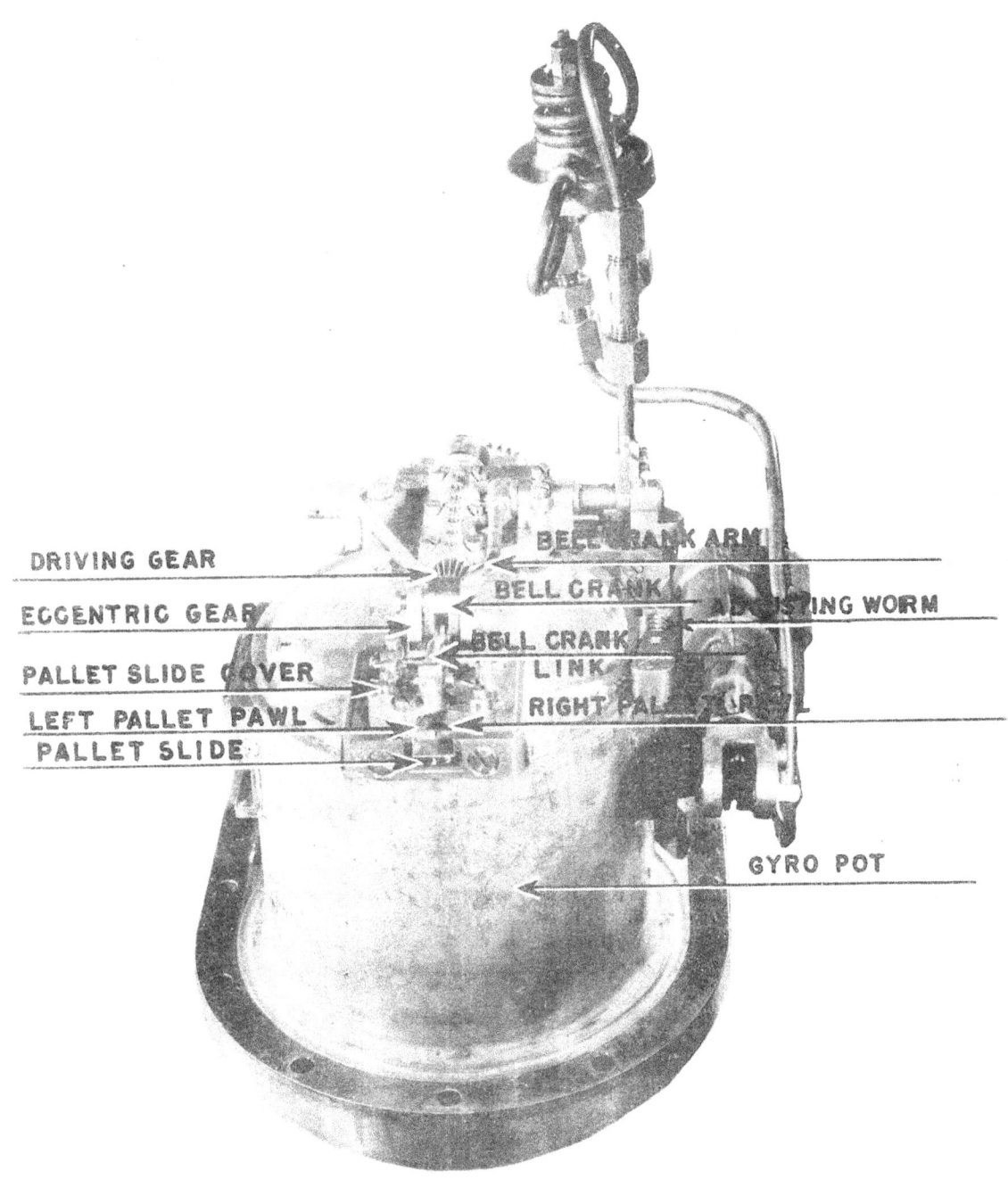

AFTER END VIEW

9008. The clamp plate cover is inserted in the gyro pot flange with the interrupted ribs on the cover passing through the lands cut away between the ribs on the pot flange. The cover is then turned in either direction until the ribs line up with those on the flange, after which the clamping plug is turned to the right, thus forcing the clamp plate down against its gasket and forming an effective seat against sea water leakage.

THE TOP PLATE.

9009. The top plate, which closes the top of the gyro pot, is a disc shaped bronze forging machined with the necessary flanges, bosses, seats and bearings for the pallet mechanism.

9010. As previously mentioned, a small worm is cut over 24° of arc around its periphery to mesh with the adjusting worm for centering the top plate. The outer circumference of the plate is suitably machined for a lap fit in the gyro pot to which it is held by two retainer plates secured with screws in such a manner as to permit rotation of the top plate on the pot without disengagement of the parts. A small adjustable pointer is located on the inside of the top plate for reference when adjusting the relative position of the pallet with the cam on the gyro, and for checking by the operator when installing the gyro. On the upper after side of the top plate are two bosses drilled and tapped longitudinally for the pallet slide plungers and springs, the space between these bosses being machined to form a channel for guiding the pallet slide, a hole being drilled through the outer end of this channel to permit the insertion of the pallet holder. The upper surfaces of the bosses are machined flat, forming seats for the pallet cover. A steel stud is riveted through the center of the top plate to form a pivot for the cam gear. Smaller studs are riveted on each side of this pivot stud for pallet slide stops.

B. **THE SPINNING AND UNLOCKING MECHANISM.**

9011. The means provided for the initial spinning of the gyro wheel to the velocity required for maintaining gyrostatic control of the torpedo's course throughout its run, and to disengage the spinning element from the wheel prior to the torpedo leaving the launching tube, is termed the spinning and unlocking gear.

9012. The complete mechanism is assembled in a spinning gear frame attached to the forward side of the gyro pot by holding screws and in a front plate similarly attached to the spinning gear frame. The spinning gear frame used on the Mark 13 and Mark 13-1 torpedoes up to and including register no. 16434 has a boss on the left side suitably bored to carry the locking gear shaft. The top surface of this frame is machined and drilled to carry the pawls and worm gear for driving the locking gear. On torpedoes after this register number the spinning gear frame has been replaced by a new frame without a boss on the side. The front plate, a brass casting, has radial holes which direct the high pressure air upon the teeth of the spinning rotor.

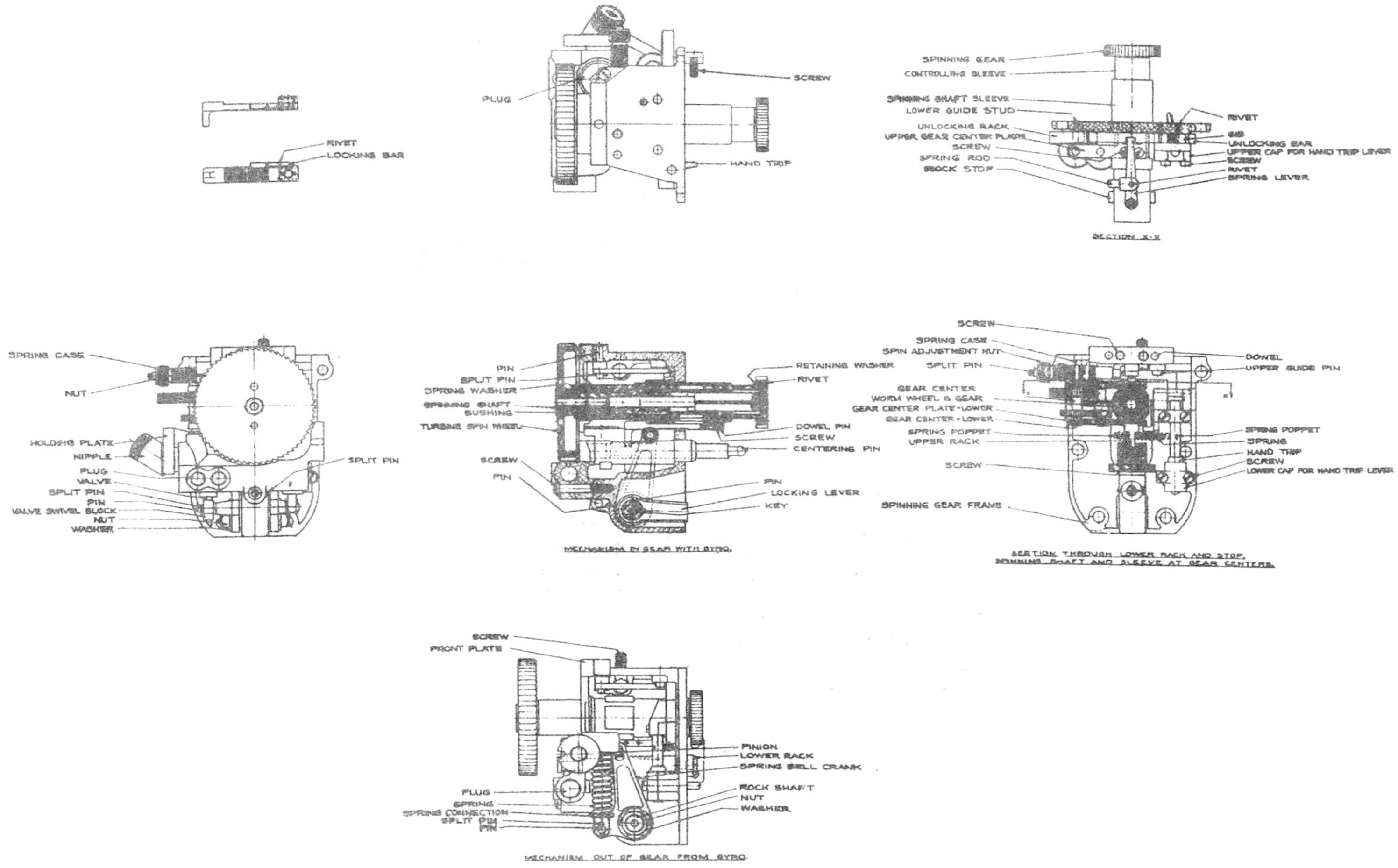

GYRO SPINNING MECHANISM

TS5
IX-4A

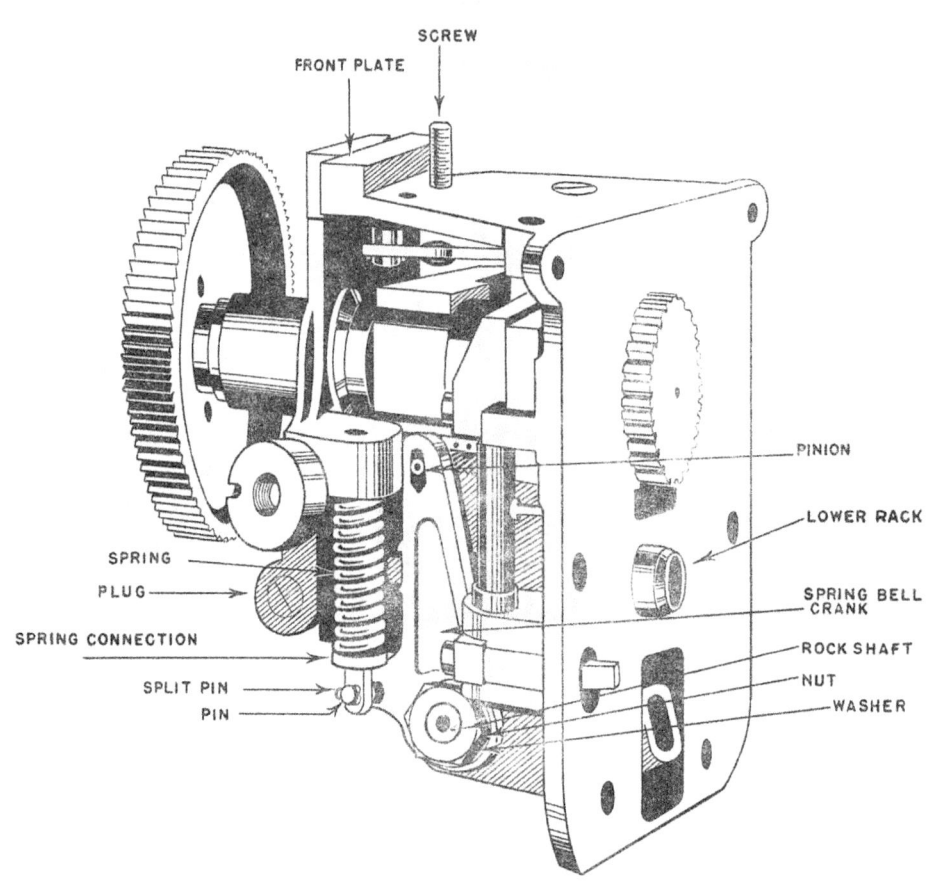

MECHANISM OUT OF GEAR FROM GYRO

TS-5   IX-4B

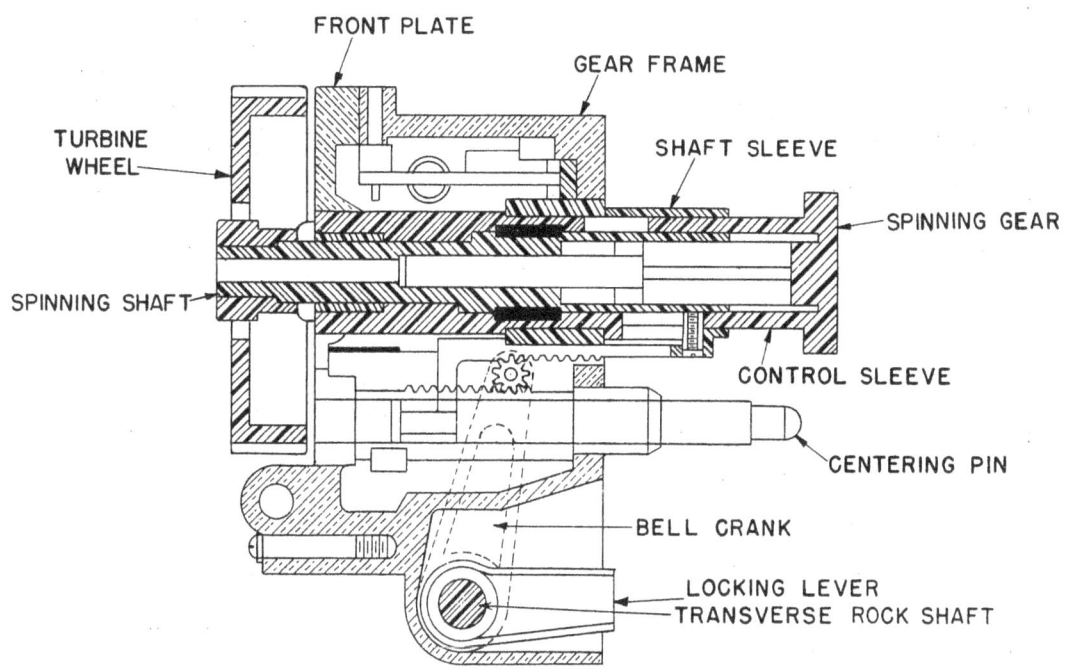

MECHANISM IN GEAR WITH GYRO

9013. The following are the essential features of the mechanism, each of which will be described in the order listed:

    (a) Spinning turbine and impulse mechanism.
    (b) Centering feature.
    (c) Locking and unlocking gear.
    (d) Spinning gear duration adjustment feature.

Spinning Turbine and Impulse Mechanism.

9014. The spinning rotor is a wheel, the periphery of which is machined with saw toothed buckets at an appropriate angle for obtaining most efficient speed. This rotor is attached to the outer end of a spinning shaft. On the extension to this shaft is machined a spur gear of similar dimensions to the spur gear machined on the gyro wheel spindle, and when the gyro is locked or in position to receive its initial spin, these spur gears are in mesh. Thus it will be seen that in this position the spinning turbine is geared for direct drive of the gyro wheel.

9015. The spinning shaft is made in two sections, both integral for rotation but the extension having independent longitudinal movement in the spinning shaft, thus permitting the travel required to engage with the gyro.

9016. The spinning shaft runs in a composition bushing and in the controlling sleeve, which parts are supported in the spinning shaft sleeve the spinning shaft sleeve in turn being supported in the spinning gear frame and front plate.

9017. The inner end of the outer section of the spinning shaft is machined with a square socket in which the outer squared end of the spinning shaft extension is inserted, thus permitting the extension with the spinning gear to move longitudinally inside the spinning shaft while at the same time being able to transmit the spinning torque.

9018. A steel tube (controlling sleeve) is loosely attached to the spinning shaft extension by means of a thin bronze retainer ring made in two halves and riveted in a recess on the face of the spinning gear so as to permit the spinning shaft extension with the spinning gear to rotate in the sleeve, when the sleeve moves longitudinally with the spinning shaft extension, acting as guide for the spinning gear moving in and out of mesh with the gyro. Due to necessity for the increased advance of the spinning gear and centering pin into the gyro pot required for non-tumble gyros, and to the use of the same angular movement of the bell cranks as used with older marks, longitudinal motion is transmitted to the spinning shaft sleeve and controlling sleeve by a pinion carried between slotted bearings on the upper ends of the valve and spring bell cranks. This pinion meshes in a fixed lower rack machined on the centering pin housing and in a movable upper rack doweled and secured to the controlling sleeve. Thus it will be seen that as the pinion is moved longitudinally upon the fixed rack due to the action of the bell cranks, and being in mesh with the movable upper rack, it will push this upper rack ahead of it. When so actuated, the linear travel of the upper rack is just twice the linear travel of the

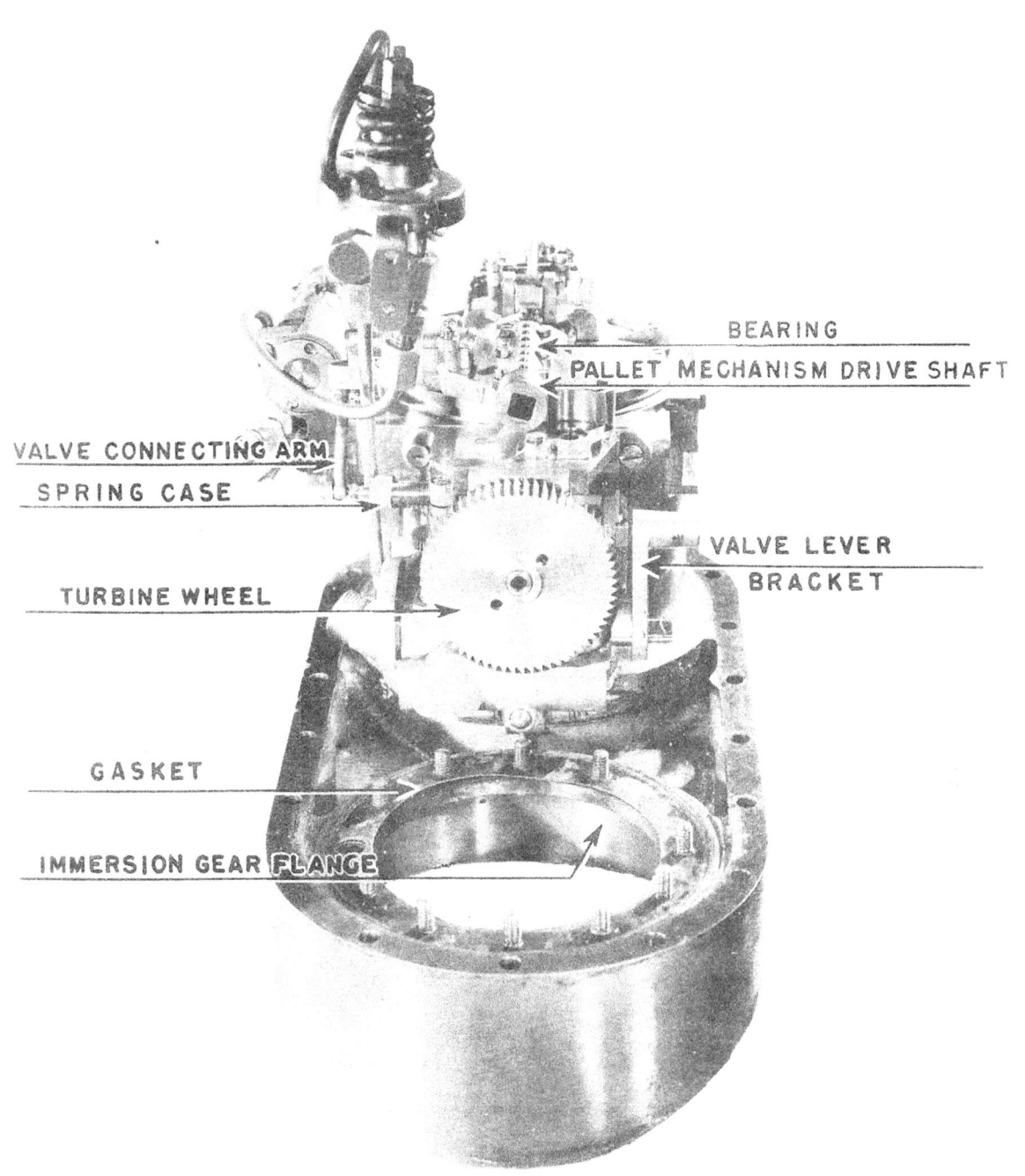

FORWARD END VIEW

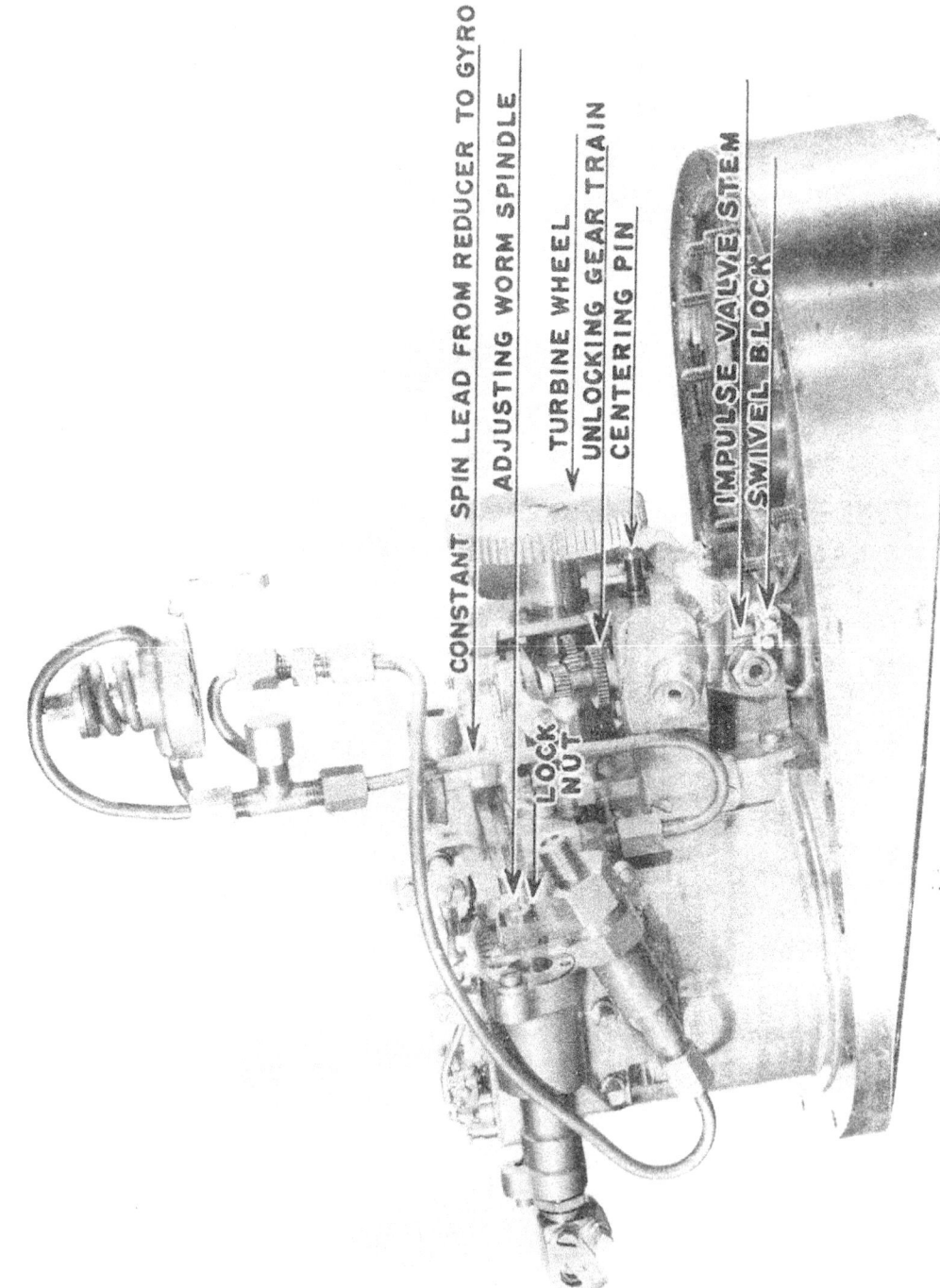

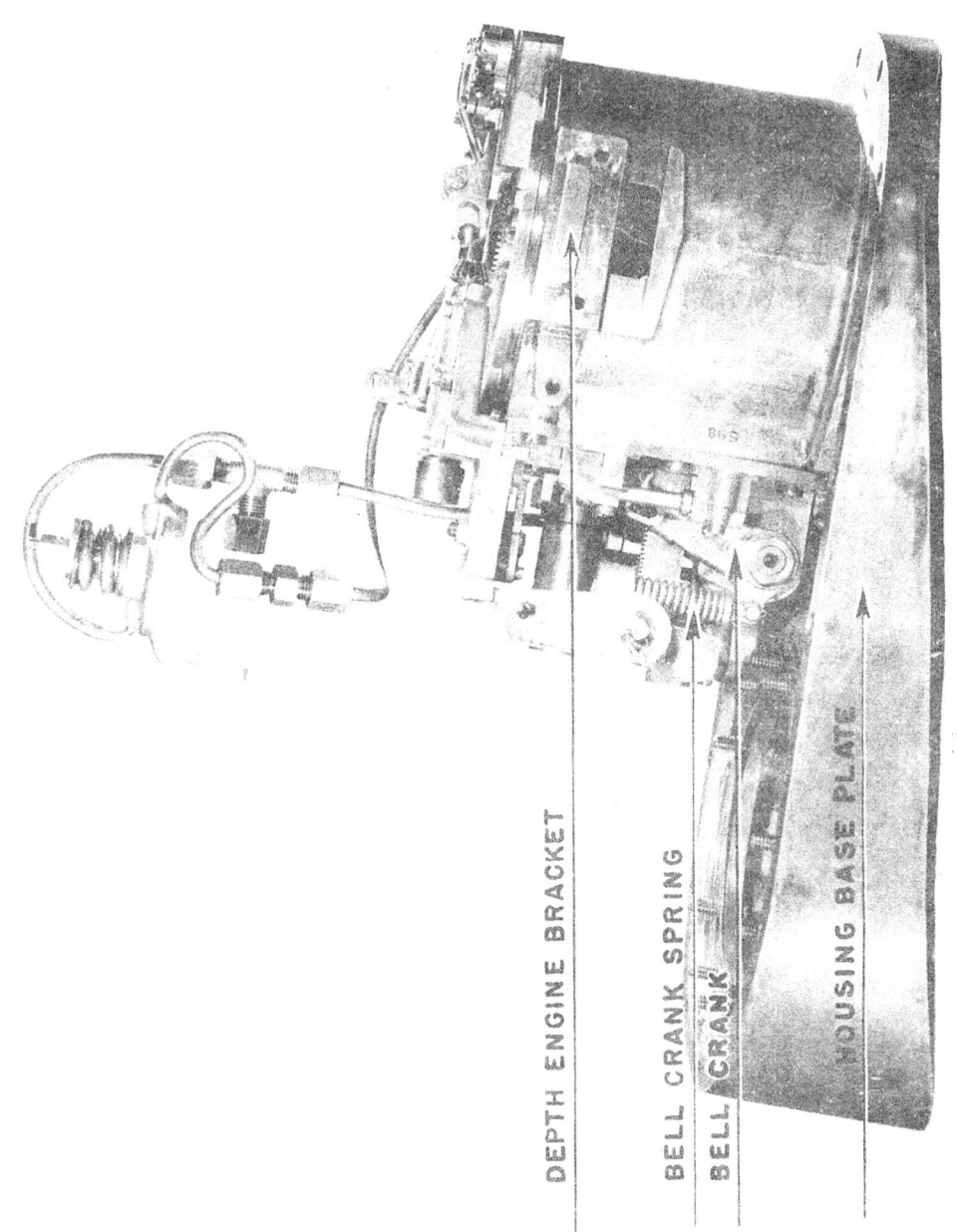

SIDE VIEW OF HOUSING

pinion itself, and it is this arrangement that permits the spinning gear to advance the additional distance required for locking the non-tumble gyro with the same angular movement of the bell cranks as obtained in previous types of spinning gears.

9019. Initial impulse is given to the gyro by air at flask pressure impinging upon the spinning turbine buckets through five holes or nozzles, drilled at a suitable angle for maximum effect, directly under the spinning rotor into a small high pressure reservoir located in the front plate. The air enters this reservoir through an impulse valve also located in the front plate. This impulse valve is normally seated, but when the gyro is locked, a swivel block attached to a projecting pin on the valve bell crank lifts the valve off its seat and holds the valve open for admission of flask pressure air for the duration of spin, when, upon unlocking the gyro, the valve will again seat and shut off the air to the turbine nozzle.

Centering features:

9020. To hold the gyro rigid with the spur gears of the timing shaft and gyro in proper mesh during initial spin, the gyro wheel is held in position by a centering pin the ball end of which engages in the hollow center of the ball bearing cup ("A" bearing) for the wheel. This centering pin is located in a separate steel housing rigidly secured between the spinning gear frame and the front plate, this housing having a dual purpose, in that the fixed rack is machined on its upper side. The centering pin is operated by a toe extending from the outer end of the upper rack on the controlling sleeve and engaging on the outer end of the centering pin.

Locking and unlocking gear:

9021. The means provided for locking the gyro in position for its initial spin and releasing the gyro from this position after receiving its impulse is termed the locking and unlocking gear.

9022. The spinning sleeve which carries the spinning element is moved longitudinally for locking and unlocking by two bell cranks, the valve bell crank and the spring bell crank, both of which are keyed to the ends of a short transverse shaft having bearings through a pocket cast in the lower end of the spinning gear frame. Keyed in the center of this shaft and accessible from the inside of the gyro pot through this pocket is a locking lever suitably machined for the insertion of a tool to lock the gyro. The upper ends of the bell cranks are machined with slotted bearings for the pinion which is carried between these bell cranks and meshes with the lower stationary and upper movable racks. Thus, it will be seen that moving the locking lever down or away from the gyro will cause the bell cranks to move the pinion longitudinally upon the fixed rack and being in mesh with the movable rack, it will push this rack ahead of it This rack being attached to the controlling sleeve will push the spinning element into the gyro pot until locked in place by the unlocking bar being pushed in front of a hardened steel insert attached to the side of the upper rack. In this position the centering pin is entered in the axial hole in the "A" bearing of the gyro wheel, holding the gyro rigid for spinning; the spur gear on the spinning shaft extension is in full mesh with the spur gear on the gyro spindle and thus in position to impart rotation; the spinning turbine buckets are lined up over the five nozzles in the front plate in line for receiving the flask pressure air impulse; the impulse valve is lifted off its seat by the swivel block attached to the projecting pin on the valve bell crank to permit passage of air to the nozzles, and the unlocking spring is compressed on the spring guide, between the spring post on the spring bell crank and its seat in a boss in the front plate, thus completing the locking of the mechanism.

9023. The unlocking mechanism is a reduction gear train operated by a single thread worm machined on the central portion of the spinning shaft and accessible for mesh in a worm wheel through a slot milled in the side of the spinning sleeve. This worm wheel is part of a gear train assembled between the upper and lower gear center plates which are rigidly attached to and move with the spinning sleeve. A gear on this worm wheel shaft meshes with a larger gear on the pinion shaft. The pinion in turn meshes with the **unlocking rack when the gyro is in the** locked positon. The ratio of gear reduction from the spinning shaft to the pinion operating the unlocking rack is 105 to 1. The unlocking rack and the unlocking bar are contained in slides machined across the upper inside portion of the spinning gear frame, the unlocking bar being in place in a retainer bracket which also forms the upper bearing cap for the hand trip. In order that the mechanism may be unlocked by hand when necessary, a small vertical shaft is mounted between bearings on the spinning gear frame. The upper end of this shaft is fitted with a cam which engages in a slot milled on the side of the unlocking bar, the lower end of the shaft having a projecting lug accessible from the interior of the gyro pot for manual operation. A spring connected between poppets on the shaft and spinning gear frame keeps the unlocking bar in its engaged position until forced out of engagement by the unlocking rack or by the hand trip.

<u>Spinning gear duration adjustment feature:</u>

9024. The duration of the spin is regulated by changing the distance between contacting points of the unlocking rack and the unlocking bar, so that more or less teeth of the unlocking rack will engage the pinion on the unlocking gear train when the gyro is locked. This is accomplished by a spring lever the outer end of which is pivoted on a stud in the upper inner side of the spinning gear frame, the inner end engaging in a slotted portion near the center of the unlocking rack. Thus, by moving this spring lever about its pivot, the unlocking rack is correspondingly moved.

9025. The position of the spring lever and consequently the unlocking rack is adjusted by means of a spring rod passing through an adjustable spring case attached to the upper end of the spinning frame through a hinged clamp threaded on its interior to fit similar threads on the spring case, which permit sufficient movement of the spring rod to take up shock and to properly mesh the teeth of the rack with the pinion of the unlocking train when locking the mechanism. By loosening a clamp screw on this hinge and turning the spring case, the unlocking movement may be adjusted to the desired number of revolutions.

9026. The operation of the complete spinning mechanism is as follows:

With the locking of the mechanism, the spinning shaft sleeve is moved in until the toe of the unlocking bar catches and holds it in the spinning position. The turbine wheel is then lined up over the nozzles; the centering pin is engaged in the forward ball cup of the gyro wheel; the spur gear on the spinning shaft engages the spur gear on the gyro wheel spindle; the pinion on the unlocking gear train is engaged with the teeth in the unlocking rack; the impulse valve is held clear of its seat and the unlocking spring is placed under a heavy compression by the respective bell cranks and the gyro is ready for a spin.

9027. Upon launching of the torpedo and when the starting valve lifts, a blast of air at flask pressure is admitted through the impulse valve and the turbine nozzles. While the turbine wheel is being spun, the unlocking rack actuated by the unlocking gear train travels toward the unlocking bar until, at the end of its travel, the toe of the unlocking rack pushes the unlocking bar clear of the spinning shaft sleeve, at which time the pressure of the releasing spring against the spring bell crank, augmented by the force exerted by the impulse valve stem against the toe on the valve bell crank, causes the instant withdrawal of the spinning shaft sleeve, disengaging the spur gears and unlocking gear pinion, closing the impulse valve, and finally withdrawing the centering pin, at which time the spinning gyro is free to take up any direction with respect to the gyro pot, thus having accomplished the purpose of the spinning and unlocking mechanism for the run.

C. DISASSEMBLY, OVERHAUL AND ASSEMBLY OF SPINNING AND UNLOCKING MECHANISM.
                                                                                     TOOL NO.

9028. Remove spinning and unlocking mechanism from gyro pot.

    NOTE: The removal of parts preceded by an asterisk is necessary prior to the removal of spinning mechanism from the gyro pot. Their removal is not otherwise connected with the spinning mechanism.

*(a) Remove steering engine valve rockshaft linkage:
    (1) Remove 2 screws for bearing cap............ 41
    (2) Remove rockshaft assembly from bearing.....
    (3) Remove spring buttons and springs under rockshaft bearings.

                                                                TOOL NO.
*(b) Remove pallet mechanism driving gear shaft and
     bearing:
     (1) Remove 6 holding screws....................   41
     (2) Remove driving gear bearing cap.
     (3) Remove pallet driving gear and shaft.

*(c) Remove driving gear bracket:
     (1) Remove hexagonal nut.......................   48
     (2) Remove screw with hexagonal head...........   48
     (3) Remove holding screw.......................   41
     (4) Remove driving gear bracket.

 (d) Remove pipe gyro reducer to nipple gyro pot....   24

*(e) Remove clamp for gyro spin pipe nipple:
     (1) Remove 2 holding screws....................   40
     (2) Remove clamp.

*(f) Remove after section of pendulum:
     (1) Lock gyro spinning mechanism...............   205
     (2) Remove 2 screws holding after section of
         pendulum...................................   39
     (3) Remove after section of pendulum.

 (g) Remove gyro spinning rotor:
     (1) Hold spinning shaft gear with a special
         spanner, pins of spanner meshing across
         teeth of spinning gear................. (special 126)
     (2) Insert pins of tool #25 in holes in rotor
         and unscrew rotor from spinning shaft.
     (3) Remove bronze spacing washer.
     (4) Unlock gyro spinning mechanism.

 (h) Remove gyro spinning mechanism frame and front
     plate assembled:
     (1) Remove 6 holding screws for gyro spinning
         mechanism.................................. 41,233
     (2) Remove gyro spinning mechanism.

9029. Disassemble:

     (1) Remove cotter pin from pin for spring guide pin. 72
     (2) Remove releasing spring and spring guide pin
         and spring button.......................... WE 178
     (3) Remove nut and washer for releasing spring bell
         crank......................................   141A

NOTE: There must be no lost motion between releasing
spring bell crank and rockshaft. The bell crank is keyed
to the shaft and if lost motion exists, it will be due
to a poorly fitting key. Remedy before proceeding.
     (4) Remove lower bearing cap and two screws for
         hand trip..................................   41

                                                                TOOL NO.
(5) Remove releasing spring bell crank (pry off
    with screw drivers if necessary)
(6) Remove pinion between upper and lower rack.
(7) Test tightness of pin for releasing spring
    guide pin with bell crank. This pin is rivet-
    ed to the bell crank and may work loose; re-
    rivet if necessary. Note that no excess play
    exists between bearing ends of pinion and slot
    on bell crank.
(8) Remove cotter pin for bottom holding screw
    and remove holding screws for front plate....... 41
(9) Remove front plate, taking care not to pry, so
    as not to bend dowels.
(10) Pull spinning shaft clear, turn, and remove
    screw from upper rack. Care should be exer-
    cised in removal of this screw so as not to
    ruin slot, due to being prick punched.
(11) Remove upper rack and locking bar. This rack
    is doweled and must be removed with great care.. 37
(12) Slip out centering pin.
(13) Remove spinning shaft sleeve and spinning shaft
    with gear train and remove spur gear.
(14) Test end play in unlocking pinion. This gear
    must be free from end play between its centers.
(15) Remove screws from upper and lower gear center
    plates... 37
(16) Remove upper and lower center plates, unlocking
    pinion, unlocking worm and pinion gear.
(17) Note that upper gear centers are tight in
    upper center plate. If not, re-rivet.
(18) Remove spinning shaft.
(19) Inspect sliding fit of unlocking bar with upper
    bearing guide cap. Clearance between bar and
    cap should be less than "001 with no binding.
    If too tight, stone the bearing surface of the
    cap; if too loose, file the lug of the frame.
    Excessive play between rack and cap may allow
    pinion to jump out of mesh with rack.
(20) Remove screws for upper bearing cap............ 41
(21) Remove upper bearing cap.
(22) Disconnect spring from hand trip lever and re-
    move hand trip lever.
(23) Remove unlocking bar and unlocking rack.
(24) Disconnect hinged clamp and remove spring case
    assembly with spring lever...................... 92,41
(25) Remove unlocking rack.
(26) Remove nut and washer on bell crank.
(27) Remove rockshaft by tapping lightly. This takes
    off valve bell crank and locking lever.
    NOTE: It is important that bell crank and locking
          lever are keyed tightly to rockshaft.
(28) Remove locking lever. Note that the side marked
    "X" is toward the hand trip lever boss, for refer-
    ence when assembling.

Disassemble impulse valve: TOOL NO.
- (1) Remove impulse valve plug and washer............ 11
- (2) Remove impulse valve.

9030. **Assembly and test:**

Before assembly of mechanism, immerse parts in clean spirits and blow dry with air. Use soft wire brush to clean gear teeth and crocus cloth on spinning shaft, if necessary. Inspect control sleeve and all movable parts for scores or burrs. Oil and assemble as follows:

9031. Assembly and test impulse valve:
- (1) Wash valve and front plate off with spirits and blow off with air.
- (2) Inspect nipple and make sure that the diameter of hole in nipple is .180".
- (3) Oil "A" impulse valve and replace valve.
- (4) Replace impulse valve plug and washer.......... 11
- (5) Connect H.P. air line to nipple, turn on air, submerge front plate with impulse valve in water, bubbles around nozzles indicate leaky impulse valve.
- (6) Other possible places for leaks are:
  - (a) Impulse valve plug.
  - (b) Around nipple.
    Remedy valve leaks by lapping or reseating valve....................................... WE164

9032. Assembly of spinning mechanism:
- (1) Replace spinning shaft in spinning shaft sleeve.
- (2) Replace lower gear center plate on spinning shaft and catch temporarily with holding screw.. 37
- (3) Replace worm gear and pinion gear on their respective pivots and hold in position by hand.
- (4) Replace upper gear center plate lining up worm gear and pinion gear to their pivots.
- (5) Tighten up screws on both upper and lower center plate. Stone heads of screws under upper rack and be sure that screw heads and rack clear each other.
- (6) See that cotter pins are in place on screw pivots in upper gear center plates.................... 92
- (7) Replace locking bar in recess in spinning gear frame.
- (8) Catch hand trip spring in poppet and replace hand trip with toe on trip engaging in locking bar.
- (9) Replace rockshaft for bell crank and locking lever. Place locking lever in position with "X" facing toward valve bell crank. Line up with keys and tap into place.
- (10) Replace valve bell crank and washer............ 141A

                                                                  TOOL NO.
  (11) Replace upper cap for hand trip and secure
     with screws.................................... 37
  (12) Note that locking bar moves freely without
     binding and without appreciable lost motion.
  (13) Replace unlocking rack. See that rack moves
     freely in recess.
  (14) Replace unlocking spring lever and adjustable
     spring case assembled. Secure with hinge screw
     and cotter pin for spring lever pin............ 92

NOTE: It frequently happens that the end of the spring rod breaks off in wake of the cotter pin hole, necessitating renewal of same. In which case proceed as follows: Remove nut from spring rod and withdraw spring rod from spring case. Remove rivet holding spring rod on spring lever and remove old spring rod. Re-rivet a new spring rod on spring lever. Remove nut on outer end of spring case and remove spring rod washers and spring. Replace spring rod washers and spring in spring case, a washer on each end of the spring. Replace nut on the end of spring rod, screwing in sufficiently to permit the insertion of a cotter pin. Replace nut in outer end of spring case and set up until all play of spring rod and spring case is removed. Replace cotter pin in end of spring rod.

  (15) Pull back hand trip lever and insert spinning
     shaft sleeve assembly.
  (16) Replace spinning shaft extension with control-
     ling sleeve and spur gear, with prick punched
     marks on spinning shaft lining up with similar
     marks on extension.
  (17) Pull spinning shaft sleeve assembly to the rear
     to clear, then turn to facilitate assembly of
     upper rack on sleeve.

NOTE: Before assembling upper rack, compare it with drawing and insure that upper shoulder has been beveled as per drawing. Shift pinion gear if necessary.

  (18) Replace the upper rack with dowels lined up.
  (19) Secure upper rack to sleeve with holding screw... 37
  (20) Prick punch screw head to prevent screw from
     turning.
  (21) Replace centering pin in its housing.
  (22) Replace front plate, make sure dowels are in line
     and tap in place.
  (23) Replace spring bell crank, nut and washer........ 141A
  (24) With the valve lever against its stop, replace
     pinion gear in mesh with the upper and lower rack.
     Measure distance from face of spur gear to face
     of spinning gear frame; this should be not more
     than 13/32". If more, change position of pinion
     until the desired measurement is obtained.

TOOL NO.

    (25) Replace releasing spring bell crank on rockshaft.
    (26) Replace washer and nut for releasing spring bell crank on rockshaft...................... 141A

NOTE: With a properly assembled gear, its own weight should cause the spinning sleeve to drop freely when held with spur gear uppermost and with trip cam released.

    (27) Replace cotter pin in lower holding screw for front plate................................. 92
    (28) Replace lower bearing cap and (2) two screws for hand trip............................... 37
    (29) Replace washer and spinning turbine............ 25
    (30) Replace releasing spring, guide and spring button. Secure with cotter pin.

NOTE: No. 1. When the mechanism is assembled, lock the mechanism and place on surface plate with the micrometer gauge placed on the periphery of the spinning gear. Rotate the spinning gear slowly and note the eccentricity of the outside of the gear. This eccentricity should not be greater than ".003. If greater than ".003, the mechanism must be disassembled and the various excess clearance between spinning shaft, control sleeve, etc., remedied.

NOTE: No. 2. With the mechanism assembled on the surface plate, place the micrometer gauge on the periphery of the spinning gear. Hold the mechanism firmly on the plate and move the spinning gear in a direct line toward and away from the micrometer gauge. The gear should not have more than ".012 lost motion. If excess lost motion exists, proceed as in NOTE No. 1.

NOTE: No. 3. When assembled in an unlocked position, the swivel block must have at least 1/32" clearance from valve stem. When in a locked position, the valve must be open. This may be tested with air.

D. **THE PALLET MECHANISM.**

9033. The system by means of which the directive force of the gyro wheel is relayed to the steering engine valve for application with multiplied power to the gyro rudders, is termed the pallet mechanism. This mechanism is so arranged in coordination with the cam on the cam plate attached to the outer gimbal ring that the rudders are operated to correct the deviation recorded by the gyro.

9034. The complete pallet mechanism is carried on the top plate and is so arranged that motion will be transmitted from the gyro to the steering engine valve without setting up a disturbing torque with consequent precession of the gyro.

9035. This is accomplished by intermittent light contact of small cam pawls with the cam or concentric ridges carried on a cam plate which is rigidly secured to the outer gimbal ring of the gyro.

9036. The cam pawls are carried on a small shaft, the pallet shaft, with bearings in a holder secured to a movable slide to which reciprocating motion is imparted by a cam gear driven from the torpedo propeller shaft and two spring loaded actuating plungers. A small pallet blade is attached to the upper end of this shaft.

9037. Thus it will be seen that when the cam pawls are moved toward the gyro and the torpedo course differs on either side of the axis of the gyro, one of the pawls will have moved into line with and will contact the cam, imparting a slight rotary movement to the pallet shaft with consequent angular movement of the pallet blade.

9038. On the other hand, when moved to position away from the gyro, the pallet blade, in position imparted through this movement, will contact one of the pallet pawls which, through a connecting linkage will move the steering engine valve to bring the torpedo course in line with the axis of the gyro.

9039. Moving the cam pawl toward the gyro when the torpedo course coincides with the gyro axis, will cause the pawls to straddle the cam, thus centering the pallet blade, in which position the blade will, when moved to position away from the gyro, pass between the points of the pallet pawls.

9040. The cam pawls are made in one piece with, and extend at right angles from, the ends of a central hub but on different planes, so that when one pawl touches the cam or concentric ridge, the opposite pawl will be on a plane with and engage in the adjacent groove, thus permitting the angular movement of the cam pawl.

9041. The ends of the pawls are machined in the form of blunt hooks, the inner sides of which are parallel to the axis of the hub and dimensioned so as to straddle the cam on the cam plate with a minimum clearance. A hole is drilled through the center of the cam pawl hub by which it is fitted and pinned to the pallet shaft.

9042. The pallet shaft is machined with bearing surfaces on each end, the surfaces being lapped into a bearing in the pallet holder. A small recess is milled lengthwise about midway of the shaft, into which is fitted a small leaf-spring, the purpose of this spring being to cause a slight drag between the shaft and the holder when assembled, and thus prevent chattering with resultant erratic deflection performance.

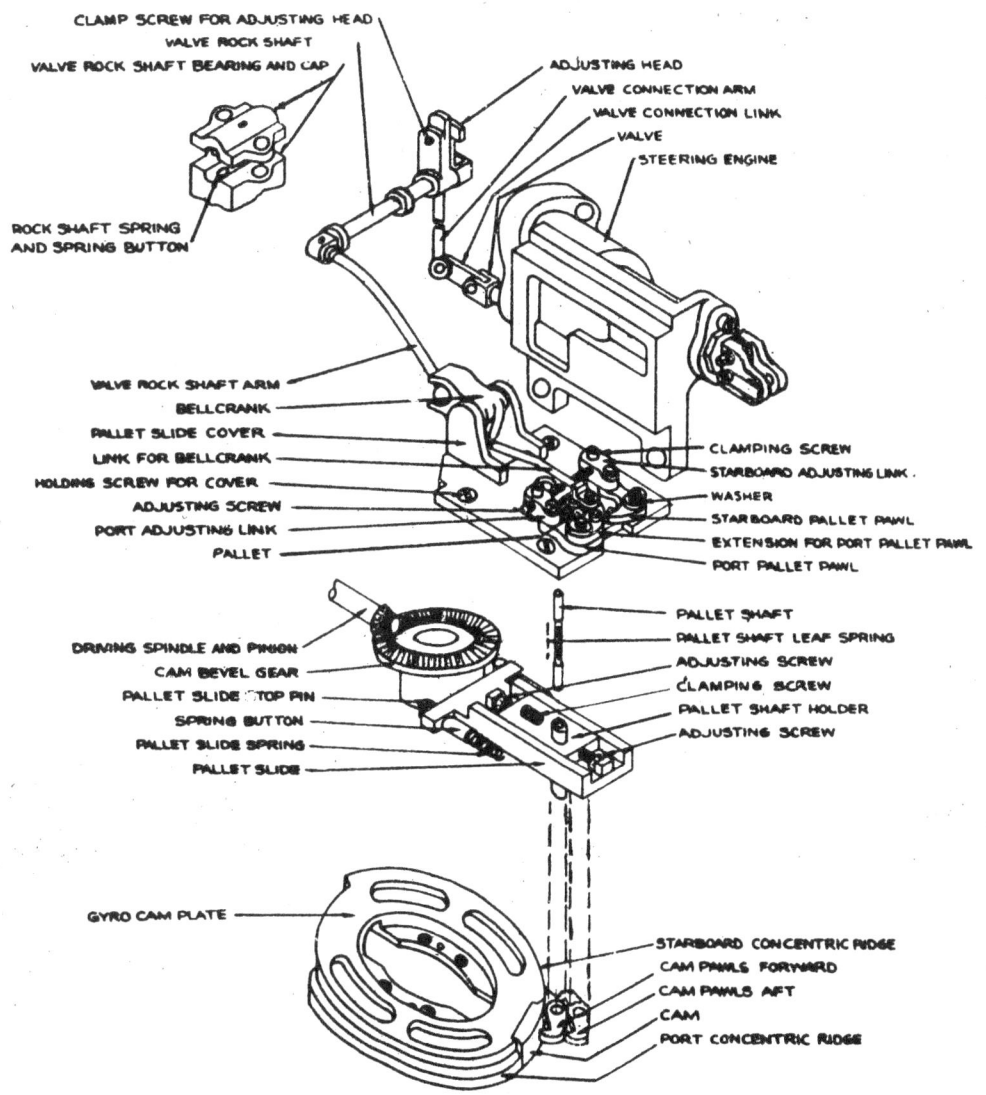

DIAGRAMATIC VIEW — PALLET MECHANISM

TS-5

IX-14A

9043. Clamped to the upper end of this shaft is the pallet. The pallet is in effect a blade extending from a hub drilled to fit over the end of the pallet shaft. It is clamped to the pallet shaft by means of a clamp screw passing through the split end of the pallet hub opposite the blade. After clamping, the screw is locked with a nut

9044. The pallet shaft with cam pawl and pallet assembled, is carried in a bearing sleeve in the pallet holder.

9045. The pallet holder is a vertical bearing sleeve machined at right angles to a rectangular bronze body, the sides of which are machined with a sliding fit for insertion in the pallet slide. Holes are drilled and tapped in each end of the rectangular portion in which screws are inserted for adjusting the position of the pallet holder on the pallet slide. An elongated slot is machined through the flat side of the pallet holder through which a screw is inserted for clamping the pallet holder in the pallet slide.

9046. The pallet holder is carried in a channel machined on the upper side of the pallet slide. The pallet slide is a rectangular steel member, one end of which is "T" shaped to afford be ring surfaces for two actuating plungers and pallet slide stops. A hardened steel insert is dovetailed into one end against which the cam gear acts in imparting motion to the slide A large hole is drilled through the bottom of the slide for insertion of the bearing sleeve on the pallet holder with sufficient clearance to permit limited adjustment of the pallet holder in the slide. Two squared projections are extended from the bottom of the slide against which the heads of the pallet holder adjusting screws bear. A hole is drilled and tapped in the bottom of the slide for the pallet holder clamp screw. The bottom and sides of the slide are machined and scraped with a sliding fit in the channel in the top plate in which it is assembled so that the "T" end is held longitudinally between two spring loaded actuating plungers and two stop pins riveted to the top plate, the slide being retained on the top plate by a cover (pallet slide cover).

9047. The pallet slide cover is suitably machined to fit over the pallet slide on the top plate to which it is secured with four screws. A slotted bracket projects at right angles from the inner end of the cover to form the support for the pallet bell crank bearing. Two posts are riveted to the outer end of the cover for the pallet pawl pivots. Three pins are riveted to the cover in appropriate locations for restricting the throws of pallet and pallet pawl, and holes are drilled through the cover for pallet shaft, for access to pallet holder adjustment clamp screw and for oil to pallet slide.

9048. There are two pallet pawls which are referred to as right hand and left hand pallet pawls. These pawls are fitted on pivot pins riveted to the pallet slide cover and are cross connected by an adjusting screw engaging adjusting links on the pawl arms. The ends of the pallet pawl arms are fitted with pins for attaching the adjusting links. The left hand pallet pawl is considerably larger across the pawl end to allow sufficient stock for drilling and tapping a hole for the extender holding screw. The extender is machined to fit on top of the left hand pallet pawl to which it is secured with one holding screw. A pin is riveted to the other end of the extender to which the link, transmitting pallet motion to the bell crank, is connected. As above stated, the pallet pawl arms are connected together by two adjusting links interconnected by an adjusting screw; the adjusting links being suitably machined to fit over pins on the pallet pawl arms and threaded right and left hand for the adjusting screw (which is similarly threaded) with facilities for clamping adjustments. The end of the adjusting screw is slotted for insertion of a screw driver when adjusting.

9049. Motion is transmitted from the pallet mechanism to the steering engine valve through a bell crank pivoted on the extended bracket of the pallet slide cover. Engaging in a slotted portion of this bell crank is the ball end of the bell crank arm extending from the valve rock shaft having bearings on the mechanism; a valve connection arm attached to the other end of this shaft is coupled to the steering engine valve link.

9050. The bell crank arm and the valve connection arm are fitted and pinned into holes drilled in the valve rock shaft and adjusting head for rock shaft respectively. The valve rock shaft is made in two parts, the rock shaft and adjusting head, each capable of limited rotary movement on the other for centralizing the valve with the pallet, with facilities for clamping the adjustments. The rock shaft is supported in a bearing machined on a bracket extending from the driving gear bracket and suitably located for engagement of the bell crank arm in the bell crank and connection of the valve arm to the steering engine valve.

9051. As previously stated, reciprocating motion is imparted to the pallet slide by a cam bevel gear and two spring loaded actuating plungers located on the top plate.

9052. The cam bevel gear rotates on a stud in the center of the axis of the top plate. Rotary motion is imparted to it by a small bevel pinion machined on a shaft running in a stationary bearing bracket attached to the pot. This shaft in turn is connected through the medium of a shaft squared on both ends with a gear driven from a pinion on the forward propeller shaft. Thus it will be seen that rotation of propeller shaft will be transmitted through gearing on the drive shaft to bevel pinion and bevel gear; the cam bevel gear giving oscillating motion to pallet slide.

E. <u>TOP PLATE CENTERING DEVICE</u>.

9053. As previously stated, angle fire devices are not provided in the Mark 13 torpedo design. However, due to tolerances in machining, etc., it is difficult, if not impossible to assemble a gyro in a torpedo so that its axis will be parallel with the torpedo axis. Therefore, in order that proper centering may be effected, means are provided for limited movement of the top plate with the pallet mechanism in azimuth about its bearing in the gyro pot.

9054. A bracket containing bearings for a small worm is located on the upper right side of the gyro pot, between the pot and the steering engine. This worm meshes with gear teeth cut over a small portion of the periphery of the top plate. The outer circumference of top plate is suitably machined for a lap fit in gyro pot to which it is held by two retainer plates. These retainer plates are secured with screws in such a manner as to permit rotation of the top plate on the pot without disengagement of the parts. A small adjustable pointer is located inside the gyro pot for reference when adjusting the relative position of the pallet with cam on the gyro cam plate.

9055. The centering adjustment is accomplished by turning the small worm. This worm, through engagement with the teeth cut in the periphery of the top plate, will cause the top plate to turn in any direction desired until proper adjustment is obtained. After adjustment, the top plate is locked in position by a clamp nut on the forward end of the centering worm spindle.

F. **STEERING ENGINE**.

9056. The steering engine is the means by which the directive force of the gyro is multiplied to give sufficient power for effective operation of the steering rudders to correct deviation recorded by the gyro.

9057. The engine has a single cylinder of 1" bore and suitably machined and lapped for the insertion and movement of a piston. The piston is operated by the admission of air at reduced pressure at either end, air being admitted through a valve operated by the pallet mechanism as described above. The engine is attached to the starboard upper side of the gyro pot with two holding screws. Recesses are machined on the cylinder to fit the rectangular lugs extending from the gyro pot.

9058. Parallel with and below the engine cylinder is drilled a hole for the valve liner. Two wide annular grooves are machined inside this hole, the raised portions thus left between grooves and on each end of hole form a seat for the valve liner. Communicating ports are drilled from each groove to the piston bore, thus leading to the forward and after ends of the piston. The valve liner has machined on its outside surface three annular grooves, forming four collars; the central collars containing the narrow or supply chamber and the end ones the exhaust chambers. Spaced evenly on the inside of liner are four narrow annular grooves; the two end ones being drilled with radial holes with exhaust cham-

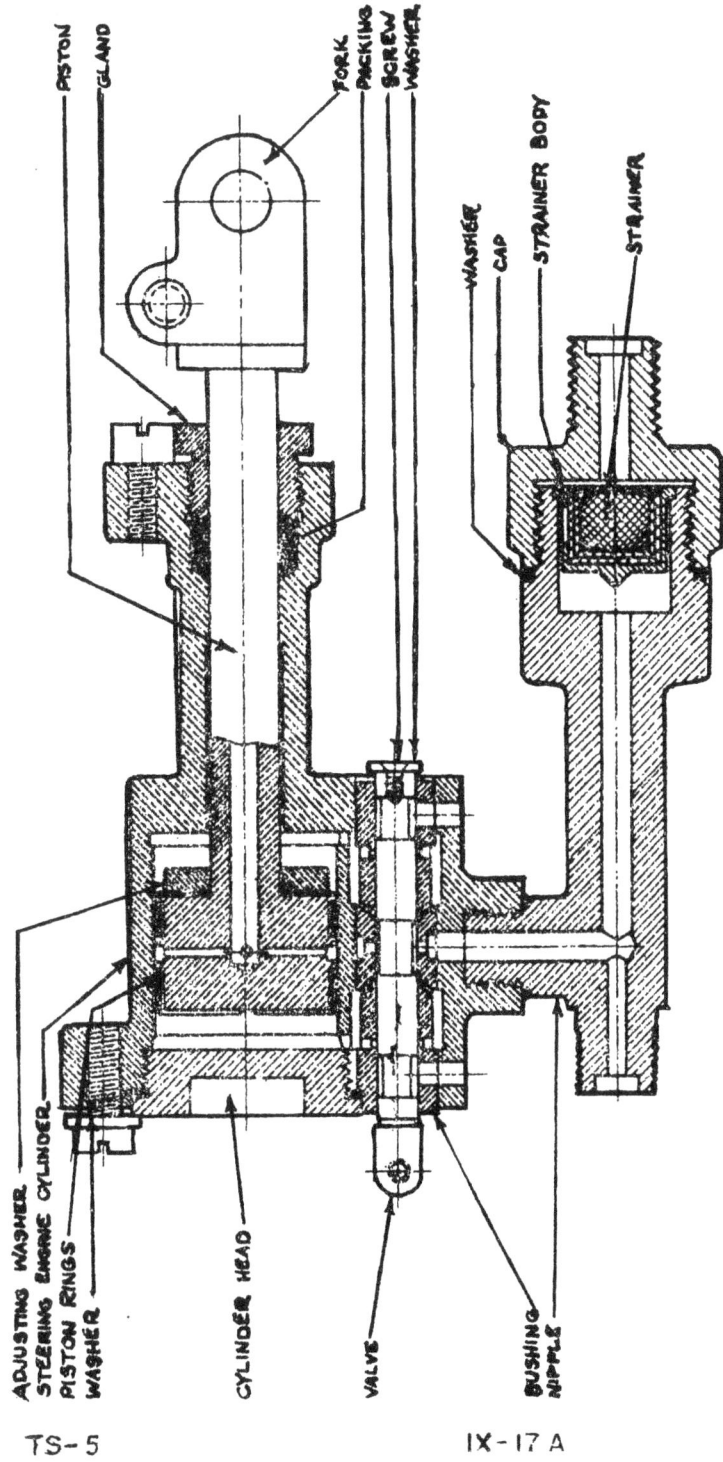

DIAGRAMMATIC VIEW OF STEERING ENGINE

ber on the outside. The two inner grooves have holes drilled through the liner terminating abaft the central collars. Holes are also drilled through the raised portions between the inner grooves terminating in outside central groove. Thus when the liner is in place, the outside collars of same are resting on the raised portions of its hole, and the supply or central chamber is lined up with the inlet nipple, which is screwed and sweated to the cylinder.

9059. The valve operating inside this liner is machined with three annular grooves, forming four collars. The two end collars are narrower than the two inside ones. The central groove is the supply and the end ones are exhaust. Drilled parallel with the inlet nipple through the cylinder body, and through the end collars of valve liner are two radial holes or exhaust ports to the afterbody.

9060. The piston and rod are machined in one piece. The end of the piston rod is threaded and screwed into the forked connection for the steering rod. Two wide grooves are machined around the circumference of the piston in which are fitted the four piston rings. A narrow groove between these grooves communicates by radial holes with an axial hole drilled throughout the length of the piston rod to relieve any air leaking by the piston rings. This leakage is thus exhausted into the afterbody.

9061. The piston rod passes through a stuffing box machined in the after end of the cylinder, packed with asbestos wicking held in place by a gland. The gland is locked from turning by a keep screw. The length of the piston stroke in the cylinder and consequently the total angle of the rudder throws is regulated by means of the adjustment of a thickness washer slipped over the end of the piston rod. This washer is placed between the piston and the after end of the cylinder, thus limiting the stroke of the piston to the desired rudder throws. The forward end of the cylinder is closed by a cylinder head, a copper washer being interposed between the head and cylinder. A lock washer is held against the cylinder head with a set screw to prevent loosening up of same.

9062. Air from the air strainer is supplied to the inlet nipple of the steering engine, and passes through the nipple to the valve liner. The central or supply chamber outside of the liner is directly under the inlet nipple and air passes through radial holes to the supply (central) chamber of the valve which operates inside the liner. When the valve, actuated by the pallet mechanism is moved forward the following occurs:

(a) The central chamber of the valve moves into line with the radial holes in the forward annular groove inside the valve liner. The air passes through these holes to the forward annular groove, thence by communicating port to the forward end of the piston, pushing the piston aft and the rudder through the connecting linkage to the left.

(b) At the same time, the after exhaust chamber of the valve moves into line with the after exhaust holes in the valve liner. Air on the after end of the piston is thus exhausted through exhaust holes in liner to exhaust chamber of valve, and thence through exhaust port in after valve liner collar and engine cylinder to the afterbody.

(c) The after middle collar of the valve closes off the holes from the supply chamber and allows the forward middle collar to open the supply holes in the liner. Thus passing air to the annular groove of the liner through the communicating port to the forward end of the cylinder. The forward middle collar at the same time closes off the forward exhaust holes of liner.

(d) Air that leaks over the piston rings passes to the central groove and then via the radial holes to the hollow stem and to the afterbody.

(e) The reversed action takes place when the valve is moved to the after position.

G. **STEERING LINE.**

9063. The steering line is that portion of the vertical rudder operating mechanism between the steering engine and rudders. The motion of the steering engine piston is transmitted through a steel rudder rod connecting from the fork on the end of the steering engine piston rod to a rudder connection passing through a stuffing box in the after bulkhead, thence through an adjustable extension to the rudder yoke on which the inner ends of the vertical rudders are secured. Thus it will be seen that with the steering engine piston moving aft, the rudder rod is being pushed in the same direction changing the angle of the rudder yoke with a consequent movement of the vertical rudders to the left. Forward movement of the piston will have the reverse effect.

9064. The Mark 13 torpedo has its vertical rudders abaft the propellers, therefore vertical rudder posts secured to the rudder yoke having lever arms attached to their outer ends connected by rudder arms to similar lever arms on the vertical rudders transmit the motion of the steering engine to the vertical rudders.

9065. The adjustable extension of the steering rod between the bulkhead connection and the rudder yoke furnishes facilities for equalizing the right and left positions of the rudders by lengthening or shortening same, in the same manner as already described for the horizontal rudder.

H. **GYRO REDUCING VALVE.**

9066. As previously mentioned, air for the steering engine is supplied from the air strainer at approximately 420 lbs., pressure. The nipple for the inlet air on engine is a "T" connection,

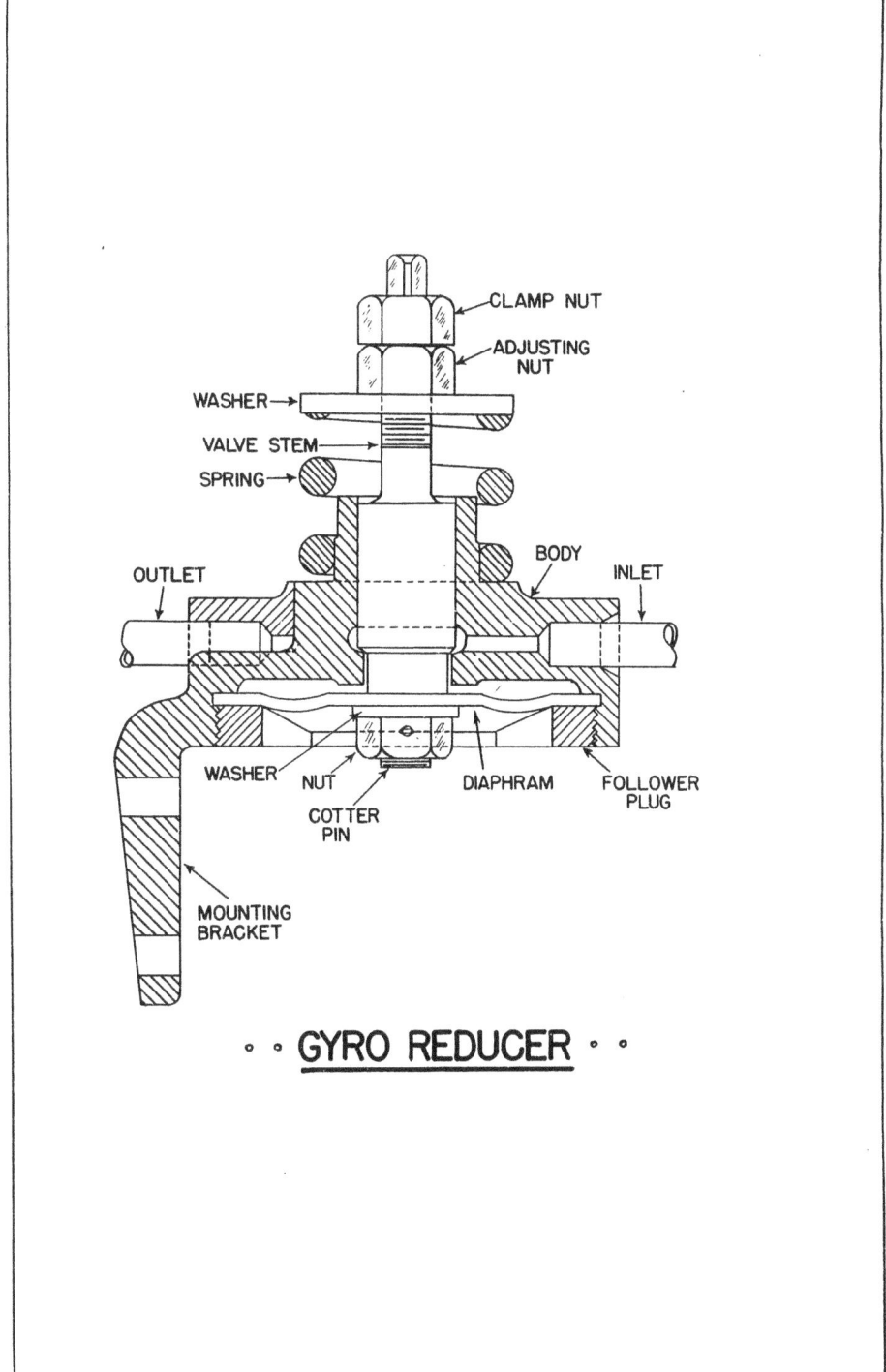

one lead going to the gyro reducing valve where it is further reduced to 125 lbs., per square inch before it is conveyed to the gyro wheel as sustaining air.

9067. The gyro reducing valve is composed of a circular composition body, on the lower end of which is an extended bracket for attaching to the gyro housing. A hole is drilled in the central portion of the body into which the valve is fitted by lapping. An annular groove machined near the lower end of this hole permits the entry of inlet air around the valve stem. A large shallow recess machined in the lower end of the valve body contains the seat for the reducing valve diaphragm. The upper end of this recess is dished out to permit sufficient space of the volume of air necessary to operate the valve and for the passage of reduced air to the outlet nipple. This recess is threaded to fit similar threads on the diaphragm follower plug.

9068. The reducing valve and stem is lapped into the hole above the annular inlet groove in the reducing valve body. The lower diameter of the valve is reduced to permit the passage of air to the space over the diaphragm. The diaphragm is rigidly secured to the valve against a shoulder of reduced diameter and held in place by a washer and nut. The outer circumference of the diaphragm is clamped to its seat in the body by the follower plug. The upper end of the valve extends through the body at a reduced diameter which is threaded for an adjusting and clamp nut. A heavy spring is inserted over the valve stem and compressed between its seat on the body and a washer on the valve stem. The compression is adjusted by means of the adjusting nut on the valve stem, after which the adjustment is locked with the clamp nut. Inlet and outlet pipes are secured and soldered into the valve body.

9069. Upon entry of air through the inlet connection of the gyro reducing valve, it passes around the valve seat down around the space above the diaphragm. This moves the valve and stem down against the compression of the valve spring, until the air pressure exerted on the diaphragm and the spring tension equalize each other. The air allowed to pass the valve when in this position is the reduced pressure used to sustain the velocity of the gyro wheel. This pressure is adjusted to 125 lbs., per square inch. The air passes from the reducing valve to a nipple on the side of the gyro pot, then through a small passage drilled in the shelf of the gyro pot to the top bearing holder.

I. <u>TEST, DISASSEMBLY AND ASSEMBLY OF STEERING ENGINE</u>.

9070. Test:
   (1) Place engine in block of leakage test stand.
   (2) Connect low pressure air line to inlet nipple. Blank off air to reducer and turn on air.
   (3) Operate valve to move piston to full "In" position.

STEERING ENGINE IN TEST STAND

(4) Fork must have a minimum clearance of from ".001 to ".002.
(5) Move valve reversely and measure length of piston stroke. Full stroke should be ".444 (7/16). If stroke is too short, it will be necessary to remove sufficient stock from adjusting washer to obtain full throw after engine is disassembled.
(6) Test for leaks around valve and through piston exhaust. The ability to adjudge excessive leakage is dependent upon knowledge of properly fitted steering engines. In general, leaks through the piston exhaust should not be audible, and the valve should fit snugly enough to prevent air from blowing into the atmosphere. There will always be a certain amount of leakage past the rings and out the piston exhaust as well as around the valve. The limiting value of such leakage must remain a matter of judgment. As an aid in determining this limiting value, operate the valve and note the drop in pressure on supply line during the stroke. This drop in pressure should be in the neighborhood of 10 lbs. per square inch (using a supply line 3/16" inside diameter and 2 feet long). Excessive leakage from the exhaust is corrected by renewal of piston rings. In extreme cases pistons or cylinder may have to be replaced.

9071. Disassembly: TOOL NO.
(1) Loosen fork clamp screw.............................. 49
(2) Screw off fork. Scribe position before removal and count turns.
(3) Remove gland lock screw.............................. 41
(4) Remove gland.. 18
(5) Remove valve. Hold forked end with pliers and unscrew valve stop screw........................... 92,37
(6) Remove locking screw and washer from cylinder head... 41
(7) Remove cylinder head................................ 25
(8) Remove piston and adjusting washer, using piece of wood to push piston out of cylinder............. 213

9072. Assembly:
(1) Immerse parts in clean spirits and blow dry with air.
(2) Clean parts, oil (A) and assemble as follows:
(3) Insert piston with adjusting washer................. 213
(4) Screw in cylinder head with gasket in place........ 25
(5) Install locking screw for cylinder head............ 37
(6) Insert valve. Secure with valve stop screw........ 92,37
(7) Pack and install gland nut.......................... 18
(8) Screw in fork to scribe mark, using same number of turns as on disassembly.
(9) Set up fork clamp screw............................. 49
(10) Test for leaks and clearance as outlined in test before disassembly. When engine passes these tests it is ready for assembly on gyro pot.

J. DISASSEMBLY AND ASSEMBLY OF PALLET MECHANISM.

9073   Disassembly: TOOL NO.

   (1) Remove cotter pin and pin for bell crank............   92
   (2) Remove cotter pin and conical washer and lift
       out bell crank and link.........................   92
   (3) Remove cotter pins from pins in pallet slide
       cover and remove pallet pawls with adjusting
       links and adjusting screw   92
   (4) Remove clamp screw and nut and pallet blade.....   37
   (5) Remove pallet shaft with spring and cam pawls.
       (Care should be taken when removing pallet
       shaft that the small bridge spring in slot on
       pallet shaft does not get lost)
   (6) Remove 4 screws, pallet slide cover and bell
       crank bracket...................................   37
   (7) Remove screw plugs, springs and actuating
       plungers for pallet slide   40
   (8) Remove 6 screws, cap for driving spindle bracket
       and driving spindle   40
   (9) Remove cam bevel gear.
  (10) Remove pallet slide with pallet holder and
       adjusting screws assembled   37
  (11) Remove 2 screws, cap for rockshaft bearing and
       rockshaft assembly..............................   37

9074. Assembly:

(a) (1) Clean and inspect pallet slide and holder Note
      hardened insert fits tight on pallet slide. Note
      that pallet shaft is a snug, lapped fit in pallet
      holder Stone out any burrs found on pallet
      slide Note that pallet holder adjusting screws
      work freely Install pallet slide with holder
      assembled on top plate   37
   (2) Clean and note that bushing is tight on cam
      bevel gear Inspect for burrs on gear teeth
      and bearing surface, and install cam bevel gear.
      Note that with pallet slide against the stops
      on top plate only a maximum of 3 5 teeth play
      exists when moving cam bevel gear to and fro
      If more play exists, it will be necessary to
      dress the pallet slide down with a stone in
      wake of stops in order to obtain the proper
      clearance.
   (3) Replace driving spindle and cap for bearing
      bracket. Secure with 6 holding screws..........   41
   (4) Clean and inspect with particular attention
      to pins and stop pins for pallet pawls being
      tightly riveted on, and replace pallet slide
      cover and bell crank bracket. Secure with 4
      holding screw After tightening the screws,
      note that pallet slide moves in and out on top
      plate without restriction.......................   37

TOOL NO.

    (5) Clean, inspect and replace pallet slide spring plugs, springs and screw plugs. Screw plugs should be flush with holes in top plate when tight. Turn cam bevel gear and note that spring plugs are moving freely.... 41

(b) Replace pallet shaft with cam pawls and spring:
    NOTE THAT:
    (1) Cam pawl is a snug fit on end of pallet shaft without any lost motion.
    (2) Pallet shaft is a lap fit with unrestricted rotary movement in bearing in pallet holder ....WE162&163
    (3) Ends of spring are free to slide in slot on shaft.
    (4) Diameter on upper end of pallet shaft extends to just below upper end of bearing in pallet holder, thus permitting pallet blade to be installed on shaft without vertical lost motion.

(c) Replace pallet blade on pallet shaft (with cut-away side down). Note that:
    (1) Pallet blade is free of burrs and that edges are sharp and square.
    (2) Pallet blade fits snugly on pallet shaft
    (3) Center pallet blade on pallet shaft by eye and set up on pallet blade clamp screw lightly (just enough to hold blade to shaft when shaft is rotated)... 37
    (4) Leave just enough vertical play between pallet blade and pallet holder to allow unrestricted rotary movement of pallet shaft.

(d) Center pallet blade on pallet shaft as follows:
    (1) Turn cam gear until cam pawls are in the (all the way out) position, and install gyro.... 246
    (2) Turn gyro until cam on cam plate is facing opposite cam pawls.
    (3) Move cam pawls "In" toward cam plate by turning cam gear, at the same time, swing the gyro in azimuth so that the cam on cam plate is brought in contact with cam pawls. At the slightest indication of cam pawls binding on cam plate, stop turning cam gear.
    (4) With feelers measure clearance between pallet blade, and right and left stop pins on pallet slide cover. Tap pallet blade gently until clearances are equal on both sides as gyro is swung in azimuth. This operation centers the pallet blade on pallet shaft, assuming that the stop pins are not bent and that the pallet slide is in its approximate full "in" position.
    (5) Move cam pawls to the "out" position and set up on pallet blade clamp screw, using two screw drivers 37

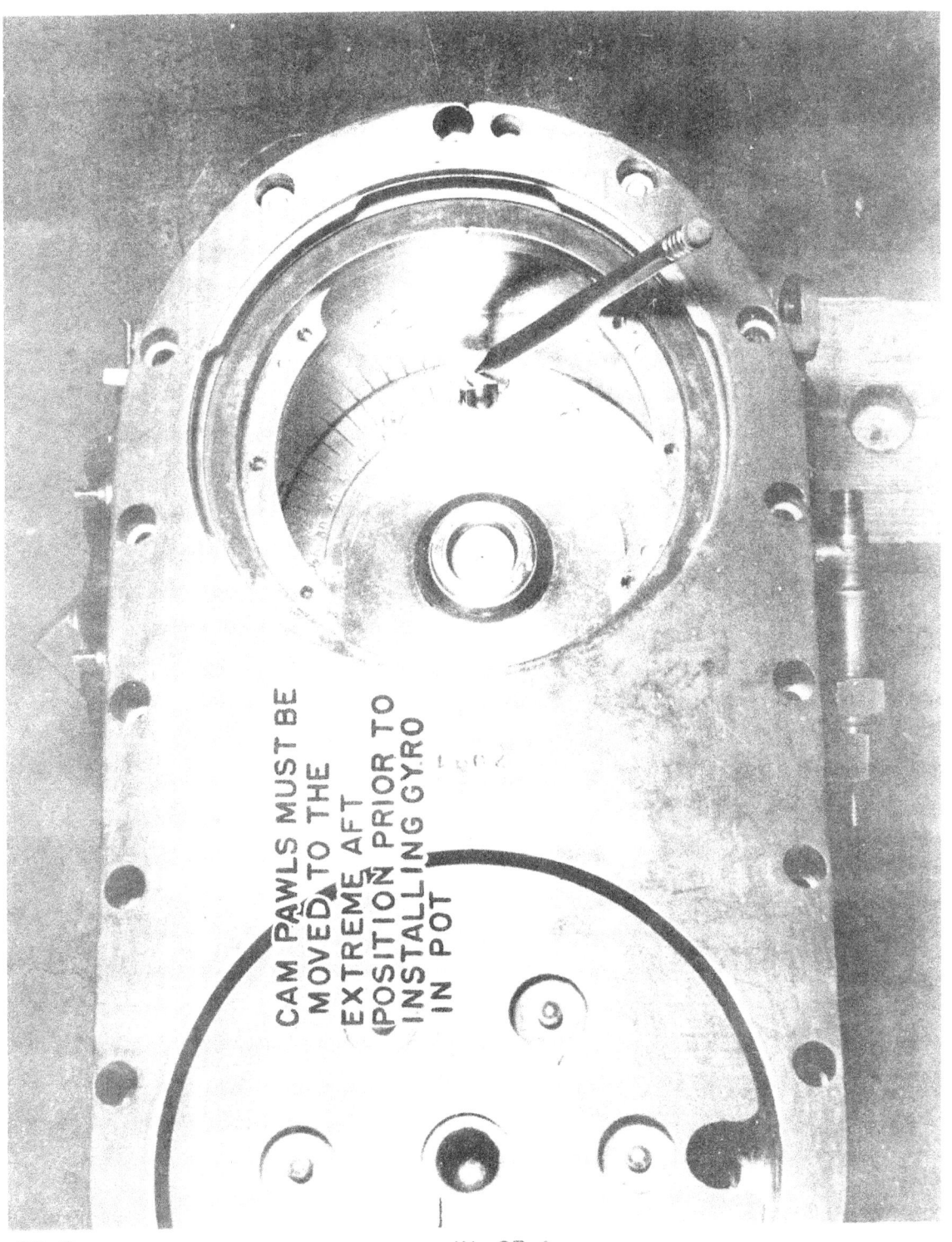

PRIOR TO INSTALLATION OF BOTTOM HEAD, CHECK TO SEE THAT SCRIBE MARKS WILL COINCIDE.

TOOL NO.

(e) Adjust clearance between cam and cam pawls:

   (1) Move cam gear until cam pawl is in the full "in" position. Swing gyro in azimuth and position pallet blade toward the right stop pin.

   (2) With feelers, measure clearance between pallet blade and stop pin. Clearance should be "010. (See Fig. #1).

   (3) If clearance is greater or less than "010, loosen pallet holder clamp screw and turn pallet holder adjusting screws, (See Fig. #2) in the proper direction, thus moving the cam pawls in the proper direction until "010 clearance is obtained.

   (4) Swing gyro in azimuth and position pallet blade toward the left stop pin. Clearance will be "010 if pallet blade is properly centered on pallet shaft, and stop pin is not bent (See Fig. #3). The object of this adjustment is to attain two ends, (a) to obtain the proper pallet movement to right and left, and (b) the cam pawl adjustment in the full "in" position. The force with which the cam pawl strikes the cam plate as the gyro is swung in azimuth should be identified by a sharp metallic "click" as the transfer takes place. A slight feeling of drag should be noticed, but no binding.

   (5) Set up on pallet holder clamp screw and recheck clearance.................................... 37

(f) Replace pallet pawls and linkage assembly on pallet pawl pins:
NOTE THAT:

   (1) Adjusting screw works freely in links and that edges of pallet pawls are sharp.

   (2) Fit of pins and holes is snug without binding.

   (3) Secure pallet pawls with cotter pins.

   (4) Move pallet blade between pallet pawls and measure clearance on either side of pallet blade. Clearance should be not less than "003 nor more than "005 (See Fig. #4).

(g) Check centering of pallet blade and zero graduation on retainer plates with line on marker in pot:

   (1) Lock gyro, and set top plate centering device on zero... 205.246

   (2) Place and hold pallet pawls in a neutral position so that the contact points are opposite each other.

   (3) Turn cam gear and note if pallet blade passes between pallet pawls. If pallet blade touches

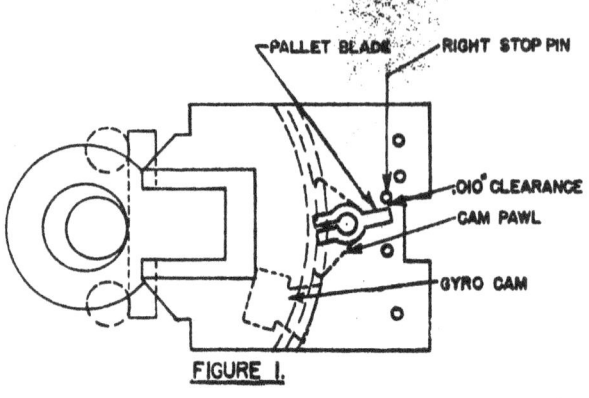

FIGURE 1.

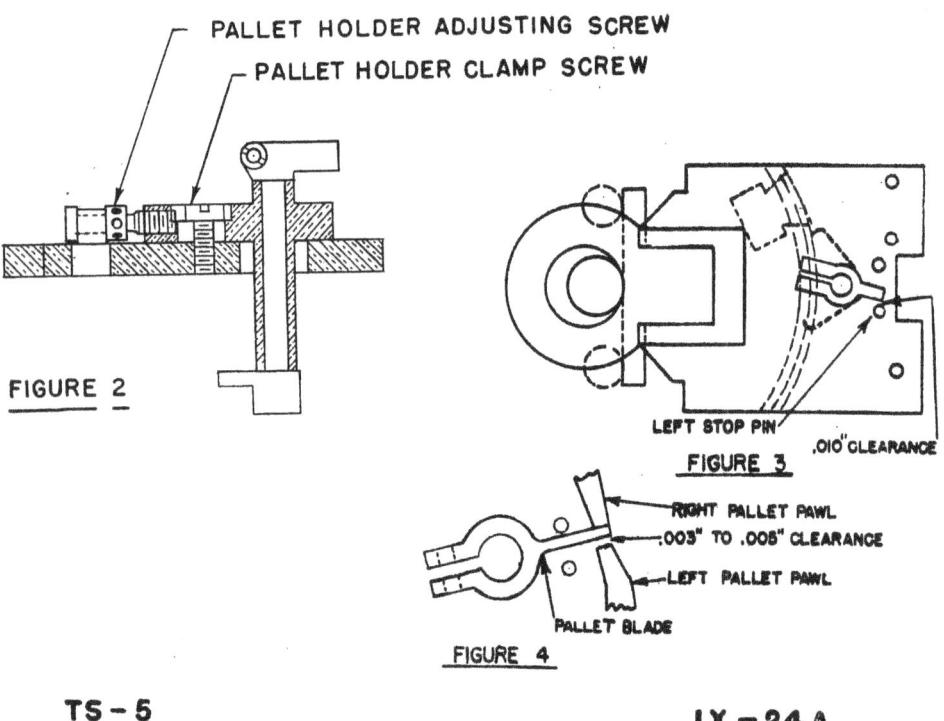

FIGURE 2

FIGURE 3

FIGURE 4

TS-5

IX-24A
ORIGINAL PAGE

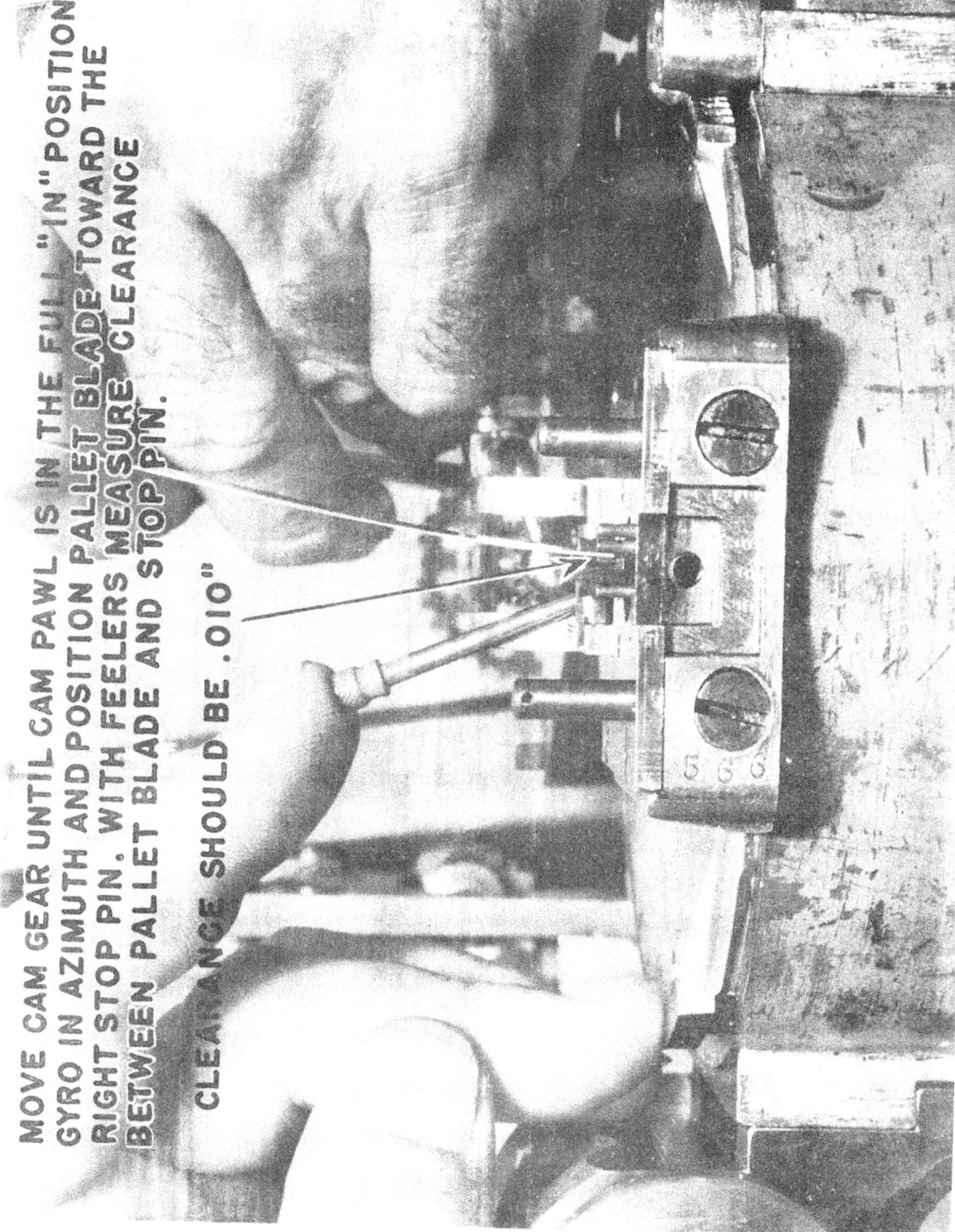

MOVE CAM GEAR UNTIL CAM PAWL IS IN THE FULL "IN" POSITION. GYRO IN AZIMUTH AND POSITION PALLET BLADE TOWARD THE RIGHT STOP PIN. WITH FEELERS MEASURE CLEARANCE BETWEEN PALLET BLADE AND STOP PIN.

CLEARANCE SHOULD BE .010"

WITH GYRO LOCKED, PALLET BLADE SHOULD PASS BETWEEN PALLET PAWLS. CLEARANCE SHOULD BE .003" TO .005"

TS-5 IX-24C

TOOL NO.

either pawl instead of passing between, the blade is off center on pallet shaft or marker for graduation on retainer is out of line.

(4) Turn top plate adjusting worm spindle slightly in a direction to cause pallet blade to run between the pallet pawls. Having obtained this position, note if pallet blade clearances are equal on both sides. If the clearances are not equal, the blade will have to be re-centered on the shaft. In such a case it will be necessary to repeat steps outlined above after re-centering blade.

(h) Install bell crank and link:
(1) Insert pin for bell crank and secure with cotter pin. Note that small conical washer is placed on bottom side of eye on bell crank link when assembled on pin of left pallet pawl extension. There should be no play in this connection after inserting cotter pin over eye on bell crank link.

(i) Install rockshaft assembly:
(1) Note that bell crank and valve connecting arms are clean and tightly riveted.
(2) Replace friction springs, buttons and valve rockshaft assembly in bearings. Note that ball end of bell crank arm is a snug fit in slotted portion of bell crank without binding. Secure cap for bearing with 2 holding screws.......... 37

(j) Replace steering engine:
(1) Secure steering engine with holding screws...... 49
(2) Move valve all the way out; move valve connecting arm in the same direction and note if moving further than valve. Move valve all the way in; move valve connecting arm in the same direction and note if moving further than valve. With valve connecting arm moved a greater distance than valve in either direction, the connection is properly lined up.

(k) Linkage friction test:
(1) (On movable top plate mechanism). Adjust the friction in the valve linkage until a force of 10 to 12 oz. on the lower end of the valve lever is necessary to operate the linkage (with valve connected). In addition to stretching the springs of spring buttons under the rockshaft it is often necessary to further increase friction in rockshaft bearing by filing material from joint surface of rockshaft bearing cap. Counterweights on rockshafts of latest mechanisms also help to eliminate flutter from linkage.

TS-5   IX-25   Original Page.

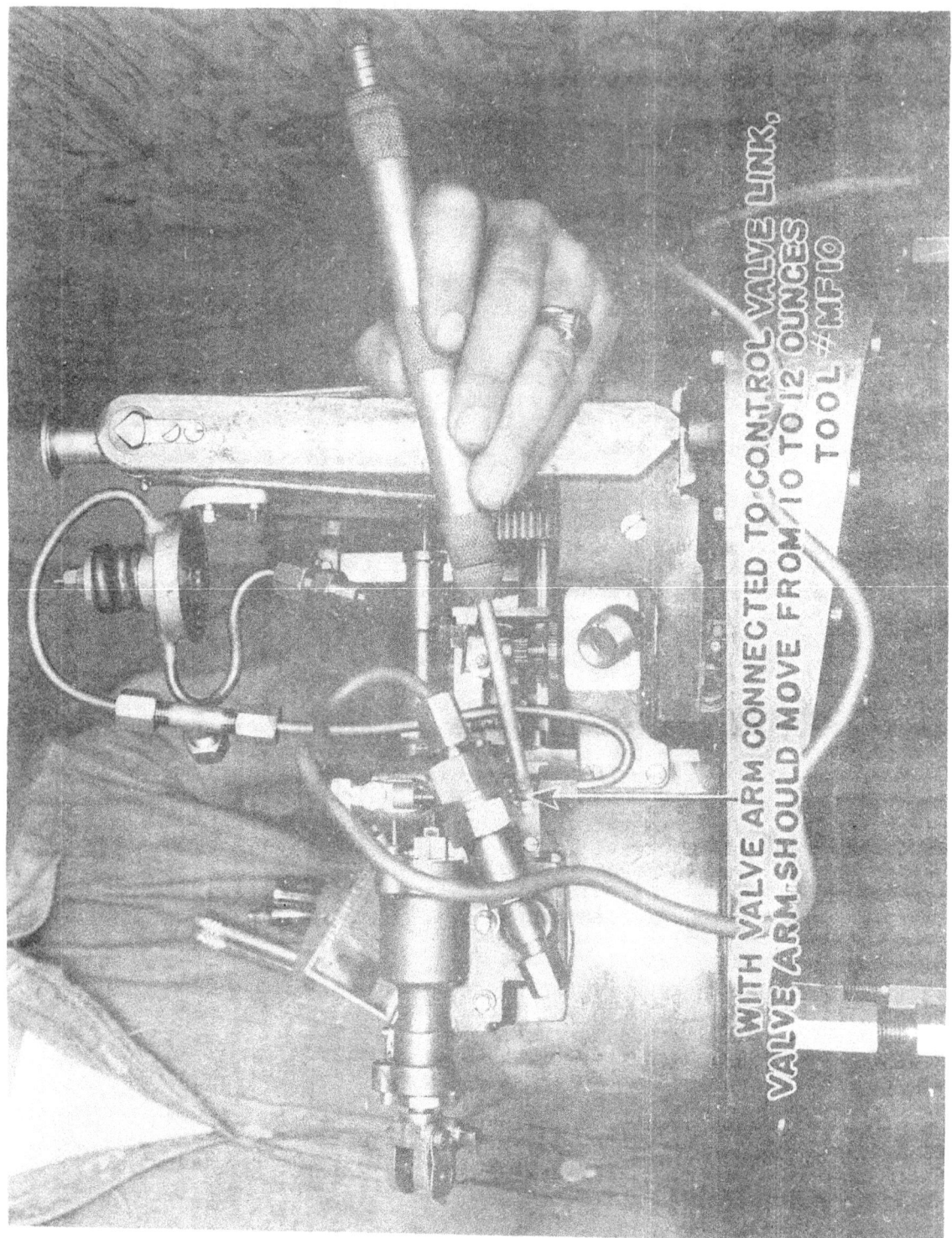

WITH VALVE ARM CONNECTED TO CONTROL VALVE LINK, VALVE ARM SHOULD MOVE FROM 10 TO 12 OUNCES TOOL #MF10

TOOL NO.

(1) Adjust clearance between pallet blade, and left and right pallet pawls:
  (1) Connect valve arm to valve connecting link........ 246
  (2) Turn driving gear shaft until can bevel gear moves the pallet slide with pallet blade to its extreme "outer" position.
  (3) Move pallet blade in line with the left pallet pawl. Clearance should be ".010. (See Fig. #6). If desired clearance is not obtained, loosen clamp screws for adjusting head on rockshaft, and rotate adjusting head on rockshaft until a ".010 feeler can be placed between pallet blade and left pallet pawl. Set up on clamp screws, at the same time holding control valve all the way forward and bell crank arm up. Thus tightening the connection with feelers in place........ WE2,37
  (4) Move pallet blade in line with right pallet pawl. Clearance should be ".005 (See Fig. #7). If desired clearance is not obtained, loosen clamp screws on right and left pallet pawl adjusting links, and rotate the adjusting screw in a direction to obtain proper clearance. Set up on clamp screws and recheck clearance between pallet blade and both, right and left pallet pawls............. WE2,37
As will be noted for pallet blade and pallet pawl clearance adjustment, (shown in Fig. 7), the clearance should be ".005; whereas the clearance of pallet and left pallet pawl, (shown in Fig. 6), should be ".010. This difference in clearance is necessary in order to obtain a uniform valve movement with a snappy steering engine throw for both pawls. The motion from the pallet to the right pawl is transmitted through the right and left adjusting links, and adjusting screw to the left pawl, with attendand lost motion due to tolerance in linkage. Whereas the motion from the pallet to the left pawl is transmitted by direct contact, and therefore, without lost motion.
  (5) Place engine valve in a neutral position so that the contact points of pallet pawls are opposite each other, and again check centering of pallet blade and zero graduation of top plate centering device. As driving gear spindle is turned, moving the pallet blade toward its "inner" position, alternate pushing the pallet blade to the right and left. When cam pawls have reached the extreme "inner" position, the pallet blade should snap back in line with the fore and aft axis of gyro top plate. on its "outer" travel should pass between pallet pawls without contacting either. See step (g)(3).
  (6) With pallet centered and top plate centering device on zero, unlock and remove gyro. Verify that adjustable index marker lines up with zero on the retainer plate. If not, loosen clamp screws and gently tap marker in place. Set up on clamp screws........... 246,37

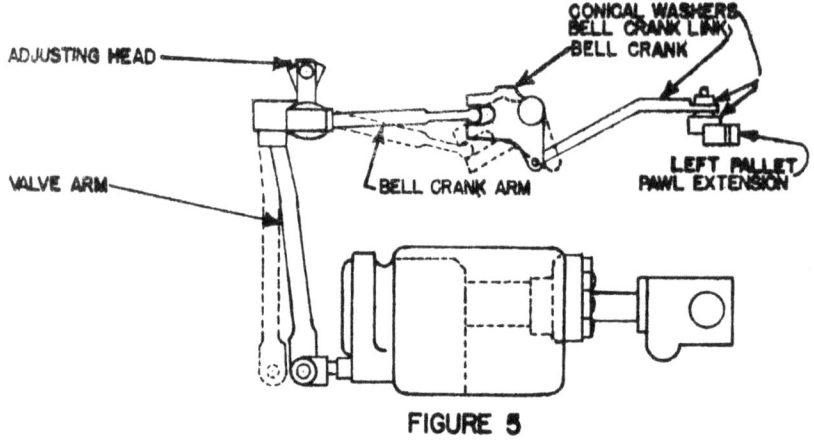

FIGURE 5

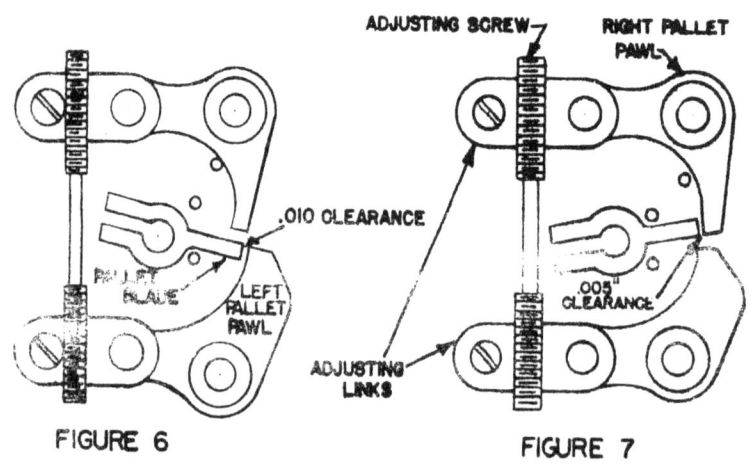

FIGURE 6

FIGURE 7

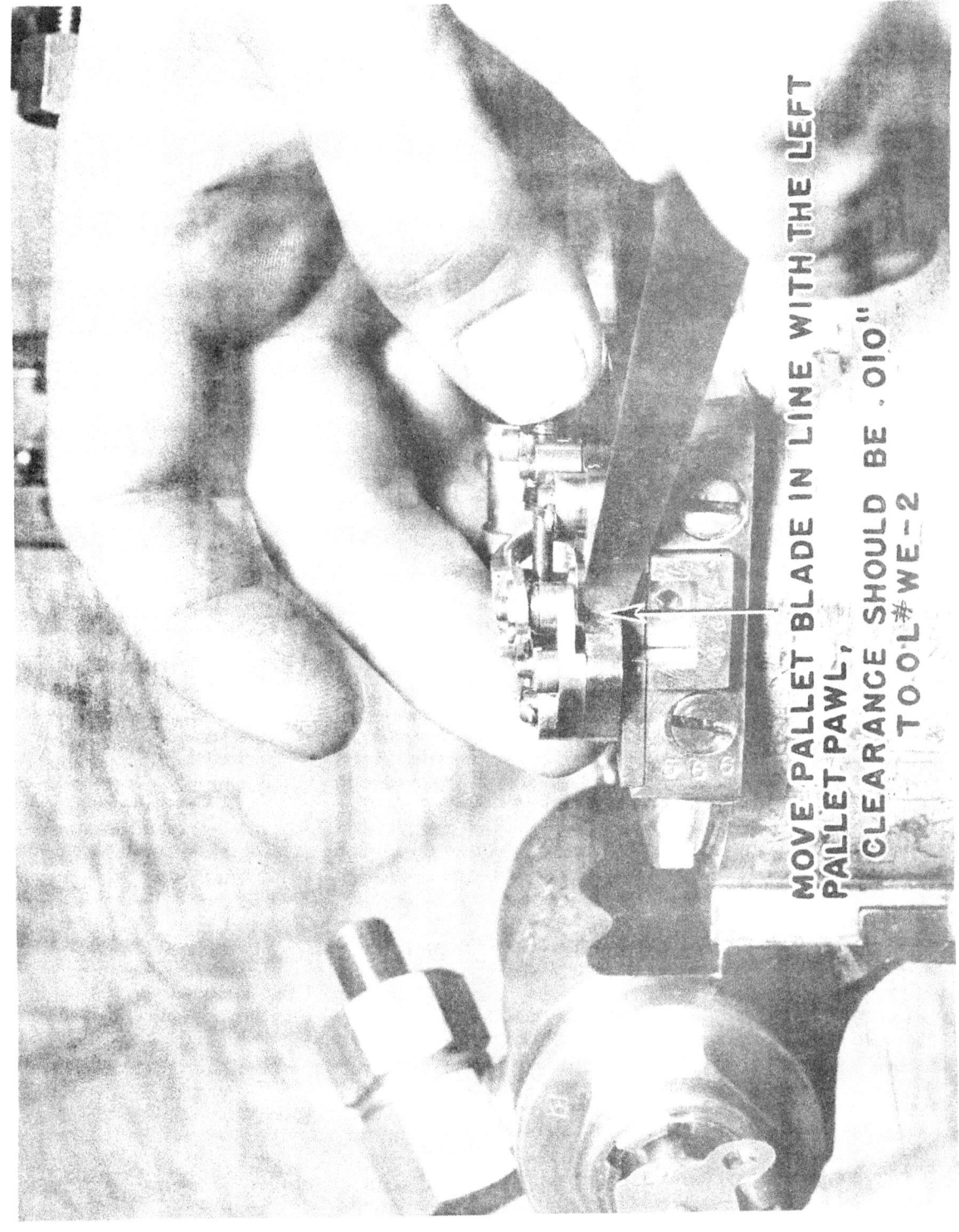

MOVE PALLET BLADE IN LINE WITH THE LEFT PALLET PAWL, CLEARANCE SHOULD BE .010" TOOL #WE-2

MOVE PALLET BLADE IN LINE WITH THE RIGHT PALLET PAWL CLEARANCE SHOULD BE .005
TOOL # WE-2

Check of Adjustments.

9075. At this point all adjustments have been made to insure proper functioning of the pallet mechanisms, and successful operation should follow. However, THE TRUE TEST OF PROPER ADJUSTMENT IS THE MANNER IN WHICH THE PALLET MECHANISM OPERATES THE VERTICAL STEERING ENGINE. At the Naval Torpedo Station at Newport, a running stand test is given as a material test and to "run in" all movable parts. This is optional in the service, but is advisable as a test of the operation of the pallet mechanism. If a running stand test is not given, at least an "air test" as described below should be given to insure proper functioning of the pallet mechanism.

9076. IT IS IMPORTANT to note that in movable top plate mechanisms, the amount of play in the linkages due to allowed tolerances and to wear and to "spring" in the lever arms make the specification of clearances between pallet pawls and blade only approximations, at best. These stated clearances should therefore be looked upon as fairly exact guides, rather than as indisputable specifications, and reasonable variations of adjustments should be made if necessary to obtain proper operation of the mechanism during the "air test".

Air Test.

9077. This test should be given if a running stand test is not given, to insure proper functioning of the pallet mechanism. For this test, it is not necessary to spin the gyro, but it should be in position in the pot and properly adjusted for height.

(1) Mount gyro mechanism with gyro in test stand. With the gyro locked, rotate the spin wheel to see if the unlocking gear functions after 60 revolutions have been made. See that the centering pin has been fully withdrawn and is clear of the gyro.

(2) Hook up a 400 lb. per sq. in. air lead to the vertical engine. Start the electric motor which drives the pallet slide, (turn pallet driving gear by hand if a test stand is not available), and turn on the air pressure. Swing the gimbal right and left so that the cam pawl will strike alternate sides of the cam on the cam plate, and note the action of the vertical steering engine. The engine should respond with prompt strokes as the gimbal ring is swung from side to side. The exhaust air explosions, if the valve is central, will ordinarily be equal in intensity, but this is not as reliable a guide as the promptness of the stroke of the piston, watch the piston. If the results appear unsatisfactory, the cause must be sought for by careful examination of the various parts in operation. No instructions on this checking can be given; the only guide is the application of a reasonable amount of thoroughness and intelligence.

(3) When the various requirements indicated have been met, the mechanism is ready for a running stand test, if such is considered advisable.

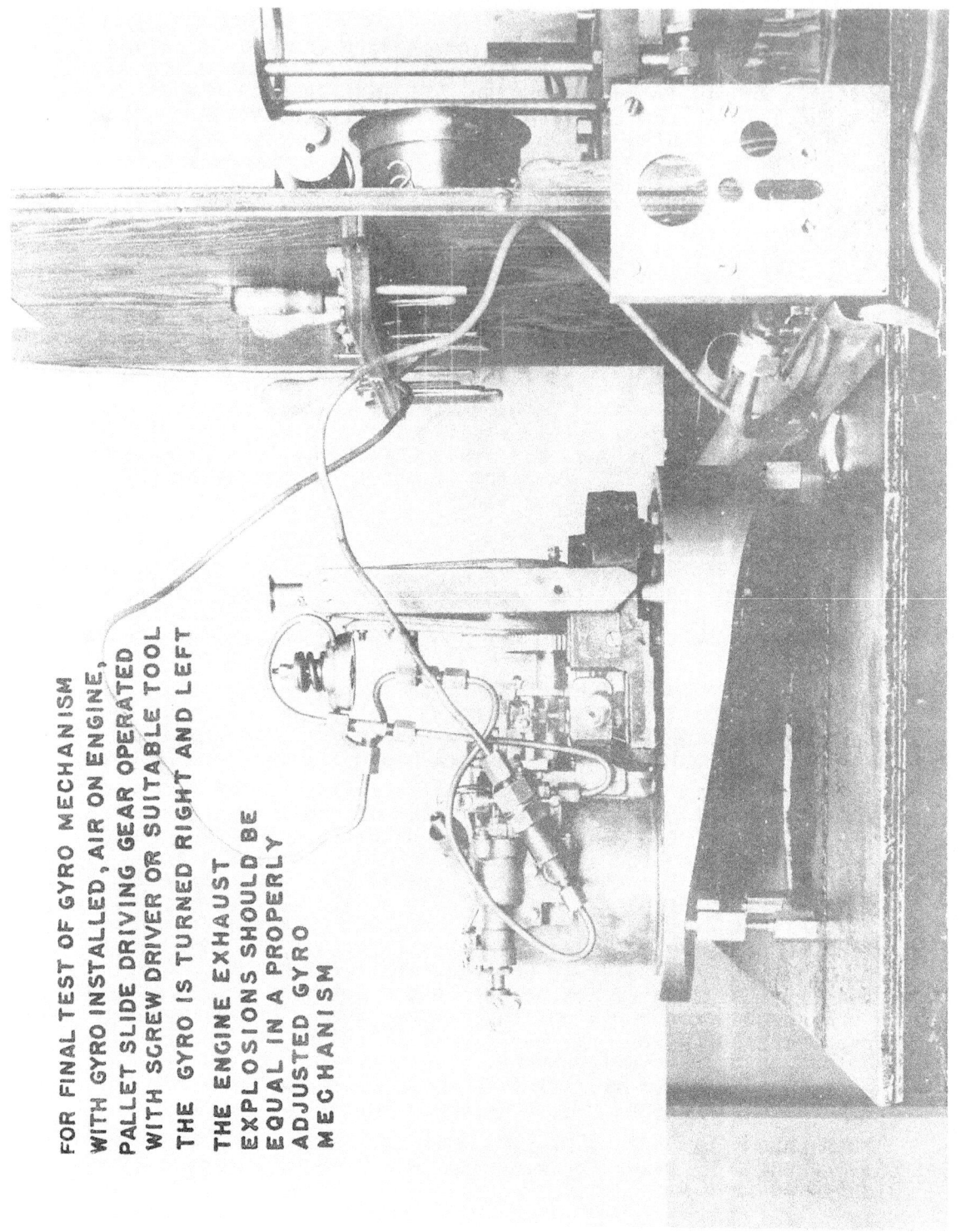

FOR FINAL TEST OF GYRO MECHANISM WITH GYRO INSTALLED, AIR ON ENGINE, PALLET SLIDE DRIVING GEAR OPERATED WITH SCREW DRIVER OR SUITABLE TOOL THE GYRO IS TURNED RIGHT AND LEFT

THE ENGINE EXHAUST EXPLOSIONS SHOULD BE EQUAL IN A PROPERLY ADJUSTED GYRO MECHANISM

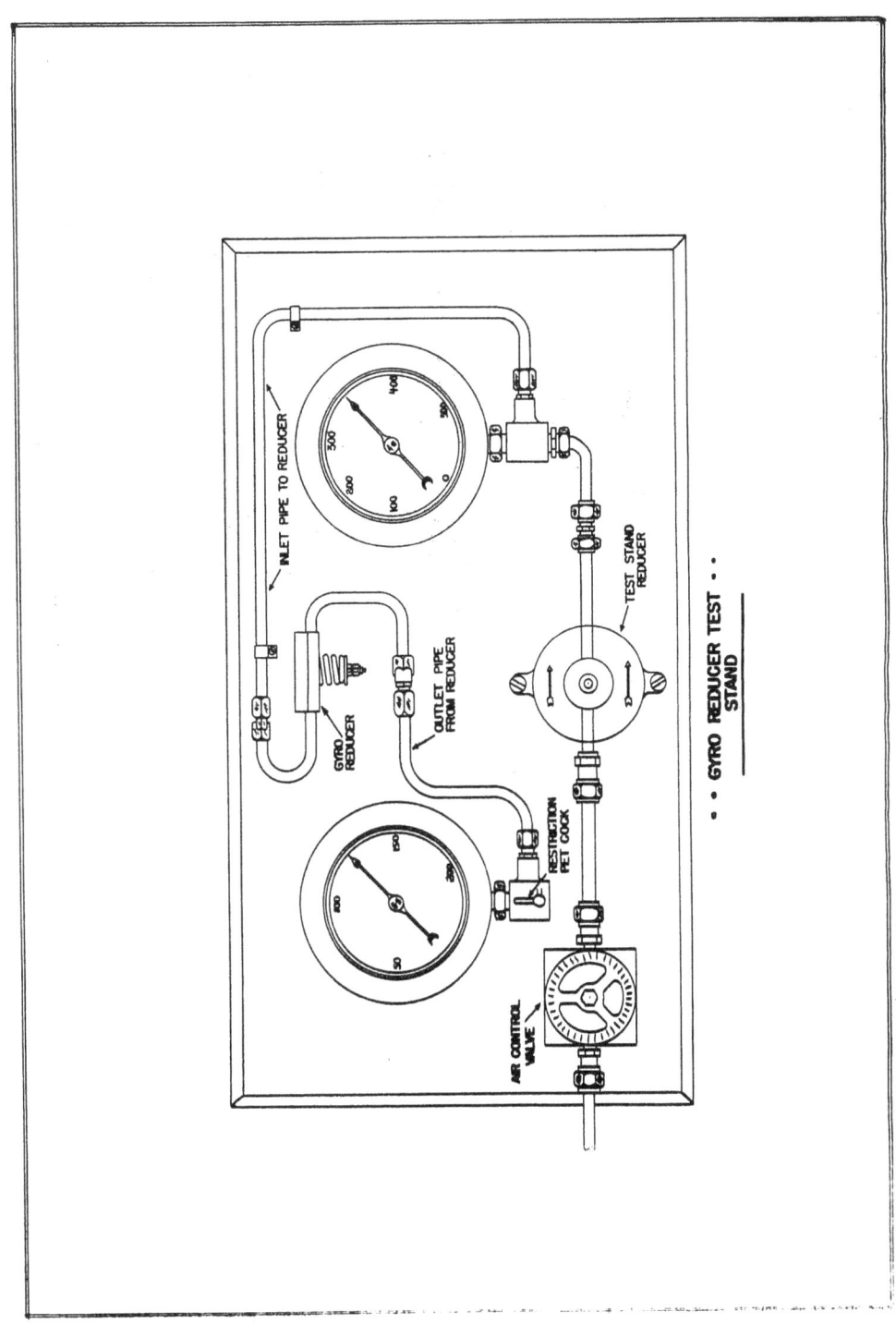

K. **TEST OF GYRO REDUCING VALVE.**

9078   To test gyro reducing valve for assembly, leakage, functioning, and proper pressure proceed as follows:

(1) Connect reducing valve inlet connection to pipe from 500 lb. gauge on test panel.
(2) Connect reducing valve outlet connection to pipe leading to restriction pet cock and 200 lb. gauge.
(3) Close restriction pet cock to 200 lb. gauge.
(4) Remove lock nut, adjusting nut, washer and spring from reducing valve.
(5) Turn on air and control pressure at 450 lbs. on the 500 lb. gauge by means of control valve.
(6) If the reducing valve does not leak, no pressure will be recorded on 200 lb. gauge. However, the valve may be considered satisfactory on this test provided that only a small leak (20 lbs. or less) is present.
(7) Close control valve, replace the spring washer and adjusting nut on the reducing valve and open restriction pet cock.
(8) Set up on adjusting nut and with 400 to 450 lbs. on the 500 lb. gauge, adjust the nut until the 200 lb. gauge registers 125 lbs.
(9) Shut off the air and turn on again several times to ascertain that valve functions properly and that the adjustment is correct. Tap reducing valve body slightly and note if pressure crawls up or down.
(10) Replace and tighten lock nut, holding adjusting nut from turning.

CHAPTER TEN

		ARTICLES
A.	General Description of Mark 12-1 Gyro	1001-1003
B.	Detailed Description of Construction	1004-1017
C.	Air Sustaining System	1018-1022
D.	Gyro Balancing	1023-1028
E.	Disassembly of Gyro - Test for Leaks	1029-1031
F.	Examination of Parts	1032
G.	Assembly and Adjustments of Gyro	1033
H.	Pot Clearances and Adjustments	1034-1035
I.	Stand Test	1036-1038
J.	Creep, or Latitude Error	1039-1052
K.	Precession	1053-1055
L.	Compensation for Creep	1056-1059

A. GENERAL DESCRIPTION MARK 12-1 GYRO.

1001. The gyro used in the Mark 13 torpedo is similar in all respects to gyros used in Mark 8-3C and 3D, 8-3, 11 and 12 torpedoes and is known as the Mark 12-1. It is of the "air-sustained" or "constant spin" type. The inner gimbal ring is machined in the form of a hood for the purpose of preventing air currents, thrown off by the wheel, from striking the gimbal rings and thereby setting up deflecting forces on the sensitive parts of the assembled gyro. The inner ring is so built that it can be rotated through 360° on the side pivots. This introduces a "non-tumble" feature, in that it is not possible for the gyro to tumble through contact with the outer ring or cam plate in case the torpedo rolls heavily. It should be noted, however, that should the torpedo roll as much as 90°, the gyro no longer has a vertical pivot (the former vertical pivot being now horizontal) and the gyro therefore has only two degrees of freedom, and may be tumbled in case the torpedo takes a sheer at this time.

1002. Gyro development in torpedoes has been very active in the past few years and has centered largely around the following features:
   (a) Necessity for constant spin at highest practicable speed (15,000 - 20,000 r.p.m.) to get constant and maximum inertia.
   (b) Hooding of the gyro with a strong serviceable hood that will reduce adverse windage effects.
   (c) Elimination of the possibility of interference between inner and outer rings at all angles of the inner ring.
   (d) Introduction of ball bearings in side and vertical pivots, with a much stronger construction to resist launchings from high speed vessels and airplanes.

1003. The function of the gyro is to maintain a torpedo during its run on the firing course. The gyro has three degrees of freedom, free to spin about its axis, free to turn about the horizontal axis at right angles to the wheel axis and free to turn about a vertical axis, the three axis intersecting at a point located at the center of gravity of the gyro wheel. Torpedo gyroscopes are so designed and built as to have three degrees of freedom when the balance nut is in the static balance position.

B. DETAILED DESCRIPTION.

1004. The gyro wheel has at each end of its hub a spur gear cut from the solid stock of the wheel. Axial recesses are drilled in the two ends of the hub and into these recesses are pressed small hardened steel centers. These centers are supported by six 1/8" steel balls which run in ball cups in a vertical plane at right angles to the hub axis. The ball cups are of special steel with hard ball recesses ground inside the cups. The ball cups are screwed into the inner gimbal ring and are located or positioned by being locked by screws to locking discs. To adjust the wheel bearing, it is necessary to slacken up on the clamping screws through the locking discs and then the locking discs can be turned and will carry with them the ball cups. In order that the torque of turning the ball cup will not be born by the clamp screws, two dowel pins carried by the locking discs are fitted in holes in the cup. The disc is turned by a small pin wrench, which fits the holes drilled in the outer face of the locking disc. The rim of the gyro wheel is machined to form a set of peripheral saw tooth buckets upon which the air jets impinge to maintain constant wheel speed.

1005. The forward wheel cup is fitted with an axial recess for the centering pin of the spinning and unlocking gear. Projecting bosses are machined on the inside of the gimbal ring to provide adequate length of screw threads for the cups. The boss on the after end of the inner gimbal ring (inside the hood) has screw threads cut on its outer surface for the balance nut. To lock the balance nut in any desired position two small screws screw into the hood and bear against the balance nut, thereby pressing it firmly against its screw threads and holding it securely in place. The outer periphery of the balance nut is numbered in tenths of a turn and the zero position is that which will be taken by the balance nut when it is screwed as far up on the boss (away from the center of the hood) as it will go. The "static balance" position for the nut of the Mark 12-1 gyro is 1.5. This means that when the gyro is in static balance the balance nut is 1.5 turns nearer the center of the hood than when in its zero position.

1006. The inner gimbal ring is a copper nickel alloy shell made to the form of the gyro wheel in two halves, dovetailed together in the vertical plane. The two halves are so machined that one accurately locates the other when they are firmly pressed together and the separating line is difficult to distinguish. Two clamping screws fit bosses machined on each side of the hood and the two halves of the hood are drawn tightly together by means of these screws. It is therefore possible to disassemble the Mark 12-1 gyro inner gimbal ring after a satisfactory performance on the range, and re-assemble without change of adjustment, providing the wheel bearings are not changed.

1007. The inner gimbal ring carries the wheel bearings and supports the weight of the gyro wheel. The amount of permissible end play is of necessity very accurately controlled by means of tool 414A which will be explained later. The inner gimbal ring is in turn supported on ball bearings which are installed in the outer gimbal ring. On each side of the inner gimbal ring and on the outboard surfaces of the clamping bosses are located the inner gimbal ring centers of hardened steel. They are secured by screws in the depressed seats in the clamping bosses. Two screw holes are drilled through the flange of each center. The inner

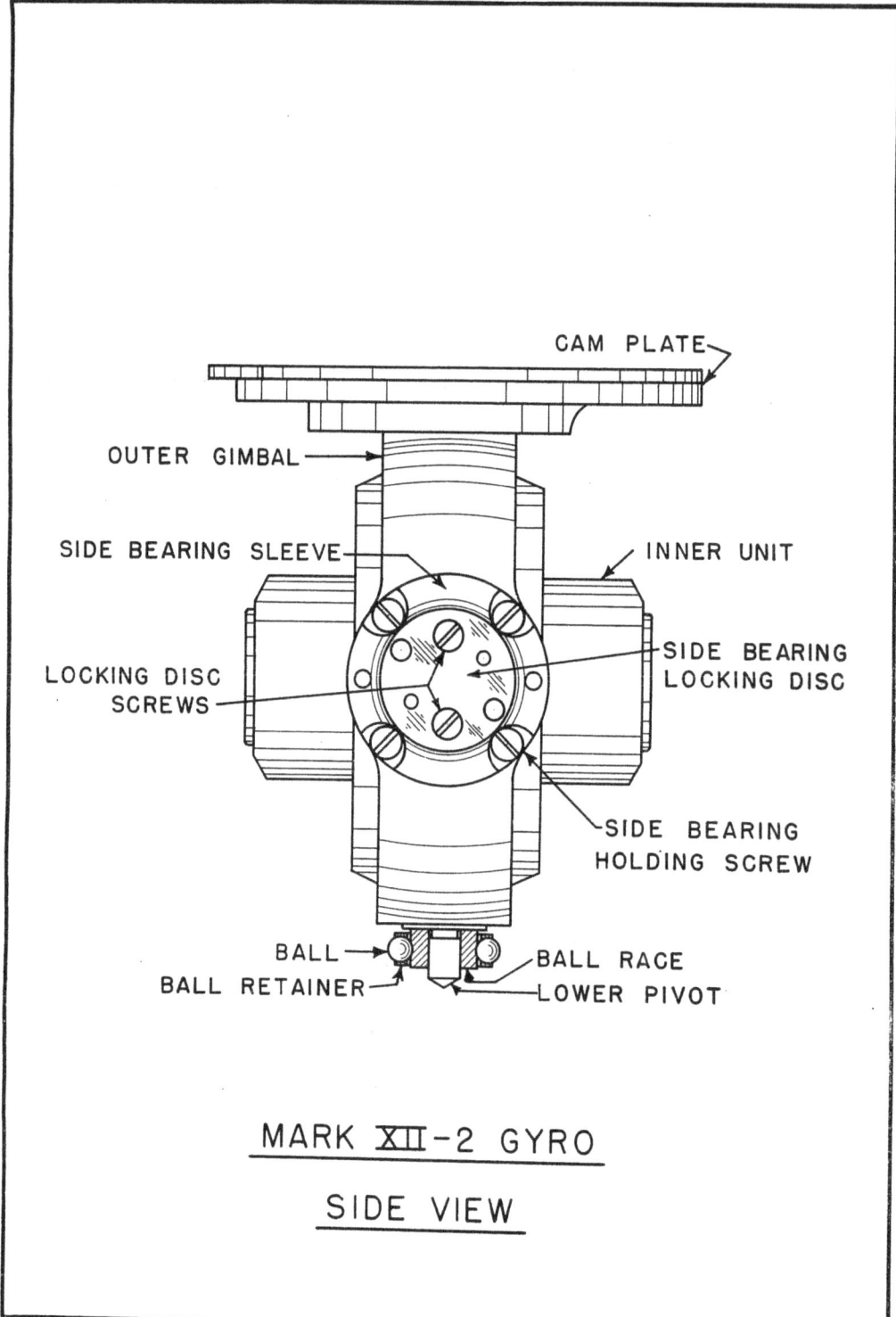

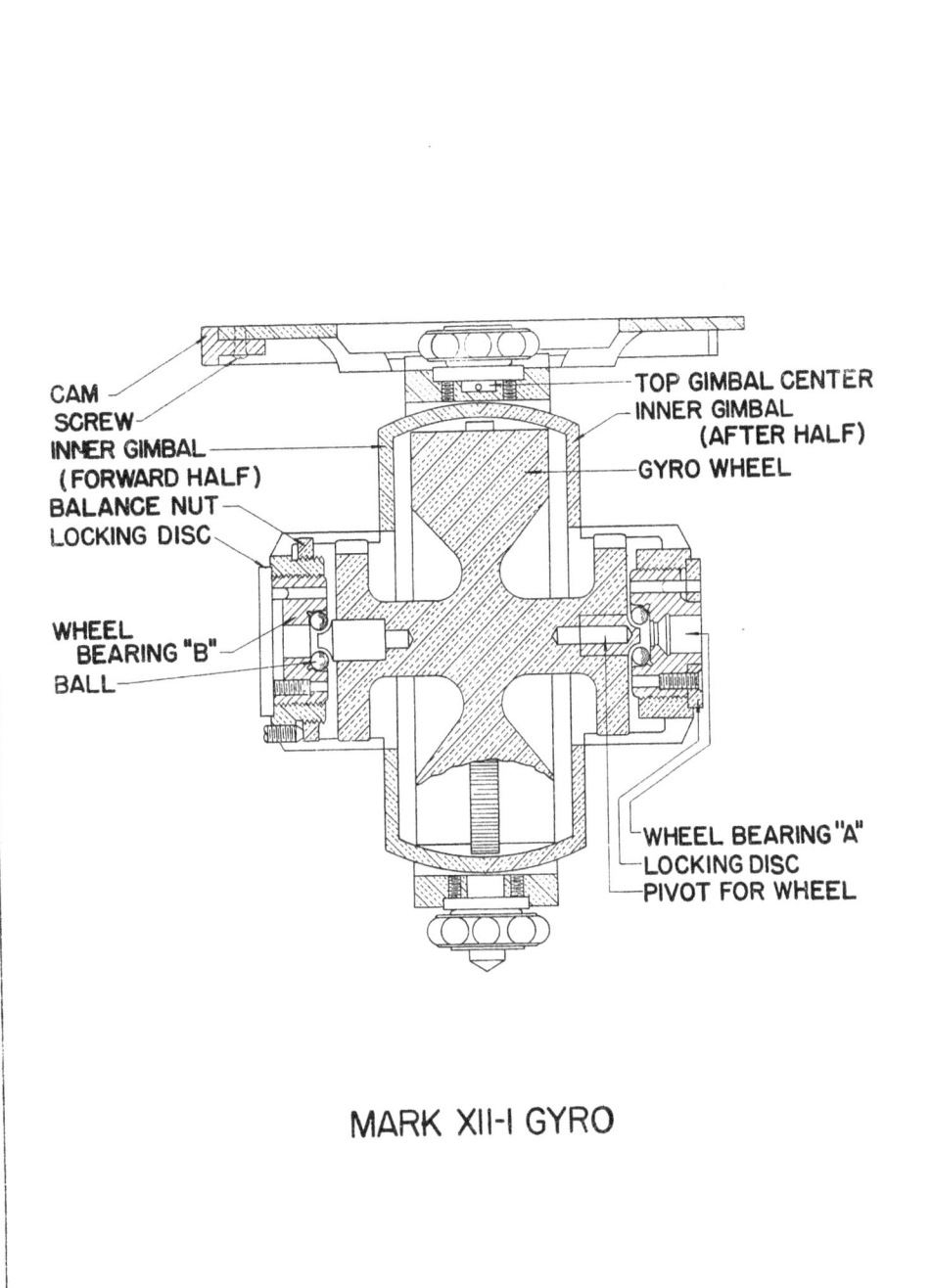

MARK XII-I GYRO

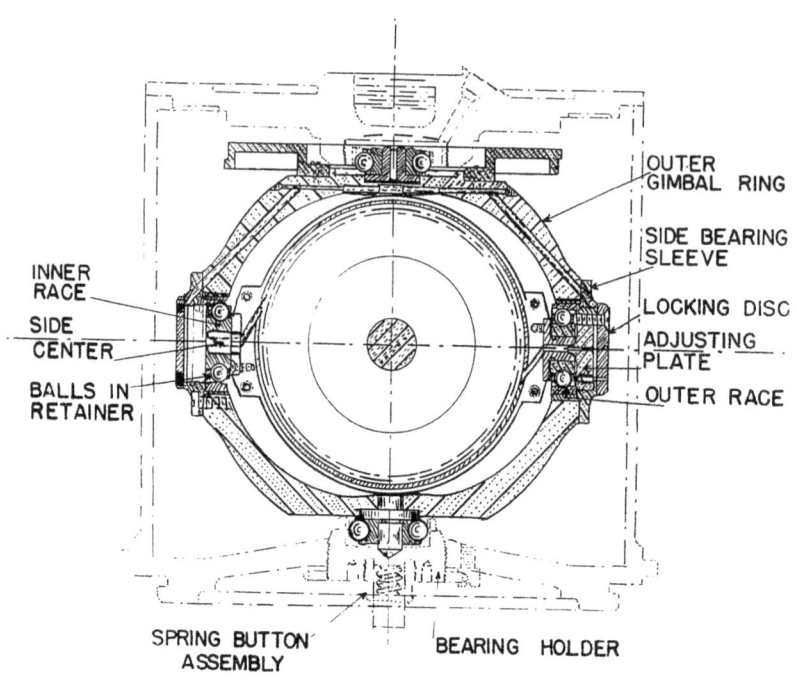

GYRO GEARS   MARK XII-1

gimbal ring centers receive the inner ball race of the commercial bearing, which is a push fit on the center. A special jack is provided for removing this inner race when necessary. The commercial ball bearing is made up in three parts; the inner race, the 3/16 balls with brass retainer ring and the outer race (a relatively thin cylindrical shell of hardened steel). Each center has a small axial air hole drilled through it, through which sustaining air is led from the outer gimbal ring. Two nozzles, one for each center are drilled from the seat of the center to the inner periphery of the hood so that the air will impinge on the buckets of the rim of the gyro wheel, thus maintaining its rotation. Each center seats on a thin paper washer to insure air tightness when the center is set up on its seat.

1008. The outer ball race of the commercial bearing (in the side bearings) fits inside the bearing sleeve (one for each side bearing) The outer ball race, however, does not screw into the bearing sleeve as in earlier types of gyros. On the contrary, it is a push fit inside the sleeve and bears against an adjusting plate which does screw inside the bearing sleeve. The adjusting plate can be moved in and out by turning the locking disc in the same manner as for the wheel bearing cups. To move the outer ball race away from the locking disc, the locking disc is turned clockwise which moves the adjusting plate, to which the locking disc is attached in the same manner as in the wheel bearing cup. Thus the movement of the adjusting plate pushes the outer ball race ahead of it. On the other hand, to move the outer ball race towards the locking disc, the adjusting plate must first be moved outward by turning the disc counterclockwise and then the outer ball race must be pushed manually against the adjusting plate. Thus, with the commercial bearings, the locking disc on the side bearing sleeve moves the outer ball race "in" only when it is turned. To move it outward, it is necessary to first turn the locking disc counter-clockwise the necessary amount, and then push the outer ball race with the fingers until it again strikes the adjusting plate. The locking discs are clamped to hold the adjusting plate rigidly against the bearing holder by two screws. Each adjusting plate has machined at its center a small nipple or extender which fits in the hole drilled in the side bearing center. The sustaining air passes through this extender from the nipple back of the adjusting plate to the center and thence to the nozzle and gyro wheel.

1009. Each side bearing sleeve is flanged and bench marked for proper assembly and is held to the outer surface of the outer gimbal ring by means of four flat headed screws. Each bearing sleeve is numbered to correspond to its location in the outer gimbal ring. Two air passages from the top vertical center are drilled inside the outer gimbal ring to seats under each side bearing sleeve. These passages are for the purpose of conducting the sustaining air from the top center seat to each side bearing sleeve.

1010. The commercial bearing is not only of heavier construction than previous types but also is a "radial" type of bearing, there being only one point of contact on the inner and and outer ball race under any condition of loading. Thus, greater sensitivity is obtained when the gyro is standing up in its side bearings, and is therefore more sensitive when a torpedo is running with some heel. The commercial bearing, as previously stated is a three piece unit. These parts are not interchangeable as different lots of bearings may be purchased from different contractors so the bearing as a whole must be replaced when any of its parts are defective.

1011. The outer gimbal ring is supported in the gyro pot by ball bearings of the same type as the side bearing. In the top vertical bearing, the outer ball race is a push fit in the top bearing holder, which screws into the pot. Vertical adjustment of this holder is obtained by varying the thickness of the washer against which the bearing holder seats when screwed home in the shelf of the pot. A special tool is necessary to withdraw the outer ball race from the top bearing holder, and also to unscrew the top bearing holder. The top bearing holder has machined at its lower center a small nipple or extender which projects into an axial hole in the top center in the same manner as in the side centers, and through which the sustaining air passes. The top center of the outer gimbal ring also seats on a paper gasket for air-tightness at this point.

1012. The installation of the top and bottom centers in the vertical bearings does not differ from that of the side bearing centers. The commercial bearing is of the same type used in the vertical or side bearings. It is necessary in installing the gyro in the pot always to make sure that the balls are well down against the grooved surface of the outer ball race. It sometimes happens that these balls jam outward in the brass retainer and in this condition they will ride in the run-out at the extreme outer end of the bearing surface.

1013. The outer ball race on the lower vertical bearing is contained in a steel bearing holder which screws into the bottom plate. The design of this holder in the Mark 13 torpedo requires two adjustments in establishing the proper vertical play of the gyro when installed. The outer bearing holder is made up of two principal parts; the first being the body which screws into the bottom plate and the second being the lower spring button assembly which screws into an axial hole in the center of the body of the bearing holder. The outer rim of the bearing holder and the outer rim of the spring button assembly are fluted to receive set screws which definitely lock these members against turning when finally adjusted. The adjustment of these two parts of the lower bearing holder accomplishes two definite steps in establishing the proper vertical play of the gyro. When adjusted, the lower bearing holder accurately controls the normal running position of the balls in the lower bearing and keeps these balls from "005 to "008 above the curved path. This construction is necessary for bearings to withstand high launching. The reason for this is that the gyro is not sufficiently sensitive about its vertical axis if the balls of the lower bearing are on the curved path of the outer ball race. Manufacturing tolerances in previous types did not permit a sufficiently accurate control of this feature.

1014. The amount of end play allowed in the wheel bearings, side bearings, and top bearings is held to very close limits, especially so in the wheel bearings. It is impossible to measure wheel bearing clearance by means of a micrometer. This is accomplished by the use of a small pendulum (tool 414A) which can be clipped to the spur gear machined on the gyro wheel hub. This tool produces a sufficient oscillating force so that if the wheel bearings have the proper clearance, this pendulum will make from 7 to 9 swings from an initial position of contact with the edge of the inner gimbal ring to the position of rest. To be reliable, this test should be made shortly after the wheel bearings have been thoroughly cleaned and supplied with fresh oil. If the oscillations are too great, the clearance is decreased by setting up on the wheel bearing cups through the medium of the locking discs. If the cups are already too tightly adjusted, additional clearance is introduced by unscrewing the cups. Even the slightest change of wheel bearing sensitivity is shown upon the number of swings the pendulum will make before coming to rest from its initial starting position in contact with the side of the gimbal ring at the spinning gear slot. This method of determining wheel bearing clearance has a decided advantage in that it permits measuring wheel bearing clearance at any time. When one considers that if the weight of the Mark 12-1 wheel is permitted to move "003 from its true mid-podition, precessive effect on the gyro is equivalent to about one full turn of the balance nut. It is apparent, therefore, that loose wheel bearings can easily introduce unsatisfactory deflection performance.

1015. The clearance or play of the inner gimbal ring in the side bearings can be directly measured with a dial indicator by setting the assembled gyro on its side and alternately raising and lowering the inner element while the contact point of the dial indicator is resting on the inner gimbal ring. This end play of the inner gimbal ring in the side bearings should never be under "005. With a little experience in adjusting gyros, it will be possible to get this side clearance very accurately by the "feel".

1016. Vertical clearance or play of the outer gimbal ring in the vertical bearings is obtained by taking a depth measurement from the bottom head to the outer gimbal ring. This vertical play should be from "0025 to "005. It is essential that when the gyro wheel is installed in the gyro pot, the axis of the wheel be in line with the centering pin in the spinning gear. This is necessary in order that when the centering pin is withdrawn after spinning, reactions will not be set up which will tend to deflect the gyro. This adjustment is accomplished in a manner later to be described but which, briefly, measures the tilt of the inner gimbal ring inside the pot by taking depth measurements from the bottom head to the two ends of the inner gimbal ring. Any adjustment which is necessary to attain this alignment is attained by adjusting the top bearing holder as previously stated. There must also be a corresponding adjustment of the lower bearing holder in order to obtain the proper end play of the outer gimbal ring in its bearings.

1017. The cam plate is secured to the top of the outer gimbal ring by 4 screws and has two dowel pins to locate it. The steel cam is secured to the after end of the cam plate and serves to throw the pallet blade to right and left as the torpedo axis changes direction with regard to the gyro wheel axis, giving right and left rudder respectively. The rim of the cam plate is further machined such that 180° of the cam plate rim from the location of the cam has a ridge above and a groove below, and the other 180° has a groove above and a ridge below. Thus when the torpedo is not on the desired course, one toe of the cam pawl will strike the ridge and the other toe will drop into the groove, rotating the pallet, causing the torpedo through the steering mechanism to turn until the pallet is deflected in the opposite direction. The continuity of these ridges and grooves is broken in the vicinity of 155° from the location of the cam.

C. AIR SUSTAINING SYSTEM.

1018. Air is by-passed from the steering engine nipple through a small reducing valve which reduces the pressure to 125 lbs. per square inch. The outlet pipe from the reducing valve connects to a nipple on the right side of gyro pot, and the nipple in turn communicates by a channel drilled through the shelf inside the gyro pot with the space over the top bearing holder. Two air passages from the top vertical center are drilled inside the outer gimbal ring to seats under each side bearing sleeve. The sustaining air passes from the space over top bearing holder through extender and top center down each leg of outer gimbal ring to each side bearing sleeve, and through a hole in sleeve to a chamber back of the adjusting plate. From the chamber, air enters the side centers through the extenders on the adjusting plates and is directed through channels (drilled in opposite directions in the inner gimbal ring) to saw-teeth buckets cut in the outer periphery of gyro wheel. The sustaining air, thus imparts the necessary force to maintain the wheel at a constant speed. The inner gimbal ring or hood closely follows the contour of the wheel and four exhaust air passages are brought as close to the center of the hub as it is possible to make them. An equalizing hole is drilled through the depth engine body, and thus, the sustaining air pressure, and consequently the speed of the gyro wheel is modified by afterbody pressure.

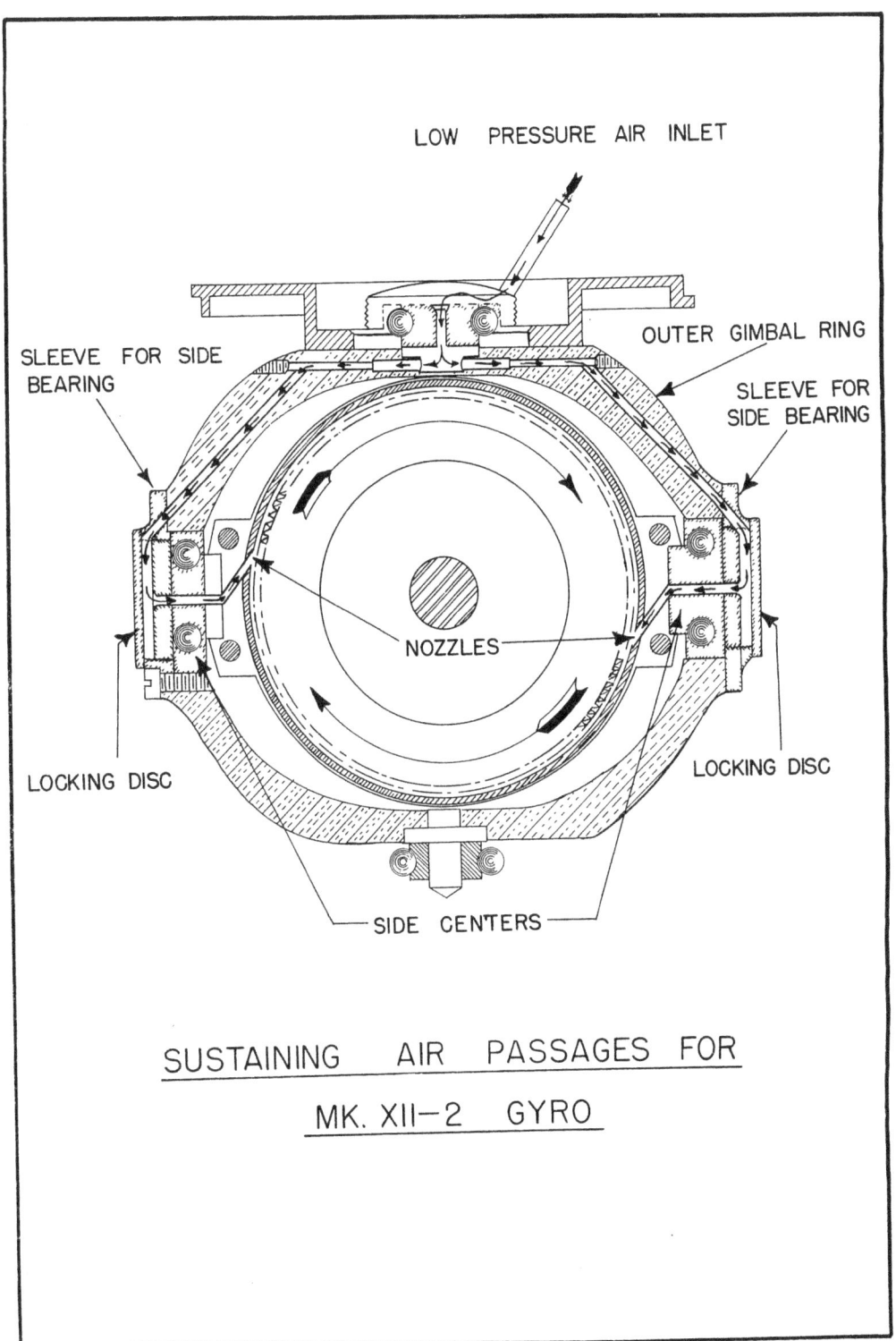

1019. Joints are kept air tight by paper washers inserted:
   (a) Under top center.
   (b) Beneath the flanges of side bearing sleeves where they contact the outer gimbal ring.
   (c) Under the locking discs on the side bearing sleeves.
   (d) Under both side centers.

1020. Air is prevented from leaking around the outside of the top bearing holder by the copper spacing washer under this bearing holder. This is the same spacing washer used in controlling the vertical adjustment of the top bearing.

1021. Paper washers should be lightly oiled to insure a tight joint. Extreme care should be taken when installing them that air passages are not blanked off.

1022. A thin film of oil between extenders and their contact surface in the center is the best safeguard against leakage and likewise prevents drag at these points.

D. <u>GYRO BALANCING.</u>

1023. The Mark 12-1 gyro is designed and built so that the three axis of rotation are at right angles to each other and intersect at a point; also that the centers of gravity of the gyro wheel and inner gimbal ring (balance nut attached and 1.5 turns from the "full up" position) taken individually or collectively, both lie in the same point, this point being the point of intersection of the three axis of rotation. The center of gravity of the outer gimbal ring lies in the vertical axis of rotation but somewhat above the point of intersection of the three axis, due to the weight of the cam plate.

1024. All balancing tests which are given to this gyro are given for the purpose of obtaining the above conditions. If the above conditions are obtained and the gyro is installed in a torpedo, the torpedo will run in a straight line at the equator only. At other latitudes it is necessary that the balance nut be moved to a new position as calculated by the latitude formula, the latitude used being known. The latitude correction at Newport is between 0.4 and 0.5 turns of the nut so that practically all Mark 13 torpedoes pass the proving range tests with balance nut positions of 1.9 and 2.0, as modified by afterbody pressure effect. In this respect, the performance of the Mark 12-1 gyro is exceptionally uniform.

1025. When gyros are issued from the torpedo station, they have been so manufactured that for static balance the balance nuts are 1.5 turns in from their hard up positions on the inner gimbal ring bosses. This condition of balance is defined as "static balance". With the proper nut correction the torpedo should make a straight run to the target. The final balance nut position found necessary to pass a torpedo on the proof range is entered in the torpedo record book for future use.

1026. With the Mark 12-1 gyro and with the balance nut 1.5 turns from the hard up position, it is sometimes difficult to tell whether the inner gimbal and gyro are nut end heavy or light. This is due to the fact that the commercial side bearings are somewhat less sensitive than the earlier types with smaller balls and no retainer ring. This can easily be overcome by attaching a small ball of putty (about 3 grains) first to one end of the hood and then to the other end. With the addition of this small ball of putty the two ends should move downward at approximately the same rate when this unit is properly balanced.

1027. It is obvious that any change in the parts of a gyro after it is issued from the torpedo station may call for a change of balance nut due to variation in the weights of new parts. When it is necessary to install new parts in the gyro, the individual parts (inner gimbal ring or outer gimbal ring) should be separately balanced to insure that the basic conditions of balance as previously outlined, will be fulfilled when the gyro is completely assembled. The torpedo stations remove metal from permissible surfaces as necessary to obtain proper balance. It appears extremely doubtful that such an expedient would have to be resorted to in the ordinary servicing of a gyro. The above outlined principles of balance should be obtained, but no person should be allowed to remove metal from any part of a gyro unless these principles are thoroughly understood. When it becomes necessary to remove metal from a gyro to obtain balance, such metal should be removed from the outer surface of the locking discs rather than the gimbal ring itself.

1028. Gyros should be kept in static balance when in storage, thus maintaining a condition of balance for easy inspection.

E. DISASSEMBLY OF GYRO - TEST FOR LEAKS.

1029. NOTE: In overhauling gyros, every effort should be made to reassemble the gyro with no change in position of parts or adjustments. Thus if no parts are renewed, the gyro should be in "static" balance when assembled, with the balance nut 1.5 turns "in" from the zero position. In order to accomplish this, each gyro worker should have a small cleaning stand to be used to hold the various parts when they are disassembled. This cleaning stand should be divided into four compartments, one compartment for parts of each wheel be ring and one compartment for each of the side bearings. All screws, etc., should be reassembled in exactly the same position they occupied before disassembly. Screw drivers for the side bearing sleeves, locking discs, and for the screws holding the inner gimbal ring halves should be ground and fitted for the particular screws they are to be used on. This prevents burring and twisting the screw heads.

Test gyro for leaks.

1030. Place gyro in balancing fixture, and adjust air so that there is 125 lbs. per square inch on the line. Turn on air and test side bearings for leaks at the following places:

TOOL NO.

    (1) Around edges of locking discs.
    (2) Around edges of side bearing sleeves.

Note position of leaks for remedy after disassembly. Leaks around edges of locking discs are remedied by the insertion of paper washers between locking discs and sleeves, also by lapping the heads of the screws for locking discs to their seats in discs. Leaks around edges of side bearing sleeves are remedied by lapping sleeves with a face lap and outer gimbal bearing surface lap, also by use of new paper washers.

Disassemble gyro.

1031.
- (1) With balance nut 1.5 turns "in" from the zero position, hold the assembled gyro in hands with inner gimbal ring in a vertical position and note if gyro is pendulous or top heavy. If such is the case, the inner unit will show a tendency to drop towards the cam plate or the bottom pivot on the outer gimbal ring. The Mark 12-1 gyro is issued non-pendulous. If a pendulous condition is found to exist, the only correction in the service is shifting of the side pivots, or minute inspection to see that pivots are properly seated (when the gyro is again assembled).
- (2) Remove holding screws for side bearing sleeves.... 41
- (3) Remove side bearing sleeves, using two #91 tools simultaneously................................. 91
- (4) Using two hardwood sticks (not supplied), lift off balls and retainers from inner races on side pivots.
- (5) Remove inner races from the side pivots (commercial name of race is outside)..................... 416
- (6) Rotate inner gimbal ring and wheel axis parallel to plane of outer gimbal ring.
- (7) Remove inner gimbal ring and wheel assembly. Inner gimbal ring pivots pass through recesses of outer gimbal ring.
- (8) Slack off holding screws for inner gimbal ring halves.. 81
- (9) Remove holding screws for inner gimbal ring pivots... 37,88
- (10) Remove inner gimbal ring pivots (there is an assembly scratch on pivot and inner gimbal to enable pivot to be reassembled in same position.......... 91
- (11) Remove holding screws for inner gimbal ring halves. 81
- (12) Start spreading inner gimbal ring halves apart. Grasp the geared wheel hub and force it up into the wheel bearing which is uppermost. This prevents balls of wheel bearing from falling out. When halves are parted, reverse position of upper half and remove gyro wheel....................... 91
- (13) Remove balls from wheel bearings.
- (14) Remove wheel bearings from inner gimbal ring halves... 37,204

		TOOL NO.
(15)	Remove balls and retainers from inner races on top and bottom pivots..	
(16)	Remove outer races from side bearing sleeves...	WE192
(17)	Remove locking discs and adjusting plates from side bearing sleeves..........................	37,204
(18)	Slack up on clamp screws for balance nut, and remove balance nut from inner gimbal ring half.	88

F. **EXAMINATION OF PARTS.**

1032

    (1) Inspect the balls, inner and outer races of top, bottom and side bearings through a two-power magnifying glass for stains, pits, and mirror finish. Parts of these bearings are not interchangeable. If one part is unfit for use, the entire bearing must be renewed.
To fit new bearings use:
      (a) Female lap for inner and outer gimbal ring pivots................................... WE195
      (b) Male lap for inside of inner races WE194
      (c) Puller for outer races of top, bottom and side bearings........................... WE192

    (2) Inspect adjusting plates in side bearing sleeves for rust, and note condition of extender for side centers.

    (3) Inspect races of wheel bearing cups for ball pits, dents, etc. In general, it is far better to renew doubtful bearings; however, the promiscuous scrapping of expensive priced and satisfactory material is to be guarded against. To fit new wheel bearings use:
      (a) Female lap for threads of bearing.......... WE176
      (b) Tap for threads of inner gimbal halves. .. WE176A

    (4) Inspect gyro wheel pivots for dents, pits, etc. Dented pivots cannot be renewed in the service. New wheels should be installed and the old wheel sent to N.T.S. Newport, R.I. for overhaul.

    (5) Inspect balls for wheel bearings for stains, pits and mirror finish. Balls must be measured for correct diameter.

    (6) Inspect cam on cam plate for rust and burrs. See that it is not loose.

    (7) Place cam plate on surface plate and see that it is not bent. Also see that concentric ridges and grooves are free from burrs.

    (8) Inspect inner gimbal ring halves for condition, giving particular attentions to dove-tailed surfaces of each half.

    (9) Inspect outer gimbal ring for condition, noting that the bearing surface for the side bearing sleeves are not burred or scratched.

G ASSEMBLY AND ADJUSTMENT OF GYRO.

1033

TOOL NO.

(1) Test top pivot for leaks. Assemble side bearing sleeves to their respective sides of outer gimbal ring with hole in sleeve for sustaining air 180° from hole in flange on gimbal ring, thus blanking off air channel. Turn on air and admit 90 lbs. pressure. Test with oil around edges and over lifting holes of pivot. When satisfactory, remove bearing sleeves ... 41

(2) Assemble inner race over top center with commercial marking outward. Place a piece of hard rubber or wood against race and tap in place with a small hammer.

(3) Assemble balls and retainers over top and bottom inner races. Oil (A) balls.

(4) Install outer races in side bearing sleeves. They are a press fit.

(5) Install adjusting plates, oiled paper washers and locking discs to side bearing sleeves.. ... 37,204

(6) Assemble balance nut on inner gimbal ring half. Screw it back until zeros line up - then screw it "in" 1.5 turns to the "static" position ... 88

(7) Assemble wheel bearing (free end) in after half of inner gimbal ring............... 204

(8) Assemble locking disc to wheel bearing..... 37

(9) Assemble wheel bearing (locking end) in forward half of inner gimbal ring................. 204

(10) Assemble locking disc to wheel bearing... 37

(11) Assemble side centers to their respective sides of one inner gimbal ring half. (The side centers are assembled at this time without paper washers and act as guides for the two halves when they are pressed together).

(12) Stand forward inner gimbal ring half on end with tool #91 protruding through hole in bearing cup. Place six balls in bearing cup. Check for slight ball clearance and oil (A)............... 172

(13) Remove gimbal ring half from tool #91 and install gyro wheel with saw teeth buckets of wheel lining up properly with machined nozzle of gimbal half.

(14) Assemble balls in after gimbal half. Check for slight ball clearance and oil (A)............... 172

(15) Grasp geared wheel hub with fingers and force it up into wheel bearing of forward gimbal half, thus preventing balls from dropping out. With the forward gimbal half in the uppermost position, assemble the two halves together with the assembly marks coinciding. Be sure wheel bearing cups permit sufficient clearance not to jam the balls

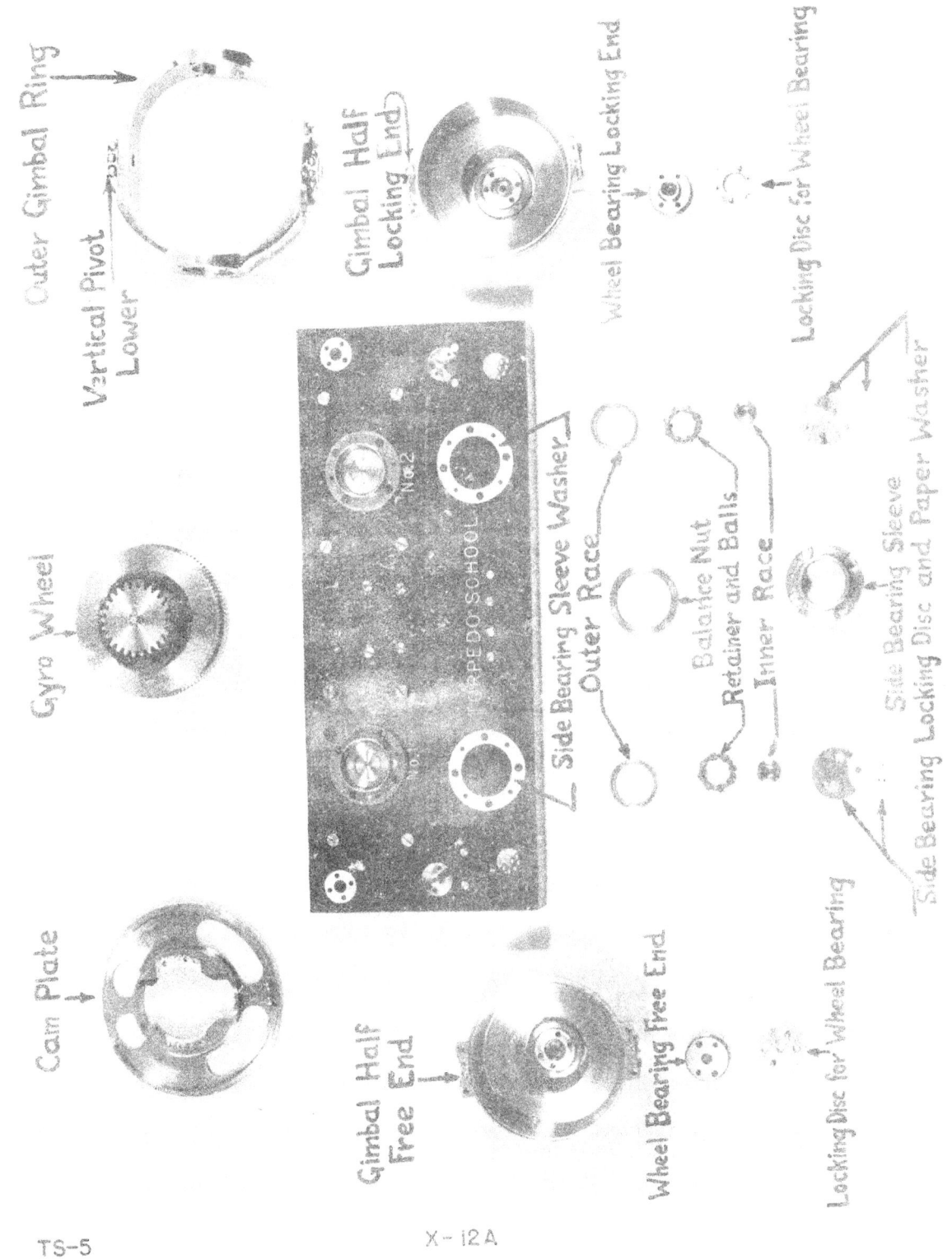

		TOOL NO.

(16) Press the halves together and replace holding screws, setting up on the screws lightly........ 37

(17) Remove side centers, install lightly oiled paper washers in the counter bore of inner gimbal and replace centers to their respective side with scribe marks coinciding...................... 88

(18) Set up on holding screws for inner gimbal ring and side centers................................ 37,41

(19) The assembled gyro wheel is now ready to be centralized in the inner gimbal ring. On the Mark 12 1 gyro, the rim of the wheel is entirely enclosed by the inner gimbal ring or hood. To center the wheel of this gyro, a centering block in the form of a short cylinder is provided. This block is used in conjunction with a surface plate and a height gauge. The assembled gyro wheel and gimbal ring are placed on the block with wheel axis vertical and the hub of the gimbal ring entering the central hole in the block. A height reading is then taken to the upper end of the wheel center. The wheel and gimbal are then reversed and a similar reading taken. If the readings to each end of wheel centers are the same, the gyro wheel is centrally located in the gimbal ring. If the readings are not the same, the wheel bearings must be moved the necessary amount to make them correspond. Wheel clearance is obtained during this operation by getting 8 to 10 swings with tool 414A. Be sure the locking discs for wheel bearings are set up tight when final clearance is obtained................. 37,414A

(20) Level knife edges on surface plate and place assembled unit so the inner gimbal pivots will rest on knife edges. It should remain in any position at any angle placed, showing no tendency to drop either way. If it drops with the wheel bearing cups initially in the horizontal position, the locking disc from the heavy end should be removed and metal ground off (first calipering the disc and not removing more than ".006). Replace disc and try for balance on knife edges again... 37

(21) The Mark 12-1 gyros are issued non-pendulous; therefore, very seldom will an assembled gyro wheel and inner gimbal ring show a tendency to drop when placed on the knife edges with wheel axis in the vertical position, when it stands up with the axis horizontal. If such a condition is found to exist, a minute inspection should be made to see if side pivots are properly seated, or if the scratch marks on pivots and gimbal coincide. If the condition cannot be corrected as a result of this examination, or by shifting

GYRO WHEEL CENTERING FIXTURE

Reading is taken by Dial Indicator to wheel center nearest to Dial Indicator. Inner Unit is then turned 180° and another reading is taken to opposite wheel center. If readings are not equal, Slack off on Wheel Bearing Locking Disc, and move Bearings until equal readings are obtained.

TEST FOR PROPER WHEEL CLEARANCE

Using tool "414 A" attached to spur gear of wheel. Pull tool to its full up position against Gimbal Boss and let go. Tool should make 8 to 10 half swings if wheel clearance is proper. Adjustment is made by moving Wheel Bearings. (in if too loose) (and out if too tight.)

TEST FOR STATIC BALANCE

With Balance Nut set 1.5 in from all the way out, Inner Unit on leveled "knife edges." Inner Unit, resting on its pivots, should remain stationary in any position. If heavy on one side, remove weight from locking disc until balanced.

TOOL NO.

the pivots, it is better to allow a slight pendulous condition than attempt to remove weight from inner gimbal ring.

(22) Place inner gimbal assembly in outer gimbal ring with scribe marks coinciding. When the cam plate is in the uppermost position and the cam facing the observer, the scribe marks on the inner and outer gimbal ring are on the "right".

(23) Lay outer gimbal ring on its side and install inner races over their respective side pivots with commercial markings outward. The inner races are a press fit and can be tapped in place by placing a piece of hard rubber or wood against the race and tapping the wood or rubber lightly with a small hammer.

(24) Assemble balls and retainers over their respective side pivots. Oil (A) balls.

(25) Assemble side bearing sleeves on their respective sides of outer gimbal ring with paper washers underneath. Be sure the hole in side bearing sleeve for sustaining air is lined up with hole in flange of outer gimbal ring and that paper washer is not covering the hole.

(26) Set up on holding screws for side bearing sleeve.................................... 41

(27) Place the assembled gyro in balancing fixture and balance the inner gimbal ring in outer gimbal ring, at the same time maintaining ".005 clearance in side bearings. To obtain balance, hold outer gimbal ring slightly off level either way and let go. Side of gimbal ring below level, should drop down slowly. Try the opposite way where similar effect should be obtained, if properly balanced. (Dropping of gimbal ring, when placed below level is caused by movement of weight due to clearance of side bearings). If one side is heavy, (the inner unit not centered in the outer unit), screw in on the locking disc of the heavy side and back out on locking disc on light side as necessary to obtain balance, maintaining the ".005 side clearance. It is to be noted that to back out on the locking disc of the side bearing moves the disc and adjusting plate only. It will be necessary to move the outer race of side bearing by pushing inner gimbal assembly against the race of bearing backed off until a "click" is heard, thus indicating that the outer race is seating against the adjusting plate. To accurately measure side bearing clearance, place assembled gyro on knife edge fixture so that the

TO OBTAIN SIDE BEARING CLEARANCE

Inner Unit of Gyro is supported on a ring fixture with locking disc against dial indicator. Press Inner Unit firmly on ring fixture with one hand, and push Outer Gimbal toward dial indicator to record clearance. Slack lock screws on Side Bearing locking disc, and move Bearings in or out as required to obtain .005"+ Side Bearing clearance.

TOOL NO.

inner gimbal ring will rest on the knife edges. With weight of gyro thus supported on knife edges, place indicator against face of side bearing locking disc on one side, and move outer gimbal ring against the bearing and note reading. It should be not less than ".005.

(28) With gyro in balancing fixture, adjust air so there is 125 lbs. per square inch on the line. Turn on air and test side bearings for leaks. It is difficult to entirely stop leaks around edges of locking discs and edges of side bearing sleeves. Every effort should be made however, to keep leakage to the minimum. See article 1029 for remedy of leaks.

(29) Remove gyro from balancing fixture and mount the assembly on a small ring support (to protect the lower pivot), with the cam plate up. A re-check is now made for balance of wheel and inner gimbal assembly previously obtained in step (20). A weight of about 5 grains is placed alternately on one wheel bearing boss and then on the other, noting whether in each instance, the inner ring shows the same tendency to turn on the side pivots. This test also checks the sensitivity of side bearings. The gyro is now in "static balance". The two centers of gravity of the wheel and inner gimbal ring are individually and collectively at the point of intersection of the side and vertical pivots, and the outer gimbal ring is non-pendulous about the centerline of the vertical pivots. The gyro wheel is exactly centralized in the inner gimbal ring, and the combined wheel and inner gimbal ring are exactly centralized in the outer gimbal ring. The balance nut is set 1.5 turns "in" from the zero pozition, thus allowing room for movement of the balance nut in north or south latitude to compensate for "creep" caused by the rotation of the earth. To correct for creep in north latitude, the balance nut is moved further "in" from its static position. To correct for creep in south latitude, the nut is moved "out" from its static position. A gyro with its balance nut so moved to correct for creep is said to be in "running balance".

H. POT CLEARANCES AND ADJUSTMENTS.

1034.

(a) Install gyro in mechanism.

(1) Move cam pawls to the extreme outer position by turning the driving gear.
(2) Install gyro in pot with cam on cam plate opposite from cam pawls to prevent fouling.

TS-5   X-15   Original Page.

TEST FOR SENSITIVENESS OF SIDE BEARINGS

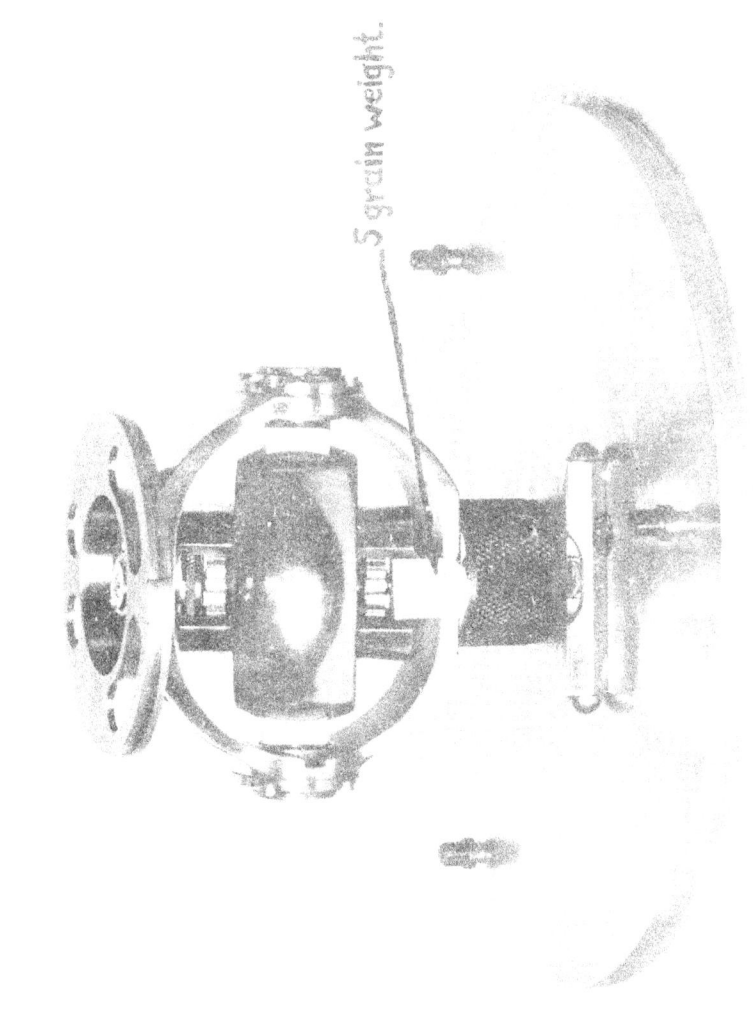

Using 5 grain weight as indicated, Inner Unit should drop at same rate of speed on either side.

Assembled Gyro in Balancing Fixture, testing Side Bearing sleeves and locking discs for leaks. Using oil syringe and 125 lbs. Pressure.

Gyro Tools and Equipment at Fleet Torpedo School, Des. Base, San Diego.

TOOL NO.

    (3) Ascertain that top pivot is centered in ball race and does not hang up. It is a good practice to cause gyroscope assembly to revolve on its upper bearing when installing bottom head, thus making certain that balls for top race do not hang up on edge of outer race.
    (4) Install bottom head with lower bearing holder removed.. 246

(b) <u>Install bottom bearing holder</u>:
    (1) Oil and screw adjusting body assembly into bottom bearing holder until it can be screwed in and out easily by hand.
    (2) With adjusting body assembly removed, install bottom bearing holder and outer race.
    (3) Set up on the bottom bearing holder until the balls of both bottom and top races are contacting the curved path of their outer races. NOTE: that balls are properly centered in their races when screwing down on bottom bearing holder. Do not jam the balls.

(c) <u>Check spacing washer under top bearing for proper thickness</u>.

NOTE: It is important that this washer be of the correct thickness to insure the alignment of axis of the gyroscope with the centering pin in the spinning gear, in order that when the centering pin is withdrawn after spinning, no reaction will be set up which will tend to deflect the gyroscope. As the relative heights of the gyroscope and centering pin in the pot cannot be measured directly on their axis, two measurements are taken, one on each end of the inner gimbal ring (See N1, N2, Fig. 8). Half the difference between these measurements will be the amount which the center of gyroscope is out of alignment with the centering pin.

    (1) Lock gyro...................................... 205
    (2) Place depth micrometer W.E. 190A base across seat for gyro clamp plate in pot. Locate so that depth gauge spindle will line up with outer edge (see location at N1) of gimbal ring locking end. Note reading carefully.......... WE190A
    (3) Reverse location of depth micrometer so that depth gauge spindle will line up with outer edge (see location at N2) of gimbal ring free end. Note reading carefully.................. WE190A
    (4) Subtract the sum of the first reading from the sum of the second reading; with allowable tolerance; the remainder should be not less than ".005 nor more than ".010.

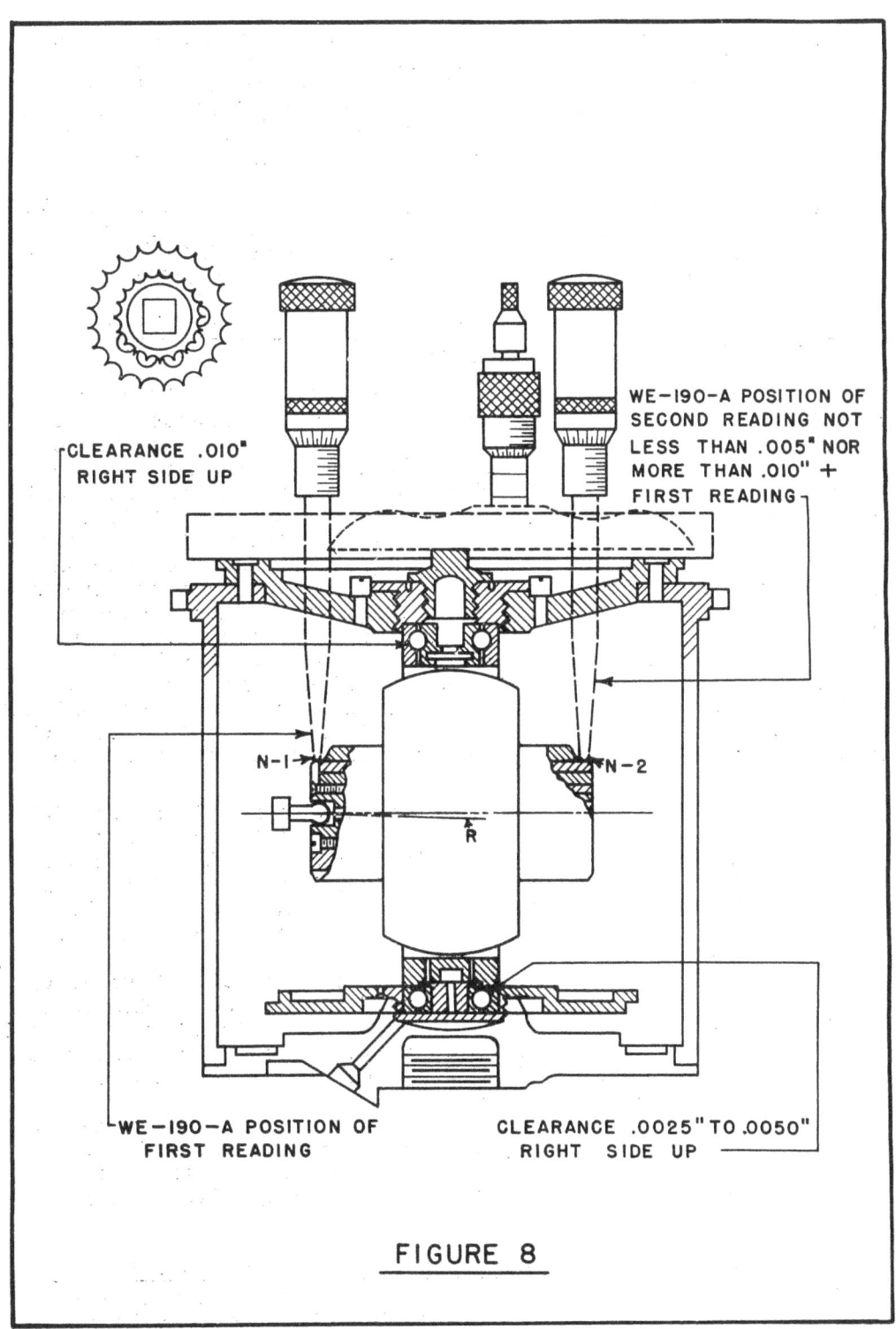

FIGURE 8

(5) With the remainder less than ".005 - for example, the washer under the top bearing holder is too thick. It will be necessary to reduce the thickness of the washer the proper amount, which is done by grinding. Care must be exercised to grind to an even thickness all over in order that the top bearing holder will seat evenly on the washer.

(6) On the other hand should the remainder be more than ".010, it is obvious that the washer is too thin. In which case, a new washer of proper thickness will have to be fitted under the top bearing holder.

(d) <u>Adjust bottom bearing holder to give clearance between upper and lower ball bearings.</u>

NOTE: The external diameter of the bottom bearing holder is threaded 20 threads to the inch to fit similar threads tapped in the bottom head. Also twenty scallops are machined around a flange on its outer end for alignment with the adjustment lock screw. Thus it will be seen that rotating the holder one scallop in the bottom head will cause a traversing movement of ".0025. For this adjustment a clearance of ".0125 is the most desirable between the balls in the lower bearing and the curved path on their outer race, provided the balls of the top race are contacting with the curved path of their outer race. This clearance will permit a ".010 clearance between balls of the bottom race and their curved path, and a ".0025 clearance between balls of top race and their curved path, after adjustments are made and gyro is right side up. However, this adjustment may vary slightly as a scallop may not line up with the lock screw hole. In such case, the scallop nearest the screw hole should be moved into line with same.

(1) See that a scallop on the bearing holder lines up with the hole for lock screw. If not, back the nearest scallop into line. If this movement is less than a half a scallop, back off 5 additional scallops and install set screw. This will give a clearance of from ".0125 to ".0137 between the balls of the bottom bearing and their curved path.

(2) If the lining up of the first scallop takes a rotation of more than ½ a scallop, the bearing holder should be backed off only four scallops and the lock screw installed. This adjustment will give a clearance of from ".0100 to ".0125 between balls of the bottom bearing and their outer race.

(3) If a scallop is lined up with lock screw hole when top and bottom balls are contacting curved path of their races, back off five full scallops and install lock screws. The clearance will be exactly ".0125, which is the most desirable clearance.

(e) <u>Adjust clearance of top outer gimbal bearing. (See Fig. 8).</u>
NOTE: In order to obtain the correct vertical alignment of the axis of the gyroscope with the centering pin, when the lower

center of outer gimbal ring is contacting the spring button in the lower bearing (torpedo right side up), the clearance in the top outer gimbal bearing must be adjusted to half of the sum of the difference obtained between first reading at N1 and the second reading at N2 of micrometer tool W.E. 190A, as taken in step (c)(2) and (3).

   (1) Ascertain that gyroscope assembly is properly seated on its upper bearing, with balls contacting curved path of outer race.
   (2) Install adjusting body assembly in bottom bearing holder.
   (3) Set up on adjusting body until no vertical play can be felt between upper bearing and lower bearing.
   (4) Place micrometer depth gauge (W.E. 9) with base resting on flat face of plug of spring button assembly.
   (5) Measure depth of flat face on bottom bearing holder. (Note position of W.E. 9 as shown in figure 8). Note the reading carefully.
   (6) Reset micrometer depth gauge, adding one half of the difference between the reading of W.E. 190A at N1 and N2, as taken in step (c) (2) and (3).

For example:

Reading at N2 - Reading at N1 = $\frac{".005}{2}$ = ".0025 to be added.

Reading at N2 - Reading at N1 = $\frac{".0075}{2}$ = ".0037 to be added.

Reading at N2 - Reading at N1 = $\frac{".010}{2}$ = ".005 to be added.

   (7) Back out the adjusting body sufficiently to permit measuring spindle on micrometer depth gauge W.E. 9 to clear flat face on bottom bearing holder, where base of gauge is resting flush on plug of spring button assembly. Replace depth gauge on flat face of plug.
   (8) Screw in on adjusting body until measuring spindle on depth gauge just contacts the flat surface on the bearing holder. This adjustment will give a clearance between the lower pivot on outer gimbal ring and spring button assembly, equal to the difference in depth gauge adjustment as reset in step (6), and consequently a similar clearance between balls of top bearing and their curved path when the torpedo is right side up.
   (9) Secure spring button assembly with lock screw.
   NOTE: It is not always possible to secure adjustment obtained in step (8), without a slight change, as a scallop in adjusting body may not line up with the hole for lock screw. Turn the adjusting body until nearest scallop lines with lock screw hole. A change thus made is considered negligible and is within the tolerance of allowable bearing clearance.

1035. If the plane of the clamp plate flanges were parallel to the axis of the centering pin, the above adjustment would definitely establish the correct gyro height. With the above conditions unknown, the gyro height as measured with W.E. 190A is approximate only, and must be checked to see if the gyro is

given a kick as the centering pin withdraws.
(a) With housing in running position and so placed that the action of the gyro may be observed, unlock gyro with hand trip. Note if gyro is given a kick. If the locking end of the gyro kicks up, gyro is high in the pot relative to the centering pin; if down, gyro is low in the pot relative to the centering pin. The allowed clearances are as follows:
   (1) Clearance between balls of top race and their curved path "0025 to "005.
   (2) Clearance between balls of bottom race and their curved path "0075 to "010.

If when adjusting gyro for correct height, the necessary movement of either the bottom bearing holder or the adjusting body alters the vertical clearance beyond their allowable tolerances, the washer under the top bearing holder must be renewed or lapped to the necessary size.

I. STAND TEST.

1036. A stand test should be given to any torpedo gyro and mechanism when new parts have been installed or there is any uncertainty in regard to adjustments of the various parts of the controlling mechanism. If the gyro and its mechanism are in satisfactory condition, it will perform best with its balance nut set at the computed value. Failure to obtain within 1 of the computed balance nut setting indicates one or more of the following:
   (a) Improper adjustment of clearance.
   (b) Improper sensitivity of bearing.
   (c) Maladjustment of pallet mechanism.

Cleaning, oiling, rebalancing and readjusting should produce the desired result rather than a change in balance nut position.

1037. Test procedure.
(a) The gyro mechanism, clean and in adjustment, is put in place in the cradle and bolted down. This can only be done with the cradle turned upright and held in that position by the pin supplied for the purpose.
(b) Put gyro in place and lock it.
(c) Swing cradle with mechanism down into normal position.
(d) Insert pin to hold cradle in position and set up on thumb nut.
(e) Make connection to "T" on vertical steering engine, and attach gauge to "T" connection in line from gyro reducing valve to gyro top plate.
(f) Have gyro reducer set for 100 lbs/in2. (This pressure for stand test only to approximate firing connection of back pressure). See that all running parts are properly lubricated.
(g) Start the motor which operates the pallet system.
(h) Open valve in low pressure line to steering engine.
(i) Throw over the lever handle of the quick opening valve in the high pressure line and throw back as soon as the gyro unlocks.

(j) Remove pin in bottom frame which prevents it from swinging about the swivel pin, and by means of shift lever, swing bottom frame, cradle and mechanism back and forth at regular intervals, going just far enough each way to cause the steering engine to operate. Note the deviations right and left as indicated on the dial plate, which is graduated in degrees and tenths of degrees.
NOTE: For computed value see Article 1056.

(k) Record the mean of these readings in tenths for every minute of the test.
NOTE: The data sheet to be used is illustrated on page X-21. The first column registers the minutes of the run. The second column registers the amount in tenths of degrees the gyro axis is to the left of the initial plane. The third column registers the amount in tenths of degrees the gyro is to the right of the initial plane. The fourth column registers the algebraic sum of the right and left deflections as represented in columns two and three. One tenth of a degree is designated as <u>one unit</u>.
  (1) Angle shots of $45°$ and $90°$ to either right or left may be made. After setting the angle, spin the gyro, remove the pin which holds the movable base plate in position and rotate the stand the set angle, completing the operation in approximately the time required by the torpedo to make the turn in question. Continue the test and taking of observations as above.
(m) Requirements for passing tests - no single deflection greater than 5 units nor a total of greater than 75 units for a 15 minute run.

1038. Explanation of Test.

The allowed limits of performance may be represented by two straight lines originating at the starting point and merging at an angle from each other by $1°$. A straight line bisecting this angle would represent the initial plane of spin. At the end of any given number of minutes, the position of the torpedo should be within these two limiting lines, never outside. Page X-21A shows a graphic picture of the sample tests recorded on page

Running Stand Data Sheet

Gyro Mark __12-1__   Torpedo Mark __13__
Gyro No. _____   Torpedo No. _____
Gyro Red. Valve Pressure __100 lbs.__ Afterbody Temp. __204° F.__
Gyro spin pressure __2500 lbs.__   Nut for static __1.5__
Gyro axis unlocked at __0°__   Nut for running __1.9__
Lat. __40° N.__ Date. _____   Oscillations __8 - 10__

Min.	L	R	Sum	L	R	Sum	L	R	Sum	L	R	Sum
0												
1	1		-1		0	0						
2	1		-2		0	0						
3		1	-1		0	0						
4		3	+2	1		-1						
5		3	+5	1		-2						
6		4	+9	2		-4						
7		5	+14	2		-6						
8		7	+21	2		-8						
9		7	+28	3		-11						
10	Failed			3		-14						
11				2		-16						
12				2		-18						
13				2		-20						
14				1		-21						
15				1		-22						
Angle	0			0								
Run No.	1			2								
Position of nut	1.6			1.9								
Dur. of spin, Min.	9			15								

TS-5   X-21   Original Page.

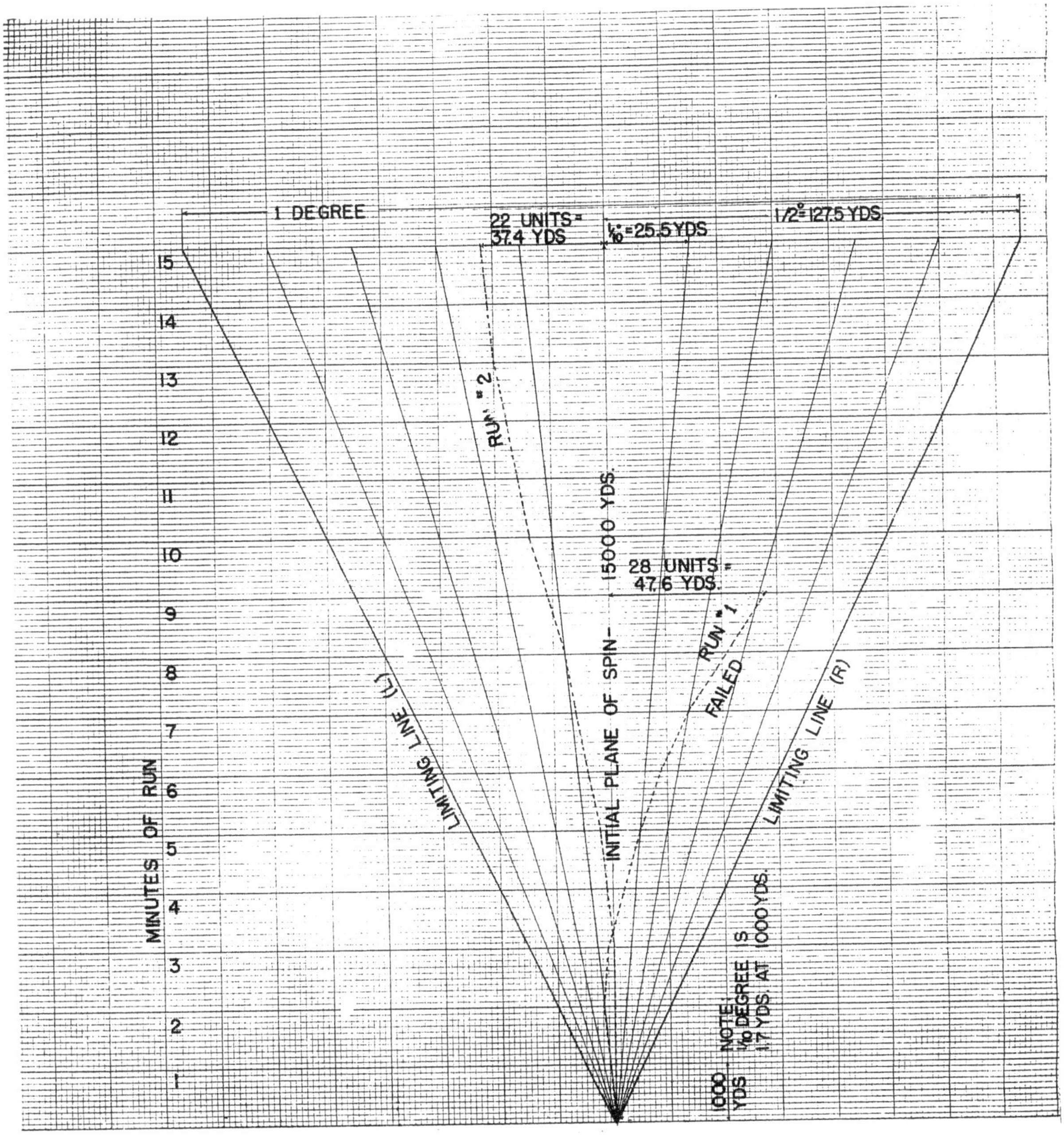

Chart Labels

- 1 DEGREE
- 22 UNITS = 37.4 YDS
- 1/10° = 25.5 YDS
- 1/2° = 127.5 YDS
- RUN #2
- INITIAL PLANE OF SPIN — 15000 YDS.
- 28 UNITS = 47.6 YDS.
- RUN #1 FAILED
- LIMITING LINE (L)
- LIMITING LINE (R)
- MINUTES OF RUN (1–15)
- NOTE: 1/10 DEGREE IS 17 YDS AT 1000 YDS.
- 1000 YDS

TS-5   X-21A

J. CREEP, OR LATITUDE ERROR.

1039. Creep is an apparent (not actual) movement of the gyro axis in azimuth, with respect to a fixed point on the earth's surface, due to the rotation of the earth alone. It is recognized as the amount the curve of the torpedoes path varies from the line of fire and may be expressed in yards or minutes of arc. The rotation of the earth on its axis carries the earth's line of fire away from the fixed celestial line of fire, in which lay the axis of the gyro when the gyro was initially spun. The deflection due to creep is to the right of the line of fire when facing the target in north latitude and to the left in south latitude. Creep varies with the sine of the latitude and is zero at the equator. The magnitude of the creep angle is independent of the direction of the line of fire on the earth's surface. The creep angle is proportional to the angle through which the earth rotates during the run of the torpedo. Therefore, since the angular speed of rotation of the earth is constant, the creep angle is proportional to the time of the torpedoes run.

1040. Consider a gyro in static balance. While a torpedo is running toward the target, the gyro is steering the torpedo toward a point in space, the direction of which, was determined by the axis of the gyro when it was spun. In the meantime the line of fire is rotating with the earth and changing its position relative to the direction that the gyro is maintaining its axis in space. The result is an apparent (not actual) precession of the gyro axis, causing the torpedo during its run to deviate constantly from the line of fire, which is rotating with the earth.

1041. In figure 1, assume that the circle Ps E Pn Q represents the earth on the plane of the Meridian with EQ as the equator. Consider point A in north latitude L the position of an observer whose line of fire is due north, directed at a target anchored at Pn. If the observer spins a gyro in static balance, with its axis along the line APn, the gyro will select a point in space beyond Pn at O, and stubbornly maintain its axis directed at that point. As an exaggerated case, suppose that the observer remained in the same position for 12 hours with his gaze fixed, due north, upon the target at Pn, and the gyro continued spinning during the twelve hour period. In 12 hours the earth has rotated through 180° and the relative positions of the point in space and the observer on the earth, will appear as in figure 2. The observer's line of fire APn, in figure 2, is still on the target, due north, and has not changed relative to the earth. The gyro axis is still directed at point O in space to right of the observer's line of fire. The amount the line of fire has moved to left of the gyro axis during the rotation of the earth or as more commonly described, the amount the gyro axis has apparently (not actually) precessed to the right of the line of fire during the rotation of the earth is the creep.

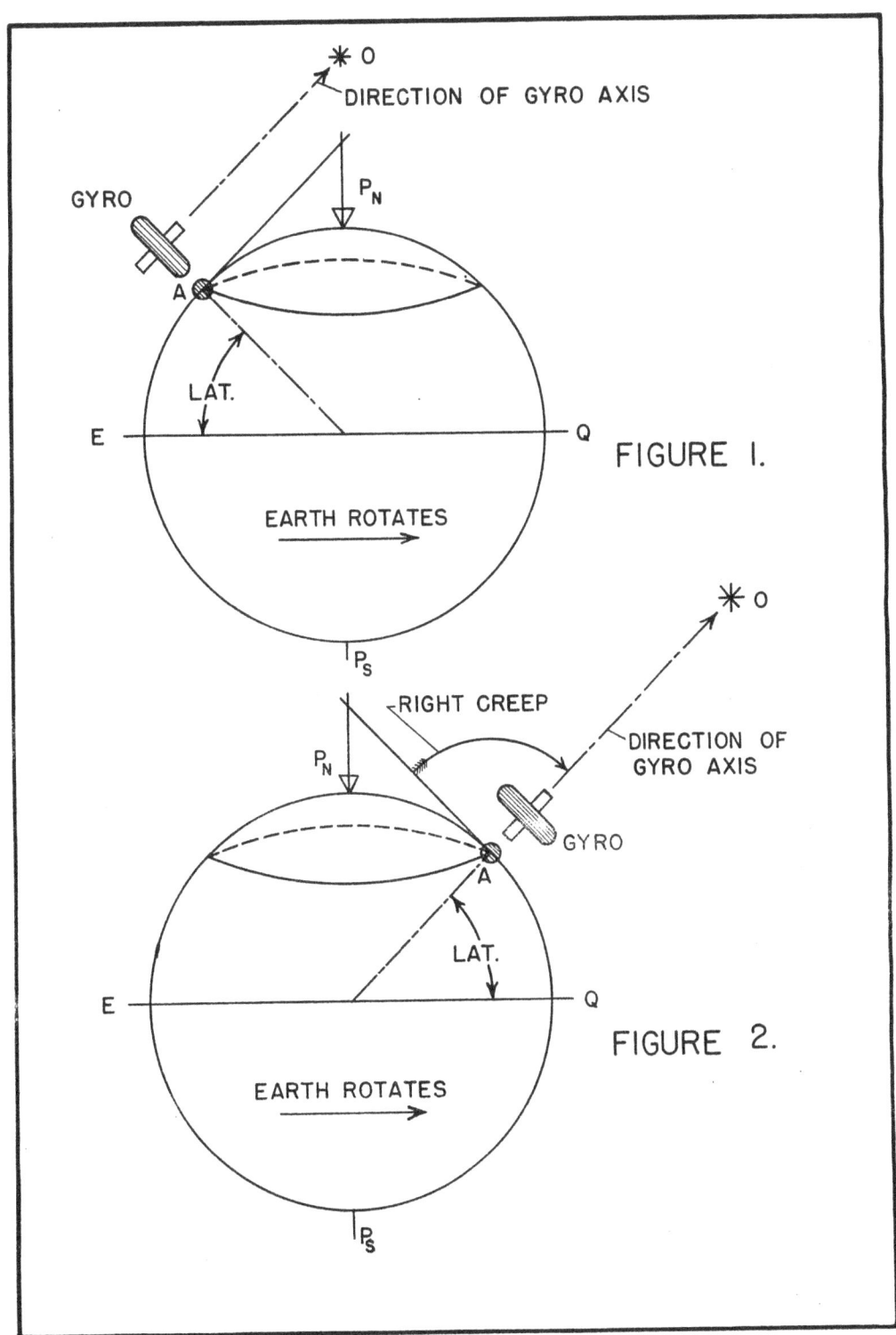

FIGURE 1.

FIGURE 2.

TS-5 X-22A

1042. Figure 3 shows an observer in North latitude whose line of fire is directed due south at an anchored target "T" and a gyro whose axis is initially spun pointed in the same direction. Figure 4 shows the relative position of the observer's line of fire and the axis of the gyro 12 hours later.

1043. Figures 5 and 6 show an observer in South latitude whose line of fire is directed at a target due south under the same conditions as described above. These last two figures illustrate that the apparent precession of the gyro is to left of the line of fire in <u>south latitude.</u>

1044. Figures 7 and 8 show an observer at the equator whose line of fire is directed at a target due north and illustrates the fact that rotation of the earth has no effect on the relative positions of the line of sight and the gyro axis, at the equator and therefore the creep is zero.

1045. The path that the torpedo makes on the surface of the earth due to the creep of the gyro, is the arc of a circle. In figure 9, OT represents the line of fire from the observer to the target on the surface of the earth and in North latitude. The arc OAN represents the curved path of the torpedo on the surface of the earth constantly deviating to the right due to the "creep" of the gyro. The distance O.T. is the range of the line of fire from the firing point and the distance O.N. is the range of the torpedo. A line R.B. tangent to the arc OAn at N forms an angle C with the line of fire equal to the creep angle. The arc TN is the distance the torpedo misses the target if creep is not compensated for, and is called the deflection. The angular deflection is the angle "d" measured by the deflection arc TN.

1046. To demonstrate that the creep angle is proportional to the sine of the latitude and to the time during which the torpedo runs (gyro spins), consider figure 10 in which the circle PNP_1Q represents the earth on the plane of the meridian. EQ is the equator and P_1CP is the earth's axis extended to O. The elipse between Ao and A6 (drawn to give it perspective) is a small circle on which all points are at a latitude L, and is parallel to the plane of the equator.

1047. Lines A1 to A6, represent the line of fire Ao, at various stages of the earth's rotation through 180°, adhering throughout to the original north and south direction relative to the earth. Since the line of fire, Ao, lies in the plane of the meridian, it will when extended intersect the earth's axis, P_1P extended, at some point O. As the earth rotates, the line of fire AoO will describe in space a right circular cone whose apex is at O and whose base is the small latitude circle at latitude L. Angle A6CQ = L. **Radius r then equals R. cos L. Considering the two similar right triangles** CMA6 and OMA6, it may be seen that angle MOA6 = L. <u>Also R cos L</u> = sin L. Then line $\overline{OA6} = \underline{R \cos L}$ =
       line OA6   Sin L
R cot **L.**

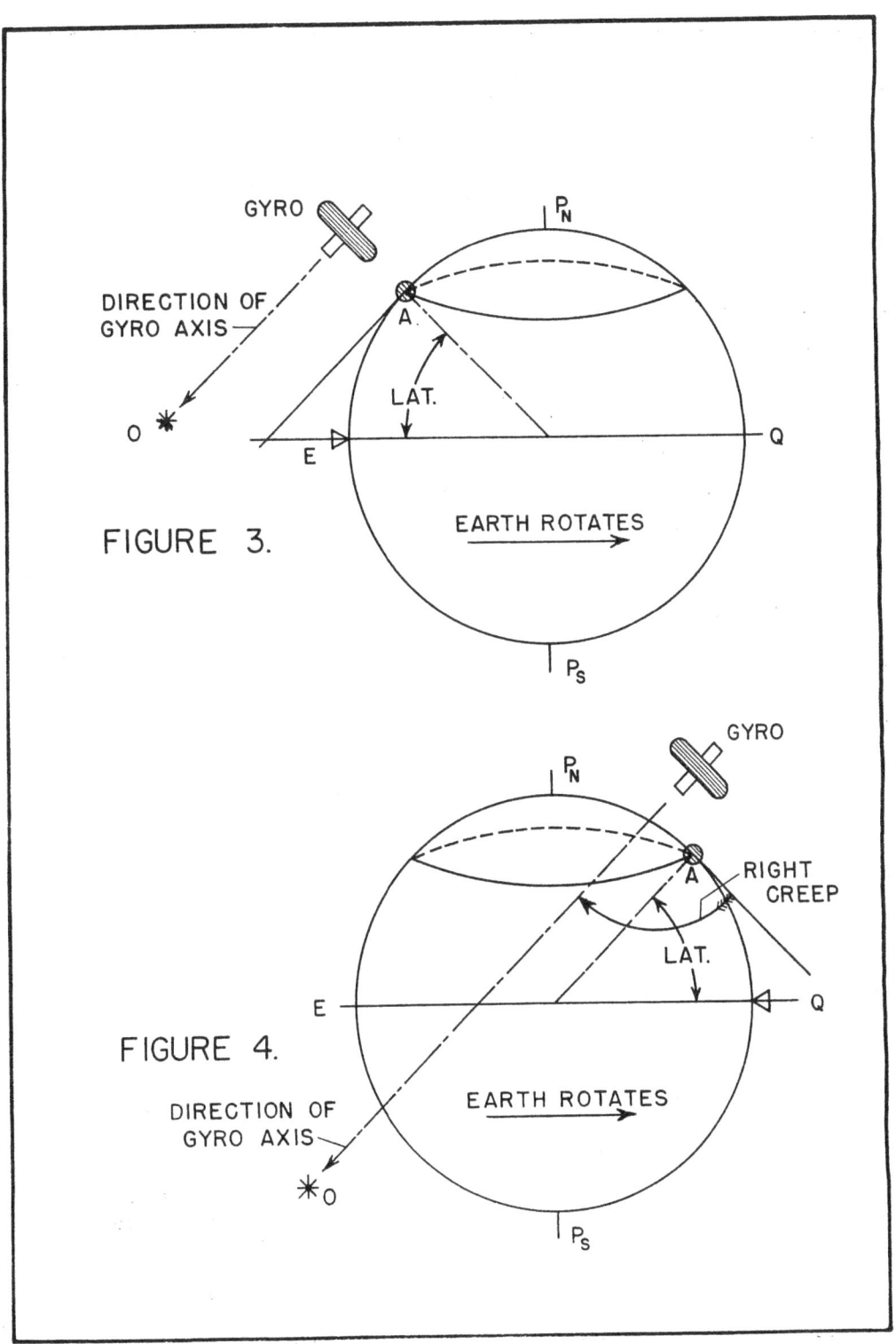

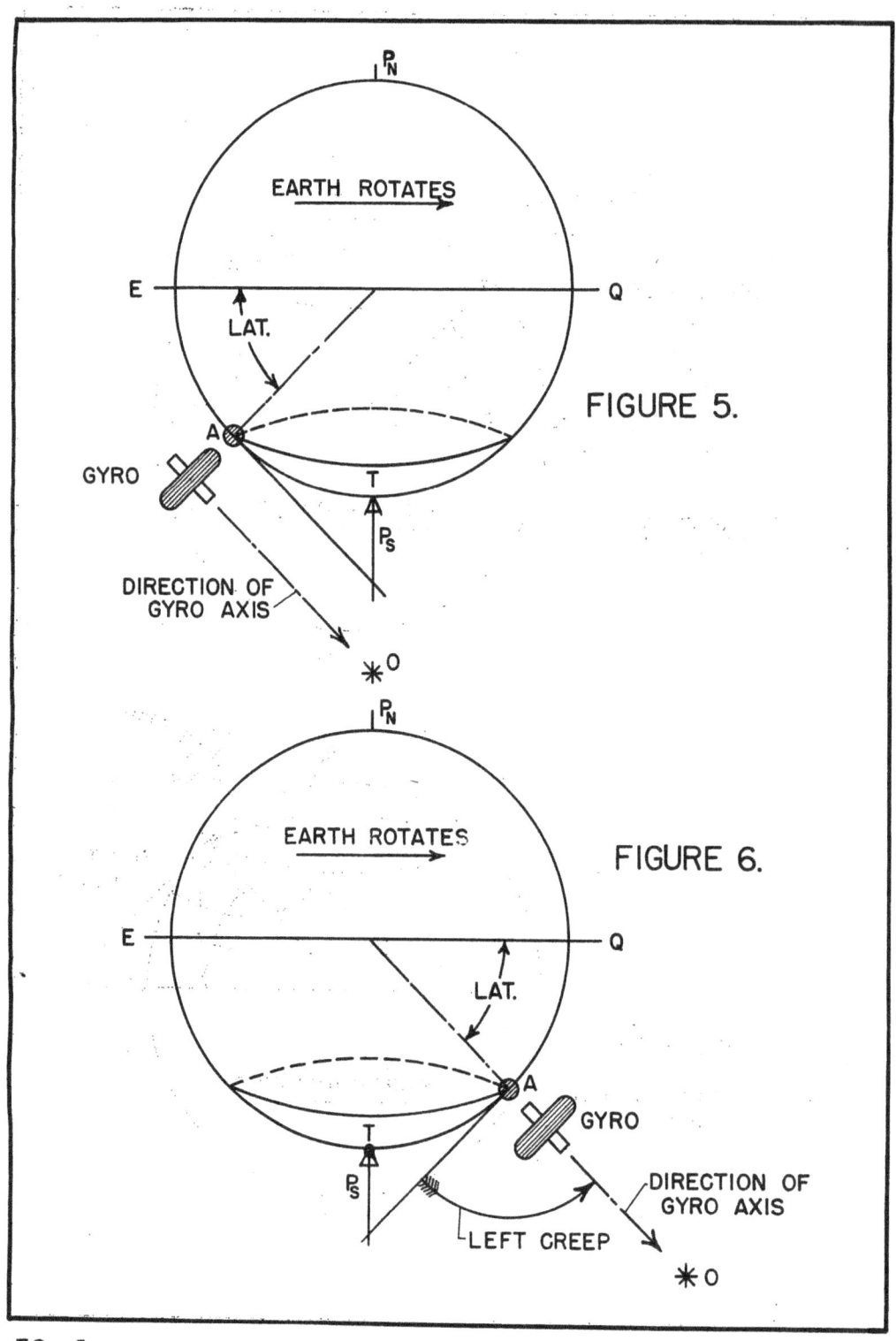

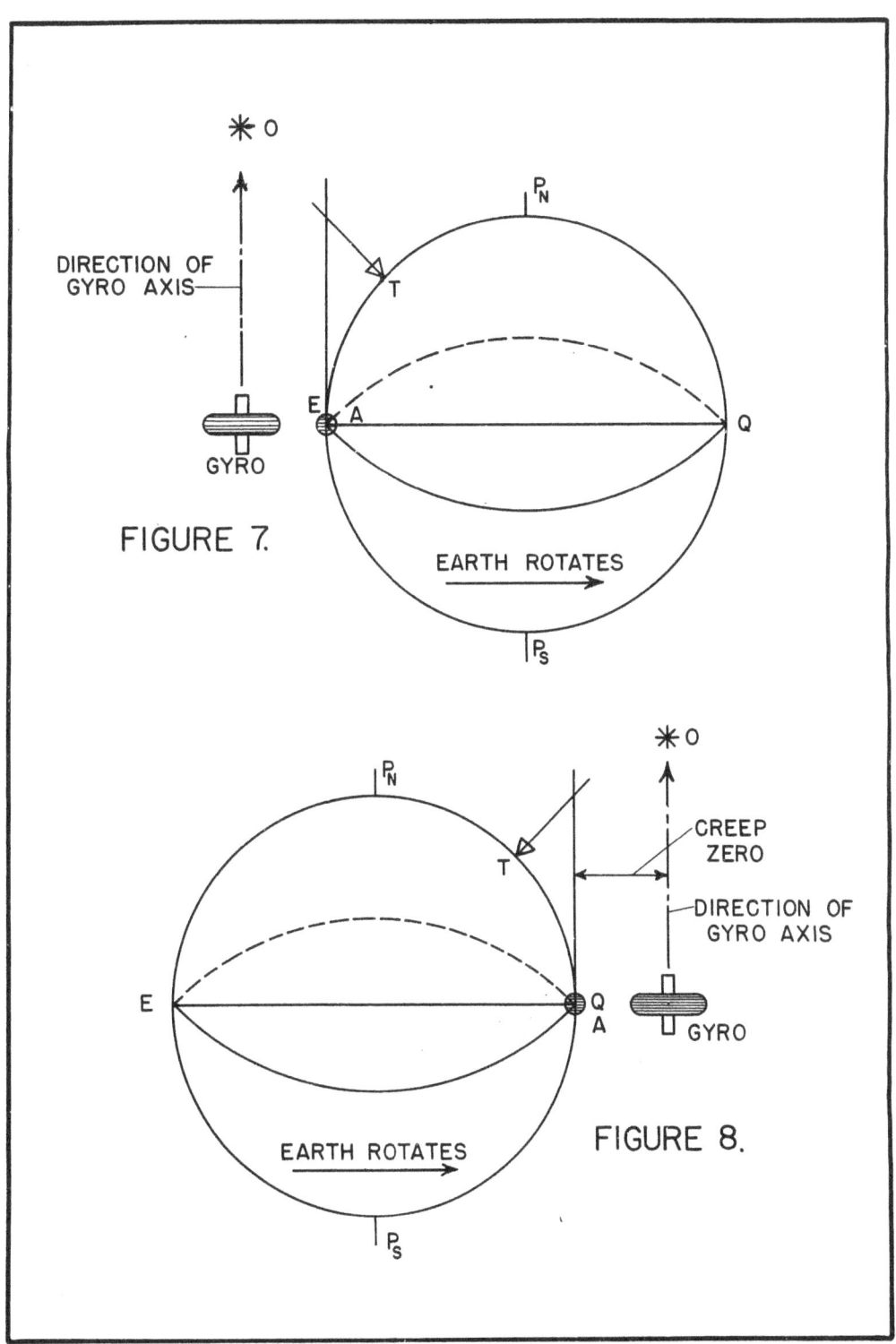

FIGURE 7.

FIGURE 8.

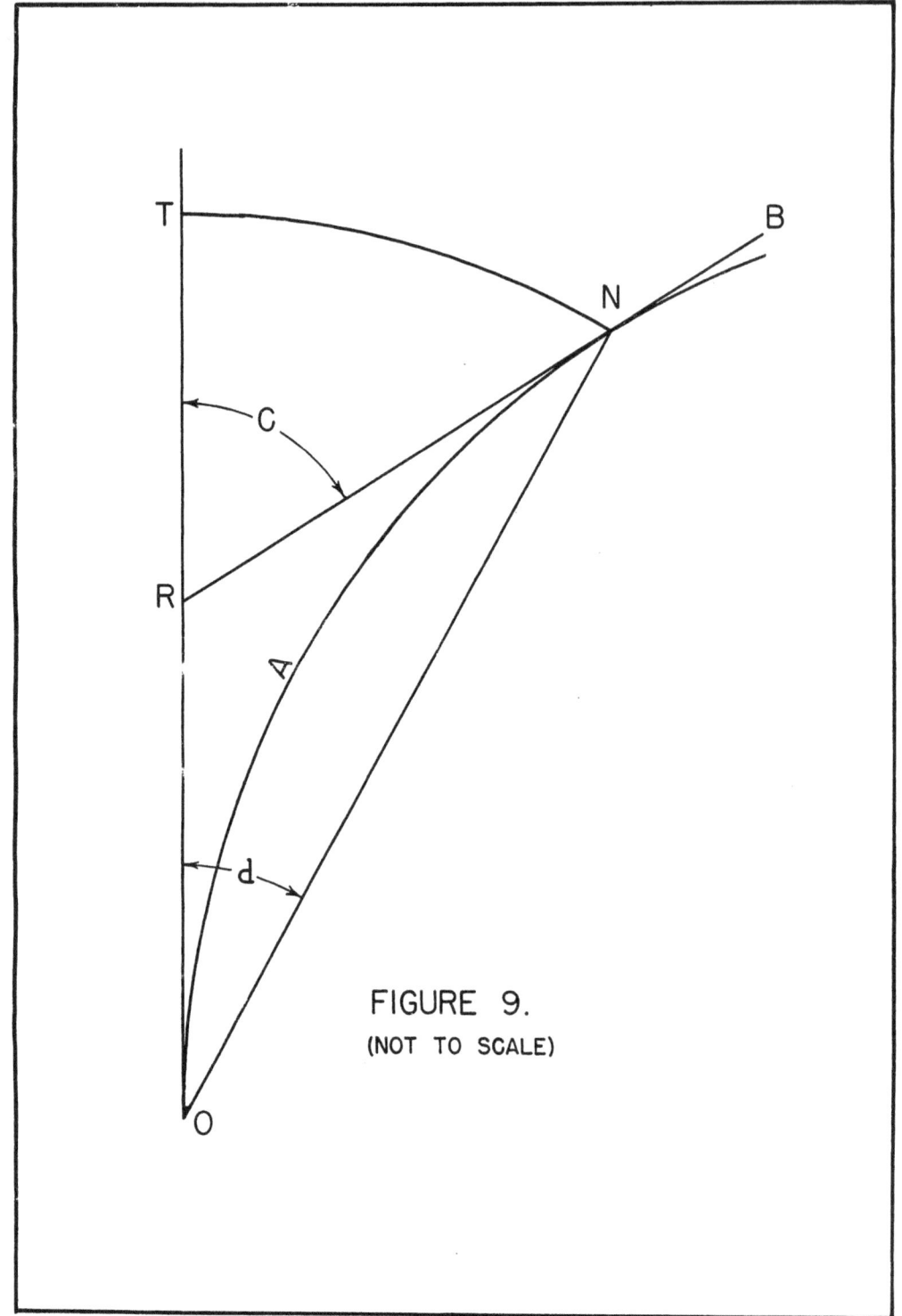

FIGURE 9.
(NOT TO SCALE)

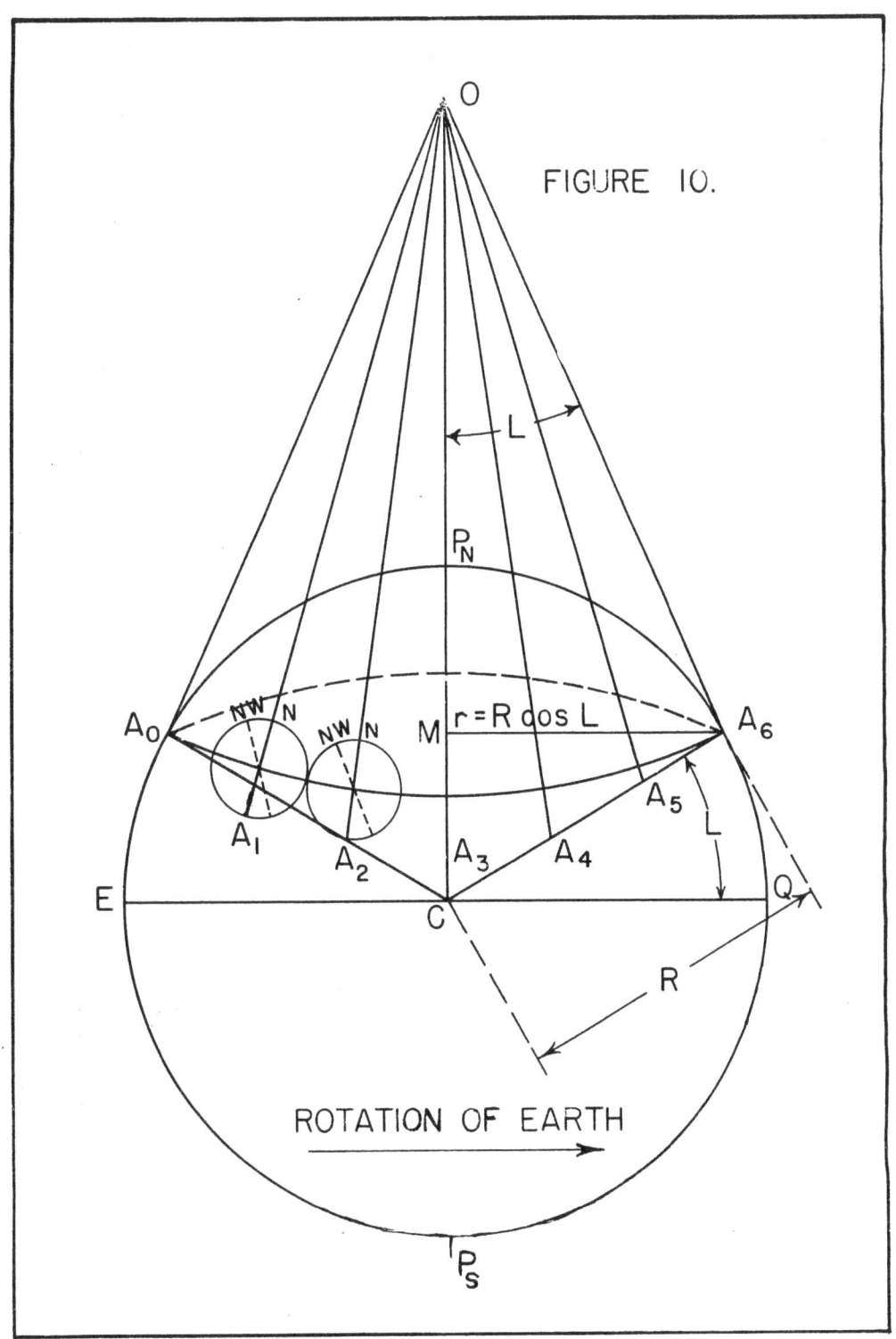

FIGURE 10.

TS-5 X-23 E

1048. From a casual inspection of figure 10 it might seem that during a rotation of the earth of 180°, the line AoO has changed its direction only through the angle AoOA6 = 2 L. This would be true if AoO remained on one plane, actually it travels over the surface of a cone, and the sum of the small angles through which it successively rotates, as the earth rotates 180°, is greater than 2 L. To see clearly the measure of the angle through which Ao rotates during a given rotation of the earth, the entire cone may be developed on a plane surface. The development of such a right circular cone is the sector of a circle whose radius equals the length of an element of the cone. In this case the radius is R cot L and the arc of the sector (considering a full 360° of rotation of the earth) is equal in length to the circumference of the small circle latitude on which Ao moves. This circumference equals $2\pi r = 2\pi R \cos L$. (See figure 11). In any sector of a circle the length of the arc divided by the radius equals, in radians the angle subtended by the arc, or $\frac{Arc}{radius} = O(radians)$. In this case, this becomes:

$$\frac{2\pi r}{R} \cdot \frac{\cos L}{\cot L} = O = \frac{2\pi \cos L}{\cot L} = 2\pi \sin L \text{ (radian)}.$$

But O is the angle through which the line Ao changed its direction while the earth rotated 2π radians (360°). The ratio may then be given:

$$\frac{\text{Rotation of Ao}}{\text{Rotation of earth}} = \frac{2\pi \sin L}{2\pi} = \sin L.$$

1049. The same relation may be proved for any angle of rotation less than 360°, and it follows that the line Ao changes its direction in space during any given length of time through an angle equal to the angle of rotation of the earth in that time multiplied by the sine of the latitude at which Ao is equal to the creep angle, and the time in which we are interested is the time during which the torpedo runs. This formula may be given: Creep angle = $c \frac{T}{4} \sin L$., in which c is expressed in degrees, and T is the duration of the torpedo's run in minutes. Since the earth rotates $\frac{1}{2}°$ per minute, the term T/4 represents, in degrees, the angle of rotation of the earth in time T.

1050. Special cases of this effect are met with at equator and poles. The formula above still applies, however, Sine L at the equator becomes zero, and at that latitude creep disappears. The line Ao drawn at the equator does not intersect the axis of the earth but is parallel to it, and so moves parallel to itself (does not change its direction in space) as the earth rotates. At the poles sin L equals unity. The cone has an apex angle of 180°. It becomes plain, with point O at the pole. The creep angle at the pole is equal to the full angle through which the earth rotates during the run of the torpedo. It may be preferable to express the formula above in terms of torpedo speed and range. The creep angle in degrees is then given by the expression: $\frac{.0074 \text{ R} \sin L}{S}$ where R is the distance the torpedo runs, in yards, and S is the torpedo speed in knots.

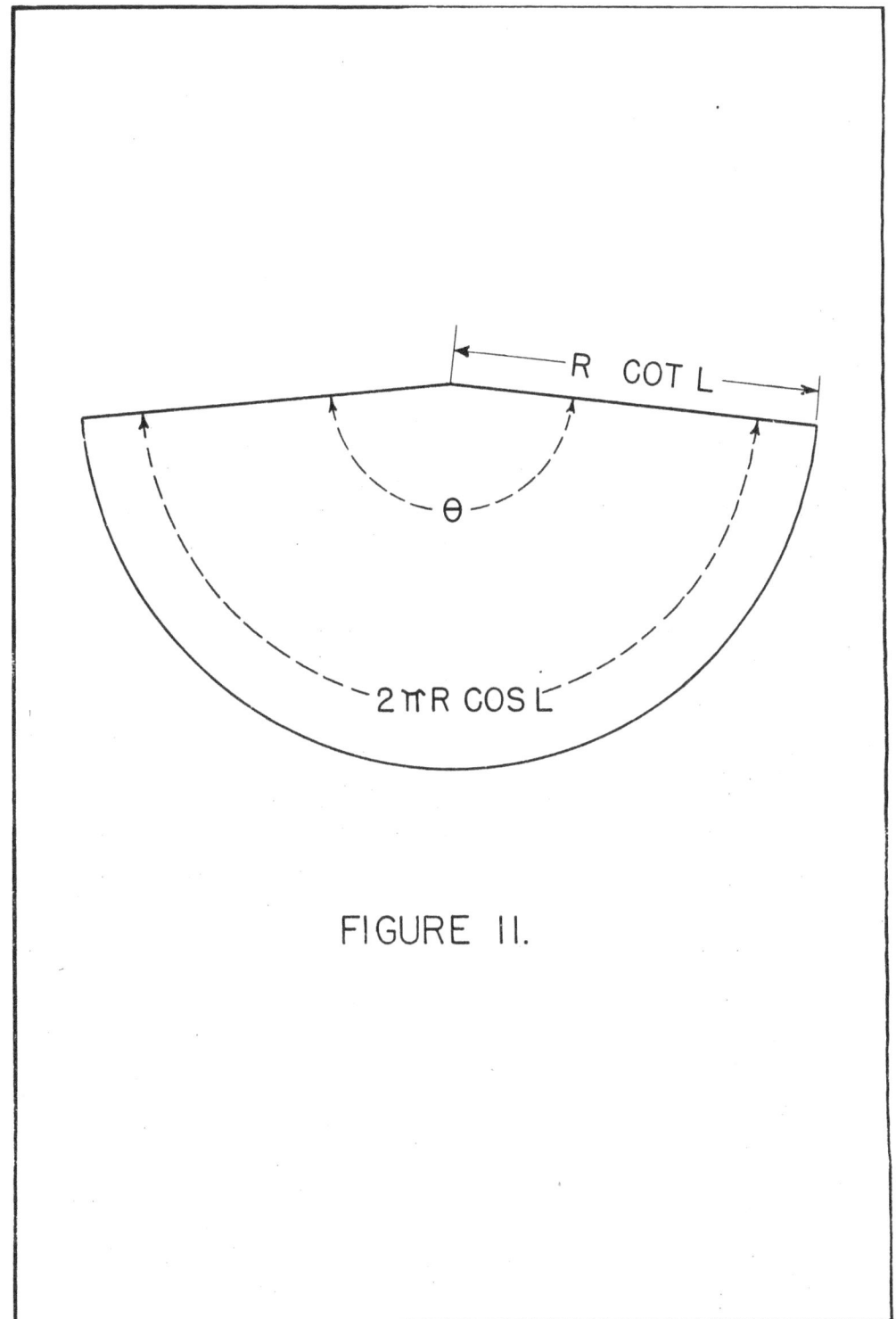

FIGURE II.

1051. In Fig. 1, the arc TN, in yards, may be found from the rather formidable expression $(TN) = \frac{.000065 R^2 \sin L}{S}$, where the terms have the same meaning as above. It would probably be simpler to compute the angle d in degrees from the expression: $d = \frac{T}{S} \sin L$ and to substitute this value in the formula
$$(TN) = \frac{Rd}{180}$$

1052. In the case of a torpedo running 10,000 yards in 10 minutes, at 40° latitude, and given an otherwise perfect run, the creep would cause the torpedo to miss the target by 140 yards.

K. **PRECESSION.**

1053. The word "PRECESSION" as used herein means movement of the gyro axis due to the application of an outside force. Such precession, may be in the horizontal plane, or the vertical plane. It may be expressed in either angular or linear units of measurement. An external force, as referred to above, is any force applied anywhere on the gyro which causes it to "precess" and thus behave in a different manner from what it would were it in pure static balance. An unbalanced force may be caused by a change in the balance nut position; by friction in any of its bearings (except wheel bearings) wherein such friction can be brought into play to oppose the wheel's maintaining its original plane of spin; permanent heel of the torpedo, by causing an inclination of the outer and inner gimbal rings, from their original position of spin, may cause the introduction of minute bearing pressures or forces; a large initial roll, or considerable variation in depth may cause precession because of the external forces of friction at the bearings, brought into play by these irregularities of torpedo performance. Too much lubricating oil on bearings may cause drag which in opposing the natural tendency of the gyro to "creep" will set up an external "precessing" force; air currents which may be allowed to impinge upon gimbal rings, or the wheel itself, may likewise constitute external forces which will cause precession.

1054. Fortunately it is possible to hold these external forces within reasonable limits if the gyro is kept cleaned, properly lubricated, and properly adjusted. The design of later torpedo gyros and mechanism (those in active service today) has practically eliminated the danger from air currents. It has been the aim of design to make all the gyro bearings as sensitive as possible and yet sufficiently rugged to withstand the shocks incident to firing the torpedo from all types of tubes and from aircraft. Ruggedness has been the first consideration because erratic performance is sure to follow failure of gyro parts due to weakness. We have had to accept a certain loss of sensitivity in order to meet the problem of ruggedness in service. Even under these circumstances, however, our torpedoes will perform

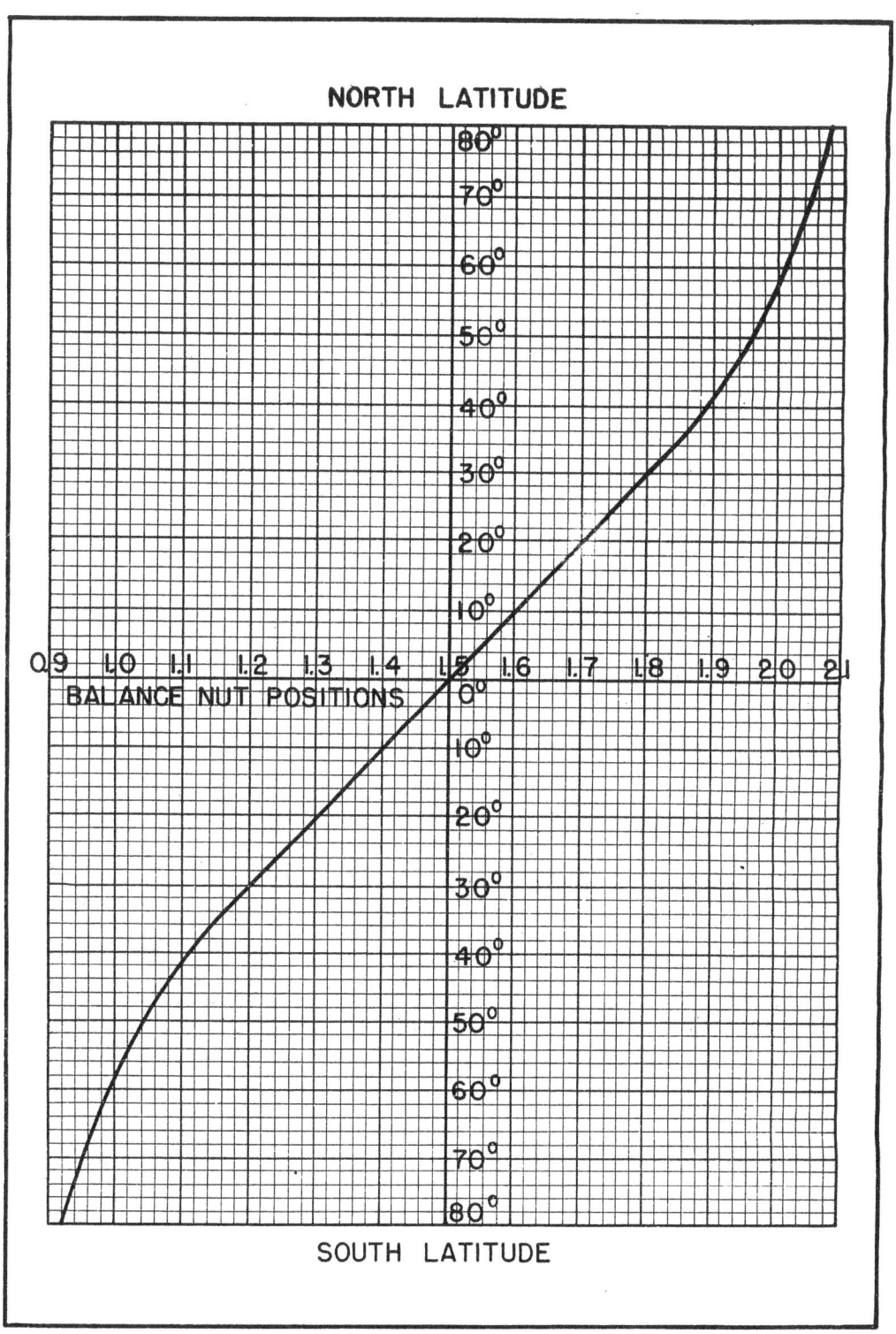

TS-5   X-25A

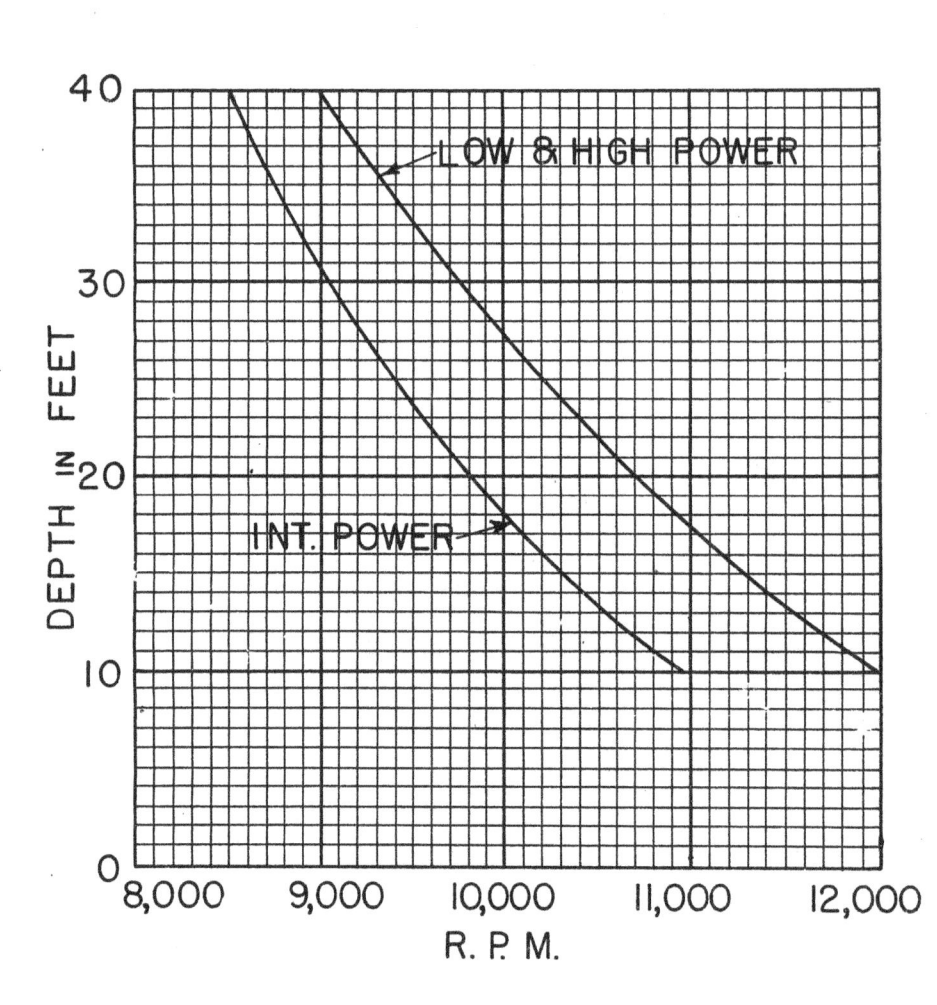

satisfactorily in deflection if the gyroscope is properly cleaned, oiled, adjusted and balanced. Precession of a gyroscope is therefore prevented by proper care of the gyroscope itself. Creep is prevented by introducing an external force of such magnitude that it will exactly neutralize it. The external force referred to is applied by a change in the balance nut position from that which it would have when the gyroscope is in static balance.

1055 Expressed in other words, this means that the purpose of the balance nut is to correct for "creep" due to the firing latitude, and for that purpose only. The balance nut therefore introduces a "precession" which should exactly neutralize the "creep" due to the latitude. If the gyroscope is properly cleaned, oiled, adjusted and balanced there is no other "precessing" force, and therefore the balance nut does not have to be given a position other than that necessary to offset the "creep" due to latitude.

L. COMPENSATION FOR CREEP.

1056. The balance nut of the gyro may be set so as to cause a precession which will equal and opposite the creep. The correction to be applied to the static balance position of the nut is given by the following formula:

   Nut correction (in turns) = 0.0000592 x N x Sin Lat.
       where N = r.p.m. of gyro wheel.
           Lat. = firing latitude

1057 Every running torpedo has an afterbody pressure in excess of the hydrostatic pressure for the depth at which it is running. This is unavoidable and is due to necessary restrictions as the exhaust gases are led from the torpedo.

1058. The afterbody pressure causes a drag on the gyro wheel because of the density of the atmosphere in which the wheel is turning. The gyro speeds which may be expected for various conditions are given on the curves on page X-25B

1059 For practical purposes a gyro speed of 10,000 may be assumed for the computation of the nut correction without resulting in serious error. The curve on page X-25A gives the balance nut setting for an assumed gyro speed of 10,000 r.p.m's.

CHAPTER ELEVEN

ARTICLES

A. Install Immersion and Gyro Mechanism............1101

B. Test Afterbody Assembly for Leaks...............1102

C. Overhaul and Assemble Tail on Afterbody
   (Mk. 13-1 Torpedo)............................1103

D. Tests conducted on Afterbody in Adjusting Stand..1104

E. Assembly of Afterbody to Air Flask..............1105

F. Test air connections in Midship section for
   Leaks...1106

G. Adjustments and Tests.........................1107-1110

H. Preliminary Adjustments........................1111

I. Attach Exercise Head...........................1112

J. Final Adjustments with War Head................1113

K. Routine for Upkeep of Fully Ready Torpedo......1114-1116

L. Treatment after a Run..........................1117

A. INSTALL IMMERSION AND GYRO MECHANISM.

1101.   TOOL NO.

(1) Place gasket on gyro door flange.
(2) Remove clamp plate and cover.................... 13,14
(3) Remove transportation pin....................... 49
(4) Depth spring should be calibrated to 10'.
(5) Set depth index on 10'.......................... 135A
(6) Insert two lifting screws for base.............. 200
(7) Install gyro mechanism assembly, meshing connecting rod for driving pallet mechanism with squared hole in spur gear on engine frame strut.
(8) Secure mechanism with holding screws and remove lifting tools................................... 49,456
(9) Replace gasket and gyro clamp plate and cover.. 13,14
(10) Replace gaskets and hand hole plates and secure holding nuts...................................... 48
(11) Replace washer and drain plug for afterbody.... 13,14

B. TEST AFTERBODY ASSEMBLY FOR LEAKS:

1102.

Blank off the following pipes:
(1) Main air pipe from stop and charging valve..... 134
(2) Fuel from check to spray....................... 141A
(3) Water from check to spray...................... 141A
(4) Reduced pressure air to air check valve........ 229
(5) With grease gun force grease (G) through after propeller shaft until it appears around bushing between shafts................................... 482
(6) Install exhaust valve gag spring............... WE228
(7) Remove drain plug and install special test fitting for low pressure air with gauge........ 141A
(8) Connect low pressure gauge and air line to this fitting.
(9) Turn air on and allow 10 lbs. pressure to accumulate in afterbody.
(10) Pour out (C) around flange and joints, and note if any air bubbles appear indicating leaks. Inspect for leaks around steering rod stuffing boxes, propeller shaft packing, and exhaust valves. In case of leaks around exhaust valves, check valve alignment to seat by inserting strips of cigarette paper dipped in oil (C) about 90° apart, across seats of valves. Hold valves against seats and try to remove paper strips. If this can be done without binding, valves are out of alignment with seat.
(11) Insert valve in a collet on bench lathe and determine if seat on valve runs true by placing indicator in tool post and against valve. If seat runs out more than ".002, it will be necessary to turn the seat true in bench lathe............. WE10

TOOL NO.

(12) If valve seat is found out of alignment, install reseating tool in valve bracket in line with valve seat to be refaced and secure bracket on studs. Place forked end of expander clamp over stem between cutter and bearing boss on bracket, set up on screw in center of expander clamp and turn reseating tool until new seat is obtained...................... 457,40,227
(13) Having obtained true seats, grind valves to seat using grinding compound sparingly......... 40
(14) All leaks so found must be remedied before proceeding further with assembly.
(15) With the test completed and leaks stopped where necessary, remove special tools and fittings used for this test.
(16) Remove hand hole plates...................... 48,200
(17) Replace depth engine and secure with holding screws, wire screw heads...................... 49A,72
(18) Connect valve connection....................... 246
(19) Replace rudder rod pin.
(20) Connect air pipe to depth engine............... 141A
(21) Replace steering engine and secure with holding screws, wire screw heads...................... 49A,72
(22) Replace valve arm connection.................... 246
(23) Connect air pipe from gyro reducer to nipple on pot.
(24) Replace rudder rod pin.
(25) Connect air pipe to gyro engine................ 141A
(26) Connect air pipe to gyro spin................. 229

C. **OVERHAUL AND ASSEMBLE TAIL ON AFTERBODY (Mark 13-1 TORPEDO).**

1103.

(a) (1) Clean and inspect all bearing surfaces of both rudder yokes. If necessary to remove burrs, use bearing scraper and fine oil stone. With aligning gauge (WET 219) check alignment of vertical rudders and with (WET 212) check alignment of horizontal rudders............... WE 212, WE 219
(2) Clean, inspect, oil (B) and replace rudder connections, washer and cotter pin on rudder yokes. 72,41
(3) Clean and inspect vertical and horizontal rudders and outboard bearing for rudders..........
(4) Oil (B) and replace rudder yokes and rudders in tail cone. Oil (B) and replace outboard bearings for rudders. Replace screws previously removed in outboard bearing in the same holes from which removed and set up tight. Note: With thickness gauge (WET 2) check clearance of rudder yokes between yokes and outer bearing surface. This should be ".015.................................. 40,WE2

TOOL NO.

IMPORTANT NOTE: When assembling rudder and outboard bearing, particular attention should be paid to the assembly numbers. The tail blades are numbered 1,2,3 and 4 on their after edges, the rudders on their forward edges and the outboard bearings on their after ends It is important that these numbers are assembled to match. When assembled, the clearance between the bearing shoulders of the rudder spindles and the outboard bearings should be ".015.......... WE 2

    (5) Clean, inspect, oil (B) and replace tail bearing in tail cone. Secure with screws, setting up tight................................... 40

(b) Assemble forward propeller sleeve:
    (1) Clean and inspect forward propeller sleeve, keys and keyways.
    (2) Replace grease packing ring in forward propeller sleeve (with beveled side of ring forward). Secure with keep screws and wire............. 41,72
    (3) Replace forward propeller sleeve on propeller shaft. Note: Assemble bench marks 0 to 0.
    (4) Clean, inspect, oil (B) and replace four (4) holding clips and secure with screws and wire..72,41
    NOTE: If previously removed, replace spider supporting straps with their reinforced pieces on forward ends of tail blades, with numbers corresponding and secure with holding screws.

(c) Replace tail on afterbody:
    (1) Note that tapped holes for joint screws in after bulkhead are clean and free of burrs.
    (2) Replace tail on afterbody and secure with sixteen joint screws........................ 184
    (3) Connect rudder connections to rudder rod eyes and replace washers and drain plugs in tail cone................................... 13,14

(d) Assemble forward propeller:
    (1) Replace two keys for forward propeller hub in forward propeller sleeve.
    (2) Replace hub for forward propeller on propeller sleeve.
    (3) Replace forward propeller on propeller hub after blocking.
    (4) Replace nut for forward propeller. Set up on nut until screw holes for lock screws are in line.185C,185D
    (5) Replace two lock screws in nut for forward propeller................................... 41

TOOL NO.

(c) Assemble after propeller:
   (1) Clean and inspect keys for keyways for after propeller sleeve.
   (2) Clean and inspect for burrs and scored surfaces four bushings for after propeller sleeve. Note that oil holes and oil grooves are clean. Replace bushings on after propeller sleeve.
   (3) Replace after propeller sleeve on shaft.
   (4) Replace after propeller sleeve holding nut....... 183
   (5) Replace after propeller sleeve hub.
   (6) Replace after propeller on hub after blocking.
   (7) Replace nut for after propeller. Set up on nut until screw holes for keep screws are in alignment.. 185C,185D
   (8) Replace two keep screws in nut for after propeller. Set up tight........................... 41

D. **TESTS CONDUCTED ON AFTERBODY ASSEMBLY, IN ADJUSTING STAND.**

1104.
   (1) Place afterbody in adjusting stand.
   (2) Blank off the following pipes:
      Reduced air to air check valve................... 229
      Check valve to fuel spray....................... 141A
      Check valve to water spray...................... 141A
      Vent pipe on bulkhead........................... 141A
   (3) Connect nipple and H.P. air line to main air pipe from stop valve to valve group............. 134
   (4) With starting gear index off zero, lift up on starting spindle arm and crack valve on main air line to test for leaks around joints of flanges and bulkhead joint............................... 135A
      NOTE: The above test is a check on afterbody leak test previously conducted.
   (5) Turn starting index to seat valve in starting gear... 135A
   (6) Remove blank on fuel or water spray pipes, loosen pipe and turn end down, place a cup of water over end of pipe and note if any air bubbles appear; if so, there must be a leak around starting gear pipe connections or the valve in starting gear; locate and repair........ 141A,12
   (7) Disconnect and remove pipe from valve group to bulkhead (gyro spin) and connect with test pipe to H.P. air line.................................. 229
   (8) Turn on H.P. air and test for leaks in gyro spin and impulse valve; if air escapes through end of pipe to fuel and water spray, there must be a leak in the line, or in the impulse valve. Locate same and remedy.
   (9) Reconnect pipe from valve group to bulkhead (gyro spin)...................................... 229

	TOOL NO.

(10) Replace blank in fuel or water spray pipe.
(11) Remove hand hole plates.................................. 48,200
(12) Remove air strainer and replace it with adapter (tool #223). Connect low pressure (450 lbs.) air line to nipple on adapter. Turn on air and note if center line of depth engine valve stem lines up with scribe mark on the valve stop. If the scribe marks do not line up, loosen clamp screw on valve connection and turn valve stem by its knurled nut until they do line up. After adjustment, leave the hole in knurled nut in a horizontal position, and set up on the clamp screw of valve connection. 372A, 377, 141A, 246
(13) Read neutral throw of horizontal rudders. The neutral throw should be $1\frac{1}{2}$ down for Mark 13-1 torpedoes. Any deviation from the above will require re-adjustment in the tail cone.
(14) Remove transportation pin. Swing pendulum by hand and get full throw of horizontal rudders. Full throws should be 1 up and 4 down for Mark 13-1 torpedoes, and both horizontal rudders should line up on zero. If the combined throws do not give a total of as above, check the following places for trouble: Fork for steering rod not screwed all the way up on depth engine piston rod; gland for steering rod through after bulkhead not screwed all the way in on packing; obstructions to pendulum, preventing its full swing.
(15) Level afterbody. Unscrew access hole plug from atmospheric chamber and install 16 lb. weight. With air on depth engine, turn depth index spindle until center line on depth engine valve stem is in line with mark on valve stop. With side gear spindle socket lifting tool, hold spindle disengaged and turn depth index to read 10 feet. Remove lifting tool and turn depth index slightly to engage square on spindle socket with square hole in adjusting socket. Turn index back to zero and back again to 10 feet. The valve center line should again line up with mark on valve stop. Swing pendulum against its stop, let go, and note that it makes 5 to 6 swings, stopping its swing with scribe marks in line............. 411B, 135A, 472
(16) Adjust pointer on cradle to zero on indicator scale on carriage of adjusting stand. Starting with afterbody level and air on depth engine, elevate and depress afterbody. Note movement of pendulum and valve for smoothness. Full throw should be accomplished at maximum inclination of $2°$ each way. Rudders should move up and down without jerk.
(17) Remove 16 lb. weight and set depth index on zero... 135A
(18) Replace access plug in atmospheric chamber......... 11

	TOOL NO.

(19) Replace transportation pin........................ 49
(20) Move the control valve of gyro engine, in and out, and take the vertical rudder throws with vernier scale. These throws should be as per record book (about 33 each way for upper and 24 each way for lower on the Mark 13-1 torpedoes). Equalize throws by rudder rod adjustment in tail cone................ 44
(21) Turn off low pressure air. Disconnect air lead and remove adapter................................ 141A,377
(22) Install air strainer and plug.................... 372A,12
(23) Turn afterbody 180° and remove gyro clamp plate and bottom head................................... 246,13,14
(24) Oil top and bottom vertical bearings of gyro pot with gyro oil.
(25) Lock spinning mechanism and unlock by the hand trip to test for proper operation................... 205
(26) Turn propellers until cam pawls are in the "all the way out" position.
(27) Oil gyro bearings with gyro oil and install gyro in pot.
(28) Install gyro bottom head with its scribe. **mark in line with scribe mark on flange** of pot..... 246
(29) Lock gyro, trip by hand and lock again............. 205
(30) Install gyro clamp plate and gasket............... 13,14
(31) Turn afterbody right side up.
(32) Check oil pot for sufficient oil to give engine a run. Turn propellers a few turns by hand.
(33) Set up on speed screw............................. 12
(34) Reconnect H.P. air line to pipe from stop valve to valve group....................................... 134
(35) Clamp afterbody adjusting stand with azimuth pointer on zero, (caution personnel not to approach close to tail during test). Seat starting piston - leave distance index off zero............. 135A
(36) Open valve on the H.P. air line. To prevent giving the gyro a premature spin the spinning turbine of the spinning mechanism can be held as the valve on the H.P. air line is opened. Opening the H.P. valve causes the starting valve to fluctuate momentarily. The spinning turbine can be released after air has equalized on the starting valve.
(37) Remove propeller lock and lift up on starting spindle arm, thus raising the starting spindle head upward and spinning the gyro. Regulate air line valve to about 700 lbs. pressure. Loosen clamp on adjusting stand and move stand each way in azimuth. Gyro rudders should operate at about 1/5 of a degree on each side of center. Note that rudders come to each side smartly as afterbody is swung in azimuth.
(38) After a satisfactory azimuth test, close H.P. air valve and replace propeller lock.
(39) Slacken up on the speed screw and remove H.P. air line.. 12,134
(40) Turn afterbody 180° and remove clamp plate, bottom head and gyro............................... 13,14,246

	TOOL NO.

(41) Replace bottom head, clamp plate, and drain plugs..246,13,14
(42) Remove blanks and air fittings from pipes.
(43) Turn afterbody right side up. Give a final inspection inside the afterbody to see that piston fork pins are properly secured and that no tools are left inside.
   NOTE: Lubricate interior, especially the main engine, with oil syringe #94.
(44) Replace hand hole plates............................ 43

E. ASSEMBLY OF AFTERBODY TO AIR FLASK.

1105.

Join afterbody to air flask:
(1) Install bail sling and hoist afterbody up about 2" above air flask joint.
(2) Join afterbody to flask holding tail higher and slightly to the right. With pipe connections lined up, lower chain fall and push afterbody in place on air flask joint.
(3) Secure with afterbody joint screws, inserting top screws first and tightening screws even around joint.. 336
(4) Connect pipes in the midship section in the following order:
   Air pipe to air check valves...................... 229
   Main air pipe to stop and charging valve......... 134
   Fuel pipe to fuel check valve.................... 141A
   Water pipe to water check valve.................. 141A
   Vent pipe to vent fitting........................ 141A
(5) Charge air flask to 1000 lbs. Remove fuel and water filling plugs when charging.
(6) Place a cigarette paper soaked in oil over water filling plug hole and note if any air leaks into water compartment while charging.
(7) Try holding screws for stop and charging valve body, air check valve body and delivery check valve body for tightness during charging.

F. TEST AIR CONNECTIONS IN MIDSHIP SECTION FOR LEAKS.

1106.

(1) Secure propeller lock in place.
(2) Turn dial on index spindle off zero, and lift starting spindle arm............................... 227
(3) Crack stop valve, squirt oil (C) around air connections in midship section, and with a lighted taper, note if oil bubbles appear around joints or if the lighted taper flickers indicating leaks.
(4) Turn index spindle to seat starting valve.

TOOL NO.

    open stop valve wide and proceed to test H.P. connections in midship as in step (3) above.

  (5) Blank off end of air pipe from blow valve to exercise head. Open air blow valve and proceed to test as in step **(3)** above.

  (6) It will be necessary to remedy any leaks which may be discovered by tests given in steps (3), (4), and (5) above before proceeding further.

  (7) Charge flask to 2800 lbs. and again go over high pressure air connections in midship section.

G. <u>ADJUSTMENTS AND TESTS</u>.

1107. Preliminary adjustments are made to ascertain that there are no leaks, and that various controlling units will function properly as received from overhaul, also to check coincidence and reference marks together with index dials and pointers for alignments and accuracy. With preliminary adjustments completed the torpedo is ready for the final adjustments to be made prior to a run.

1108. It will be noted that symbols (capital letters) are used to designate the oils and greases having application in the various steps of the adjustments.

1109. Preliminary adjustments should be made with the torpedo resting on a truck or chocks, with the exercise head removed, located under a chain hoist.

1110. Equipment tools and supplies required for use in making preliminary adjustments: Torpedo truck or chocks; hoisting strap (short); safety strap; *torpedo testing set with wing nut; two (2) propeller locks; pressure gauge (portable).
Tools required: Supply and ready tools, numbers of which are listed in the right hand margin opposite each step.
Supplies required: tail packing compound - (G); hot running torpedo oil - (B); light lubrication oil - (C); gyro oil - (A); small quantity of spirits (cleaning strainers).
The above material (except spirits) is listed in the torpedo allowance for vessels.
* - General use test panels if available.

H. <u>PRELIMINARY ADJUSTMENTS</u>.

1111. Charge Air Flask:
  1. Place torpedo right side up on truck or chocks.
  2. Put on propeller lock, secure lanyard. Install clip lock or lanyard on starting spindle arm.
  3. Remove water filling plug............................ 11

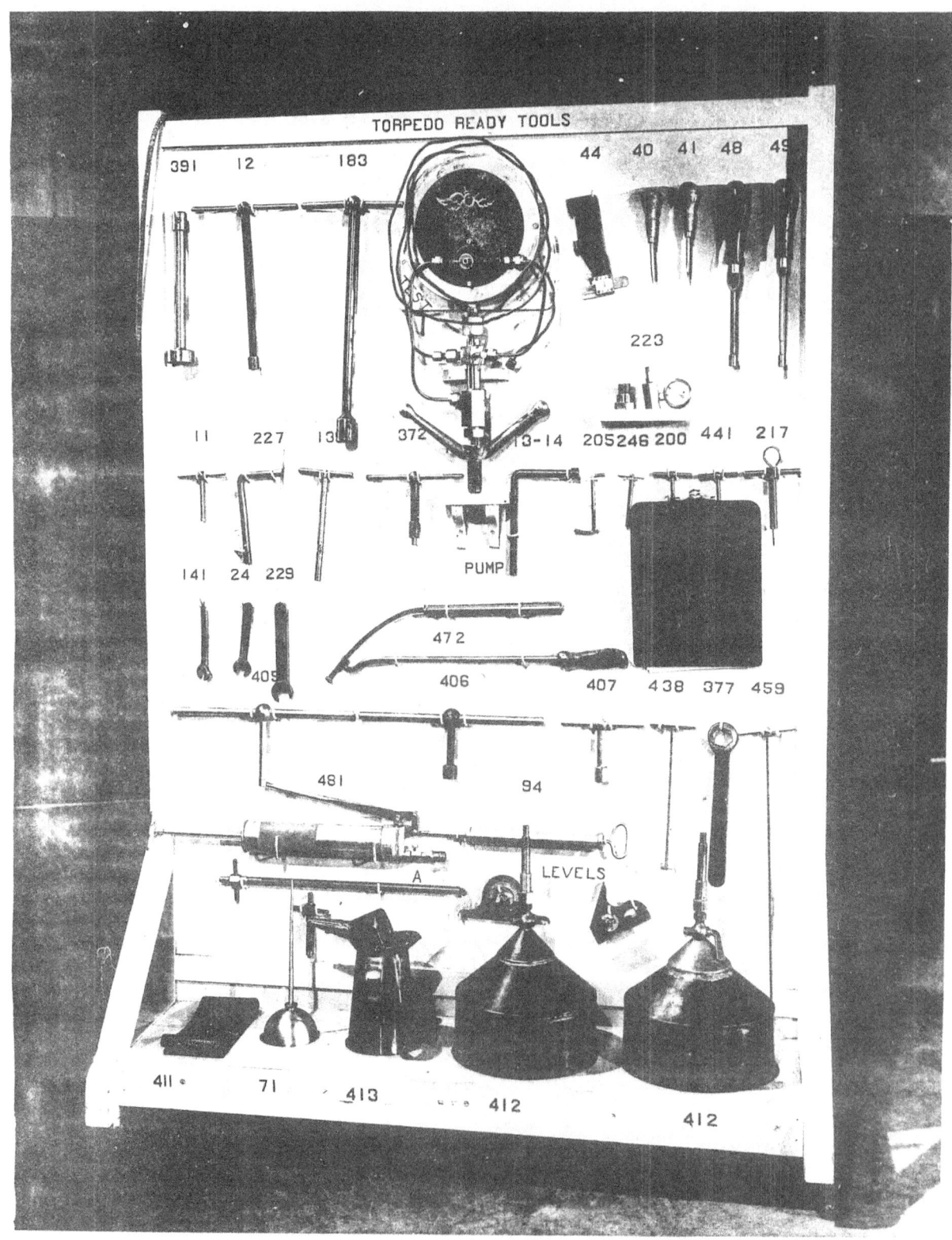

	TOOL NO.
4. Remove fuel filling plug.................................	217.14
5. Remove charging valve and plug washer.............	13.14
6. Rotate distance index spindle to seat starting piston, line up scribe marks........................	227
7. Close flask stop and blow valve....................	49
8. Open stop valve.......................................	227
9. Install charging line and safety strap. Note: do nut on wing nut.	
10. Open air inlet valve to charging line, crack main air inlet valve from charging source, slowly, charge torpedo to at least 1500 lbs. air pressure. (If final adjustment, follow charge to desired pressure).	
11. Close main air inlet valve.	
12. Close stop valve.....................................	227
13. Open bleeder valve on inlet valve to charging line and bleed air.	
14. Remove safety strap, wing nut, and charging line. Note: While torpedo is being charged frequently bleed moisture from air through separators.	
15. Disconnect air pipe from flask stop and blow valve. Not used on war shot................................	24
16. Place sluing strap and bar on air flask, station man at end of flask and tail of torpedo for safety and turn torpedo over until the pipe lead from air flask to flask stop and blow valve is at its lowest point. Open flask stop and blow valve and drain water from air flask, close valve and turn torpedo upright..	49
17. Remove oil tank filling plug.........................	13.14
18. Fill oil tank with hot running torpedo oil. (Capacity of tank 8 pints)...........................	413
19. Remove propeller lock and rotate propellers and note that oil pump takes suction. Refill tank and replace plug......................................	13,14
20. Remove after propeller sleeve holding nut, after propeller and sleeve.................................	183,468
21. Remove grease packing screw from inner propeller shaft...	184A
22. With grease gun fill after bearing with tail packing compound until compound begins oozing out around propeller shaft grease ring (Approximately 1 oz.)...	481,481A
23. Replace grease packing screw.........................	184A
24. Inspect four part bushing on after propeller sleeve. Replace sleeve, note that assembly marks on end of sleeve and after propeller shaft are in line; replace holding nut and set up tight........	183
25. Remove grease plug on tail cone and fill tail bearing with tail packing compound, watch for compound to ooze out of tail bearing. Replace plug and note bottom of plug is flat.............	13,14,481,481A

	TOOL NO.

26. Remove grease screw from forward propeller and fill propeller nut with compound, watch for compound to ooze out between forward and after propeller. Replace screw.......................... 481,481A
Clean and oil check valves and strainers.
27. Remove fuel and water air check plugs.............. 405
Remove one air check valve at a time, clean, inspect, oil (A) and replace. Replace leather washers and plugs; it is imperative that air check valve 245,74 plugs be started in place at least two threads by 12 hand set up....................................... 405
28. Remove fuel and water strainer plugs. Remove 406 strainers, clean, inspect, replace................ 372A
29. Replace strainer plugs, note copper washers in place. Again it is imperative to start plugs in by hand to prevent cross threading.................... 406
30. Remove fuel delivery check valve plug and valve. Clean, inspect, oil (A) and replace. Note copper washer in place and start plug in by hand.......... 405,407,74
31. Remove water delivery check valve plug and valve. Clean, inspect, oil (A) and replace. Note copper washer in place and start plug in by hand.......... 405,407,74
32. Remove and check thickness of speed ring, should be in accordance with record book. Set up on speed 12 screw... 1" micrometer
Center depth steering line. Calibrate depth spring - tilting test.
33. Remove air strainer plug, install adapter...... 377,12,223
34. Remove access plug from air chamber............... 11
35. Install testing set (Dr.No. 79646) in charging valve plug bushing and connect air lead adapter.... 141A
36. Check alignment of depth rudders with zero lines on tail cone, moving rudders by hand. (Top edge of rudder should be read against tail cone).
37. Remove hand hole plates........................... 48,200
38. To check depth steering line locked by transportation screw in pendulum; open stop valve and build pressure on gauge to 450 lbs....................... 227
39. Sight depth engine valve stop and valve connecting rod through left hand hole. Scribe marks should be in line; if not, slack up on valve connecting rod clamp screw and line up scribe marks by turning knurled nut, leave hole in knurled nut in the horizontal, set up on clamp screw....................... 246
40. Check mid-position of depth rudders (neutral throw) should read 1½ increments of depth rudder down. If rudders are off remove drain plugs on left side of tail cone... 13,14
41. Loosen clamp screw, through after drain plug on rudder adjusting rod................................ 49
42. Remove rudder rod connecting screw and turn rudder adjusting eye in direction desired (left - up for up - down for down - right)....................... 13,14

TOOL NO.

Note: One half turn is the least amount this eye can be turned for alignment with the forked end of steering rod connection.

43. Line up eye with fork on rudder rod extension, insert connection screw and clamp rudder adjusting rod with clamp screw............................ 13,14,49
44. Replace drain plugs, note copper washers in place. Close stop valve................................... 227,13,14
45. Put on hoisting strap on flask over water compartment bulkhead, being mindful of the low pressure test set installed. Hoist torpedo about four feet from the ground or deck.(2 ton chain hoist)
46. Remove transportation screw........................ 49
47. Set depth index to 10 feet......................... 135A
48. Install screw hook in diaphragm plate lock nut and hang 16 pound weight on hook..................... 411B
49. Place spirit level on top center of air flask and maintain torpedo level.
50. Open stop valve, build and maintain pressure on gauge to 450 pounds air pressure.................. 227
51. Sight depth engine valve stop and valve connection rod scribe marks are in line, if not, turn depth index spindle in desired direction to bring scribe marks in line. Swing pendulum from both directions to check alignment of scribe lines. Note **depth rudders 1½ down**.. 135A
52. Check depth index, if it does not register ten feet disengage socket from head on spring adjusting spindle through right hand hole and turn index to read 10 feet. Re-engage socket with spring adjusting spindle head................................. 472A,135A
53. Verify the adjustment by removing tension on depth spring and resetting to 10 feet.................. 135A
54. Place bevel protractor on tail vane, offset ½ degree from zero. Starting with torpedo level tilt torpedo until the bubble in protractor comes to center. Place one man at the left handhole and one at the tail to observe the movement of valve connection rod and depth rudder, they should "Mark" at the slightest amount of movement. Reverse protractor and tilt in opposite direction.
55. Offset bevel protractor 2 degrees from zero.
56. Set on tail vane. Starting with torpedo level tilt torpedo until bubble in the protractor is center. Read the depth rudder, should read full throw either 1 up or 4 down. Place the hand through the left handhole and try to move pendulum in the same direction as the tilt of the head. Pendulum should be up against the stop on immersion gear casing. Reverse the protractor and tilt until bubble is center and try to move pendulum by hand.
57. Remove 16 pound weight and hook................... 411B
58. Set depth index to 10 feet......................... 135A

	TOOL NO.

59. Lower torpedo on truck or chocks. Remove level and protractor.

Check Vertical Rudder Adjustment.

60. Remove propeller lock and turn propellers and at the same time have one man reach through right handhole and move control valve link of vertical steering engine until a sharp throw to the left and right is obtained on the vertical rudders. Replace propeller lock.

61. Move control valve link forward and hold. With vernier take throw of upper vertical rudder. With the movable scale set to zero place the after rest on scale holder on left side of upper vertical tail vane and the movable scale on the rudder, move scale in firmly until the forward rest is flush against the tail vane. Remove scale and read. (The fixed scale is graduated in tenths of an inch; the movable scale in one hundreths of an inch). The reading should be the same as the record book. Move the control valve link all the way in and shift the scale to the right side and carry out the same procedure.

Note: Should the readings vary from those given in the record book but are equal and approximately "35 right and "35 left, no adjustment need be made. Take the readings of the lower vertical in the same manner but as the length of the lower rudders are smaller the readings will be smaller approximately 25 right - 25 left.. 44

62. Should the rudders throw for example: read - "38 right upper and "33 left upper; remove the drain plugs on the right side of the torpedo tail. Loosen up on clamp screw. Remove the rudder rod connection pin and turn the adjusting eye on the rod to the left one half turn equal to "03 - this will lengthen the rod to the left and the rudders will read, "35 right - "35 left. Reconnect the rudder connection, set up on clamp screw, replace drain plugs, noting copper washers in place........ 13,14,49,44

63. Close stop valve.. 227

64. Disconnect and remove test set and adapter, replace air strainer plug.. 377,12

65. Replace charging valve washer, charging valve and plug.. 11

Test Gyro Mechanism.

66. Replace water filling plug.. 11
67. Tie down starting spindle arm.
68. Open stop valve one full turn................................ 227
69. Turn torpedo 180 degrees, observe step #16.
70. Remove air chamber.. 48,409

		TOOL NO.

71. Remove diaphragm plate cotter pin, lock nut and plate..................................(pliers) 461,407
72. Check diaphragm lever and depth spring. Renew diaphragm.
73. Replace diaphragm plate, lock nut and pin.........461,407
74. Replace air chamber............................. 409,48
75. Install adapter in access plug hole and pump air into chamber with hand pump to 15 pounds. Must hold pressure for 5 minutes. Note: Use adapter that is countersunk to fit about the diaphragm plate lock nut as the issued adapter may bend diaphragm lever. Test, successful, remove adapter and replace access plug, note copper washer is in place... 11
76. Remove gyro cover, clamp plate and washer......... 13,14
77. Remove gyro bottom plate......................... 246
78. Wipe gyro pot dry with lint free rag.
79. Note that pointer on shelf and top plate are at zero.
80. Remove propeller lock and rotate until cam pawls are all the way out from the center of the pot. Replace propeller lock.
81. Inspect and oil (A) gyro top bearing......... Hypo Syringe
82. Lock and unlock spinning and unlocking mechanism by hand, oil (A) and note freeness................ 205
83. Inspect and oil (A) and install gyro. (Align the wheel axis with the vertical axis, the horizontal axis of the gyro with longitudinal axis of the torpedo; care should be exercised in setting the top center over the extender in top bearing, balls in retainer should be all the way in to insure seating in the outer race of top bearing).
84. Replace gyro bottom plate and bearing lining up zero marks, do not force outer race over balls, check to see balls are all the way in to inner race, replace six holding screws and set up evenly. 246

Lock Gyro.. 205
85. Turn torpedo upright.
86. Hoist torpedo from chocks about four feet from deck or ground.
87. Level torpedo. Unlock gyro by hand, note centering pin is withdrawn without kick to inner unit up or down. Note: The shock of the spinning mechanism unlocking is sometimes sufficient to vibrate the inner unit and care should be exercised in determining a kick from vibration.
88. Check vertical clearance by hand - should between ".0025 to ".005.
89. Lock gyro....................................... 205
90. Replace gyro cover and clamp plate - note good gasket in place.................................. 13,14

                                                                    TOOL NO.
91. Place gyro engine control valve in approximate
    mid-position, remove propeller lock and rotate
    propellers. Sight through right hand hole and
    note if valve connection arm moves, if it does
    gyro top plate must be moved to center gyro axis
    with torpedo axis.
92. Install dummy igniter.................................... 391
93. Connect up air lead to igniter............................ 141,438
94. Install replacement screw................................. 49
95. Place tools for operating stop valve and index
    spindle in starting gear................................. 227
96. Remove clip lock or lanyard from starting spindle
    arm, with scale attached to cam release toggle,
    pull toggle clear, note weight required to remove
    toggle; should not exceed 10 pounds. After gyro
    is spun, close stop valve................................ 227
97. Remove both propeller locks.
98. Clear vicinity about propellers and open stop
    valve and build up engine speed.......................... 227
99. Swing torpedo up and down noting action of depth
    rudders for full throw in a smooth action.
100. Move torpedo in azimuth and vertical, rudder action
    should be obtained within 1/5 of a degree.
101. Satisfactory results from test obtained rotate in-
    dex spindle to seat starting piston. Close stop
    valve.. 227
102. Lower torpedo in chocks or truck......................... 227
103. Replace propeller lock.
104. Remove replacement screw and replace with trans-
    portation screw.. 49
105. Slack up on speed screw.................................. 12
106. Turn torpedo upside down.
107. Remove gyro cover and clamp plate........................ 13,14
108. Permit gyro to run down by itself.
109. After gyro stops, remove gyro bottom plate, remove
    gyro, inspect and check for binding, oil (A) and   246,414A
    wrap in oil paper and stow in container..... Hypo Syringe
110. Replace gyro bottom plate, clamp plate and cover,
    noting good gasket in place.............................. 246,13,14
111. Replace handhole plates.................................. 48,200
112. Turn torpedo upright.
113. Drain afterbody by removing drain plug, replace
    drain plug; note good copper washer in place............. 13,14
114. Remove water filling plug, a puff of air indicates
    tight check valves, replace water filling plug........... 11
115. Remove sluing bar and strap. Note: When torpedo
    is turned listen for unusual sounds, as foreign
    matter can be removed from interior if personnel
    are alert.
    Make haste slowly, observe safety precautions.

I. ATTACH EXERCISE HEAD.

1112.   TOOL NO.

1. Disconnect and remove air releasing mechanism...... 48,141A
2. Hoist exercise head into alignment with the forward joint of the air flask and connect pipe from air flask blow valve to nipple on exercise head bulkhead.. 141A
3. Blank off lead to air release mechanism in head. Crack air flask blow valve and test pipes and connections for leaks. Note that no air escapes from relief valve in buoyancy compartment indicating a leak in the piping. Close blow valve and remove blank.   141A 49
4. Join exercise head to air flask and secure with holding screws. NOTE: Care should be exercised not to distort the pipe from air flask blow valve to head when joining the exercise head to air flask.
5. Exercise discharge valve by hand. Note that the valve seats properly and that the spring is not broken.
6. Fill forward compartment of exercise head with fresh water.
7. Calibrate air release mechanism to prescribed pressure.
8. Crack air release mechanism blow valve and blow through piping. (Note that there is no restriction in pipes).   49
9. Attach air release mechanism to air line in head... 141A
10. Open air flask blow valve and note that blow valve is seated on its outboard seat..................... 49

    NOTE: The Mark 26 exercise head uses the Mark 2 air releasing mechanism that has to be cocked by hand with tool number 441. Care should be taken not to apply side force when lifting the cocking tool, which has a tendency to bind the valve stem. This possibly would result in the failure of the exercise head to blow and the consequent loss of the torpedo.

11. Test air release mechanism and connection by immersing in water. If no leaks are found around stem, put on air release protection cap. Note that washer is in place and vent hole in cap is clear.
12. Note that forward bulkhead is tight by observing if bubbles are present around bulkhead or piping.
13. With air release mechanism cocked, secure with leather gasket and cover plate. (Note that gasket is evenly in place).................................... 48
14. Install torch pot in torch case, using leather gasket. Secure cover evenly in place............... 48
15. Inspect washer of air relief valve and note that it is in good condition and even on its seat.

TS-5   XI-16   Original Page.

NOTE: It is particularly important in preparing the exercise head for firing to be sure that the leather gaskets under the air release mechanism and torch case are tight. If they are not, the head may not blow or the air buoyancy chamber may be flooded and the torpedo will be lost.

J. **FINAL ADJUSTMENTS WITH WAR HEAD.**

1113.

1. Remove protecting ring from war head. REASON: To inspect joint ring and install war head.
2. Place torpedo on truck or chocks. Close air flask stop and blow valve. REASON: To isolate the air in the air flask.
3. Place hoisting strap on war head at center of gravity, 31.42 inches from joint ring; (for Mk. 13) hoist war head and fit to air flask. Line up hole in joint ring with hole in air flask and secure head to flask with joint screws. (Screws on bottom can be put in first time torpedo is turned over). #49.
4. Put on propeller lock, secure lanyard. REASON: To keep propellers from turning; personnel safety.
5. Rotate distance index spindle to seat starting piston, line up scribe marks. #227. REASON: To keep the main starting valve seated when torpedo is being charged. To insure the unlocking cam is parallel to the unlocking lever.
6. Remove fuel and water filling plug...#217, 74. REASON: To prevent damage to fuel flask while torpedo is being charged.
7. Open stop valve...#227. REASON: To permit the entry of air into the air flask.
8. Remove charging valve and plug and leather washer. 13,14,74. REASON: To permit inserting of the wing nut to charge air flask.
9. Install charging line wing nut and secure safety strap. REASON: (1) No tit on wing nut so that air pressure will unseat the charging check valve. (2) To prevent injury to personnel while torpedo air flask is being charged.
10. Open air inlet valve to charging line, then open main inlet valve from charging source. Charge torpedo slowly through banks and separators if available. Bleed moisture from air through separators frequently while torpedo is being charged.
11. When torpedo is charged to 2800 pounds or desired pressure, close the main air inlet valve <u>FIRST</u>. REASON: To stop the flow of air from the main supply source.
12. Close stop valve...#227. REASON: To isolate the air in the air flask.
13. Remove safety strap wing nut and charging line. NOTE: Secure charging line and place cap on wing nut thread for protection.
14. Install high pressure air gauge in charging valve plug bushing and check air pressure in flask by opening stop valve slowly...#227. If flask charged to required pressure close valve and remove air gauge...#227.

15. Replace charging valve washer, charging valve and plug. #13, 14.
16. Remove dummy igniter. #433, 391.
17. Remove oil tank filling plugs; fill oil tank with hot running torpedo oil (B). Capacity of tank 3 pints. 13,14,413. REASON: To lubricate the working mechanism in the afterbody.
18. Remove propeller lock, rotate propellers by hand. Note that oil level drops. Refill tank and replace plug on copper washer. 13,14,413. REASON: Turning the propellers causes the oil pump to operate. The oil level dropping indicates oil line between tank and pump is clear and pump is taking suction.
19. Remove after propeller sleeve holding nut, propeller, and sleeve bushings. 183, 468. REASON: The nut holds sleeve to inner propeller shaft.
20. Remove grease packing screw from inner propeller shaft. #184A. REASON: With grease gun fill after bearing with tail packing compound until it oozes out around grease ring. (Approximately one ounce is used).
21. Remove grease gun, 481, 481A, and fitting, replace grease packing screw. REASON: To lubricate the after bearing and propeller shafts.
22. Inspect four part bushing on propeller sleeve, grease and place on sleeve. Line up scribe marks on sleeve with inner shaft and install. Replace holding nut for sleeve on shaft. 183, 468. REASON: To insert keys on sleeve in fitted keyways on shaft.
23. Remove grease plug in tail cone. Insert grease gun fitting, with grease gun fill after bearing with compound until it oozes out between tail bearing and forward propeller hub. Replace plug. REASON: To lubricate the tail bearing and bushings.
24. Remove grease screw from forward propeller and fill propeller hub with compound until it oozes out between propellers. Replace screw. 40, 481, 481B. REASON: To lubricate bushings between propeller sleeves.
25. Remove plug on reducer oil line and lubricate reducer. Oil (A). 13, 14, 94. REASON: To lubricate the valve sleeve and cylinder.
26. Remove and check thickness of speed ring. Speed ring size should be in accordance with torpedo record book. Replace and set up on speed screw. 1" mic., 12. REASON: The size of the speed ring controls the amount of compression applied to the reducing valve spring by setting up on the speed screw.
27. Open stop valve slowly and turn. REASON: The stop valve is opened at this time so that the main starting valve will be seated. This step must be completed prior to filling fuel and water compartment, installation of gyro and igniter. If stop valve were first opened after the above steps are done, the fuel and water would be forced into combustion flask, the gyro prematurely spun and the igniter fired.

28. Remove fuel and water air check valve plugs. Remove one air check valve at a time. Clean, inspect, operate and oil (A) and replace. See good leather washer in place and start plugs in body by hand to prevent cross threading. 245, 74, 12, 403.
29. Remove fuel and water strainer plugs. Remove strainers, clean, inspect and replace. Replace strainer plugs, note that the copper washers are in place, start plugs in by hand to prevent cross threading. 406, 372A.
30. Remove fuel and water delivery check valve plugs. Remove one valve (one delivery check valve at a time). Clean, inspect, operate, oil (A) and replace. Note copper washer is in place and again start the plug in by hand to prevent cross threading. REASON: The check valve bodies are contoured to the torpedo shell and are easily cross threaded if plugs are started in with the tools for setting up. 405, 407, 74.
31. Screw fuel filling funnel in fuel flask; fill flask with alcohol. Note: Exercise care that splashing alcohol does not injure fellow workman. Fuel flask is filled when liquid squirts out of air vent holes in funnel. Close valve on funnel and remove funnel. Sight alcohol level in fuel flask and replace fuel plug. 412, 74, 217.
REASON: To provide the fuel for heating the air.
32. Fill water compartment with distilled water. Note: Water from land bases should not be used because where torpedoes are in the fully ready condition over prolonged periods fungus in the water grows and strainers are clogged, affecting the normal operation of the superheating system. Replace water filling plug, note copper washer in place. #11.
REASON: To cool interior of combustion flask and after heated becomes steam, mixes with other gases and assists in driving the turbines.
33. Place sluing strap around torpedo, attach bar and rotate torpedo 90 degrees to the left, hold finger over igniter lead and drain combustion flask. REASON: To prevent explosion in combustion flask. (2) The two men keep the torpedo from sliding while it is being turned.
34. Rotate torpedo 90 degrees more and bring bottom side up. REASON: To prepare gyro pot for reception of gyro.
35. Remove gyro clamp plate and cover. #13, 14.
36. Remove gyro bottom plate and bearing. #246.
37. Wipe interior of gyro pot with clean rag, note top plate and pointer on zero to align gyro and torpedo axis. Inspect top bearing and oil (A). Hypo syringe.
38. Remove propeller lock, rotate propellers until cam pawls are all the way out from the center, replace propeller lock. REASON: To clean cam plate.
39. Lock and unlock spinning mechanism. #205. REASON: To check proper operation.
40. Check gyro balance nut and set for latitude correction. (Not necessary on short range torpedo). Oil gyro (A). Align the wheel axis with vertical axis, horizontal axis with torpedo longitudinal axis and insert gyro on top bearing. Do not force gyro in place. Rotate gyro easily

in top bearing. Replace gyro bottom plate and bearing note scribe marks in line. Gyro screw driver No. 88, 414A, 246

41. Line up cam with cam pawls, locking and bearing with centering pin and lock gyro. #205. REASON: So that gyro is not damaged and properly locked for spin.
42. Unlock and relock gyro. Note: Hand trip lever flush with gyro pot wall. #205. REASON: To insure proper withdrawal of spinning gear and centering pin.
43. Replace gyro clamp plate and cover, note good gasket in place. #13, 14. REASON: To insure all openings are sealed and joints set up tight.
44. Check afterbody joint screw. #49.
45. Check tail joint screws. #49.
46. Check drain plugs and washers in afterbody and tail. #13, 14.
47. Replace handhole plates and gaskets. #43, 200.
48. Install remaining joint screws from war head. #49.
49. Inspect live igniter and install. REASON: To ignite the fuel after torpedo is launched for a hot shot. (Note: The air lead to igniter is not made up until the torpedo is attached to plane).
50. Remove base plate from war head. MF2, 49. REASON: To install exploder.
51. Inspect interior of exploder casing for dirt and corrosion; wipe dry. Inspect base plate gasket for tears and wear.
52. Remove impeller guard. Inspect impeller and anti-countermining plate for corrosion and burrs. REASON: Access to impeller for tests which follow.
53. Inspect inside of base plate. REASON: Corrosion and dirt. Removal of three holding screws and lock washers.
54. Insert grease gun, MF-1, into hollow impeller shaft and force in grease. REASON: To lubricate impeller shaft bearings.
55. Place tool MF 6 on castlelated nut and back off on impeller shaft packing gland nut. REASON: Easier to get to before the exploder is mounted. Check freeness of movement by hand.
56. Check serial numbers on base plate, exploder, vertical shaft, vertical shaft upper gear and push rod. Set base on legs. REASON: Same number on each unit.
57. With MF-11 compress push rod spring on push rod; insert rounded end of rod in free end of the diaphragm lever. Check worm gear on bottom of vertical shaft for arm which indicates thin tooth and engages with worm wheel on transverse shaft.
58. Engage removable gear of vertical shaft with gear train on top plate. Slots in gear should be placed parallel to pins on vertical shaft. REASON: To install exploder on base plate without damage to any part.
59. Place exploder on base plate, inserting vertical shaft through gear to top plate; push rod through hole in bracket in bell crank bracket. Line up dowel pins with base plate. If exploder does not seat itself readily, turn impeller until mechanism seats in line with base plate and removable gear falls in place over pins on vertical shaft.

60. Install three holding down screws on lock washers Set up evenly with tool #64. REASON: So that exploder is not canted on base plate.
61. Remove MF-11. REASON: So that push rod spring will force push rod down into cup of diaphragm lever.
62. Connect base plate to exploder test set No. 3. Turn on motor and arm exploder. Place step of plunger of tool MF-11 on anti-countermining rack and force in. Pressure required should be between 1 to 2 pounds; time for withdrawal, 1 to 2 seconds. REASON: To test for sensitivity and to ascertain that scribe on rack returns clear of bell bracket.
63. Fire exploder. Force forked rack in all the way until it contacts trigger cap. With feeler gauge between top of push rod and heel of bell crank. REASON: To gauge clearance between fork and trigger cap when exploder is fired. Clearance should be between ".002 to ".005. If proper clearances are not obtained, check for corrosion, alignment.
64. Hang 3 pound weight in outer notch of lever arm bearing in anti-countermining diaphragm plate. Turn on motor and arm exploder. (Note: Dummy detonator is installed with Winchester caps). REASON: Attempt is made to fire exploder and the rack should prevent this by inserting the forked rack between the exploder top plate and trigger cap.
65. Move the 8 pound weight to the inbound notch in the lever arm. Force the firing ring until the mechanism fires. Remove dummy safety chamber; cock and unarm exploder. Replace MF-9 on guide posts. REASON: The live caps explode and is indicative that the firing spring is of sufficient strength. The anti-countermining gear will not operate before a depth of 55 feet is reached.
66. Place flat spring gauge, MF-7, on edge of impeller blade. Impeller should rotate at not more than 3 ounces pressure. REASON: Friction test.
67. With MF-6 set up packing gland nut on impeller shaft until the pressure required to turn impeller lies between 12 to 20 ounces. REASON: Places enough drag on impeller to prevent the exploder from arming below air speed of 270 miles or 234 knots.
68. Lock castlelated nut with safety wire. REASON: To lock the packing gland in place.
69. Remove idler gear from gear train in top plate. REASON: To facilitate the arming and unarming of the exploder.
70. Arm exploder by hand. REASON: To remove the safety devices so that the exploder can be fired.
71. Place MF-10 in center of firing ring between two supporting studs. Exploder must fire between $3\frac{1}{2}$ to $5\frac{1}{2}$ pounds. REASON: Sensitivity. The MF-10 can introduce a drag if not used correctly. Using the step on the rod on the top of the firing ring as the mechanism fires, the tool is usually caught by the edge of the trigger plate denting it. If mechanisms fail to fire or require more than $5\frac{1}{2}$ pounds, friction must be found and eliminated.

72. By hand, cock and place exploder in the unarmed position. Replace the idler gear and set up on lock screw. Note that scribe mark on arming screw and arming gear are flush. NOTE: Disregard the raised ridge for centering the safety chamber. REASON: To rearrange the component parts of the exploder so that the safety devices are operative.
73. The detonator box, containing 3 detonators assembled in their safety chambers, is then opened. One detonator is removed from its own sealed container, protected by corrugated paper and cork. REASON: The detonators are usually stored above the water line in sealed containers and protected from heat. (Fulminate of mercury is a very unstable explosive and deteriorates rapidly at temperatures of 104 degrees F. or over).
74. Test the detonator holder for ease in operation. Wipe with lint free rag. REASON: If burrs are present, remove. NOTE: Never disengage detonator holder from safety chamber; to do so the safety chamber's purpose is defeated.
75. Oil the thread of detonator holder and the exploder mechanism. Oil (A). REASON: lubrication.
76. Screw detonator holder in flush with safety chamber, scribe marks in line. REASON: It requires 781 turns of the impeller to fully arm the exploder, that is, to disengage the thread of the detonator holder from the thread of safety chamber.
77. Remove MF-9 from guide posts. REASON: To install the detonator, holder and safety chamber over guide posts.
78. Line up scribe marks of safety chamber with scribe mark on arming gear. REASON: To align detonator holder with guide posts and arming gear.
79. Secure safety chamber on arming gear. REASON: So that safety chamber will rotate with arming gear.
80. Remove tetryl booster from sealed metal container and unseal from its individual container. Inspect and install base plate gasket. REASON: Six boosters are stored in metal boxes and usually with war heads.
81. Install booster in booster recess with recess facing man installing. REASON: So that detonator holder can be inserted in recess.
82. Screw MF-2 in base plate and install assembled exploder in the cocked-unarmed position in the exploder casing on the base plate flange. The anti-countermining diaphragm plate is aft.
83. Secure base plate on base plate flanges with monel joint screws. Set up evenly. REASON: To prevent sea water from entering.
84. Remove testing plug from base plate. REASON: To test for tightness.
85. Install adapter in test plug hole. Pump interior to 5 pounds air pressure for 5 minutes. Test all joints with soapy water. Fill cavity around impeller housing above the level of impeller shaft and look for bubbles. REASON: If air can come out, sea water can enter and cause corrosion.

86. Thread arming wire through nose piece, impeller and plug in exploder base plate. Connect with two Fahnstock clips. Pressure to pull free not to exceed 20 pounds.
87. Replace impeller guard.
88. At arrival at plane with torpedo, open stop valve wide and back off ¼ turn. REASON: To insure full delivery of flask pressure.
89. Check starting gear that scribe marks align. REASON: To prevent the accidental seating of starting piston.
90. Set depth. REASON: Pilot usually knows type target anticipated, set as instructed. NOTE: If not instructed, set to 10 feet.
91. Check tightening of speed screw. REASON: To insure torpedo speed.
92. Check stabilizer secured to tail vane for type plane. (Note: If TBF, secure lanyard in bomb bay for dual action).
93. Place lifting band at center of gravity. REASON: For hoisting torpedo.
94. Connect hoist cable to band. CAUTION - keep strain on cable. DO NOT KINK OR HOIST TORPEDO WITH KINK IN CABLE. IT WILL BREAK.
95. Hoist torpedo. It is imperative that ONE MAN give commands. REASON: To reduce overlapping of orders and personnel safety.
96. Position torpedo to receive guide bolt or pin secured in plane, in hole in midship ring. REASON: To prevent lateral movement of the torpedo.
97. When torpedo is two-blocked in cradle secure release strap to torpedo release rack. Lock with lock nuts.
98. Back off on bomb hoist and see if release strap and torpedo racks support weight of torpedo. REASON: Personnel safety.
99. ONLY EXPERIENCED PERSONNEL SHOULD COMPLETE THE FOLLOWING STEPS:
100. Connect tripping latch on plane to cam release toggle.
101. Under no circumstances connect igniter lead before step 100. Connect igniter lead.
102. Remove lanyards holding down starting spindle arm. REASON: If torpedo should start the propeller, lock still affords safety to torpedo and personnel.
103. Remove propeller lock. Insure that auxiliary lock is in place. REASON: If slip stream turns propeller oil pump will drain tank before torpedo is launched and torpedo will probably burn up due to lack of lubrication.
104. Above all use check off lists. REASON: To eliminate chances of a slip.
105. A slip means a ship. BE THOROUGH, CONFIDENT, CHECK ANY DOUBT, REGARDLESS. IT IS TOO LATE AFTER THE PLANE IS IN THE AIR.

K. ROUTINE FOR UPKEEP OF FULLY READY TORPEDO.

1114.   DAILY   TOOL NO.
  (A) If attached to plane:
    (1) Secure toggle lock to toggle and starting spindle arm.................................. 92
    (2) Lower torpedo to truck or chocks.

	TOOL NO.

    (3) Gauge and boost air flasks as necessary......
    (4) Check oil level in oil tank................... 13
    (5) Attach torpedo to plane..
  (B) Reload torpedoes:
    (1) Gauge and boost as necessary.
    (2) Check oil level in oil tank................... 13

WEEKLY

1115.
  (A) If attached to plane:
    (1) Secure toggle lock to toggle and starting spindle arm..................................... 92
    (2) Lower torpedo to truck or chocks.
    (3) Turn propellers 50 turns, refill oil tanks.
    (4) Repack propeller shafts and sleeves with grease.. 462,481A
    (5) Operate depth mechanism from 0-50, reset to required setting............................... 135A
    (6) Attach torpedo to plane.
  (B) Reload torpedoes:
    (1) Turn propellers 50 turns, refill oil tanks.
    (2) Repack propeller shafts and sleeves with grease.. 462,481A
    (3) Operate depth mechanism from 0-50, leave tension off depth spring...................... 135A

MONTHLY

1116.
    (1) Secure toggle lock to toggle and starting spindle arm..................................... 92
    (2) Lower torpedo on truck or chocks.
    (3) Close stop valve................................ 14
    (4) Disconnect and remove igniter. Note condition of end seal. Replace in container and put in stowage space........................ 141A,391A
    (5) Remove afterbody drain plug and one tail cone drain plug....................................... 13
    (6) Turn torpedo bottom up draining combustion flask on way over.
    (7) Remove exploder mechanism, detonator, and booster.
      (a) Remove base plate holding screws........ 49
      (b) Remove safety chamber and detonator from exploder mechanism, replace in container, seal tightly and replace in stowage space. 40
      (c) Remove booster from warhead, replace in container and replace in stowage space.
      (d) Place base plate and exploder mechanism in safe stowage space.

	TOOL NO.

(8) Turn torpedo upright.
(9) Remove check valve plugs, work and oil check valves, and replace plugs............. 405,74,12
(10) Remove strainer plugs, remove and clean air, fuel, and water strainers and replace; replace strainer plugs........................ 405,372A
(11) Check action of starting gear.
   (a) Remove toggle lock from toggle and starting spindle arm......................... 92
   (b) Trip starting spindle arm.
   (c) Rotate starting gear index spindle through several complete revolutions.... 227A
   (d) Replace toggle lock to toggle on starting spindle arm.
(12) Operate depth setting mechanism from 0-50 and return to required setting.
(13) Remove either hand hole plate and swing pendulum back and forth a few times by hand.... 48
(14) Work horizontal rudders a few times up and down by hand, using one hand on each rudder, with the same amount of force on each.
(15) Work vertical rudders by hand using same method as step 14.
(16) Remove propeller lock, turn propellers by hand at least 50 turns. Listen for unusual sound. Check action of pallet and slide to see that pallet does not engage pallet pawls with gyro locked.
(17) While propellers are being turned over move vertical steering engine control valve by hand. Note that pallet blade does not strike pallet pawls.
(18) Replace propeller lock.
(19) Lubricate torpedo.
   (a) Fill oil tanks.
   (b) Fill tail bearing with tail packing compound................................... 462,481A
   (c) Fill forward propeller grease cavity.... 462,481B
(20) Crack stop valve............................. 13,14
(21) Turn torpedo bottom up draining combustion flask of water.
(22) Inspect gyro and gyro mechanism.
   (a) Remove gyro cover plate, unlock gyro by hand and remove bottom head............. 246A,13,14
   (b) Remove and inspect gyro, noting that balance nut is tight. Apply two (2) drops of gyro oil in each bearing using syringe furnished.
   (c) Lock and unlock spinning mechanism by hand and by rotating spinning shaft..... 205A
   (d) Remove propeller lock, turn propellers til cam pawl is in extreme after position, replace propeller lock.

                                                                TOOL NO.
  (e) Install gyro, bottom head, lock gyro,
    install gasket and cover plate....... 246A,13,14
 (23) Replace hand hole plate and afterbody and
   tail drain plugs.................................. 48,13,14
 (24) Install booster, detonator, and exploder
   mechanism in accordance with instructions in
   Preparation for a War Shot.
 (25) Turn torpedo upright.
 (26) Gauge and boost air flask.
 (27) Check level of fuel, oil and water.
 (28) Load torpedo on plane according to instructions in Final Adjustments.

L. **TREATMENT AFTER A RUN.**

 The following treatment should be carried out immediately after firing to prevent deterioration of material:

1117.
 Immediate treatment:
  (1) Close stop valve and air flask blow valve..... 227A,49
  (2) Put on propeller lock.
  (3) Rotate starting gear index spindle until click
    is heard....................................... 227A
  (4) Place torpedo on truck, wipe shell dry and
    slush.

 Drain Afterbody and Tail:
  (5) Remove replacement screw..................... 49
  (6) Put in transportation screw 49
  (7) Remove drain plug in afterbody.............. 13,14
  (8) Remove drain plugs in tail.................. 13,14
  (9) Remove gyro clamp plate cover............... 13,14
  (10) Remove igniter 438,391A

 Fill Oil Tank:
  (11) Fill oil tank with hot running torpedo oil
    (B) Note quantity of oil required to fill
    tank in order to determine the quantity of oil
    remaining after a run........................ 13,14

 Drain Combustion Pot:
  (12) Turn torpedo over sufficiently to drain com-
    bustion pot.
  (13) Drain combustion pot. Turn torpedo bottom up
    and install the dummy igniter, connect air   391A
    lead... 438

 Remove Gyro:
  (14) Remove gyro bottom head..................... 246A
  (15) Remove gyro, wipe pot dry, clean and oil top
    and bottom gyro bearings and replace gyro bot-
    tom head and clamp plate cover............... 13,14

                                                                TOOL NO.
Drain Fuel and Water Compartment:
  (16) Remove water compartment filling plug. Measure
       quantity of water remaining after run by drain-
       ing into a can... 11
  (17) Remove fuel filling plug. Measure quantity of
       fuel remaining after run by draining into a
       can.. 217,74
  (18) Replace water compartment filling plug.......... 11

Overhaul and Clean Tail (Mk.13 torpedo).
  (19) Remove holding screws for top, bottom and side
       rails.. 39
  (20) Remove side rails with rudder arms.
  (21) Remove depth rudders.
  (22) Remove bottom rail with rudder arms.
  (23) Remove lower vertical rudder.
  (24) Remove upper rail with rudder arms, upper verti-
       cal rudders and rudder support body assembly.
  (25) Remove holding nut, after propeller and sleeve
       with bushings... 183A
  (26) Remove forward propeller lock nut set screws(2). 40
  (27) Remove forward propeller lock nut............... 185C
  (28) Remove forward propeller, use lead maul, if
       necessary. Remove keys from sleeve............. 40
  (29) Remove rudder connection screws................. 13,14
  (30) Remove tail joint screws....................... 49,184A
  (31) Remove tail.
  (32) Remove forward propeller sleeve locking screws. 40,72
  (33) Remove forward propeller sleeve.
  (34) Insert cotter pins in holes on after end of
       exhaust valve stems...................................... 72,40
  (35) Remove cotter pins and nuts on exhaust valve
       bracket.. 72,408
  (36) Remove exhaust valves and bracket intact.
  (37) Examine exhaust valve springs and seats and oil
       (compound steam cylinder oil).
  (38) Renew exhaust valve spring and reseat valves if WE202A
       found necessary... 72,408
  (39) Replace exhaust valve bracket and valves....... 72,408
  (40) Remove cotter pins from holes in exhaust stems. 72,40
  (41) Replace forward propeller sleeve "zero" marks
       to coincide.
  (42) Replace forward propeller sleeve lock screws
       and wire all screws....................................... 72,40
  (43) Replace tail.
  (44) Replace tail joint screws....................... 49,184A
  (45) Replace forward propeller sleeve keys.
  (46) Replace forward hub and propeller "zero" marks
       to coincide.
  (47) Replace forward propeller lock nut.............. 185C
  (48) Replace forward propeller lock nut set screws.. 40
  (49) Wipe clean and apply tail compound thoroughly
       and replace bushings on after propeller sleeve.

	TOOL NO.

(50) Replace after propeller sleeve, hub and propeller ("zero" marks to coincide) and replace after propeller sleeve lock nut.................. 183A

(51) Replace rudder support body assembly and vertical rudders with rudder arms and rails. Secure rails with holding screws..................... 39

(52) Replace depth rudders, side rails and rudder arms, secure rails with holding screws......... 39

(53) Replace rudder connection screws............... 13,14

(54) Replace tail drain plugs....................... 13,14

(55) Grease tail bearing with tail bearing compound until it shows.

(56) Pack shaft with tail bearing compound until it shows.. 462,481A 184A

(57) Grease forward propeller until it shows (tail packing compound)............................... 40,481B

Overhaul and Clean Tail (Mk. 13 Modification Torpedoes)

(58) Remove holding nut, after propeller and sleeve with bushings.................................. 183

(59) Remove four (4) bronze bushings from sleeve for after propeller.

(60) Remove two (2) lock screws from nut for forward propeller and remove nut..................... 185C

(61) Remove forward propeller and hub from sleeve for forward propeller.

(62) Remove keys from sleeve. Use screw driver in milled slot in end of keys.................... 41

(63) Remove two (2) pins for rudder connections..... 13,14

(64) Remove sixteen (16) joint screws and remove tail.. 459,184

(65) Remove wire from four (4) screws for locking clips and remove screws....................... 92,41

(66) Turn sleeve for forward propeller until locking clips are in alignment with milled slot in after bulkhead and remove four (4) locking clips. Pry out with screw driver.

(67) Remove sleeve for forward propeller.

(68) Remove wire and two (2) screws for grease packing ring and remove grease packing ring from forward propeller sleeve.

(69) Insert cotter pins in holes on after end of exhaust valve stems. 72,40

(70) Remove cotter pins and nuts on exhaust valve bracket studs. 72,408

(71) Remove exhaust valves and bracket intact.

(72) Examine exhaust valve springs and seats and oil with compound steam cylinder oil (D).

(73) Renew exhaust valve spring and reseat valves if found necessary................................ WE202A 408,72

(74) Replace exhaust valve bracket and valves....... 408,72

(75) Remove cotter pins from holes in exhaust stems. 72,40

(76) Clean, wipe dry sleeve for forward propeller...

TS-5   XI-28   Original Page.

	TOOL NO.

(77) Clean, grease (G) and replace packing ring in forward propeller sleeve. Replace holding screws and wire.................................... 72,41

(78) Grease (G) sleeve and replace on forward propeller shaft.

(79) Clean, oil with compounded steam cylinder oil (D) and replace four (4) holding clips for forward propeller sleeve. Replace screws in holding clips and wire. Try to pull sleeve off as a check to see that pins are properly seated in the holes in the shaft....................... 72,41

(80) Clean interior of tail cone and wipe dry. Grease (G) and replace on afterbody. Oil and replace joint screws.................................. 459,184

(81) Replace rudder connection screws............... 13,14

(82) Replace four (4) copper washers and four (4) drain plugs in tail cone...................... 13,14

(83) Clean, oil with hot running torpedo oil (B) and replace two (2) keys for hub on forward propeller sleeve.

(84) Clean, grease (G) and replace hub and propeller on forward sleeve.

(85) Clean, oil with hot running torpedo oil (B) and replace nut for forward propeller. Replace two (2) eep screws for nut....................... 185D 41,185C

(86) Clean, wipe dry and grease (G) after propeller sleeve.

(87) Clean, grease (G) and replace four (4) bronze bushings for after propeller sleeve.

(88) Replace sleeve, hub and after propeller on after propeller shaft.

(89) Replace after propeller shaft lock nut......... 183

(90) Turn torpedo upright.

(91) Remove grease plug in tail cone. With grease gun force grease (G) into tail bearings to grease bearing surfaces. Remove grease gun and replace grease plug in tail cone............... 13,14 462,481B

(92) Remove grease plug in after propeller shaft. With grease gun force grease (G) between propeller shafts and grease packing ring. Replace grease plug in after propeller shaft........... 184,481B

Make Check Run to Check and Oil the Engine.

(93) Remove charging valve plug..................... 13,14

(94) Put on safety strap and charging lead.

(95) Open stop valve................................ 227A

(96) Charge torpedo to 1000 pounds.

(97) Close stop valve............................... 227

(98) Bleed and remove charging lead and safety strap.

(99) Replace charging valve plug and washer......... 13,14

(100) Put tool 227 on starting gear index spindle as a safety measure............................. 227

	TOOL NO.

(101) Remove propeller lock and turn propellers over by hand to see if engine turns freely.
(102) Lift starting spindle arm to open starting piston.
(103) Open stop valve sufficiently to give the torpedo a check run at reasonable speed........ 227
(104) Close stop valve... 227
(105) Rotate starting gear index spindle to close starting piston.
(106) Put on propeller lock.
(107) Remove water compartment plug, replace fuel   11,217
plug and replace water compartment plug........ 74

NOTE: If torpedo is to be made ready for firing immediately proceed with the final adjustments. If torpedo is not to be fired immediately, proceed as follows:

Reducing Valve:
(1) Oil reducing valve.. 386,14,94
(2) Back off speed screw to relieve compression on reducing valve spring................................. 12

Remove Gyro and Immersion Mechanism:
(3) Set depth index on zero................................ 135A
(4) Turn torpedo bottom up.
(5) Disconnect air leads to horizontal and vertical steering engines..................................... 141A
(6) Remove pipe from steering engine to reducer valve.. 141A
(7) Remove valve connection clamp screws for both engines... 246A
(8) Remove engine screws, detach engines from gyro pot and lay aside in afterbody leaving engines   92
connected to rudder rods................................ 49,49A
(9) Disconnect gyro spin lead.............................. 229
(10) Remove gyro clamp plate cover and bottom head.. 13,14,246A
(11) Remove transportation pin, and holding screws for gyro and immersion mechanism base......... 49
(12) Put lifting screws in gyro housing and remove.. 200
(13) Wipe gyro pot dry.
(14) Blow off gyro top and bottom bearings with low pressure air; with gyro syringe drop 2 drops of gyro oil (A) on each bearing.
(15) Blow moisture off gyro spinning and locking mechanism with low pressure air. Wipe dry all parts and oil well. (Compound steam cylinder oil (D)).
(16) Wipe dry all parts of gyro housing and immersion mechanism. Oil (B) well for preservation (hot running torpedo oil).

Oil Engine Gearing and Ball Bearings:
(17) With hot running torpedo oil (B) use oil gun and lubricate and spray all parts of main engine,

	TOOL NO.

starting gear, etc., paying particular atten-
to ball races, washers and steel parts which
would become rusted from lack of attention.
This can be accomplished through the hand holes
and gyro and immersion mechanism housing openings.. 94

Replace Gyro and Immersion Mechanism:
(18) Replace gyro and immersion mechanism housing
 with gasket, replacing connecting rod for driving pallet bevel pinion....................... 200
(19) Secure housing with clamp screws and replace
 transportation pin............................ 49
(20) Attach gyro spin lead.......................... 229
(21) Attach engines to gyro pot and wire screws together.. 72,49,49A
(22) Replace valve connection screws................ 246A
(23) Replace pipe from reducer to vertical engine... 141A
(24) Attach air leads to steering engines........... 141A
(25) Replace gyro bottom head....................... 246A
(26) Replace gyro clamp plate cover and gasket...... 13,14
(27) Turn torpedo upright.

Replace Fittings:
(28) Drain afterbody and tail and replace plugs..... 13,14
(29) Replace hand hole plates and gaskets........... 18

The above will place a torpedo in a state of preservation pending periodic overhaul.

CHAPTER TWELVE

		ARTICLES
A.	General Information on Torpedo Loading..........	1201-1202
B.	Safety Precautions and Causes of Torpedo Failures.......................................	1203-1204
C.	Air Trajectory.................................	1205-1211
D.	Initial Entry and Underwater Trajectory.........	1212-1231
E.	Dropping Conditions and Restrictions, BuOrd C/L T 7-42.......................................	1232-1247
F.	Biplane Stabilizers............................	1248-1267
G.	Cold Weather Operations........................	1268-1274
H.	O.D. 3818, 3820, 3825; BuOrd C/L's T-145, 150, 155, 162, 165, 173, 2-42, 8-42...........	

A. GENERAL INFORMATION ON TORPEDO LOADING.

NOTE: In this Letter of Instruction it will be assumed that the torpedo will be delivered to the airplane with all preliminary and final adjustments completed. Also, it will be assumed that the airplane is completely equipped to receive a torpedo.

1201. INTRODUCTION.

1. These instructions are meant to afford instructions for loading torpedoes on any and all Army Air Forces airplanes that are equipped to carry them. It must be remembered, however, that each airplane model has different installations insofar as details are concerned. Yet, the loading procedure in most cases is precisely the same. In general, the torpedo is held in a nearly horizontal position against suitable sway bracing. It is held by two steel cables around the air flask (one either side of the c.g.). One end of each cable is held fast in the airplane and the other end is attached to a bomb shackle so that it may be released in a similar manner to a bomb.

1202. LOADING PROCEDURE.

1. Make the desired depth index setting for attack against the expected target. It is important that this setting be made at this step in the loading procedure because it may be found inconvenient to change the setting later on. It is easiest to make the setting before the torpedo is positioned on the airplane. In the A-20B and A-20C airplanes, the depth setting cannot be changed once the torpedo is hoisted in place.

2. Cock the proper bomb release unit or units and place the bomb rack or racks in the locked position in order that accidental release during the torpedo installation may be avoided.

3. Open the turn-buckles located on one end of the suspension cables to about three-quarters of maximum. This procedure will allow sufficient slack in the cables during the preliminary steps of the loading.

4. Attach the turn-buckle end of each suspension cable to the proper bomb rack hook. This end of the cable is therefore held fast in the airplane when the torpedo is released.

  (a) If the particular airplane torpedo installation calls for release of the suspension cables from alternate sides of the bomb bay, the turn-buckle end (fixed end) of each cable should be hooked on the first station above the release shackle for the other cable. This procedure is usually necessary where long suspension cables are used, in which case the cables must be released from alternate sides in order to avoid an initial roll of the torpedo on release. The B-26 airplane has such an installation.

5. If hinged sway bracing is a part of the installation, it should be released from its retracted position and swung into place to receive the torpedo.

6. Place the 2000 lb. bomb hoisting sling around the torpedo and attach the ends to the hoisting cables. The hoisting sling should be positioned so that the torpedo will be in static balance while being hoisted.

NOTE: The center of gravity of the torpedo is located at approximately the center of the air flask (19 inches back from the forward end of the air flask).

7. Attach the starting lanyard to the starting mechanism by placing the thimble over the hook and rotating the same into its locked position while holding the spring locking pin in a retracted position. This can be done by pushing upwards with a screwdriver. Then allow the locking pin to spring down, thereby retaining the hook.

8. Begin hoisting the torpedo, being careful to take up evenly on the hoists. Hoist the torpedo into position against the sway bracing chocks being sure that the stop-bolt enters the hole in the torpedo during the process.

CAUTION: Do not pull the torpedo too tightly against the sway-bracing. The hoisting straps or cables may give way as a result of overload.

9. Attach the free ends of the suspension cables to the bomb shackle or shackles, and bring the cables around the torpedo. Place the shackle or shackles on the proper station or stations.

10. Tighten up on the suspension cable turn-buckles until the torpedo fits snugly against the chocks. Excessive tightening will overload the bomb shackle or shackles.

11. Attach the retracting cables to the FIXED ends of the suspension cables.

12. Remove the hoisting slings and hoisting cables.

13. If the torpedo is equipped with an adjustable stabilizer, attach the loose end of the stabilizer lanyards to the specified position on the airplane. Eight feet of slack should be provided for on each one of these lanyards. Care should be taken to arrange this slack so that the entanglement will not take place when the torpedo is released.

NOTE: Adjustable stabilizers are usually used where the tail or the entire torpedo is enclosed in the bomb bay.

14. Attach the loose end of the starting lanyard to the proper place on the airplane (usually vertically above the starting toggle on the torpedo).

XII-3B

THE STABILIZER IS PUT ON THE TORPEDO AT A 45° ANGLE, SIGHTED, BROUGHT BACK IN PLACE AND FITTED INTO SLOTS OF THE STABILIZER.

PLACING THE BELLY BAND OR HOISTING BAND ON THE TORPEDO.

MEASURE DISTANCE FROM GUIDE STUD HOLE TO CENTER OF GRAVITY 33¼" FOR HOISTING BAND TBF TYPE PLANE.

THE FIRST STOP.

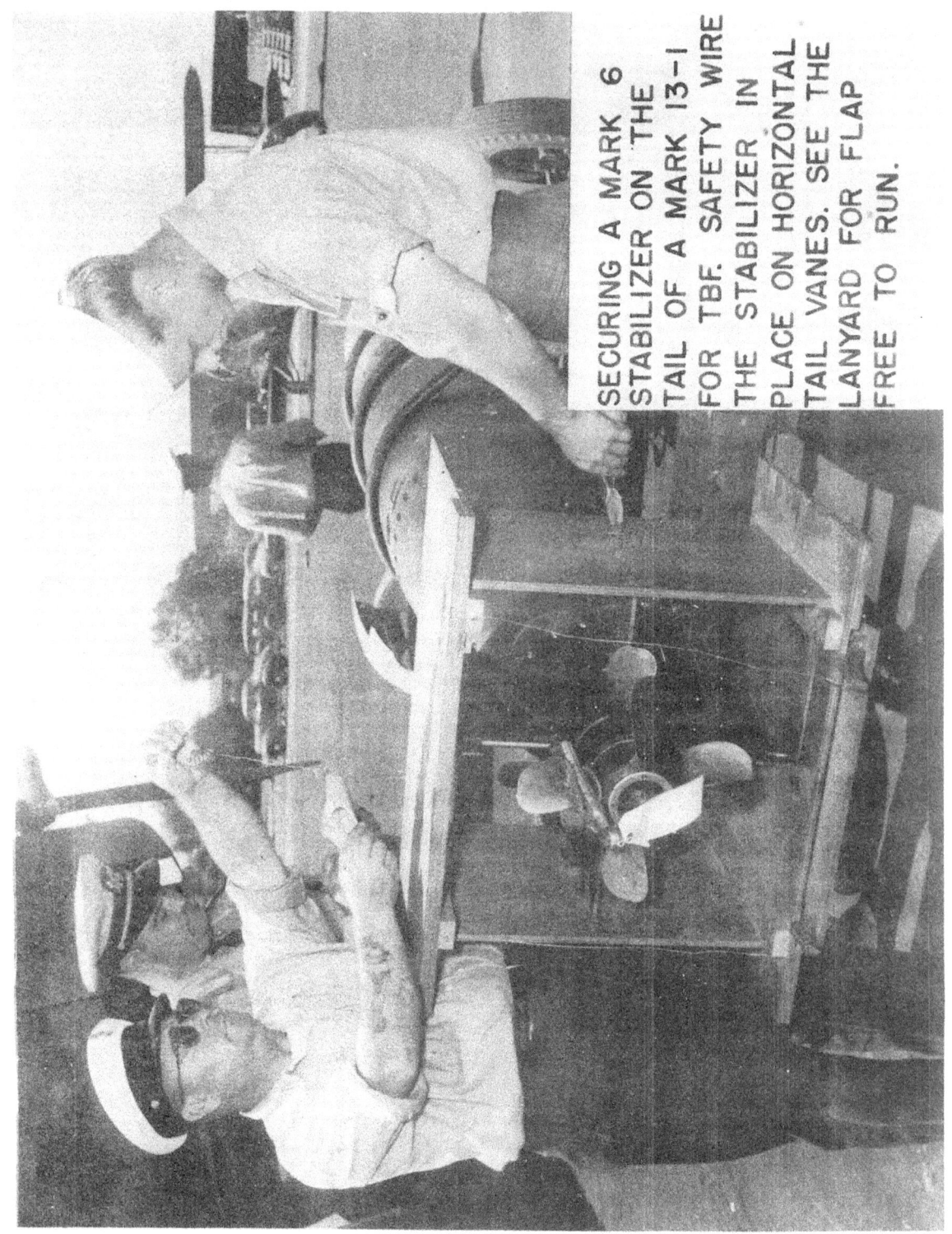

SECURING A MARK 6 STABILIZER ON THE TAIL OF A MARK 13-1 FOR TBF. SAFETY WIRE THE STABILIZER IN PLACE ON HORIZONTAL TAIL VANES. SEE THE LANYARD FOR FLAP FREE TO RUN.

CHECKING THE STARTING GEAR TO SEE MARKS IN LINE.

SETTING THE DEPTH.

OPEN THE STOP VALVE WIDE AND BACK OFF 1/4 TURN.

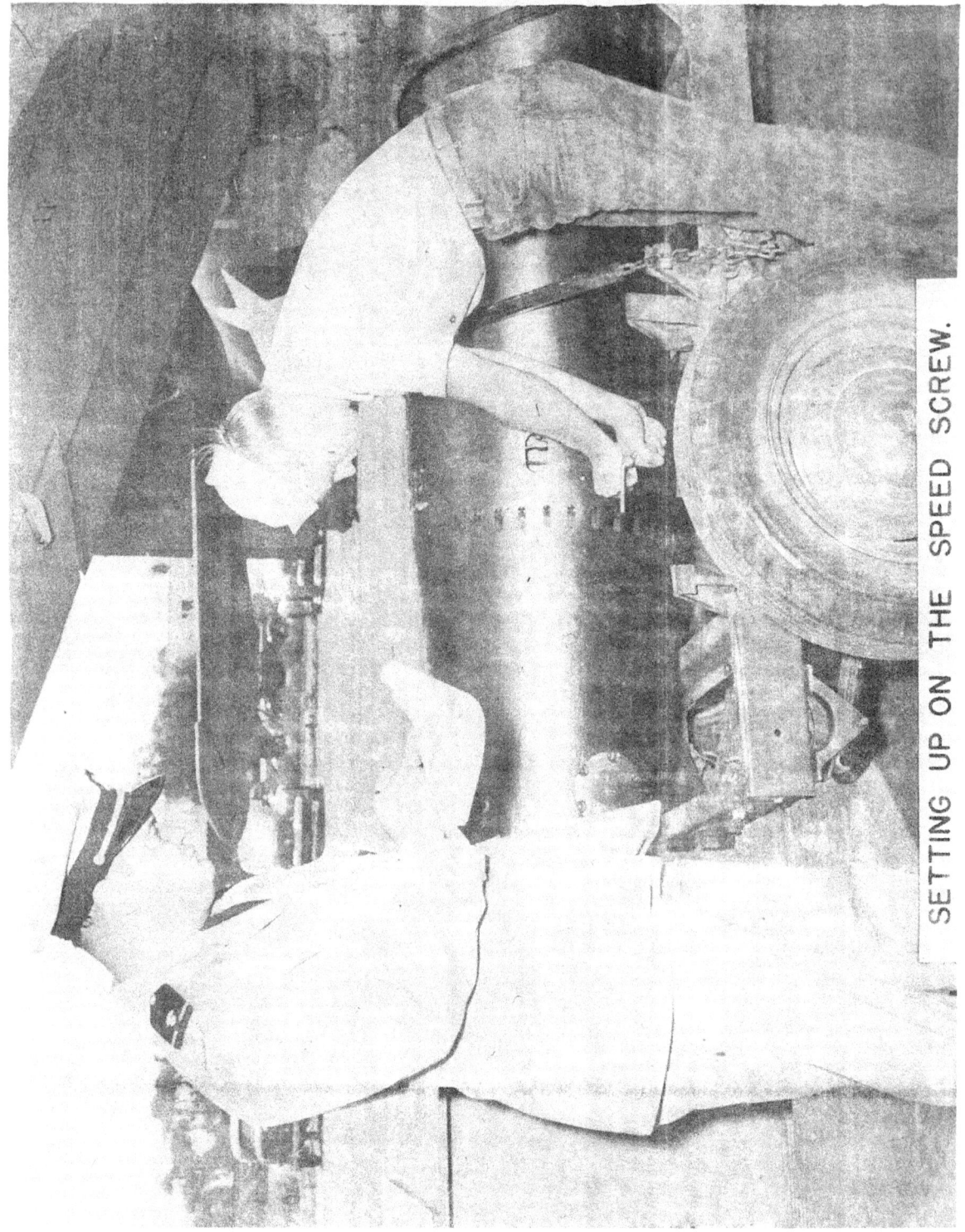
SETTING UP ON THE SPEED SCREW.

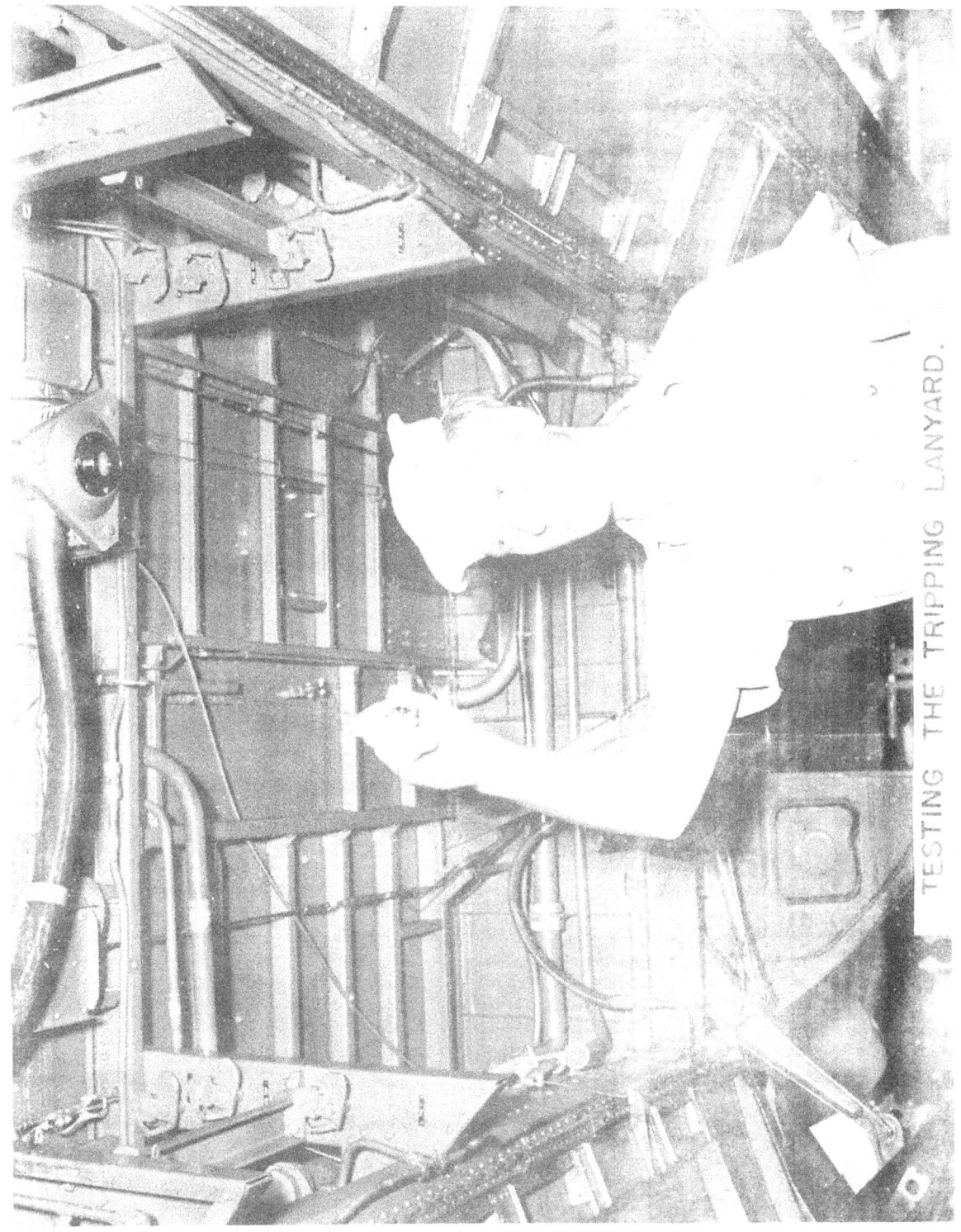

TESTING THE TRIPPING LANYARD.

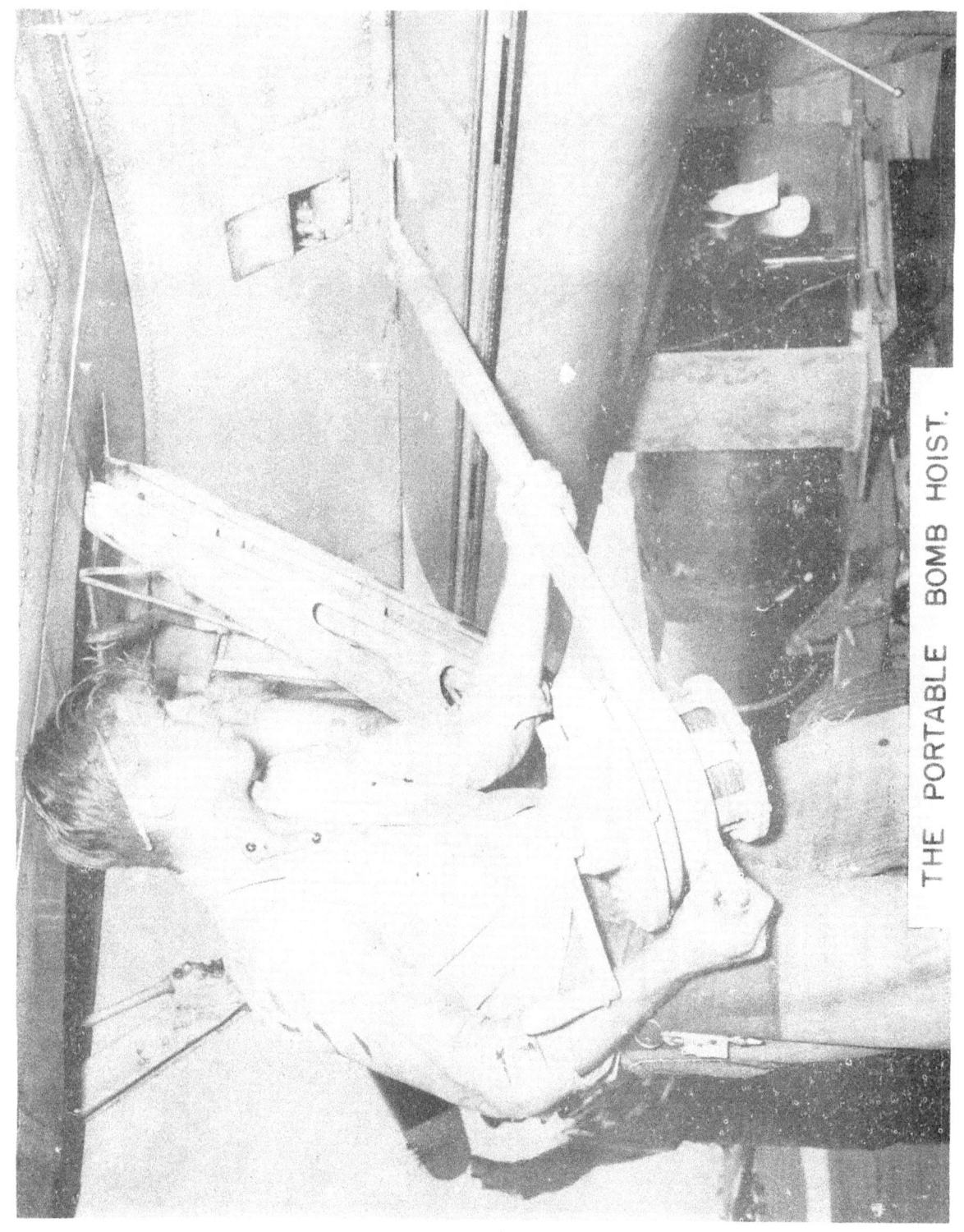

THE PORTABLE BOMB HOIST.

VIEW OF MARK 7 TORPEDO RACK. PORT RACK NOT HOOKED UP FOR RELEASE.

LOOKING FORWARD IN TBF BOMB-BAY. NOTE: STOP BOLT, RELEASE LANYARD, HOISTING CABLES & RELEASE CABLES.

THE BOMB-BAY ALL SET TO RECEIVE THE TORPEDO.

ROLLING THE TORPEDO UNDER THE PLANE.

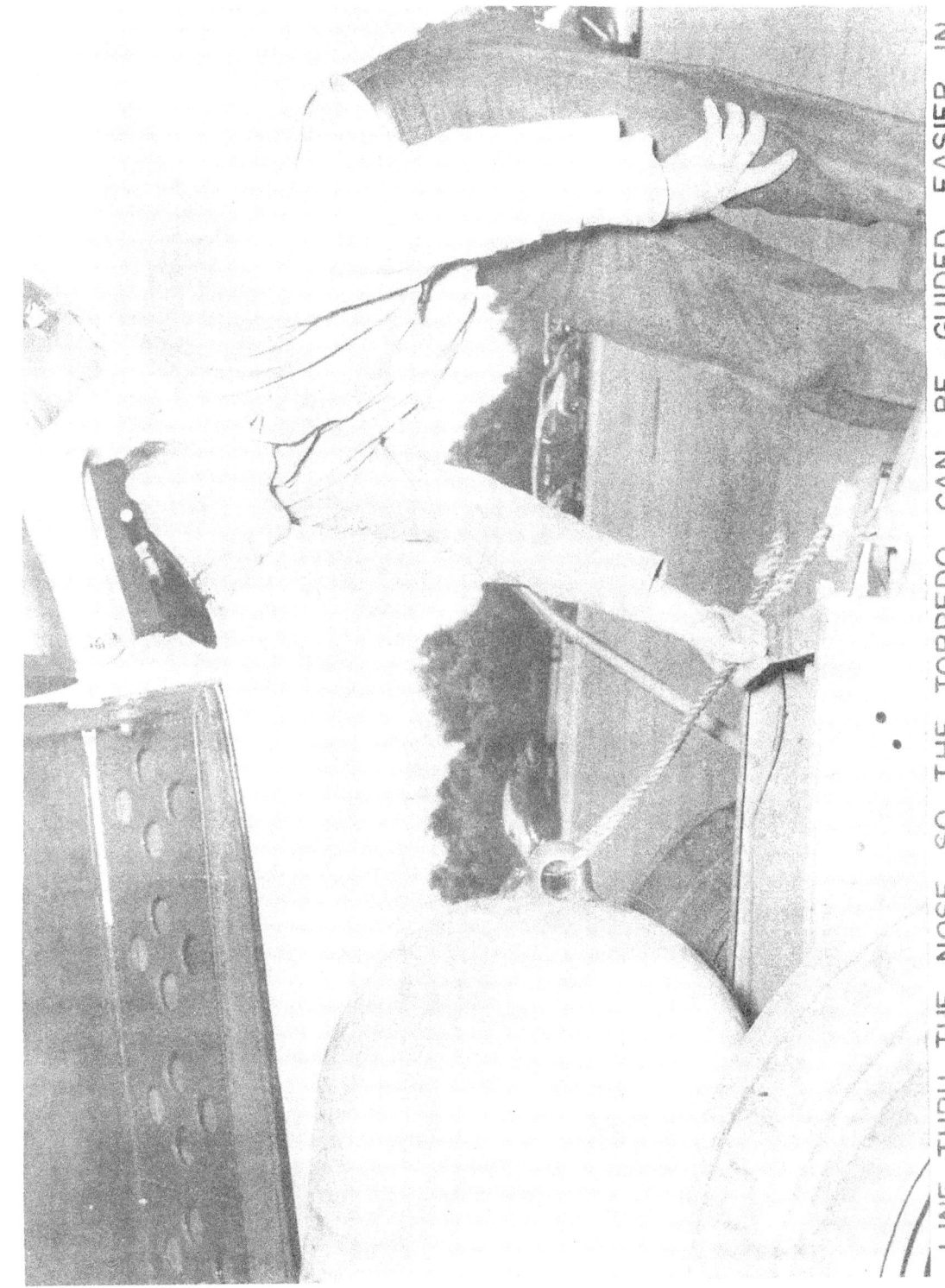

LINE THRU THE NOSE SO THE TORPEDO CAN BE GUIDED EASIER IN HOISTING.

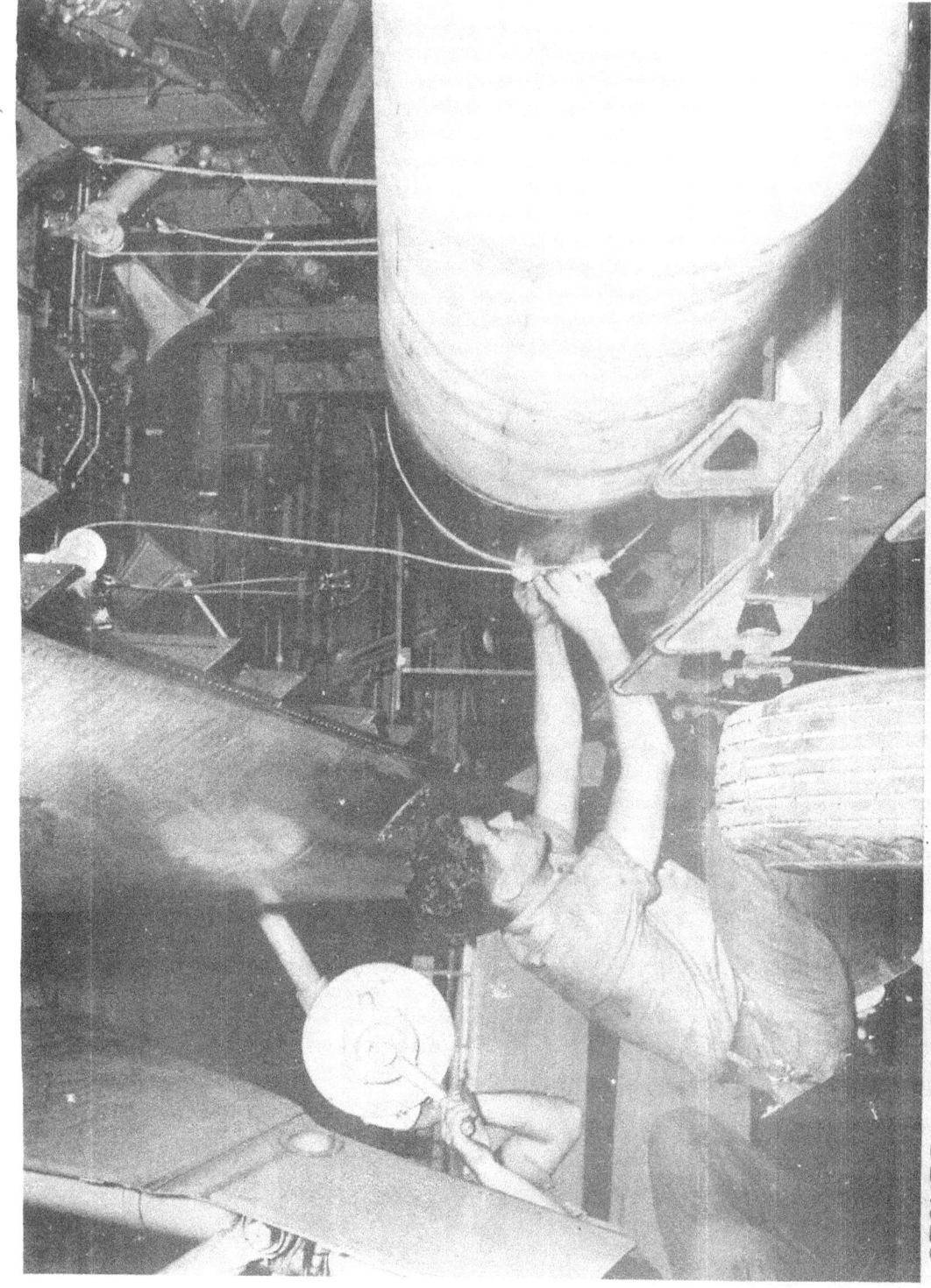

SECURE HOISTING CABLE TO HOISTING BAND, PUT IN PIN & SECURE WITH SPLIT COTTER PIN.

ONE MAN DIRECTING THE TEAM. TWO HOISTMEN, ONE MAN INSIDE GUIDING THE STOP BOLT IN HOLE AND THE MAN ON THE WAR HEAD HOLDING DOWN AND PUSHING AS DIRECTED.

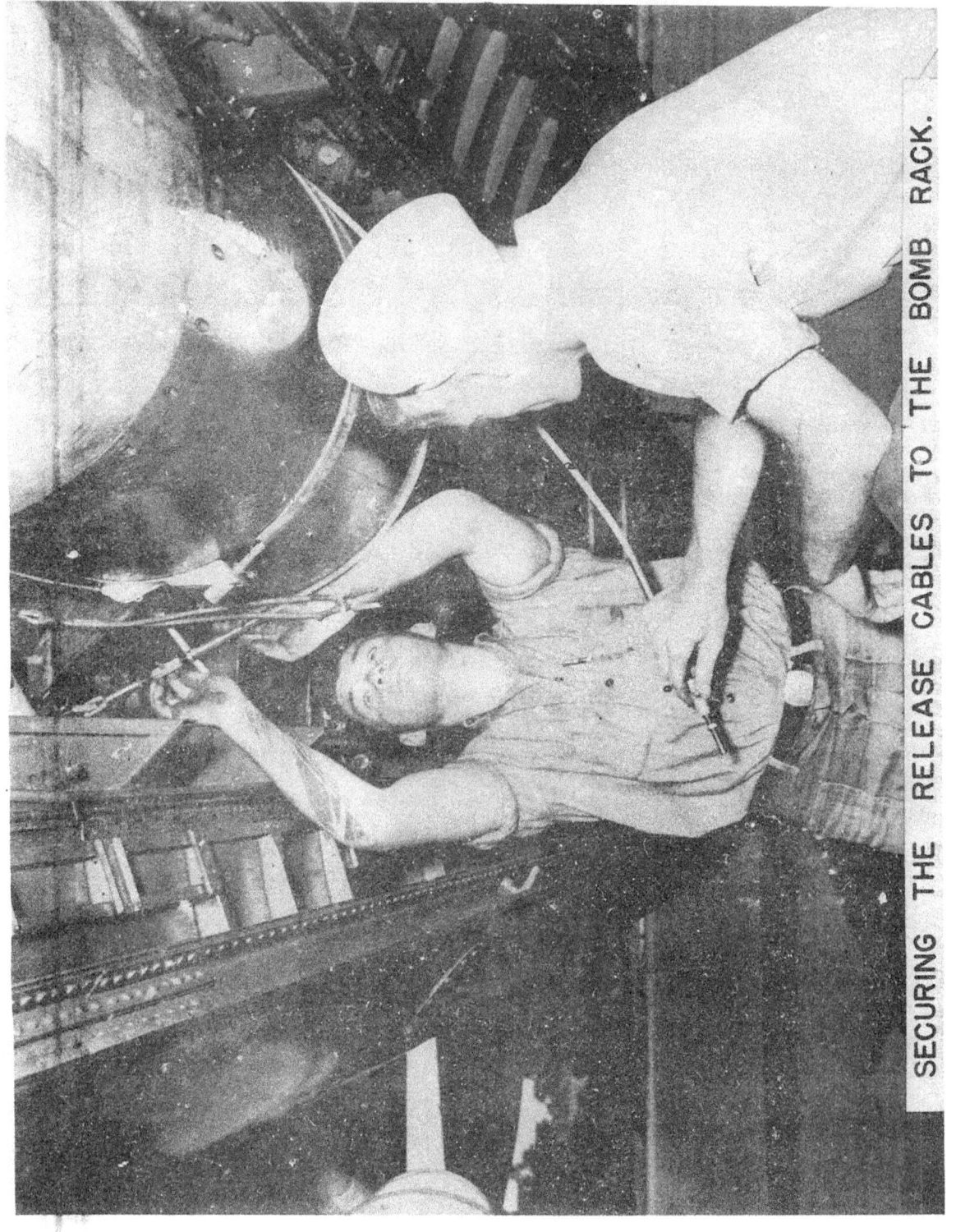
SECURING THE RELEASE CABLES TO THE BOMB RACK.

REMOVING THE HOISTING BAND.

THE BONGEE LANYARDS INSURE A SNAPPY RELEASE OF THE RELEASE CABLES.

TORPEDO HOISTED IN BOMB-BAY, & EMPTY TRAILER READY TO BE REMOVED.

HOOKING UP STARTING LANYARD TO CAM RELEASE TOGGLE.

CONNECTING UP THE IGNITER LEAD.

REMOVING THE SAFETY WIRE FROM THE STARTING SPINDLE ARM.

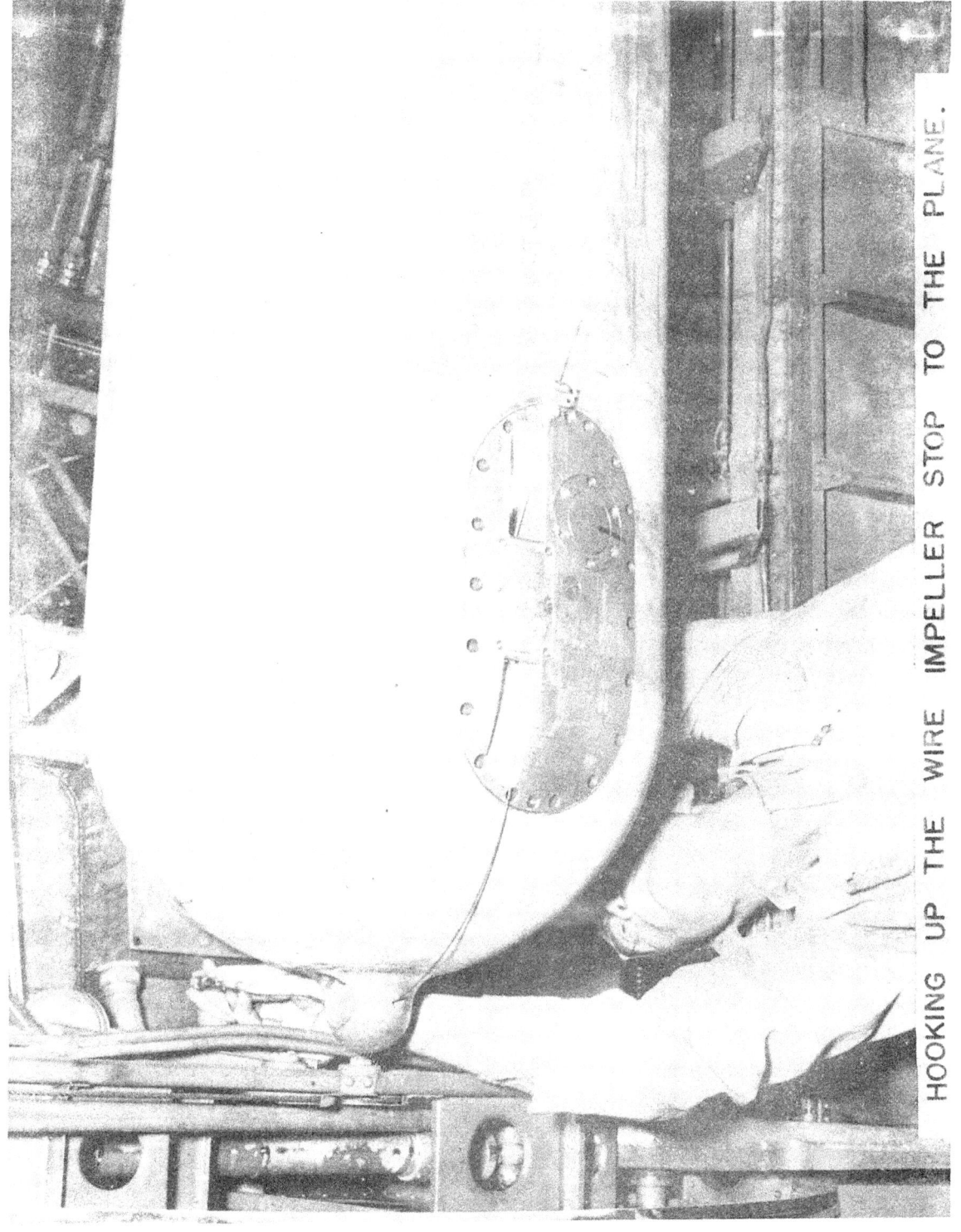

HOOKING UP THE WIRE IMPELLER STOP TO THE PLANE.

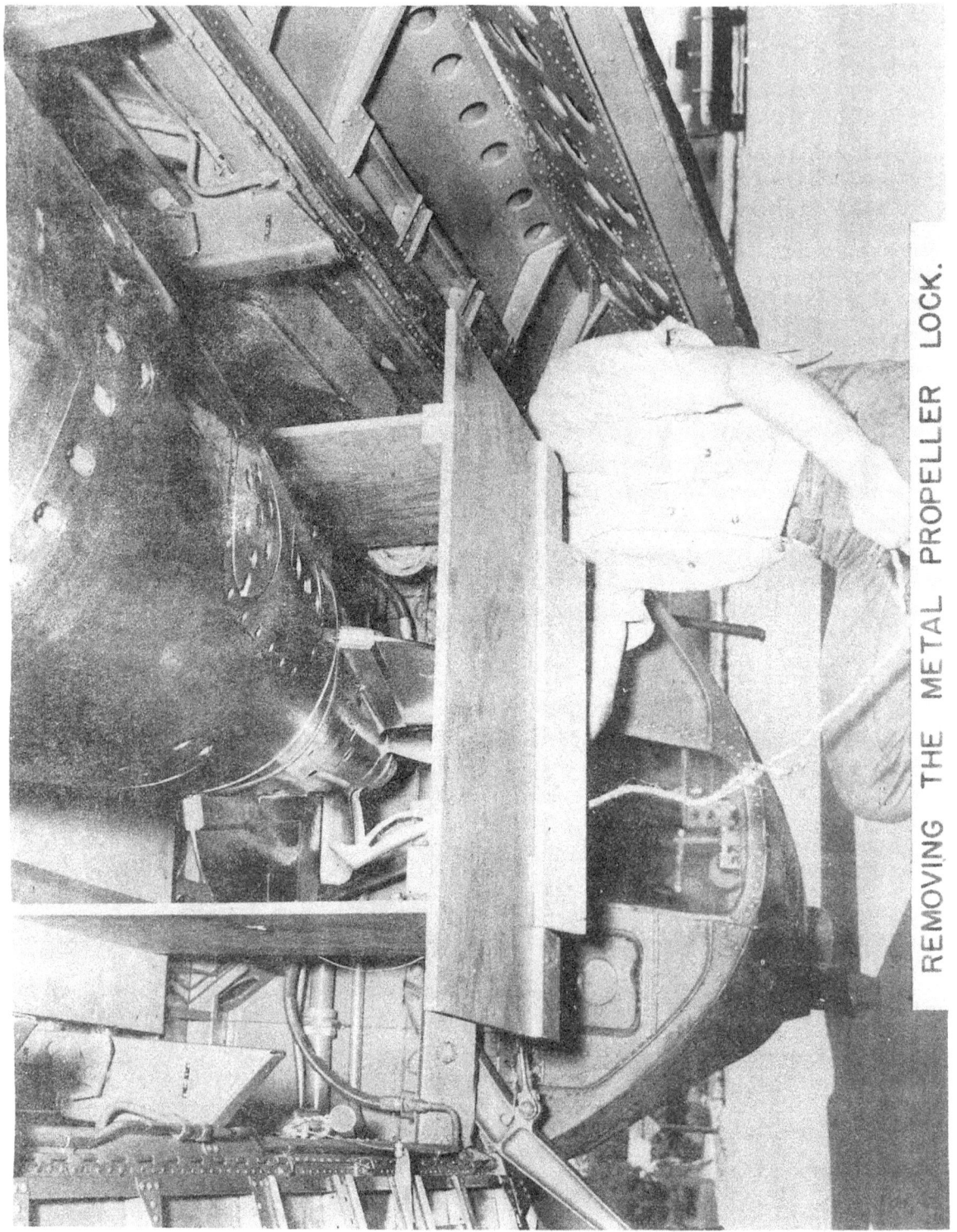

REMOVING THE METAL PROPELLER LOCK.

THE BOMB-BAY CLOSED READY FOR THE TAKE-OFF.

15. __IMPORTANT__: Remove the starting toggle locking wire that has been wrapped around the torpedo to prevent accidental release of the starting mechanism.

16. Remove the bronze safety propeller lock from the torpedo and wire the propellers to eliminate any rotation that may be caused by the slip stream. This locking procedure is of great importance since any rotation of the propellers would serve to exhaust the oil supply of the lubrication system within the torpedo. A more detailed description of wiring the propellers is given in Figure 1.

    (a) Figure 1 shows a suggested method of locking propellers during flight by means of a small wire strap which is sufficiently strong to prevent rotation, and yet breaks easily when the torpedo starts, even though the shot be a cold one (air driven only). It is to be noted that the wire propeller lock is not a subsitute for the conventional bronze lock which is still to be used as a safety device during the preparation, adjustment and loading of the torpedo, and __which must be removed prior to take-off of the plane__.

    (b) For each wire propeller lock the following material is required:

        $12-\frac{1}{2}$" (approximately) Insulated Soft Copper Stranded Wire (American Standard No. 14 -0.064 inch diameter).

        2   50 ampere Single Hole Lugs.

        1   3/16" Bolt and Nut.

    (c) To make the lock, the insulation is stripped from each end of the wire and the bare ends are soldered into the lugs to form an assembly $13-\frac{1}{2}$ inches from center to center of the bolt holes as shown in Figure 1. This is then looped about the propeller blades in a figure eight fashion as shown in Figure 1. The lugs are then bolted together, and the threads on the bolt deformed to prevent the nut from backing off.

    (d) If the recommended type of wire is not available, suitable substitutes may be used. Bare stranded wire of the same wire size, or even solid wire of equivalent strength are perfectly suitable. The Naval Torpedo Station, Newport, Rhode Island, has used without failure the usual soft copper safety wire, 0.0319 inch diameter (American Standard No. 20), but with this wire two separate figure eight loops of two turns each on opposite pairs of blades have been used. No lugs were used with the safety wire, the ends being secured by twisting together.

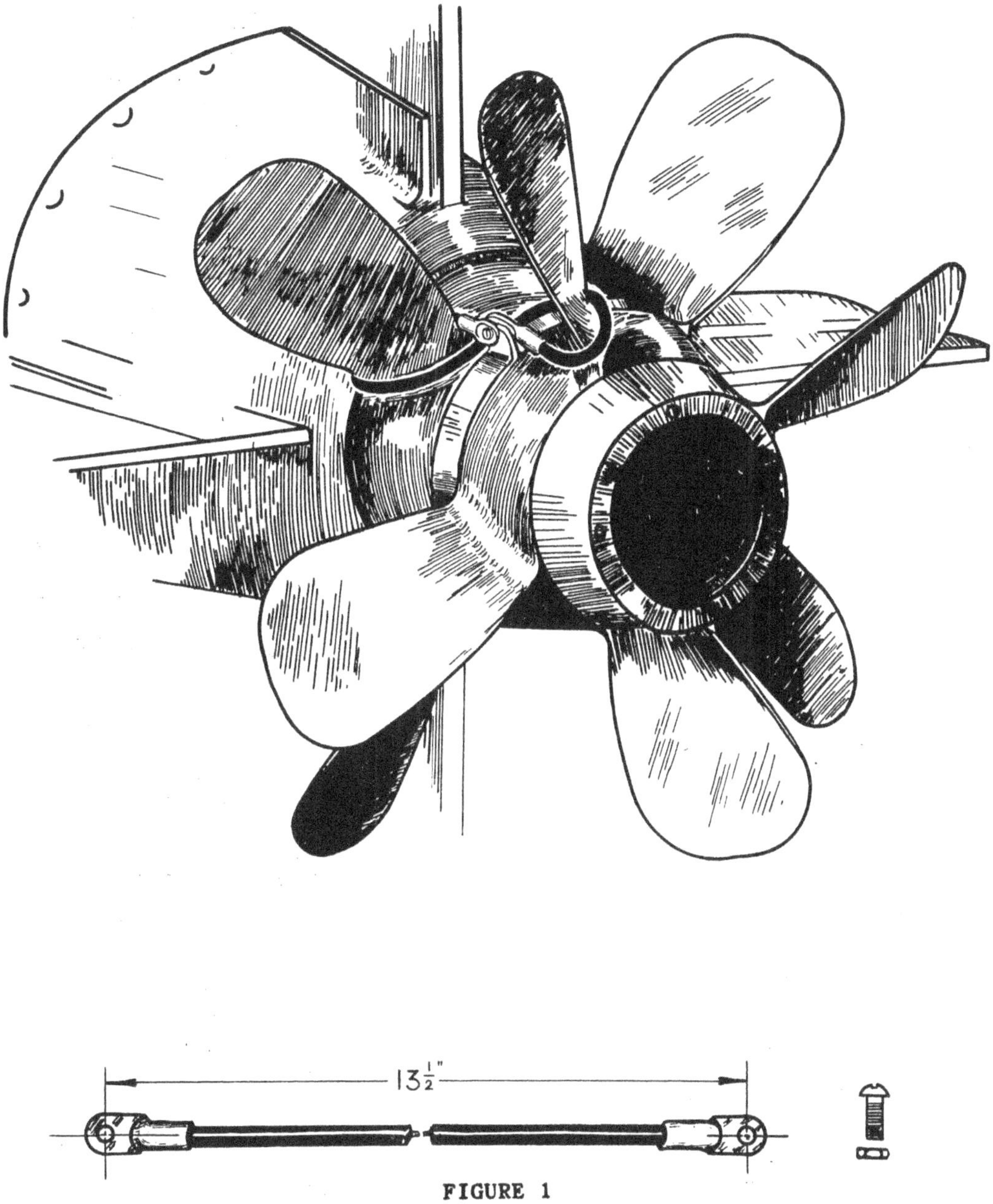

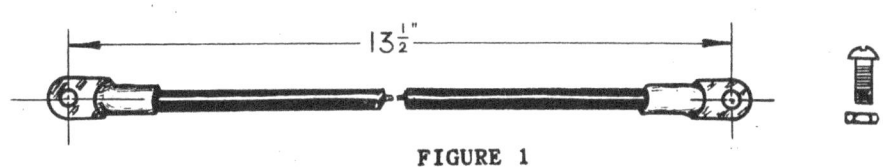

FIGURE 1

TS-5   XII-4A

B. **SAFETY PRECAUTIONS AND PILOTS' CHECK-OFF LISTS.**

1203.

1. Safety precautions are thoroughly covered in the various Bureau Manuals, Navy Regulations, Bureau Letters and Squadron Orders. The necessity of observing these safety precautions should be impressed on all personnel who are in any way connected with torpedo work. Pilots must know the safety precautions so that they can intelligently supervise the handling and loading of torpedoes.

2. Due to the number of various torpedo installations and the various conditions under which torpedo planes may operate, it is difficult to compile a pilot's check-off list which will apply in all circumstances. A check-off list is used in preparing the torpedo for firing and it is desirable to have a check-off list for use when loading the torpedo on the plane and one for the pilot to check before taking off. The loading and pilot's check-off lists should be prepared by the loading activity and by the Squadron concerned.

1204. **CAUSES OF TORPEDO FAILURES.**

1. Not all the aircraft torpedoes that are dropped will make good runs. Some of the causes of failures can be avoided by careful preparation of the torpedo and by observing the restrictions on launching speeds and altitudes. The following list includes the known causes of failures of torpedoes dropped during the past year at the Naval Torpedo Station. The reason for some erratic runs could not be determined:

  (a) Igniter failures:
    (1) Failure to hook up igniter air connection after torpedo was installed. Due to carelessness in following check-off list.
    (2) Faulty igniter that failed to ignite.
    (3) Faulty igniter that failed to burn completely.

(A cold shot runs so much slower and for such a short distance that it constitutes a failure).

  (b) Damage to torpedo due to poor entry into water. With present stabilizers entrance angle is generally good.
  (c) Bent propeller blades. With the heavier blades installed no failures of propellers have occurred when launching restrictions have been complied with.
  (d) After propeller coming off in the air, due to securing nut not being tight; carelessness.
  (e) Main stop valve not open; carelessness.
  (f) Starting piston failed to lift. Piston too short.
  (g) Starting toggle inserted backwards; carelessness.
  (h) Starting toggle not connected to starting lanyard; carelessness.

(i) Failure of starting lanyard; poor method of connecting lanyard to toggle or to plane.
(j) Tumbled gyro. Gyro tumbled shortly after impact causing erratic run.
(k) Damaged pallet mechanism. Probably due to water impact on initial entry or broach.
(l) Dives to bottom. Torpedo failed to recover from initial dive in 100 feet of water.
(m) Failure of exercise heads. Due to exceeding launching restrictions or not filling head completely with water. Exercise heads have been strengthened and tests with war heads indicate that the latter are satisfactory in strength.
(n) Bent rudders. Rudders, horizontal or vertical, may be damaged on impact with water.
(o) Cold shot believed to result from igniter being extinguished by torpedo rolling in the water.
(p) Bent tail vanes, caused by impart with the water.
(q) Sheared pins in the rudder linkages.

C. <u>AIR TRAJECTORY</u>.

<u>THEORETICAL FLIGHT IN AIR</u>.

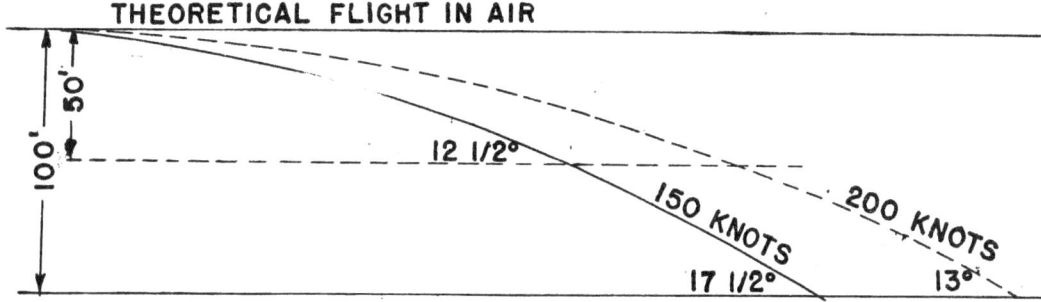

AIR TRAJECTORY — FIG. I

THEORETICAL FLIGHT IN AIR

1205. If the aircraft torpedo could be made to follow the vacuum trajectory, it would follow a path similar to one of those illustrated in Figure 1. The center of gravity of the torpedo would travel on a parabolic path and its longitudinal axis would be tangent to the trajectory at all times. The torpedo would therefore enter the water at a predetermined angle to the horizontal which would depend upon the altitude and air speed of the plane at release. Figure 1 shows the vacuum trajectories for two air speeds and indicates the theoretical entrance angles for the given altitudes and speeds

1206. Actually, the aircraft torpedo cannot be made to follow the vacuum trajectory, for reasons to be discussed later. However, for purposes of illustration, it is assumed that the torpedo does follow the vacuum path. Plate 1 is a nomogram calculated for vacuum conditions and is used to determine theoretical entrance angles. Since it is based on vacuum conditions, the entrance angles are not correct for actual conditions, but they are a good index of the entrance angles which should be obtained with a well stabilized torpedo.

1207. In order that the aircraft torpedo may follow the vacuum trajectory, the following conditions must be fulfilled:
- (a) No air resistance.
- (b) The torpedo must be aerodynamically stable.
- (c) The torpedo must be released so that it is horizontal at the instant of release.
- (d) The torpedo must have no angular accelerations but must have angular velocity at release.
- (e) To obtain the correct entrance angle, there must be no wind.

PRACTICAL FLIGHT IN AIR.

1208. Actually the torpedo does not and cannot follow the vacuum trajectory for the following practical reasons:
- (a) Air Resistance: Air resistance, even though its effect is small, does make the torpedo depart from the vacuum path.
- (b) Angular Acceleration: To follow the vacuum path, the torpedo should have no angular acceleration at the instant of release. If the plane is turning up or down at the instant of release, an angular acceleration will be imparted to the torpedo which will cause it to depart from the vacuum path.
- (c) Angular Velocity: It is difficult if not impossible to give the torpedo consistently the correct angular velocity at release.
- (d) Inclination of the torpedo at release: The torpedo is seldom released with its longitudinal axis horizontal, which is one of the conditions which must be fulfilled if the torpedo is to follow the vacuum path. For certain conditions of speed and plane loading, the torpedo may be horizontal when the plane is in level flight. However, any departure from these conditions results in the torpedo being launched either nose up or nose down, even though the plane is in steady, level flight.
- (e) Inclination of the Aircraft at release: If the aircraft is not in level flight but is gaining or losing altitude at a constant rate, the torpedo will not follow the vacuum trajectory.
- (f) Instability of the Torpedo: Wind tunnel tests show that the torpedo, without auxiliary stabilizing vanes, is unstable in the air. The center of pressure is

forward of the center of gravity so that if the torpedo axis is displaced from the tangent to the trajectory or from the relative wind, the air stream, acting on the center of pressure, will set up a couple which tends to further displace the torpedo from the tangent to the trajectory. The Mark 13 torpedo without auxiliary stabilizing vanes, is unstable until the longitudinal axis has been displaced about 18° from the relative wind. From this it can be seen that the torpedo must be stabilized, since a departure of 18° from the trajectory would result in entrance angles that might cause serious damage to the torpedo.

(g) <u>Effect of Wind</u>: Wind has no effect on the trajectory but it does have an effect on the desired entrance angle. The best that any practical method of air stabilization can do is to stabilize the torpedo for airspeed. This, under no wind conditions, will produce the desired entrance angle. With wind, the entrance angle will not be correct, since entrance angle should be based on ground speed, not air speed. For example, from Plate I, the entrance angle for a torpedo dropped from 200 feet at an air speed of 100 knots would be 34°. Now if the plane were flying down wind, with wind force 40 knots, the entrance angle for a ground speed of 140 knots should be 26°, or if flying up wind the entrance angle should be 48°. Since the torpedo is stabilized for air speed only, its entrance angle will be 34°, and on the down wind run the torpedo would enter 8° steep to the ground speed entrance angle and on the up wind run 14° shallow to the ground speed entrance angle.

USE OF AUXILIARY STABILIZERS.

1209. From the above it can be seen that some means must be provided to stabilize the torpedo in the air so that its center of gravity will follow some constant trajectory and so that the longitudinal axis of the torpedo will remain tangent to this trajectory. It is not essential for the torpedo to follow the vacuum trajectory, but it is essential for the torpedo to follow some constant trajectory. Depending upon the type of air stabilizers used, this trajectory may be a curve that approximates the vacuum trajectory, or it may be one which departs radically from the vacuum parabola and approaches a straight line.

1210. As mentioned above, the entrance angle should be based on the ground speed of the torpedo, since this is the factor which determines the trajectory relative to the water. If any fixed type of auxiliary stabilizer is used, the trajectory relative to the air is determined by air speed and the torpedo will not enter the water with the correct entrance angle except under no wind conditions. Even with controlled auxiliary stabilizers, it would be difficult to compensate for the effects of wind on entrance angles. Fortunately, the entrance angle is not too critical for the Mark 13 torpedo and fixed auxiliary stabilizers have been used with good results in winds up to about 20 per cent of the air speed at release.

1211. Two types of fixed auxiliary stabilizers have been developed at the Naval Torpedo Station and are described in detail in **Section F.**. One type is for use on the TBD-1 airplane and the other on the PBY series. Either one may be used on the TBD-1 due to space limitations of the airplane. These auxiliary stabilizers cause the Mark 13 torpedo to follow a trajectory which closely approximates the vacuum trajectory. A modification of one type of stabilizers with a two position trailing edge has been developed for use with the Mark 14 and Mark 15 torpedoes, and a modification of the TBD type is being used on the Army A-20C airplane.

D. INITIAL ENTRY AND UNDERWATER TRAJECTORY.

ENTRANCE VELOCITIES.

1212. Plate II is a nomogram showing the relationship based on vacuum conditions between altitude, speed at release and entrance velocity. Since air resistance can be neglected during the short time of fall from the relatively low altitude used, this nomogram can be used without much error for calculating entrance velocities and the kinetic energy of the torpedo when it hits the water.

1213. From Plate II it can be seen that air speed (or ground speed) has much more effect on entrance velocity than has altitude of release. For example a torpedo released at 100 knots from 100 feet has an entrance velocity of 187 feet per second, while if it were released from 200 feet at the same speed the entrance velocity would be 203 feet per second or an increase of only 8.5%. But if the torpedo is released from the same altitudes at 200 knots the entrance velocities would be 347 and 357 feet per second, or increases of 86% and 76%.

1214. Since the kinetic energy varies as the square of the velocity ($1/2 MV^2$), from the above examples it can be seen that any increase in air speed has a marked effect upon the energy which must be absorbed when the torpedo strikes the water. If this velocity of impact is doubled, the energy is four times as much.

INITIAL ENTRY.

1215. Anyone who has ever done much diving has probably had the painful experience of hitting the water too flat from the high board. A clean entry is essential for a good dive, and especially for the comfort of the diver. The same is true for the aircraft torpedo. The torpedo must enter the water with its axis along the flight path, or close to it; otherwise the torpedo may be damaged and a poor run result. It would be de-

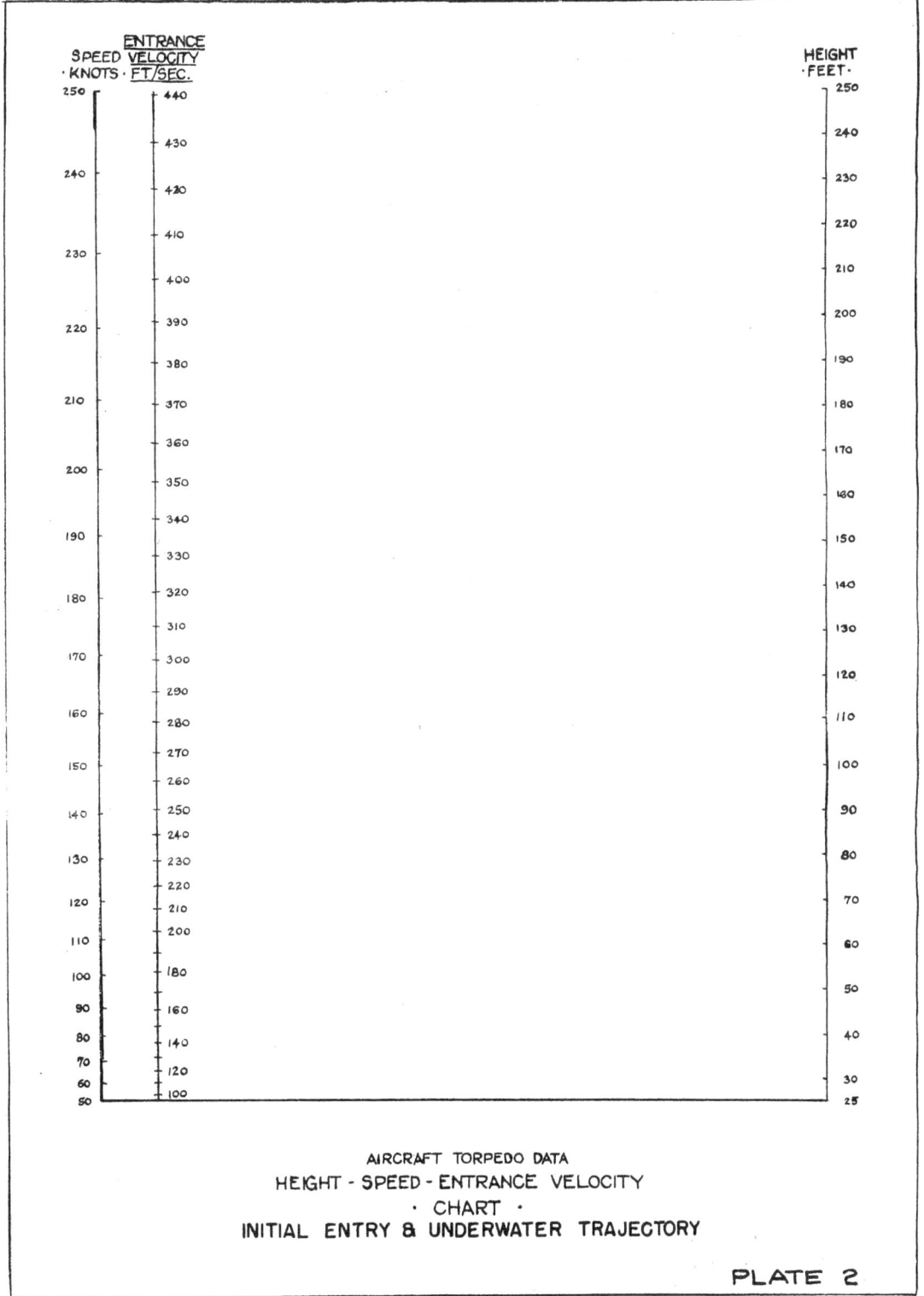

AIRCRAFT TORPEDO DATA
HEIGHT - SPEED - ENTRANCE VELOCITY
· CHART ·
INITIAL ENTRY & UNDERWATER TRAJECTORY

PLATE 2

sirable for the torpedo to hit the water slightly nose high to the flight path, so that the force of impact would tend to turn the head upward and cause a shallow initial dive. As stated previously, it is very difficult, if not impossible to stabilize the torpedo to obtain an entrance angle based on ground speed. Therefore, it is not possible to have the torpedo enter at the desired angle but generally if properly stabilized, the Mark 13 torpedo will have a sufficiently good entrance angle to insure good runs.

1216. Most of the kinetic energy of the torpedo is absorbed during the first fraction of a second after impact. Some studies indicate that possibly 80 per cent of the energy has been absorbed by the time the torpedo has entered about six inches into the water. If the torpedo enters the water nose first, high stresses are set up in the head, but the torpedo is able to stand up under impacts of this type. If the torpedo lands flat, the total energy is absorbed more rapidly and although the stresses are distributed over a larger area, the construction is such that the torpedo is not able to withstand these impacts, and fails at the joints or midship ring.

1217. During the past year it has been necessary to reinforce the exercise head in order to prevent damage to the head and torpedo. The standard exercise head is satisfactory for drops at about 115-125 knots from 75 feet but will not stand up under repeated drops from higher altitudes or speeds. In two cases where the standard exercise heads were badly damaged, probably because they had been incompletely filled, the deceleration forces were so great that most of the mechanism in the afterbody was torn loose and the torpedoes were complete wrecks. This happened on drops from 80 to 130 feet at 125 knots, and the torpedoes appeared to have good entrance into the water.

1218. Some data has been collected concerning the magnitude of the decelerations on impact. Accelerometers were mounted in the exercise head to measure the decelerations along the three principal axis. For the Mark 13 torpedo, dropped from 200 feet at 125 knots and with good entrance angles, these decelerations averaged about 40 g along the fore and aft axis and about 30 g perpendicular to this axis. This gives a total of about 50 g which is in close agreement with results obtained with a torpedo similar to the Mark 7. However, when the entrance angles were not close to the ideal, decelerations as high as 400 g were obtained when the torpedo entered slightly tail first, or about 18° flat to the trajectory. This torpedo was not recovered in one piece. From these data it can be seen that good entrance angles are essential, for it is very difficult to design a torpedo to take 400 g.

If we accept the theory that most of the energy is absorbed on impact and knowing that the energy varies as the square of the velocity, it can be seen that there is a practical limit to the speed at which the torpedo can be dropped. We have seen that altitude has little effect on entrance velocity, for example, at 125 knots, the Mark 13 torpedo has been dropped from as high as 450 feet. Air speed of release is therefore the controlling factor. If we have decelerations of 50 g at 125 knots we should expect to get decelerations of much higher magnitude at increased speeds. Until recently, planes have not been available at the Naval Torpedo Station for dropping torpedoes at speeds in excess of 150 knots, so it has not been possible to determine the speed at which the Mark 13 torpedo will fail. It is expected that in the near future dropping speeds can be increased so that the maximum speed at which the torpedo can be launched will be determined. The strength of the torpedo and the inability to design for extremely high decelerations may limit the dropping speeds to those considerably below the maximum of present day airplanes.

1219. The decelerations given above were for the spherical head. Limited investigation with an ogive head indicates that the decelerations are considerably less with this type. The ogive head has many disadvantages to balance against this fact, but it may have to be used for high dropping speeds, if some of its disadvantages can be overcome.

UNDERWATER TRAJECTORY.

1220. Very little is known concerning the underwater trajectory during the first few seconds after impact while the torpedo is decelerating to its controlled running speed. Some information has been gained by observing small models which were shot into a glass tank and some from observing actual torpedoes both from the air and from a boat. The model tests were not conclusive since a small moddel was used. Observation of an actual torpedo is unsatisfactory since it cannot be seen after entry and only a surmise can be made as to what actually happens.

1221. From these observations the so called "bubble theory" has been advanced to explain the erratic action of the torpedo during the initial dive. Briefly this theory is as follows: When the torpedo enters, the water is displaced with such force that it moves beyond the circumference of the torpedo. Thus the torpedo is supported on its nose by the water but it is not supported along its length. The torpedo is therefore free to pivot about its nose and may change directions radically until the water closes in around the torpedo and the rudders take charge.

1222. This bubble has been observed in the tank with models and may follow the torpedo for two to three times the length of the torpedo. Observation from a boat and from a plane also indicates that the bubble travels with the torpedo for some distance. Frequently a large bubble and boiling of the water appears 30 to 40 feet from the point of entry of the torpedo. On a few occasions parts of the plywood stabilizers have been seen to come up with this bubble, although most of the stabilizer is shattered on impact and remains at the point where the torpedo entered the water.

1223. This theory has not been definitely established, but it does explain the well extablished fact that the aircraft torpedo is definitely erratic during the initial dive. Initial dive covers that part of the underwater run from the instant of impact until the torpedo has assumed normal running speed and the vertical and horizontal rudders are controlling. The Mark 13 torpedo, because of its short length, large diameter and consequent small moment of inertia, is more erratic during the initial dive than the long, thin torpedoes such as the Mark 7, 14, and 15. It appears to form a larger bubble and therefore has more opportunity to turn than the longer torpedoes.

1224. Although the Mark 13 torpedo is erratic during the initial dive, it generally makes a satisfactory run. Instead of continuing to follow the trajectory after impact, the torpedo frequently turns one way or the other. If the torpedo lands a little steep to the air flight path, it stubs its toe and takes a deep dive, the extreme case, which is a rare occurance, being that it comes up on a reverse course. When the torpedo lands flat, it tends to make a shallow dive, and may broach or porpoise before taking depth. It is not unusual for the Mark 13 to come 15 feet out of the water, land tail first and still make a good run. Frequently the torpedo tends to hook right or left during the initial dive. This may be caused by cross-wind or by yawing of the plane at the moment of release. Although this hook may sometimes be more than 200 yards, generally it is much less and the average is under 30 yards.

1225. The Mark 13 torpedo is more stable during the initial dive than the Mark 13 Mod. 1, or the Mark 13 Mod. 2. The Mark 13 has the rail tail which offers a greater drag and this may account for its better performance. The rail tail Mark 13 seldom broaches and the average hook is about 15 yards, but it tends to make a deeper initial dive. Although the rail tail gives better underwater performance during the initial dive, it has high drag which reduces the running speed of the torpedo about 3.5 knots. The rail tail design was superseded by the straight tail for this reason. From the results with the rail tail as opposed to the straight tail, it appears that it would be desirable to give the torpedo high drag aft, if this drag could be eliminated as soon as the torpedo has recovered from the initial dive.

DEPTH OF INITIAL DIVE.

1226. The depth of the initial dive becomes an important factor when torpedoes are dropped in shallow water in a harbor or near the coast. To date no practical way has been found to limit the depth of the initial dive. In theory, if the torpedo lands shallow to the flight path, the impact force should tend to push the head upward causing the torpedo to take a shallow dive. This appears to hold true in practice in the majority of cases. On the other hand, a shallow initial dive generally produces a broach and the torpedo may take a deep dive after the broach.

1227. The depth of the initial dive of the Mark 13 torpedo with the spherical head seldom exceeds 100 feet, although a few do go below this depth. The Mark 7, 14 and 15 torpedoes and the Mark 13 with the ogive head tend to make deeper initial dives. Sufficient data are not available on these torpedoes to determine average depths of initial dive, but results to date indicate that they should not be dropped in much less than 100 feet of water.

1228. The table below gives the results for 245 proof Mark 13-1 torpedoes dropped during September to December 1941 at the Naval Torpedo Station. These torpedoes were dropped under the following conditions:

PLANE	AIR SPEED	ALTITUDE
PBY-3	125 knots	50-200 feet
TBD-1A	110-115 knots	50-100 feet
TBD-1A	115-125 knots	100-165 feet

Sea - smooth to choppy.
Wind - 0 to 20 knots. Drops made up, down and cross wind.

1	2	3	4	5	6
Depth of initial dive - ft.	% to given depths-249 torpedoes dropped from 50-200 feet	% to given depths-72 torpedoes dropped from 50-75 feet	% to given depths-74 torpedoes dropped from 76-100 feet	% to given depths-44 torpedoes dropped from 101-125 feet	% to given depths-59 torpedoes dropped from 126-200 feet
10	12.1	12.5	5.4	15.9	13.6
20	34.0	27.8	29.8	36.4	45.8
30	52.2	51.5	43.2	52.2	67.8
40	71.5	73.6	66.3	75.0	77.9
50	91.1	89.0	89.2	95.5	95.0
60	94.6	91.6	93.3	97.8	98.5
70	98.5	98.6	96.0	100.0	100.0
80	98.8	100.0	96.0		
100	100.0		100.0		

1229. Since this table represents a fair sample of drops made under varying conditions, it can be used as a guide in estimating the percentages of torpedoes which will not hit the bottom when dropped in shallow water. Thus about 70 percent of the torpedoes dropped in 40 feet of water should recover from the initial dive without hitting the bottom and about 95 percent should recover in 60 feet of water. These figures are open to some question, since the depth recorder used is not too reliable or accurate. However, this is the best information available.

1230. It might be supposed that shallow dives would result if the torpedo were launched from low altitudes. The entrance angle from low altitude is small and if the torpedo continued along the line of entry the dive would be shallow. In practice this is not correct, as a study of the table will show. In general, the torpedo recovers more quickly when dropped from altitudes above 100 feet than from below and in dropping in shallow water it is best to launch from altitudes of 125-150 feet. A few drops have been made from altitudes of 400-450 feet with the initial dive not exceeding 60 feet on any drop.

1231. No practical method has so far been developed to limit the depth of the initial dive. The Mark 13 torpedo is fairly satisfactory in this respect, but it would be desirable to increase the percentage which does not exceed 40 feet on the initial dive.

E. <u>DROPPING CONDITIONS AND RESTRICTIONS</u>.

<u>LAUNCHING RESTRICTIONS</u>.

1232. Up to the present time nearly all torpedo drops at the Naval Torpedo Station have been made from the PBY-3 and TBD-1A airplanes. The TBD-1A is a TBD-1 on twin floats, and has a somewhat lower speed than the landplane version. A few drops have been made from the PBM-1, the Army B-26, and the Army A-20C airplanes. Nearly 500 drops have been made during the past year, and the following restrictions on speed and altitude have been established as a result of this work.

1233. During the past year several changes have been incorporated in the Mark 13-1 torpedo. Mark 13-1 torpedoes having register numbers higher than 17620 have had these changes incorporated in them during manufacture. Mark 13-1 torpedoes with register numbers below 17621 may not have had these changes made. Unless the torpedoes with register number below 17621 have been modified in accordance with Change Lists No. 22 and 23 they cannot be dropped under the same conditions as those with register numbers above 17620.

1234. Mark 13 Mod. 1, and Mark 13 Mod. 2 torpedoes having register numbers above 17620 and Mark 13 and Mark 13 Mod. 1 torpedoes which have been strengthened may be dropped under the following conditions of speed and altitude:

    (a) PBY airplane - 50 to 150 feet, at air speeds up to 125 knots.
    (b) TBD airplane - 50 to 100 feet at air speeds up to 110 knots.
    (c) Experience with the PBM-1 airplane is very limited. These restrictions are therefore tentative but should be satisfactory at air speed 125 knots and altitude between 100 and 150 feet.

1235. For Mark 13 and 13-1 torpedoes with register numbers below 17622 which have not been strengthened, 75 feet is the maximum altitude with the same speeds as listed in paragraph 1234 above for the PBY and TBD airplanes.

1236. The Mark 7 torpedo is not very strong, but can be used with fair results if launched from less than 30 feet at air speeds below 80 knots.

1237. The Mark 14 and Mark 15 torpedoes at the present time are limited to an airspeed of 110 knots and altitudes from 50 to 75 feet.

1238. Until each new type of airplane has been given a thorough test it is difficult to predict the launching restrictions which may have to be imposed on each type. However, the following general considerations apply, and in the absence of specific tests to determine launching restrictions, they may be used as a guide for launching torpedoes from new types of aircraft:

    (a) Maximum safe speed for launching is 125 knots airspeed for the Mark 13 torpedo, 80 knots for the Mark 7 torpedo, and 110 knots for the Mark 14 and Mark 15 torpedoes.

    (b) The torpedo should be horizontal at release, that is, it should be parallel to flight path when the plane is in horizontal, level flight at launching speed. Maximum launching speeds can be used within the altitude ranges given above if the torpedo is horizontal or within 3 degrees of horizontal. If the torpedo is carried at an angle of more than 3 degrees to the flight path at launching speeds, as for instance by the TBD-1 airplane, launching restrictions similar to the ones for this plane may be necessary. Since the torpedo at release from the TBD is displaced several degrees from the flight path of the plane, rather violent air oscillations may be set up at the higher speeds.

1239. The Mark 13 torpedo has been launched at air speeds in excess of 125 knots and a few drops have been made at ground speeds of 175 knots. Although it has not been definitely established, it appears that the Mark 13 torpedo should give satisfactory runs when the GROUND speed does not exceed 150 knots. Therefore, for upwind runs the air speed may be increased above 125 knots but for downwind runs it is essential that the ground speed does not exceed 150 knots.

1240. The Mark 13 torpedo has been launched at 125 knots airspeed from altitudes up to 450 feet. Only a few drops have been made at 450 feet but enough have been made at 200 to 250 feet to establish this as a safe altitude for strengthened torpedoes. The number of drops above 250 feet are not sufficient to establish the upper limit. However, 150 has been established as the upper limit for operating units for the following reasons:

    (a) The present altimeters are subject to errors of more than 100 feet, especially if the torpedo plane has come several hundred miles from point

of take-off. If the torpedo is launched at 150 feet
by altimeter, it may be considerably higher.
- (b) 150 feet is high enough to avoid shell splashes, masts
of ships and so forth.
- (c) 150 feet allows reasonable freedom of movement for
the torpedo plane.
- (d) 150 feet is a comfortable altitude and the pilot does
not have to worry about flying into the water.

1241. Exercise heads which have not been reinforced will not withstand repeated drops under conditions of maximum specified speeds and altitudes. Care should be taken that exercise heads are completely filled with water and that there are no leaks. After each drop, exercise heads should be carefully inspected for dents or for any looseness of the nose piece. Any dents should be removed and any looseness of the nose piece should be remedied before the head is used again. Tests with the war head indicate that it will withstand drops at 245 knots ground speed from 300 feet.

FLIGHT CONDITIONS.

1242. lthough good runs have been obtained from torpedoes dropped in moderately bumpy air, it is essential that every effort be made to drop the torpedo when the plane is in steady flight. Any departure from steady flight conditions at release will cause errors in the run of the torpedo in the water.

1243. The course of the torpedo in the water is controlled by a small gyro. This gyro is spun when the torpedo starting valve is opened and is unlocked about 0.4 seconds later. Thus the torpedo will drop about 4 feet before the gyro is unlocked and the torpedo heading in the water will be the heading on which the gyro was unlocked. From this it can be seen that it is essential for the plane to be steady about the yaw axis at the instant of release. The angular velocity of the plane in yaw will be imparted to the torpedo, and by the time the gyro is unlocked, may cause a considerable change in heading of the torpedo. The water run of the torpedo will therefore differ from the plane heading by the amount the torpedo yaws from release to the unlocking of the gyro. Slips, skids, and yawing of the plane must be avoided at the instant of release.

1244. The gyro should be unlocked when the torpedo is on an even keel, that is, with its longitudinal axis horizontal and the tail fins vertical and horizontal. If the gyro is unlocked in any other position, deflection errors will result in the water run. Under normal conditions this error is small, but every effort should be made to keep the plane steady about the pitch and roll axis. It is difficult to have the gyro unlock when the longitudinal axis of the torpedo is horizontal. When the torpedo leaves the plane it gains angular velocity and will be nose down several degrees when the gyro unlocks. Carrying the torpedo about 3 degrees nose up at release will make the torpedo about horizontal when the gyro is unlocked. However, from the standpoint of air stabilization it is desirable to have the torpedo horizontal at release.

1245. It is not essential to maintain constant speed at release, provided the plane is in steady flight. Constant altitude during the approach is not required, but it is desirable to have the torpedo about horizontal at release. Thus the torpedo may be dropped in a glide or in a climb, provided neither is steep and speed and altitude restrictions are not exceeded at release. However, level flight at the instant of release is desirable.

1246. Since the air stabilization depends upon air speed only, up, down, or cross wind runs have no effect upon the torpedo in the air. However, with respect to ground speed, the entrance angle will not be correct but if the wind does not exceed about 20 percent of the air speed, the entrance angles should be satisfactory for good runs regardless of the direction of the relative wind during the approach.

1247. With the TBD-1 airplane, the clearance between the stabilizers and the sides of the torpedo bay is very small. There has been no evidence of the stabilizers hitting the plane, but because of the small clearance there is some possibility of interference. Therefore, it is important that the TBD-1 and similar airplanes be held as steady as possible from the instant of release until the torpedo is well clear.

BUREAU OF ORDNANCE CIRCULAR LETTER NO. T 7-42.

Subject: Data Concerning Release of Aircraft Torpedoes by all Types of Airplanes.

Enclosure: (A) Curves giving "Angle of Entry vs Height of Drop".
(herewith)
(B) Curves giving "Air Travel After Drop".

1. The following information is being forwarded for service use and represents the latest and best practice in aircraft torpedo work as of this date:

AIRCRAFT TORPEDOES.

The two sets of curves attached hereto will be found of great interest and value to aircraft torpedo personnel. One set of curves shows the horizontal travel of the torpedo when dropped at various heights while flying at various air speeds. The second curve shows the angle of entry of the torpedo to the water when released from various heights at various flying speeds, and both contain an area extending across the several curves (hatched area) which shows the range of heights within which torpedoes must be released at the various air speeds in order to obtain satisfactory entrance and subsequent torpedo run.

These curves presuppose the use of stabilizers as directed by the Bureau for the various types of planes; they presuppose that for the higher air speeds only strengthened torpedoes are used and with heads specifically designated to be used at these higher speeds. Bureau of Aeronautic despatch 091700 of July covers in general the handling of torpedoes with the new TBF type plane; Bureau of Ordnance despatch 222130 of July covers the use of torpedoes from the B-25, B-26 and A-20C. The two sets of curves are applicable to all types of planes within the speed restrictions imposed by current instructions provided properly strengthened torpedoes are used.

As an example of the use of the curves of horizontal travel, suppose you are flying a plane at an air speed of 150 knots and you wish to release your torpedo at 250 feet. These conditions give point A on the curve and by projecting vertically downward it will be seen that the horizontal travel is about 340 yards. On the other hand, if you are flying at 175 knots and decide to release your torpedo at 300 feet, this will be point B on the curve, and by projecting vertically downward it will be found that the horizontal travel is 430 yards. If the arming distance of the exploder be added, then in the latter case the torpedo should not be dropped closer than about 800 yards from the target. The cross hatched area on this set of curves corresponds to the cross hatched area on the set of curves giving angle of entry; i.e., it delineates the correct height of drop for each flying speed within the range of 100K-225K. Attention is particularly invited to the fact that in the horizontal travel curves the scale is different in the case of the vertical and horizontal graduations, thus the angles of entry are not correct, so these curves should not be used for measuring angle of entry.

For angle of entry corresponding to various heights of drop and flying speeds, the second set of curves should be used. It will be noted that regardless of air speed $20°$ is considered the minimum angle of entry which should obtain; on the other hand, angles of entry up to $34°$ may be had at the lowest air speed, but as air speeds are increased the maximum permissible angle of entry falls off very rapidly.

To use the set of curves giving angle of entry, suppose it is desired to release the torpedo at 125 knots at 300 feet. This will be point C on the angle of entry curve and by projecting vertically downward it will be seen that the angle of entry is $34°$; since point C is outside the hatched area, it is a forbidden condition as at 125 knots the maximum height of drop is about 240 feet which gives an angle of entry of $31°$.

Angle of entry is controlled by the plywood stabilizers and the forces necessary for this control are obtained from the apparent wind; thus, air speed is the important speed for stabilizers and thus for angle of entry. If the torpedo is stabilized to give a satisfactory angle of entry (as when stabilizers are used as directed by the Bureau for the particular type of plane in use), then torpedo damage is unlikely and a satisfactory water run may be expected.

It will be noted that in prescribing maximum air speeds for torpedo release the Bureau has consistently authorized 200 knots only in emergency and provided the war head is used. We have no exercise heads which are strong enough to permit the large amount of test dropping at 200 knots needed to establish this speed as a permissible speed except in emergency; therefore, it is preferred that torpedoes be dropped at air speeds of not over 175 knots for best results. Later, when a satisfactory exercise head has been developed it may be possible to prescribe 200 knots under all conditions. An emergency condition as defined above is considered to exist when a successful attack on an enemy vessel can be made only by flying at the higher speed.

An exercise head which can stand dropping at 200 knots air speed has been manufactured and is being tested at Newport at the present time.

The following table gives information concerning the various types of stabilizers to be used on various types of planes:

MARK	DRAWING	SHEETS	LATEST REVISION	USE
1	226107	1	A	Mark 13 torpedoes on PBY-3,4,5, 5A; PB2Y-1; PBM
2	226108	1	A	Mark 13-1 torpedoes on the planes listed for Mark 1 and also for use on B-25 and B-26 airplanes.
3	226109	1	A	Mark 13 torpedoes on TBD-1 airplanes.
4	226110	1	B	Mark 13-1 torpedoes on TBD-1 planes.
5	(226116)	2	B	Mark 14-3, 15 and 15-1 torpedoes for PBY's specially fitted.
	(226117)	1	B	
	(226118)	1	B	
6	226256	2		Mark 13-1 torpedoes for TBF-1
7	226289	2		Mark 13-1 torpedoes for A-20C
8	226258	2		Mark 13-1 torpedoes for A-20C airplanes.
9	226259	2		Mark 13-1 and 13-2 torpedoes for British Bristol Beauforts.
10	226290	1		Mark 13-1 torpedoes for TBD-1 airplanes (designed to prevent fouling arresting gear)
11	226332	1		Mark 13-1 for PBM-3, 3C planes.

So far as stabilization is concerned the same stabilizer may be used on either Mark 13-1 or Mark 13-2 torpedoes. The Mark 5 stabilizer was designed for special use on Mark 14 and Mark 15 torpedoes on certain patrol planes (at Naval Air Station, San Diego). Attention is invited to the fact that either Mark 7 or Mark 8 stabilizers may be used on the A-20C airplane. The Mark 10 stabilizer was designed for use on TBD-1 planes to meet certain objections to the Mark 4 having to do with possible fouling of arresting gear and interference with the center bomb rack on this type airplane.

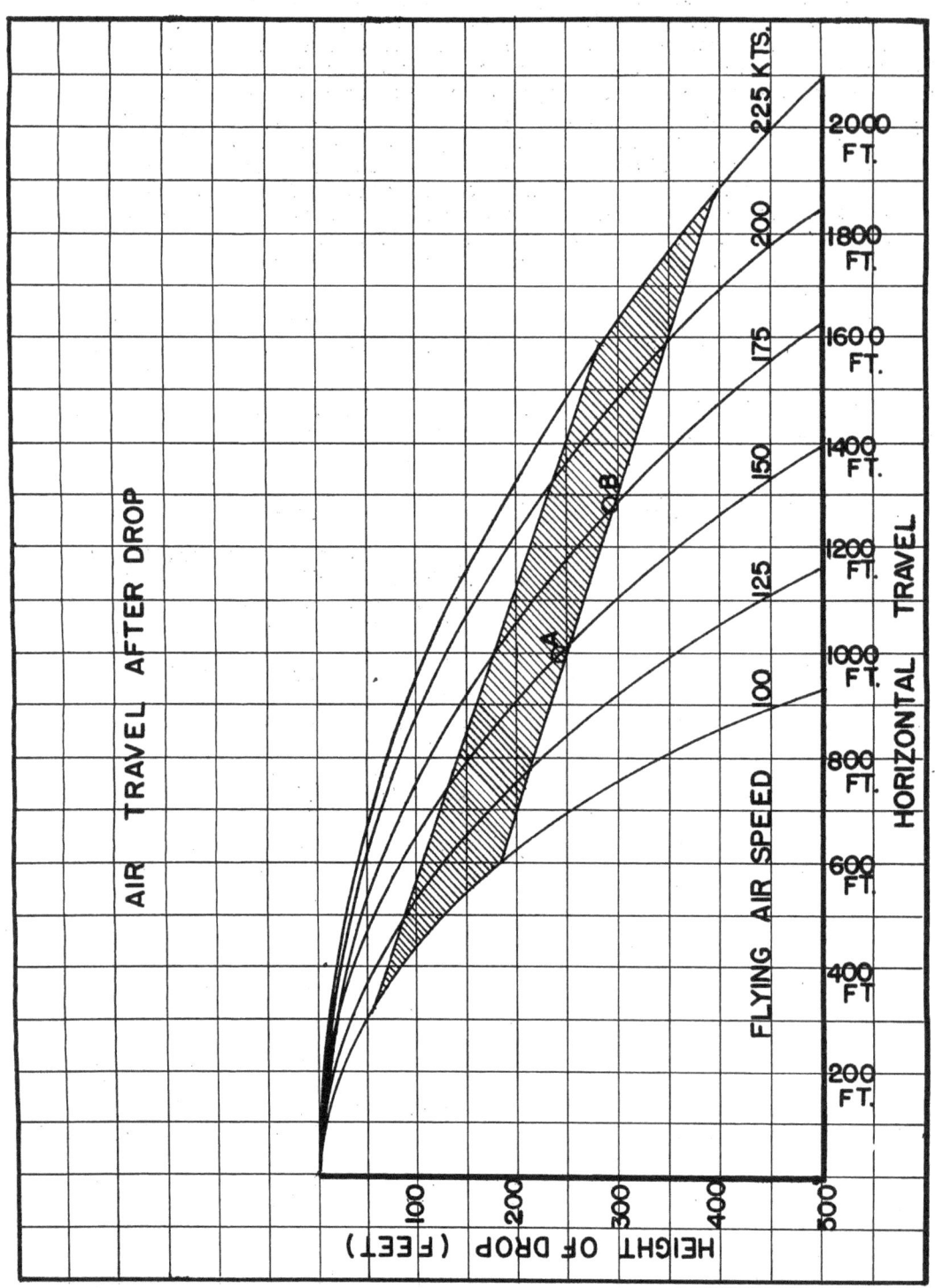

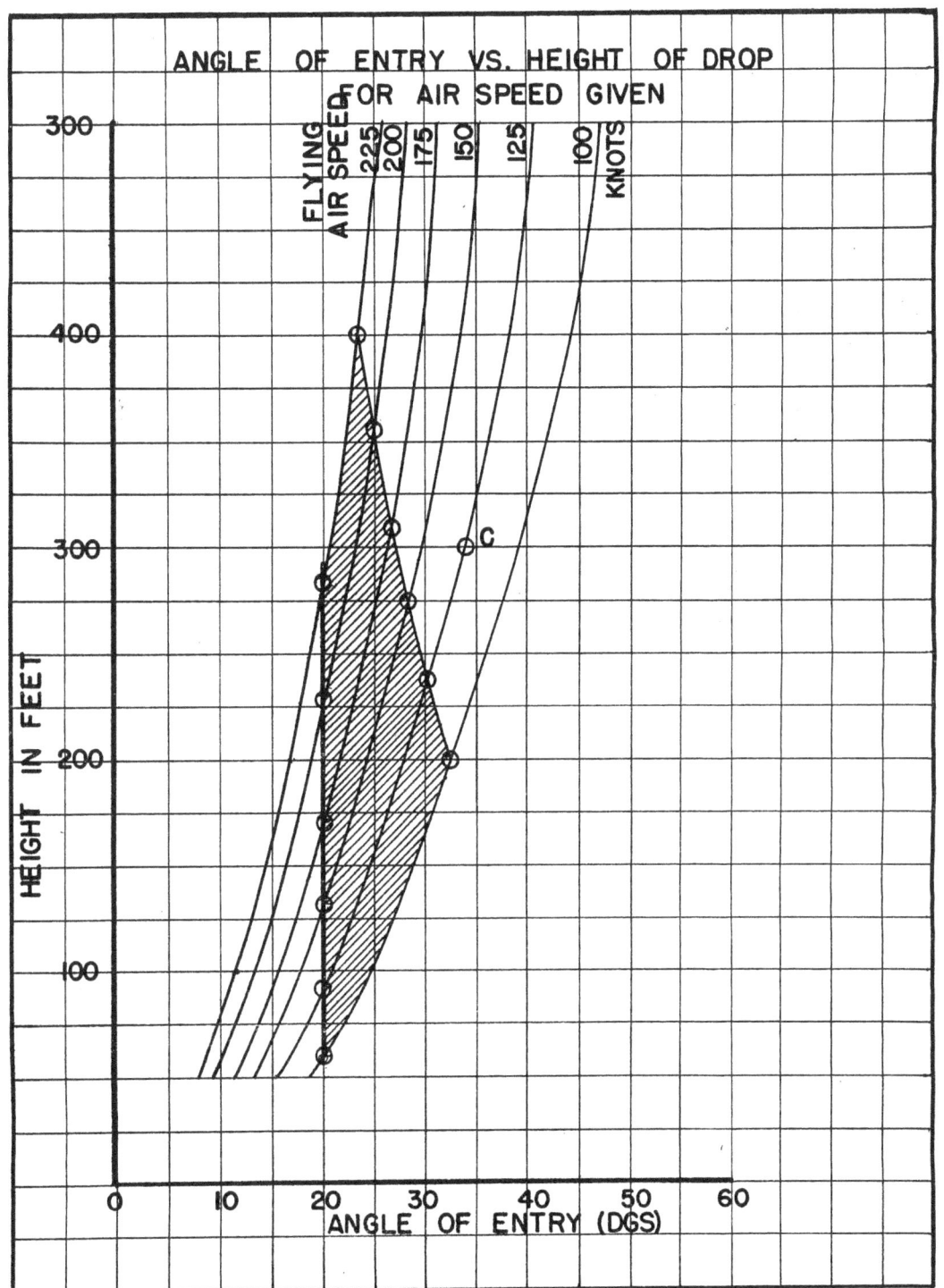

F. **BIPLANE STABILIZERS.**

STABILIZERS FOR THE MARK 13 TORPEDO.

1248. Plates 3 and 4 give the details of construction of the biplane flat board stabilizers for the Mark 13 and Mark 13-1 and 13-2 torpedoes. These stabilizers differ only in the construction of the blocks which hold them on the fins of the torpedo tail. The Mark 13 torpedo has a rail tail and therefore requires slightly different blocks than the 13-1 and 13-2 which have a straight tail.

1249. The construction of these stabilizers is simple and all materials are readily available at most stations. The stabilizers can be made in various combinations of partial assembly so that they can be easily stowed and if drilled in a jig the final assembly using wood screws will take only a short time. Fully assembled they require considerable space for stowing but in the partially assembled stage little room is required. It is recommended that all parts be given two coats of shellac, varnish, or other water proof coating and then painted to conform with the color of the plane. When the stabilizers or any of their component parts are stored, they should be stowed in a dry place and in such a manner that warpage will be prevented. Any warping of the flat boards will tend to cause the torpedo to roll, and although a roll in the air may not cause a poor run, it is not desirable.

1250. The assembled stabilizer can be put on the torpedo by rotating it $45°$ and sliding it forward over the tail fins of the torpedo. The stabilizer is then rotated to the proper attitude and pulled aft until the blocks come up against the tapered portion of the rail tail or the rudder bearing blocks of the straight tail. If the fit is a bit snug, which is desirable, it may require some tapping with a hammer or maul to move the stabilizer to the proper position. The blocks, if they have the proper snug fit, are sufficient to hold the stabilizer on the straight tail. The rail tail requires a piece of safety wire around the vertical part of the stabilizer and through the hole in the horizontal tail fin to prevent the stabilizer from sliding forward.

1251. Although it may appear that the stabilizers can be put on either way, there is a definite difference for the one used with the Mark 13-1 and 13-2 torpedoes. If the stabilizer is assembled according to the drawings and put on the torpedo properly, all blocks on the stabilizer will bear against all four rudder bearing blocks of the torpedo. If the stabilizer is reversed or not properly assembled, all blocks will not be up against the rudder bearing blocks.

1252. Plates 5 and 6 show the details of construction of the biplane wedge board stabilizer for the Mark 13 torpedo used with the TBD-1 airplane. Since the bomb bay of the TBD-1 is

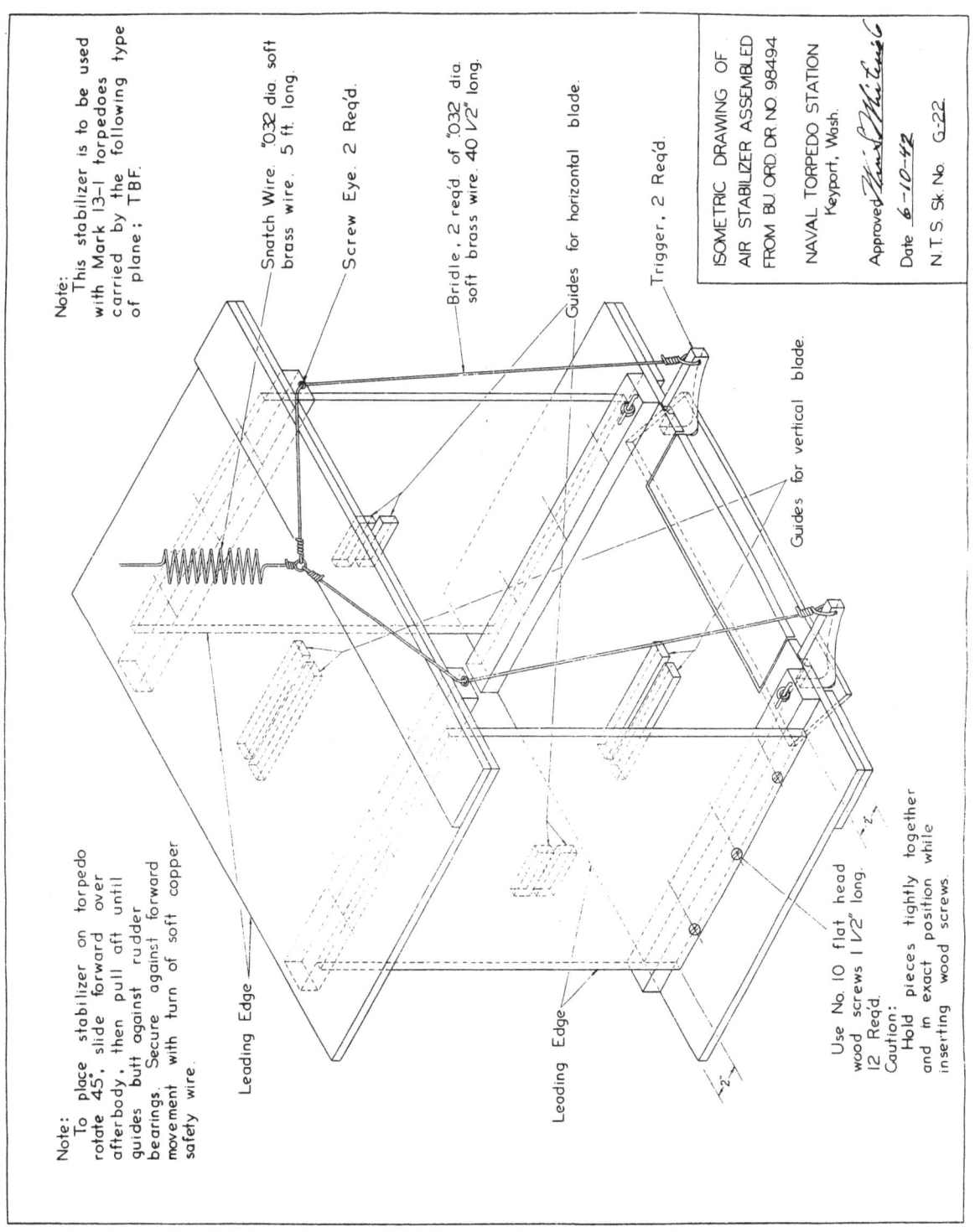

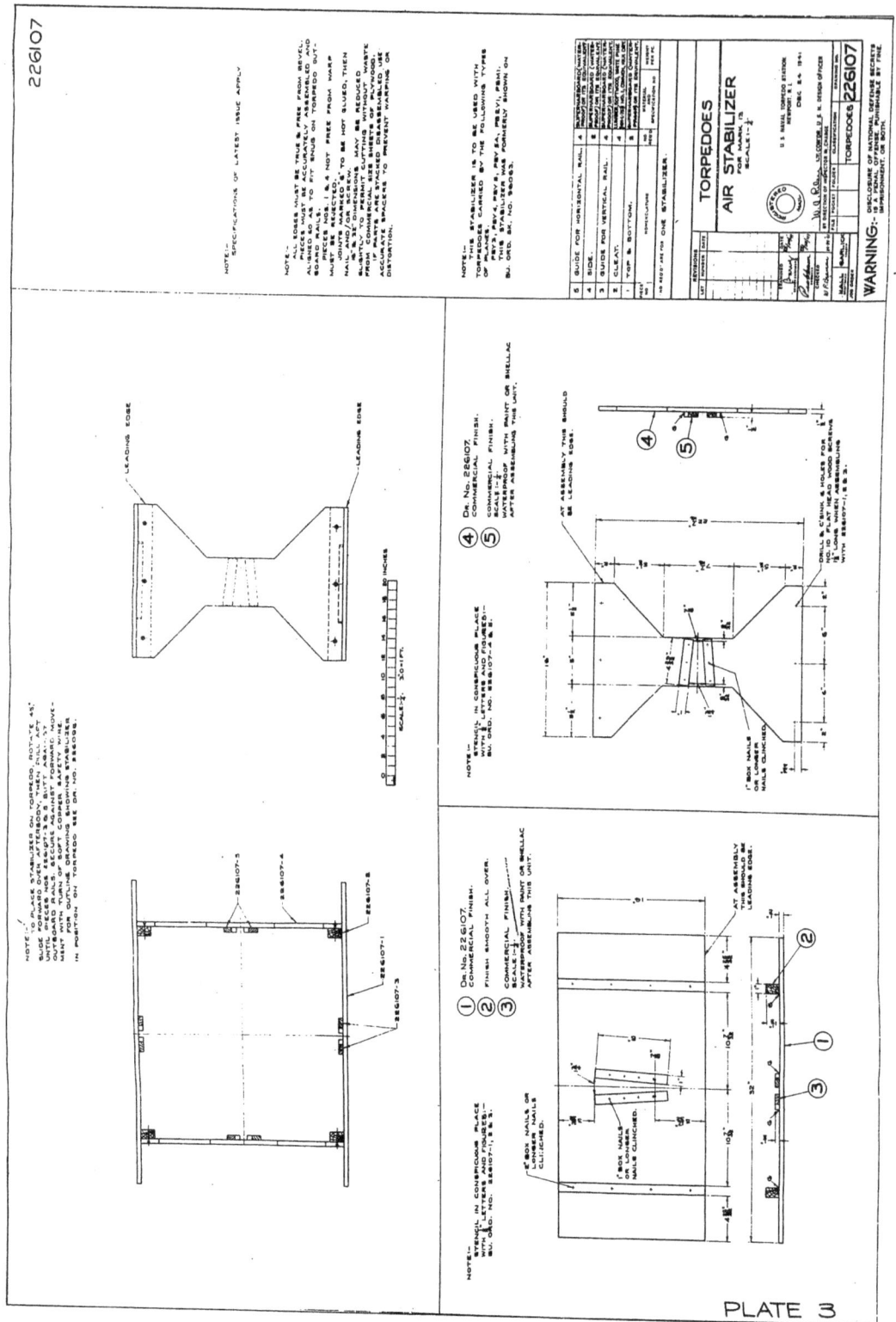

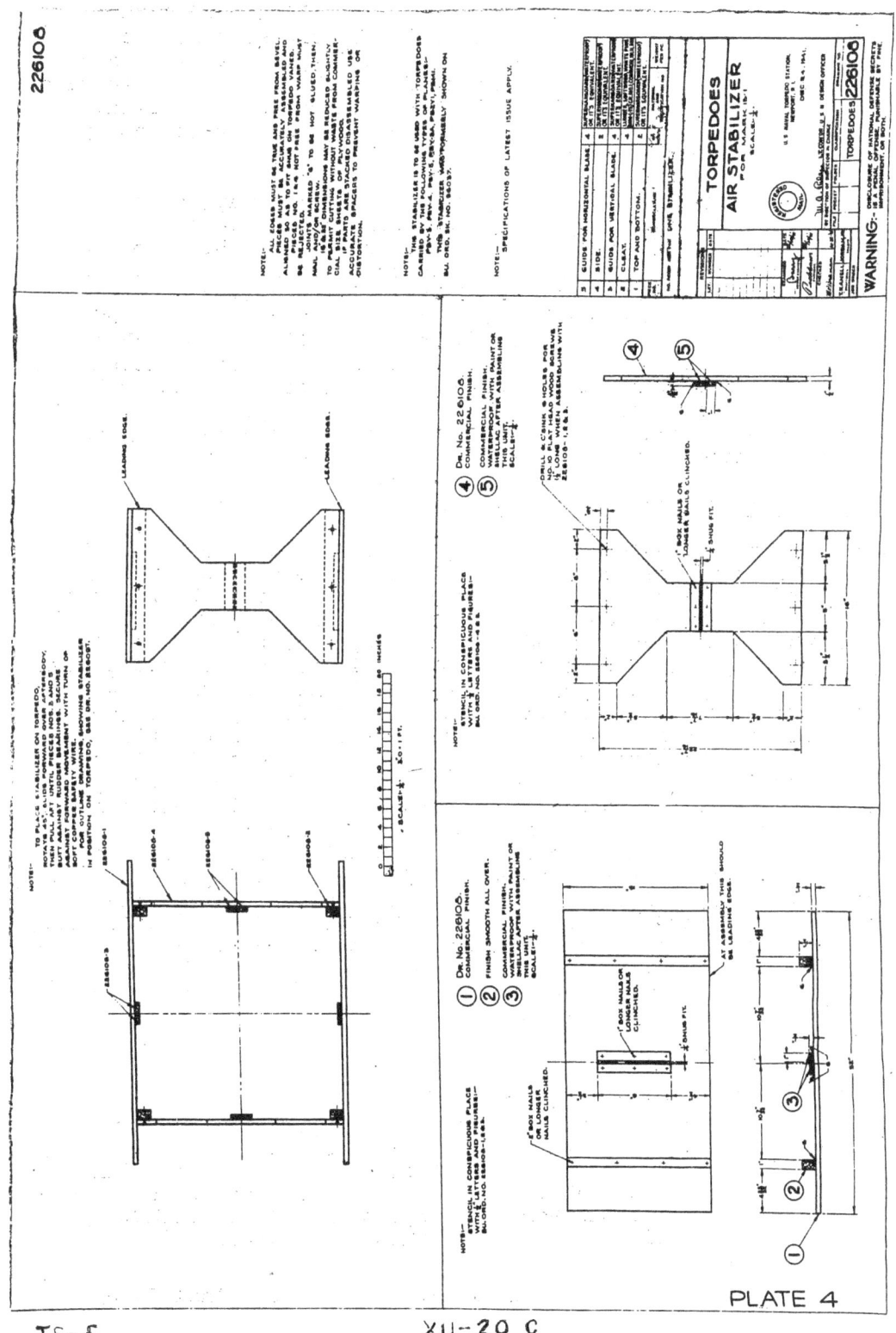

only 24 inches across, the span of the stabilizer had to be reduced to the minimum. These stabilizers are somewhat more difficult to construct than the flat board type, but the materials are readily available and a squadron should be able to make them without any difficulty. The same remarks as to partial assembly, stowage, painting and assembly on the torpedo apply to the wedge board as well as the flat board type.

1253. With the TBD-1 airplane, some difficulty may be experienced in loading when the stabilizers are attached to the Mark 13 torpedo. In order to get the torpedo to the hoisting position beneath the plane it is necessary to lift the tail of the plane as the torpedo is moved aft under the plane. Also, if the torpedo chocks on the plane are not properly adjusted there may be some interference between the after bomb rack and the top of the stabilizer. If the chocks cannot be adjusted so that the stabilizer will clear both the deck and the bomb rack, it may be necessary to slot the top wedge so that it will clear the bomb rack. This slot should not be any larger than absolutely necessary.

STABILIZERS FOR MARK 14 AND 15 TORPEDOES.

1254. Plates 7, 8, and 9 show the details of construction of the stabilizers used on the Mark 14 and 15 torpedoes. The main part of the stabilizer is the same as the flat board type for the Mark 13 torpedo except that it is larger. The two position flap on the trailing edge of the upper board is controlled by a tripping lanyard. The Mark 14 and 15 torpedoes have a high moment of inertia and tend to remain horizontal after release. Therefore, the flap is initially set in the down position so that the torpedo starts to gain angular velocity soon after release. When the torpedo has dropped about 12 feet, the flap is reversed so that the torpedo will remain flat to the trajectory at entry into the water. If the Mark 14 or 15 enters close to or steep to the trajectory, the initial dive will probably exceed 100 feet on all drops while if stabilized so that the entry is slightly shallow to the trajectory, the depth of dive is considerably reduced. This stabilizer gives good results at 110 knots from altitudes of 50 to 75 feet. Other stabilizers for use with these torpedoes are under development.

1255. The same remarks concerning assembly, painting and so forth apply to this stabilizer as well as that for the Mark 13 torpedo.

EFFECTIVENESS OF BIPLANE STABILIZERS.

1256. The wind tunnel reports for the flat board and wedge type stabilizers are contained in the supplement. Practical experience with the two types from several different airplanes indicates that these stabilizers, or some modification of them are satisfactory for the Mark 13 torpedo and are satisfactory for the Mark 14 and 15 torpedoes for limited ranges of altitude

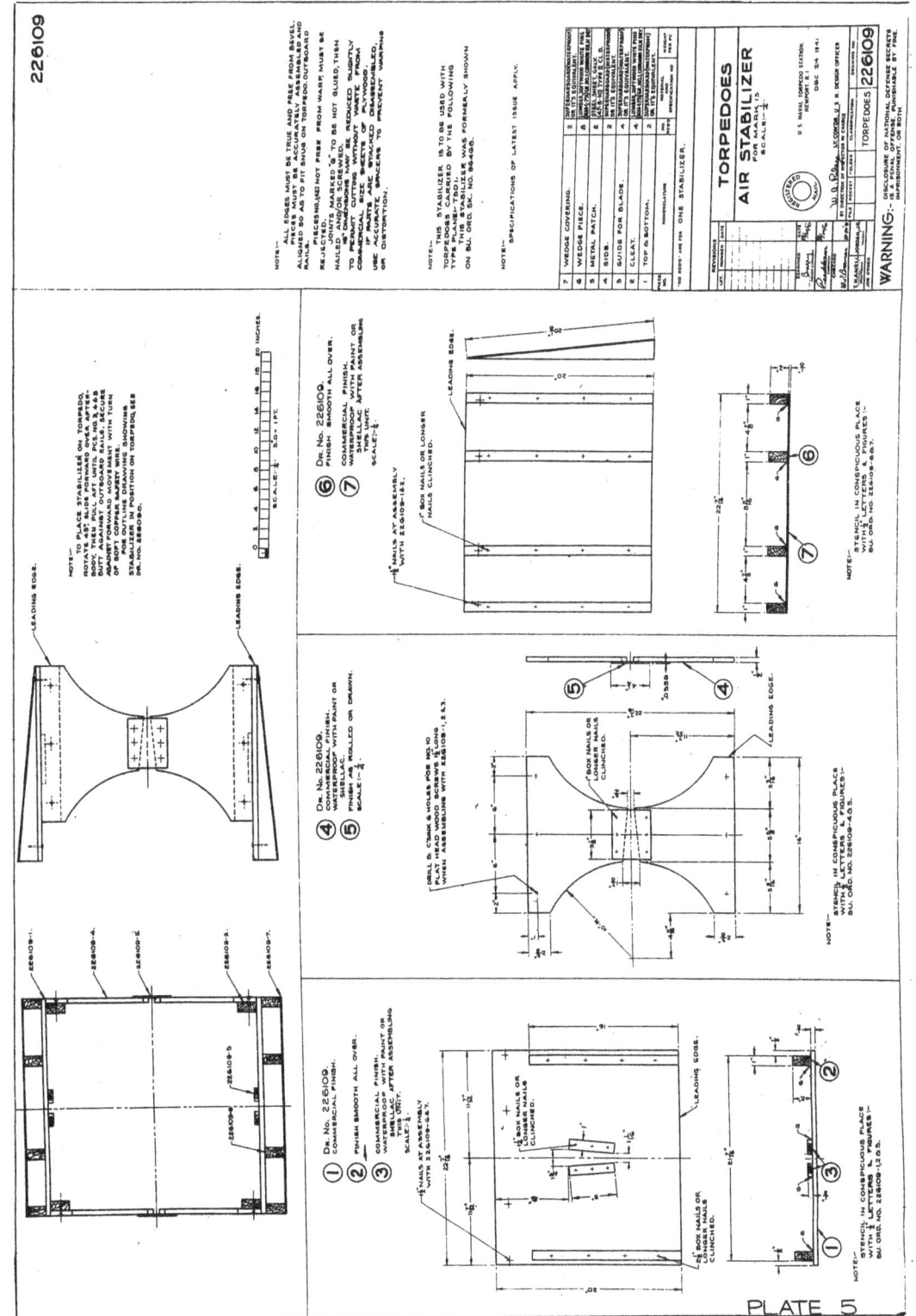

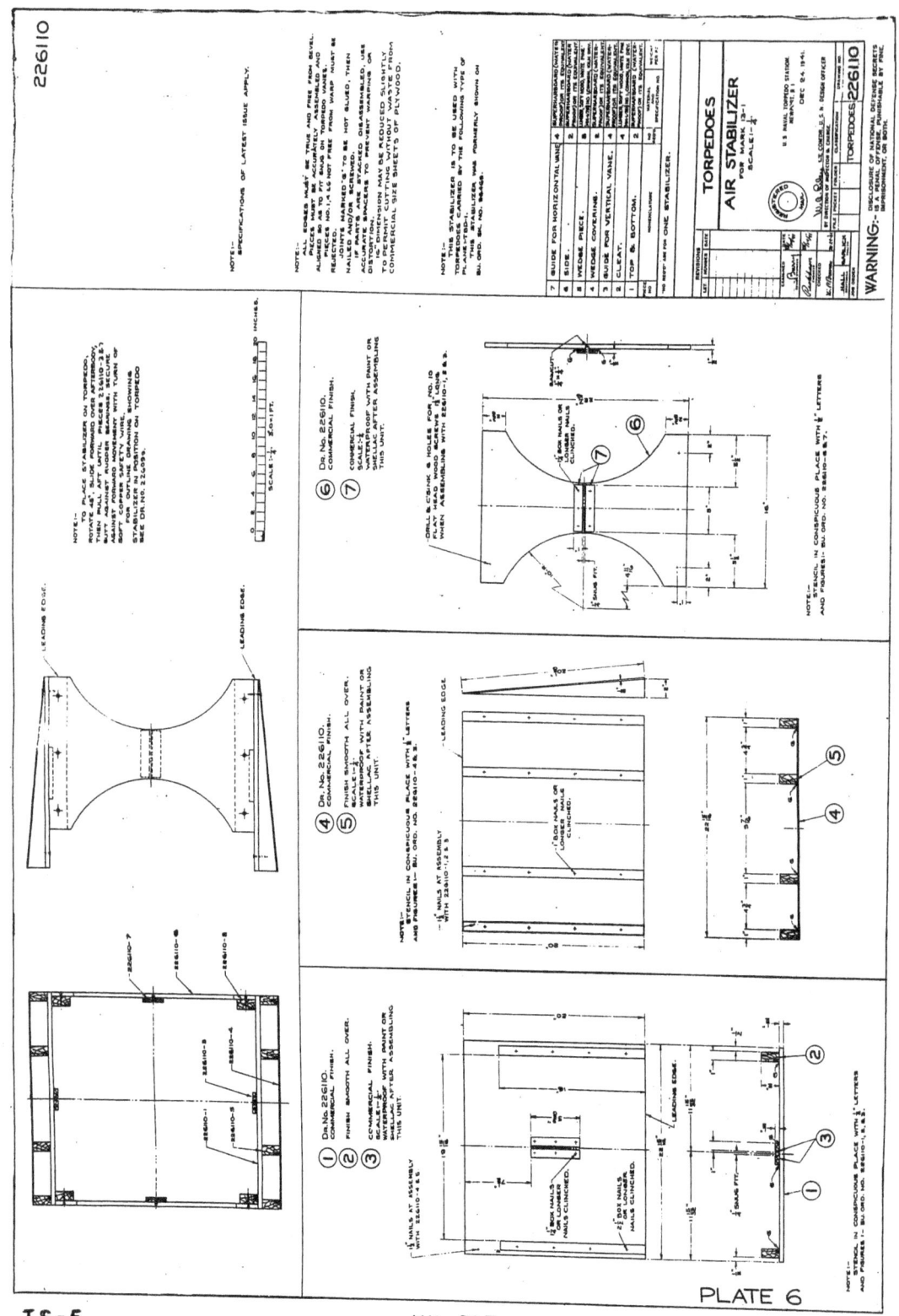

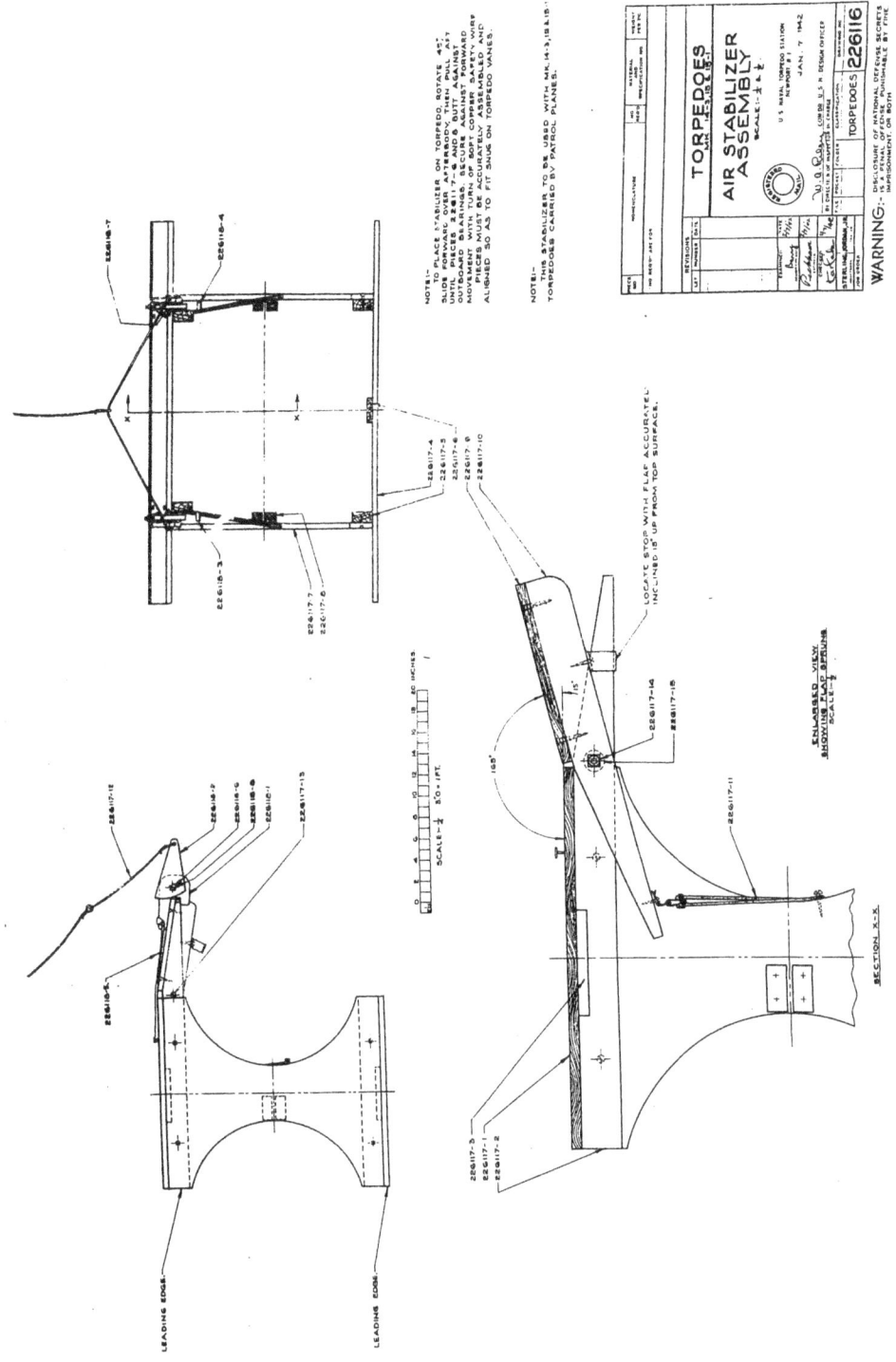

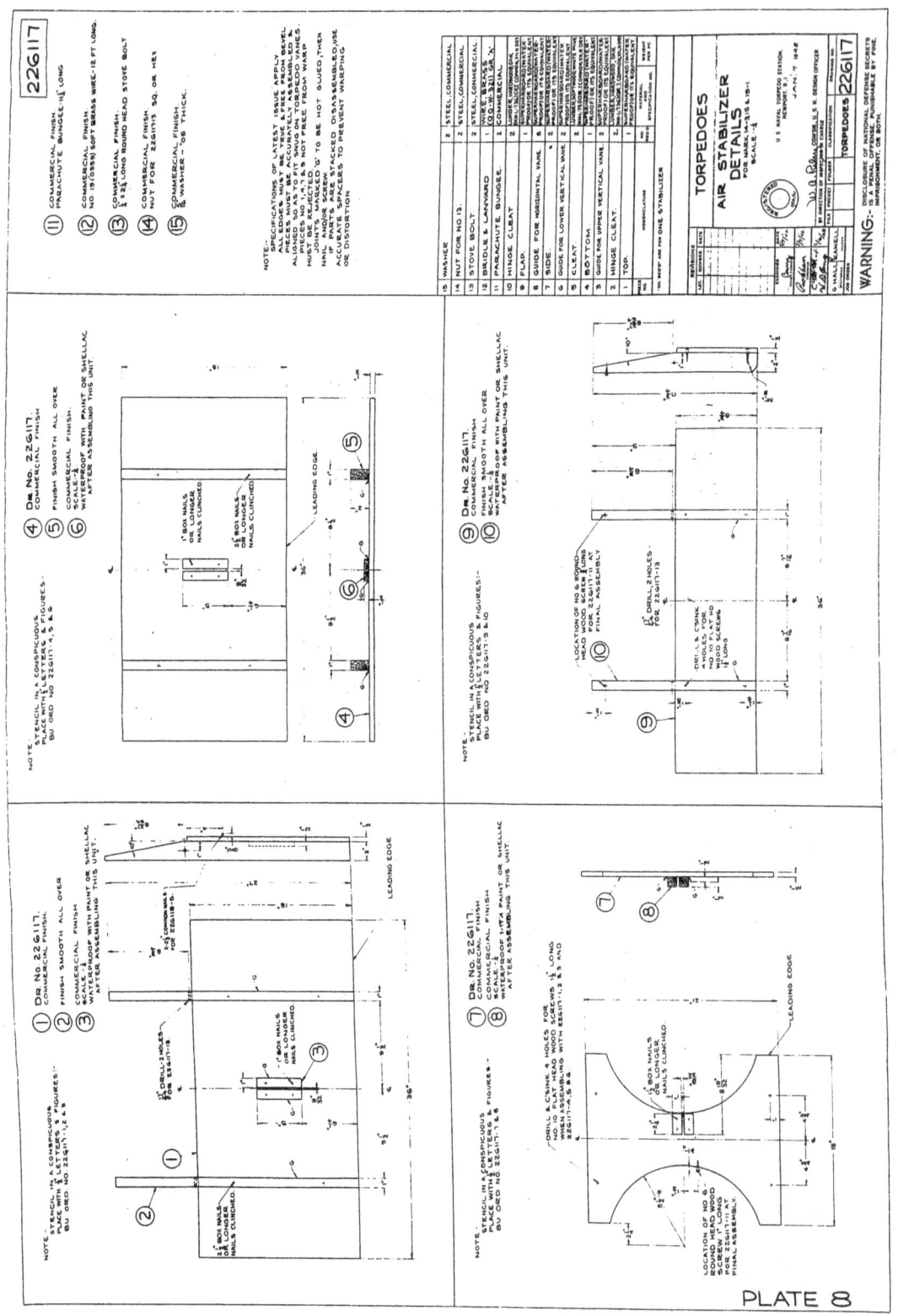

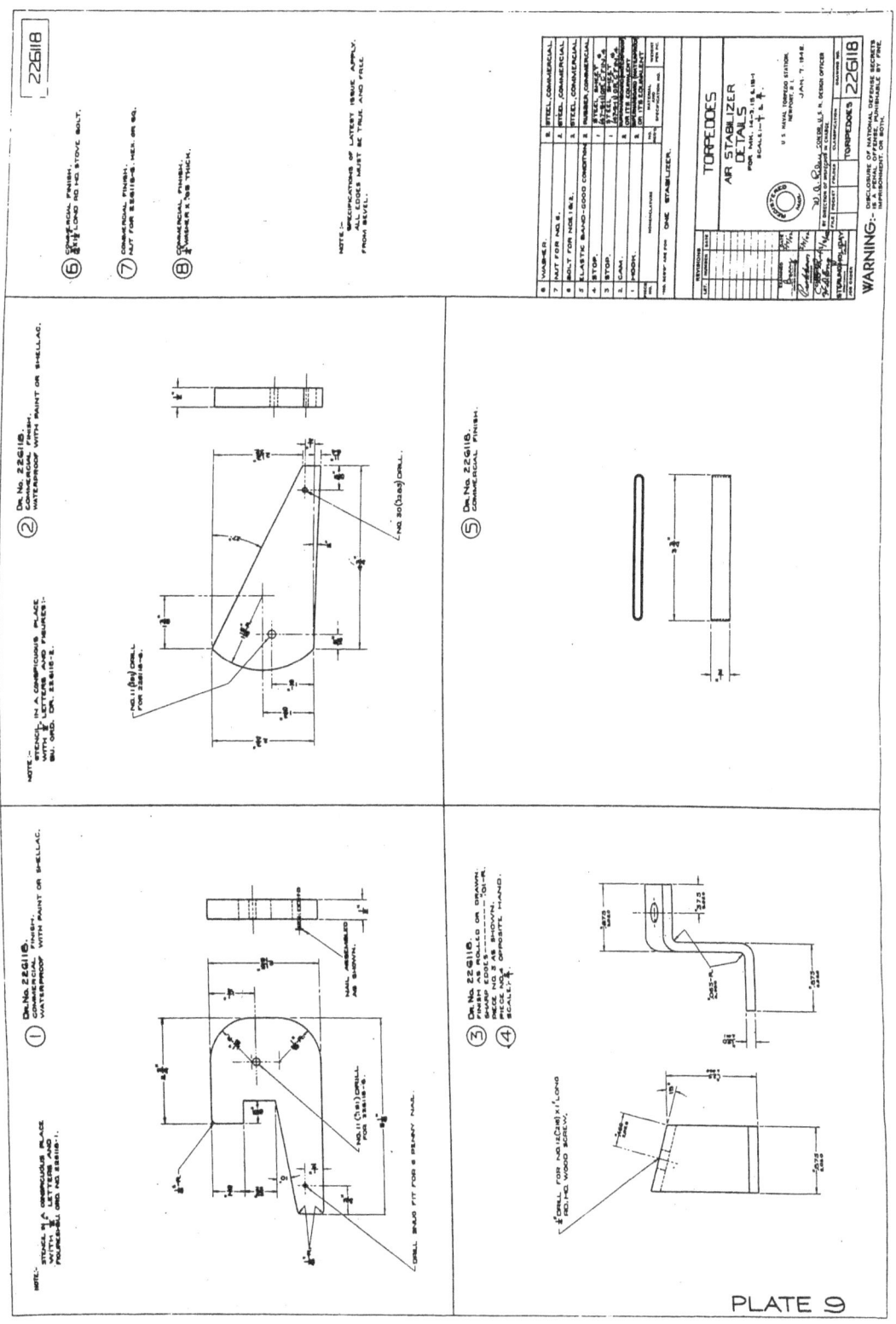

and plane speed. The type of stabilizer to use depends upon the plane, but present experience indicates that these stabilizers or some modification of them can be made to work successfully with the Mark 13 torpedo from any type of plane in which sufficient room is allowed for its installation.

1257. Plate 10 is a sketch made from a photograph showing the trajectory of a Mark 13-1 torpedo dropped from the PBY-3 airplane using the flat board type of stabilizer. This is an example of good air stabilization. The wedge type stabilizer is equally good from the PBY-3.

1258. Plate 11 is a sketch showing the trajectory of a Mark 13-1 torpedo dropped at moderate speed from the TBD-1A plane using the wedge type stabilizer. This clearly shows that the torpedo is nose down to the trajectory for some time after release. However, the torpedo does not oscillate much and the damping is such that the torpedo is very close to the trajectory before entry into the water.

1259. Plate 12 is a sketch showing a drop from the Army A-20C using the wedge type stabilizer. This shows definite air oscillations. Plate 13 shows the results with a modified wedge type and indicates good air stabilization.

1260. Each type of airplane designed to carry torpedoes must be given a complete series of tests to determine what stabilizer should be used to obtain satisfactory air performance of the torpedo. Only two types of planes have been given a thorough test, the PBY-3 and the TBD-1A. The stabilization of the Mark 13 torpedo from the PBY-3 using the flat board stabilizer is very satisfactory and it is believed that equally good results should be obtained with the PBY-4, PBY-5, PBY-5A airplanes. The stabilization of the Mark 13 from the TBD-1A is satisfactory and a limited experience indicates that the results from the TBD-1 are also satisfactory at low speeds (100-110 knots). Six (6) drops from the PBM-1 showed definite air oscillation when using the flat board stabilizer, and a plane of this type has not been available for a sufficient period to determine definitely the proper stabilizer to obtain satisfactory results. Four (4) drops from the Army B-26 airplane indicated good air stabilization using the flat board stabilizer. Tests are now being conducted with the Army A-20C airplane using a modified wedge board with good results to date. Tests with the TBF-1 have been started, but are sufficiently advanced only to say that a special stabilizer will probably be necessary for this airplane.

SUMMARY OF RESULTS.

1261. During the past year some 500 research and proof drops have been made with Mark 13 and Mark 13-1 and 13-2 torpedoes from the PBY-3, TBD-1A, PBM-1, B-26 and A-20C airplanes. Only

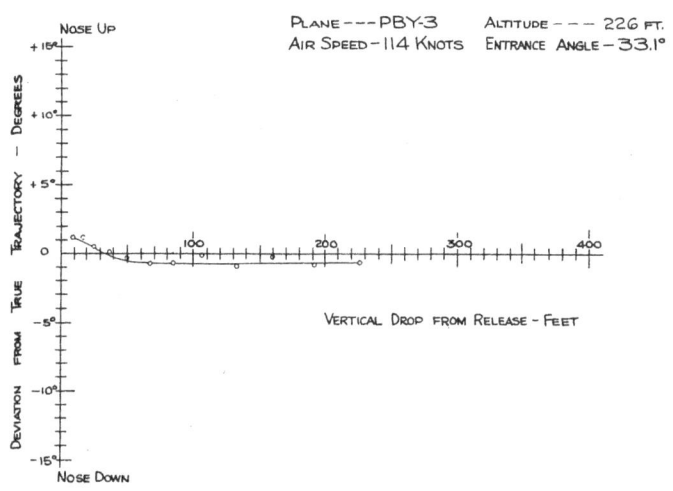

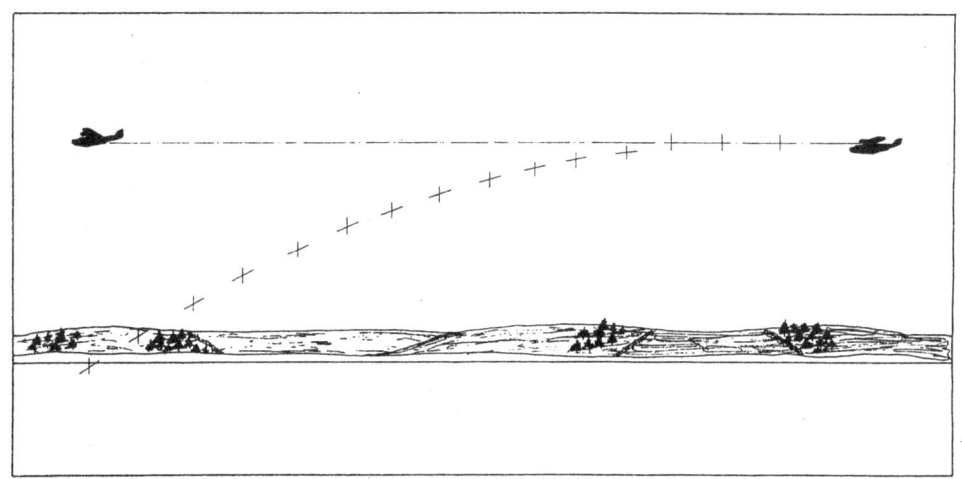

TRAJECTORY OF MARK 13-1 TORPEDO FROM PBY-3 AIRPLANE

PLATE 10

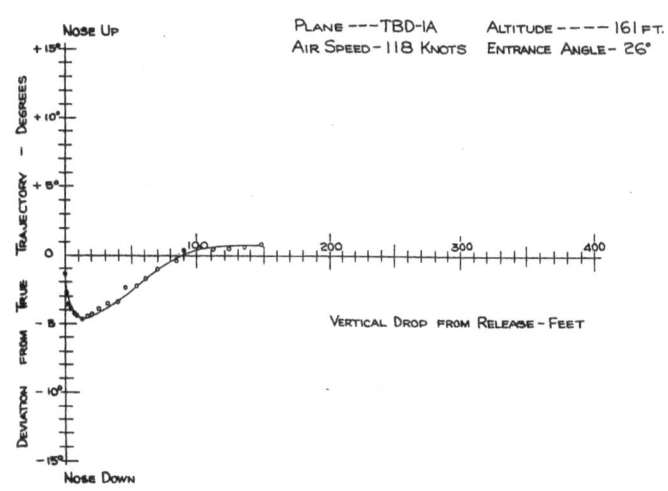

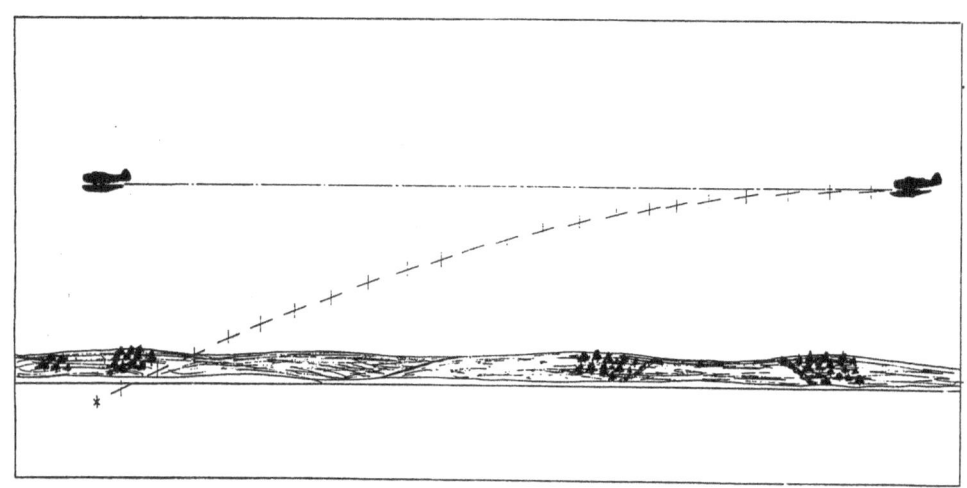

TRAJECTORY OF MARK 13-1 TORPEDO FROM TBD-1A AIRPLANE

PLATE 11

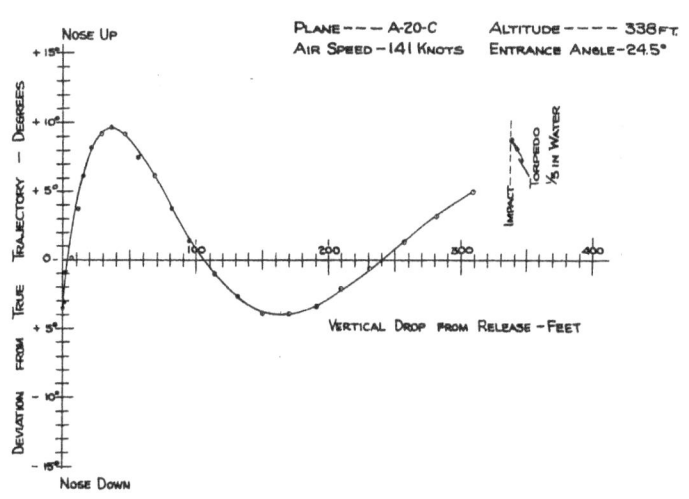
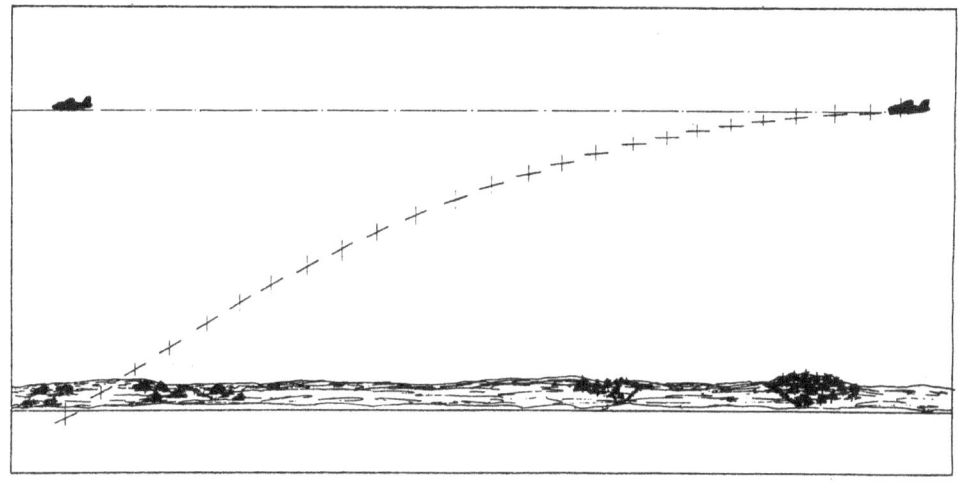

TRAJECTORY OF MARK 13-1 TORPEDO FROM A-20-C AIRPLANE
UNSATISFACTORY STABILIZATION

PLATE 12

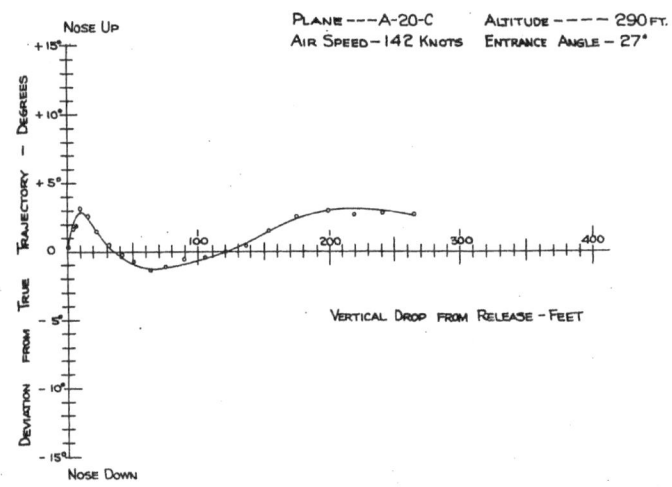

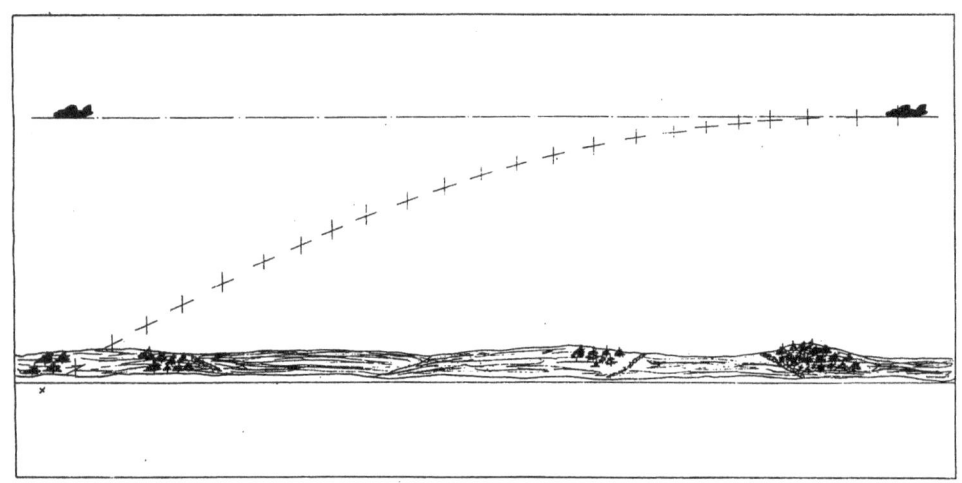

TRAJECTORY OF MARK 13-1 TORPEDO FROM A-20-C AIRPLANE
SATISFACTORY STABILIZATION

PLATE 13

a few drops were made from the PBM-1, B-26 and A-20C planes. Most of the drops were made from the PBY-3 and TBD-1A airplanes. Therefore the results obtained apply essentially to these two planes, but it is believed that similar results will be obtained from any plane for the same ranges of speed and altitude provided the air stabilization is satisfactory.

1262. The aircraft torpedo tests conducted at the Naval Torpedo Station, Newport, are of two classes, research and proof. All aircraft torpedoes manufactured must make a passing run from the torpedo testing barge at Newport. For this test the torpedo is dropped from a special rack on a stationary barge and must make the required minimum speed, maintain the proper depth and must be within certain deflection limits. After making a passing run from the barge, at least one-third of the aircraft torpedoes must be launched from aircraft. Present specifications require that these torpedoes be launched at 125 knots airspeed and from altitudes of 50 to 150 feet, with at least 25% launched from the maximum altitude. These are referred to as proof drops. All other aircraft torpedo drops come under the heading of research drops.

1263. Normally, after an aircraft torpedo makes a passing run from the barge, it is either sent to the Torpedo Station for overhaul prior to being stored, or it is prepared for dropping from a plane. Thus the proof torpedoes dropped do not have the benefit of an overhaul prior to being launched from a plane at the Torpedo Station. On the other hand, all torpedoes delivered to operating units have undergone a complete overhaul after the last time they were fired at the Torpedo Station. The torpedoes received by operating units should therefore be in better condition than when dropped from aircraft for proof at the Torpedo Station, and should show better performance.

1264. During the past few months a program of proof drops were completed and the results of these drops should be a good guide upon which to base expected performance in service, if the torpedoes are launched from the same type of plane and under similar launching conditions. These conditions were as follows:

PLANE	ALTITUDE	AIR SPEED
PBY-3	50-200 feet	125 knots
TBD-1A	50-100 feet	115 knots
TBD-1A	100-150 feet	125 knots

Of 205 torpedoes dropped, 155 or 76% made passing runs, with 50 or 24% failing. Most of the torpedoes that failed were returned for repair or overhaul and on the first repeat drop 28 out of 47 or 60% made passing runs. Most of the failures were due to excessive heel, erratic depth or erratic course. Many of these required only minor adjustment to correct the faults. It is believed that if the torpedoes were given an overhaul after being fired from the barge and before launching from the plane that 85 to 90 percent would make passing runs from the plane. Thus torpedoes launched by operating units should make good runs about 90 percent of the time, since they do receive a complete overhaul prior to issue by the Torpedo Station. A few failures were caused by avoidable errors, such as failure to connect the igniter and failure to connect the starting lanyard. Other unsatisfactory runs caused by igniter failures, starting failures, and failure to recover from the initial dive in 100 feet of water. All these failures amounted to about 5% of the total number of torpedoes dropped.

1265. The table in paragraph 1228 gives the best available information concerning the expected depth of dive for the Mark 13-1 torpedo. From this table the number of torpedoes that may be expected to run when launched in water of a given depth can be estimated. For instance, if it is expected torpedoes will be launched in a harbor having a minimum depth of 40 feet, assuming that 90% of torpedoes launched should make good runs and that 75% should not exceed 40 feet on the initial dive, 90 x .75 or 67% of the torpedoes launched should be expected to make good runs.

1266. About 50 percent of the torpedoes broach or porpoise after recovery from the initial dive. Although this is undesirable, it does not appear to have much effect upon the run of the torpedo. For the research drops the percentage of good runs was somewhat higher for torpedoes that did not broach than for torpedoes that broached. For the proof drops, the percentage of good runs was about the same whether the torpedo did or did not broach. If the torpedo hooks before broaching there is a fair chance that an erratic run will result. Some means of preventing broaching is desirable but a practical solution has not yet been found.

1267. The results with the Mark 14 and Mark 15 torpedoes have not been as satisfactory. Only research work has been done with these torpedoes and no proof of production torpedoes of these marks has been attempted from aircraft. These torpedoes tend to take a deep initial dive and should not be launched in less than 100 feet of water. About 75% good runs can be expected if the Mark 14 or 15 is launched at 110 knots from altitudes of 50 to 75 feet in at least 100 feet of water, providing certain alterations have been made to prepare these torpedoes for aircraft use.

G. COLD WEATHER OPERATIONS.

1268. Cold weather tests with torpedoes indicate that the operation should be satisfactory at air temperatures as low as -10°F., (-23.3°C). At about -15°F (-26.1°C) the gyro is slow in unlocking and the control is somewhat sluggish. Delivery of lubricating oil to certain parts of the afterbody may be slow but normally there should be sufficient oil on all parts to lubricate them until the afterbody becomes heated and the supply of oil is normal.

1269. The water in the exercise head and the water in the water compartment must be treated with an anti-freeze if the torpedo is to be carried into freezing temperatures. At present alcohol is used as the anti-freeze in both the exercise head and the water compartment. The following tables give the amount of alcohol that should be added to the exercise head and water compartment:

WATER COMPARTMENT - PINTS.

Torpedo Mark	13	14	15
Capacity (total)	46.0	80.0	121.0
Plus 20° F.	8.0	13.5	21.0
Plus 10° F.	12.0	20.5	31.5
0° F.	16.0	27.0	41.0
Minus 10° F.	19.5	33.0	51.0

EXERCISE HEADS - GALLONS

Torpedo Mark	13	14	15
Capacity (total)	40.0	43.8	43.6
Plus 20° F.	7.0	7.5	7.5
Plus 10° F.	10.5	11.5	11.5
0° F.	13.5	15.0	15.0
Minus 10° F.	17.0	18.5	18.5

1270. Torpedoes that do not have anti-freeze should not be carried to altitudes where freezing temperatures may be encountered. When the surface temperature approaches freezing, it is essential that anti-freeze be added since freezing temperatures will be encountered shortly after take-off. For instance, if the surface temperature is plus 2° C (35.6°F), at 2000 feet the air temperature will be -2° C (28.4°F) and if the torpedo remains at this altitude very long the exercise head and water compartment will freeze unless alcohol is added before take-off. Torpedoes should never be carried to altitudes where temperatures are lower than the temperature for which the torpedo has been treated with anti-freeze.

ANTI-FREEZE SOLUTIONS FOR BELOW FREEZING TEMPERATURES.

1271. Planes which operate in freezing air, the result of either altitude or cold climate, should protect both the water compartment and the exercise head against freezing by the addition of alcohol to the water.

1272. Possibility of freezing also makes additional precautions necessary to guard against the presence of water in any of the torpedo tubing.

1273. The method described below should be strictly adhered to for mixing anti-freeze solution:

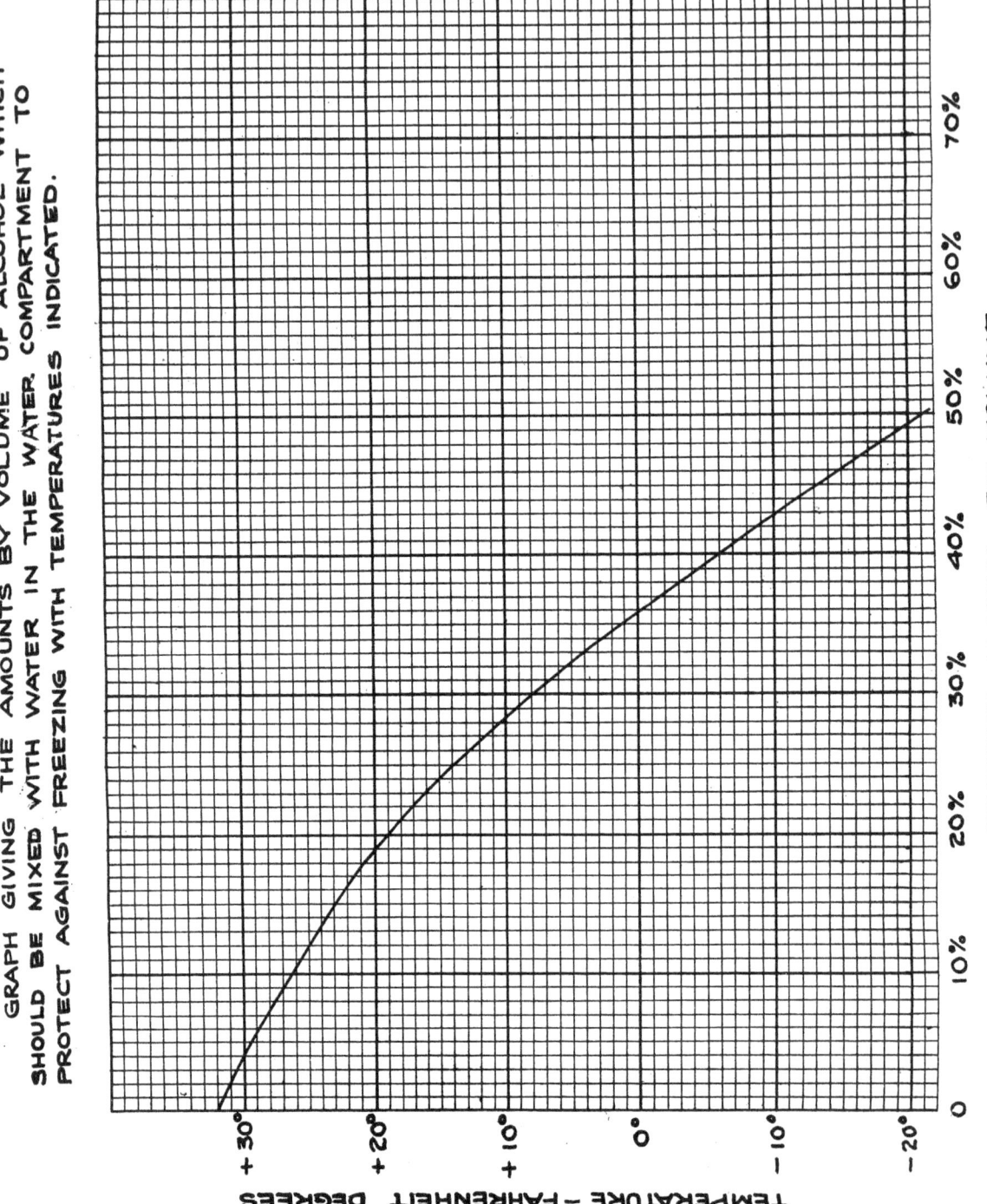

TS-5 XII-25A

(a) Alcohol and water should be thoroughly mixed before pouring into the water compartment.

(b) Whenever practicable the percentage of alcohol in the water compartment should not exceed that necessary to protect against freezing at a temperature of 30° below the existing atmospheric temperature, otherwise power plant damage may result from the resulting excessive combustion flask temperature.

<u>INSTRUCTIONS FOR MIXING POUR POINT DEPRESSANT WITH HOT RUNNING TORPEDO OIL FOR TORPEDOES, (Santopour, Monsanto Chemical Co.)</u>

1274. Six (6) oz. of Pour Point Depressant should be used for each five (5) gallons of hot running torpedo oil. The hot running torpedo oil must be heated to between 140-150° F, the Pour Point Depressant added and the mixture agitated until thorough blending is assured.

H. O.D. 3818, 3820, 3825; BuOrd Circular Letters T-145, 150, 155, 162, 165, 173, 2-42, 8-42.

O.D. No. 3818: DESCRIPTION OF AND INSTRUCTIONS FOR INSTALLING WIRE IMPELLER STOPS ON MARK 13 WARHEADS AND IN MARK 4-1 EXPLODER MECHANISMS.
Plates 1 and 2
Dwg.No. 226242

1. To prevent the arming of Mark 4-1 exploder mechanism during flight in TBF-1, B-26, and A-20C planes until torpedo is released from plane, an arming wire similar to that used in bomb fuses is installed and arranged on the warhead and exploder mechanism so that it will effectively lock the impeller in flight regardless of the speed or maneuvers of the plane and be withdrawn intact from warhead and remain attached to plane upon release.

2. Instructions for fitting and installation:

(a) Drill a 1/8 inch diameter hole through warhead nose piece perpendicular to the warhead and on its vertical centerline. It will be noted that plate 1 shows this hole drilled forward of centerline of the nose piece hole which is incorrect. The 1/8 inch hole should be drilled through the vertical centerline of the nose piece hole.

(b) Tap hole (1/4" pipe tap) adjacent to after center securing screw hole port side for the insertion of a $\frac{1}{4}$ inch pipe plug, Dwg. No. 226242-1.

(c) Drill a 1/8 inch hole in squared end of the $\frac{1}{4}$ inch pipe plug directly over threaded portion of plug, file metal off sides of the plug parallel with drilled hole and round off ends to streamline plug and cut down water resistance.

(d) Install plug in hole tapped in (b) above and screw in flush with the exploder base (in order to do this, it may be necessary to cut off the lower threaded portion of the plug sufficiently for the plug to screw in flush as it is impossible to thread hole to the required depth.)

TS-5   XII-26   Original Page.

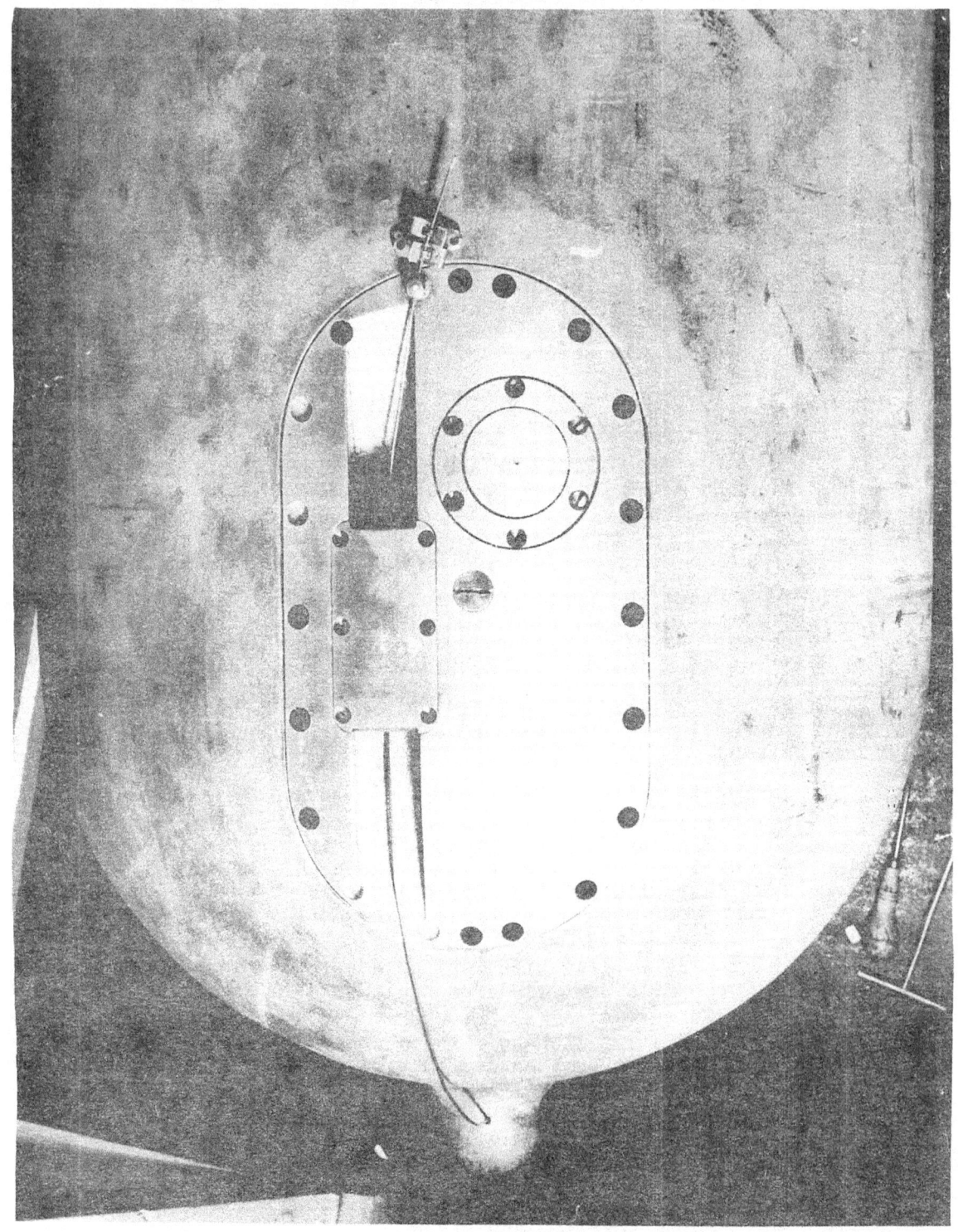

TS-5 XII-26A

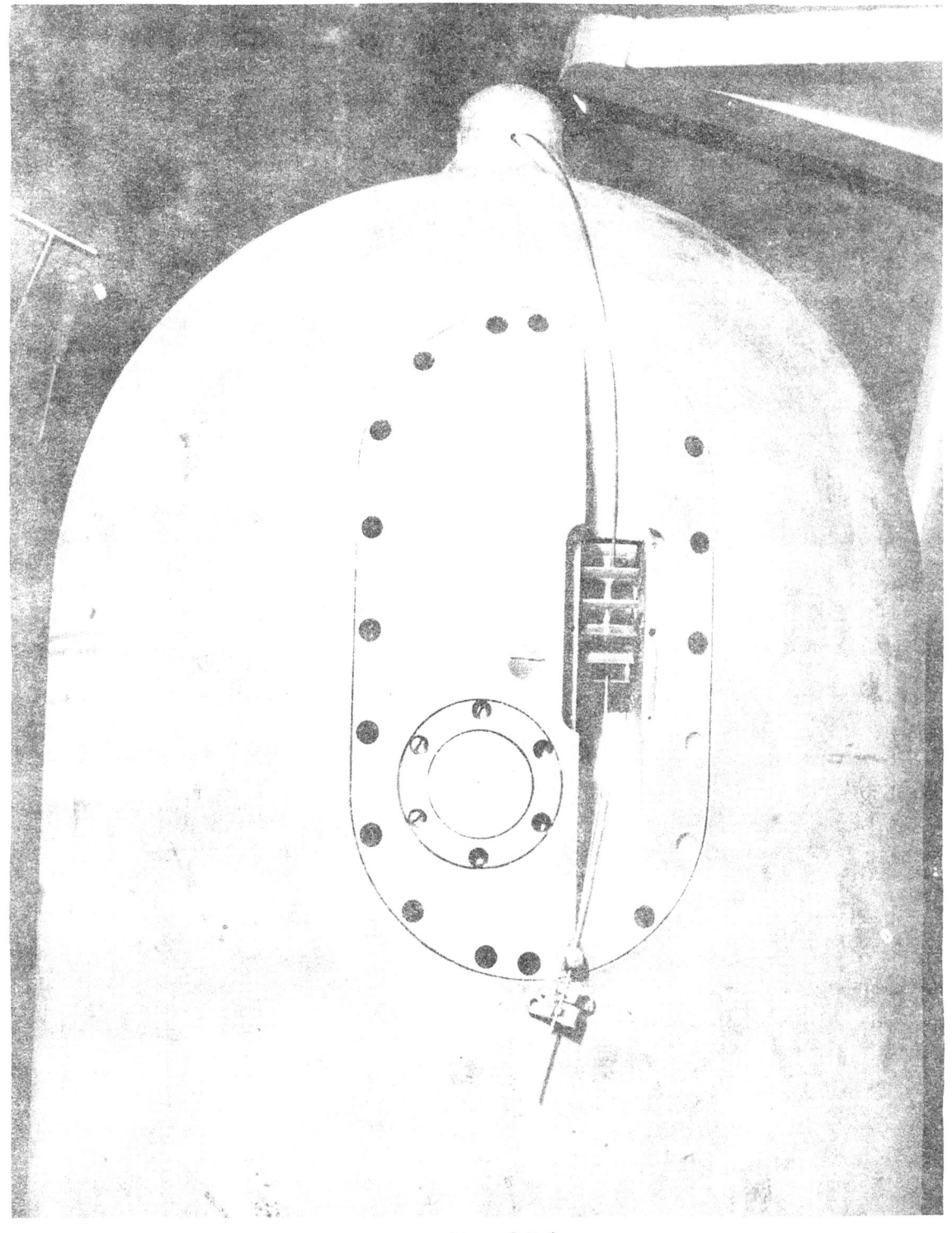

TS-5   XIII-25B

(e) Loop one end of wire, Dwg. No. 226242-3, into an eye for attaching to lanyard securing to fuselage of plane and thread the other end of wire through 1/8 inch hole previously drilled in nose piece. Pass end of wire under the port side (outboard) of four of the impeller blades and thread through the $\frac{1}{4}$ inch pipe plug previously attached to base plate. Take in the slack of the wire on after side of plug and attach two Fahnstock clips, Dwg. No. 226242-2, around wire against the afterside of the pipe plug in such a manner that the wire cannot become slack and be disengaged from the impeller blades. The arming wire should be cut off about 2 inches from the second Fahnstock clip and any burrs made should be filed off.

3. The above will complete installation of wire impeller stop on warhead and exploders. **NOTE:** It will not be necessary to remove impeller guard to effect this installation.

4. A wire lanyard should lead from the arming wire to some part of the plane directly above the nose of the torpedo. It requires a pull of approximately 20 pounds to disengage the wire stop from the exploder mechanisms.

5. THIS STOP DOES NOT PERMIT DROPPING OF TORPEDOES AT PLANE SPEEDS IN EXCESS OF 270 M.P.H. AS GOVERNED BY THE LOAD ON THE IMPELLER SHAFT.

O.D. No. 3820. INSTRUCTIONS FOR MANUFACTURE AND INSTALLATION OF LOCK FOR WRENCH OF IMPELLER GLAND NUT MARK 4-1 AND 6-1 EXPLODER MECHANISMS.

1. To prevent the impeller gland nut in Mark 4-1 and 6-1 exploder mechanism from backing out with possible leakage of water into the exploder mechanism casing, means for locking the impeller gland nut wrench has been approved for manufacture and installation by the service personnel.

2. The lock for Mark 4-1 exploder mechanism is shown in figure 1 and for Mark 6-1 exploder mechanism in figure 2 of the attached plate.

3. These locks are made of sheet brass in the form of a "U" to the dimensions shown on plate and consists of a series of locking tongues disposed about the radius of the lock in such a manner that three tongues 90° apart may be bent over and engaged in the cut-a-way recesses of the outer diameters of the wrench, the lower end of the lock resting on the flat surface of the exploder base directly under the impeller shaft stuffing box.

4. To simplify manufacture of these locks a templet of thin sheet brass may be made to the exact dimensions as shown on plate and used as a guide in cutting out the locks. Rough cutting may be done with tin snips and finished to size by filing.

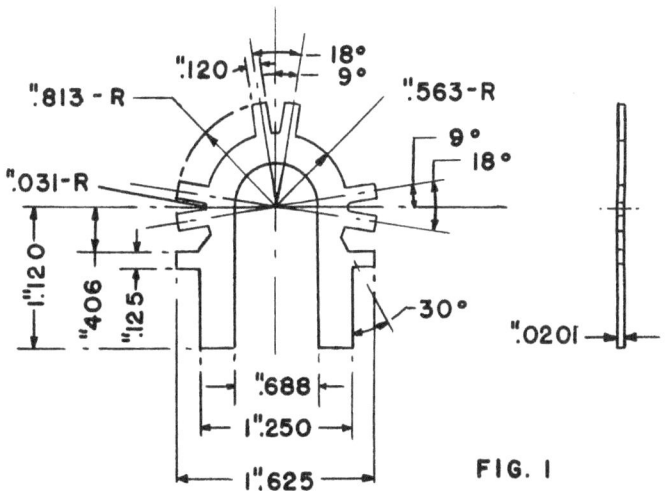

MARK 4-1 EXPLODER

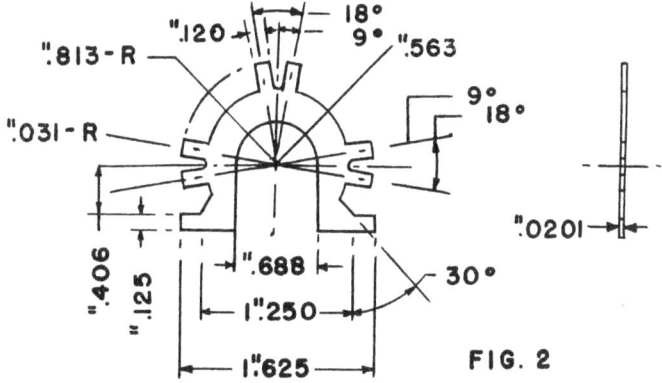

MARK 6-1 EXPLODER

O.D. No. 3825: CHANGE IN DESIGN OF MARK 13 TYPE TORPEDO STARTING GEAR CONNECTION LEVER, DWG. 173669-4.

1. An analysis of starting gear failures of Mark 13 type torpedo indicates that it is highly desirable to change the design of connection lever catch for the following reasons:

    (a) In addition to the forces tending to close the connection lever catch, namely, the high pressure air acting against the starting piston head, and the starting spindle spring; the surface of the connection lever catch when engaging the last step in the unlocking lever inclines 12° in a direction tending to force its disengagement with this lever when fully open. The only force tending to oppose this disengagement being the unlocking lever spring.

    (b) It is therefore probable that a force such as when a torpedo strikes the water when firing from aircraft, or even vibrations in torpedo when running, may, due to this incline of the surface of the connection lever catch, cause its disengagement from the last step of the unlocking lever and thus throttle or close starting piston with consequent loss of torpedo (see figure 1 attached sketch).

2. In order to remedy this, the upper or engaging face of the connection lever catch is being machined to parallel the last step in the unlocking lever when fully open, thus eliminating the condition outlined in paragraph 1(b) above. (See figure 2 attached sketch).

3. This is accomplished as follows:

    (a) Remove and disassemble starting mechanism sufficient for the removal of connection lever with catch (see O.P. 629 or 629A for procedure).

    (b) Lay out, scribe, and grind upper face of connection lever catch to the contour and dimensions shown in figure 3 of Plate No. 1 attached.

    (c) Reassemble connection lever in starting gear.

    (d) With starting gear tripped and connection lever catch entered in the last step of the unlocking lever, note that the engaging surfaces of the catch and unlocking lever are approximately parallel. Turn index to seat starting piston.

    (e) Connect starting gear to high pressure air line (2800 p.s.i.) to try operation of tripping with toggle latch.

    (f) A quick withdrawal of the starting toggle in any direction between 45° forward or abaft the vertical should consistently result in fully engaging the unlocking-lever catch on the last (bottom) step of the unlocking lever and should lift the starting piston at least 0."060 off its seat. The necessary pull on the lanyard vertical to axis of the torpedo should not require a force greater than 30 pounds to extract the toggle. Forward pulls may require more force to clear

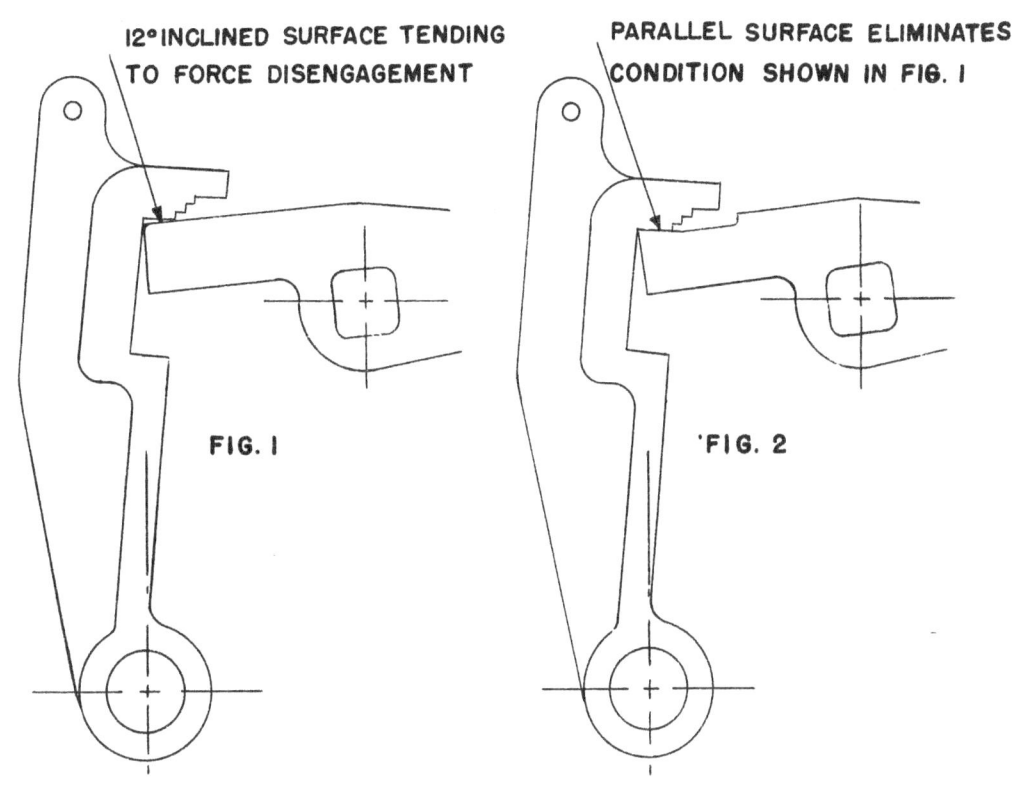

FIG. 1 — 12° INCLINED SURFACE TENDING TO FORCE DISENGAGEMENT

FIG. 2 — PARALLEL SURFACE ELIMINATES CONDITION SHOWN IN FIG. 1

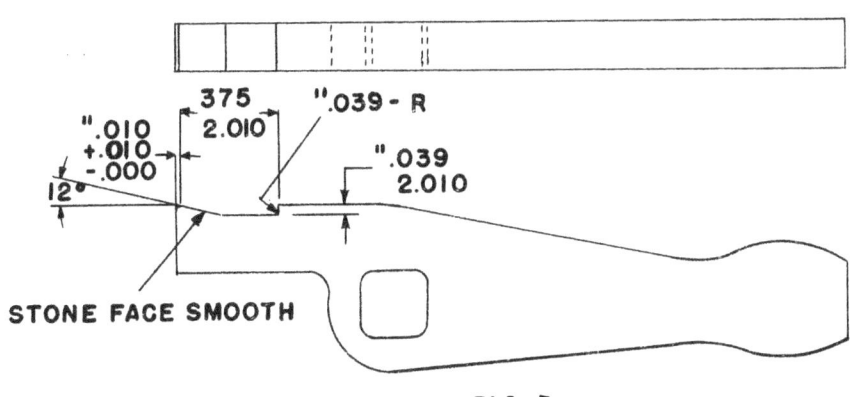

FIG. 3

PLATE I

the toggle and to fully open valve with connection lever catch engaging in the last (bottom) step of the unlocking lever.

(g) To measure lift scribe a light pencil mark on the starting piston stem against the valve body when the valve is in open position, turn index to close valve and with air still on, measure distance from pencil mark on stem to valve body, this should not be less than 0."060 in a properly adjusted mechanism.

(h) Reinstall starting gear in afterbody.

INSTRUCTIONS FOR FITTING NEW TOGGLE (Dwg.No. 226237-1) IN STARTING GEAR OF MARK 13 TYPE TORPEDOES.

1. With the adoption of the new toggle for Mark 13 type torpedo, certain changes were necessary to eliminate interference, these changes are being made in torpedoes now in production.

2. However, as it appears possible that the new type toggle may be used with starting gears in torpedoes now in service in which interferences will be encountered tending to prevent full throw of the starting piston and free release of the toggle from the spindle arm, instructions for procedure to remove these interferences are given below:

(a) Check starting gear to note existing interferences (see fig. 1 attached sketch).

(b) If interference exists between the forward end of the toggle and the forward upper end of the toggle slot (see (A) Fig.1), Plate 2, remove stock to dimensions shown in (A-1) fig. 2 by filing.

(c) If interference exists between the upper end of the starting spindle and the end of its slotted hole in the spindle arm (see (B) fig.1) increase the length of this slotted hole by filing off an equal amount on each end to the dimension shown at (B-1) fig. 2.

(d) If interference exists between the forward end of the spindle arm and the starting gear flange (see (C) fig. 1), remove stock from the forward end of the spindle arm by filing to the radial dimension shown at (C-1) fig. 2.

(e) If interference exists between the starting spindle head and the lower end of the starting spindle bushing (see (D) fig.1) remove stock from the lower end of this bushing to the dimension shown at (D-1) fig. 2. (Use counterbore if available). If using file, great care must be exercised to keep the surface of the end of the sleeve level.

3. After obtaining the above clearances, test starting gear in accordance with paragraph 3(e) and (f) on sheet 3 and note that gear will function in accordance therewith.

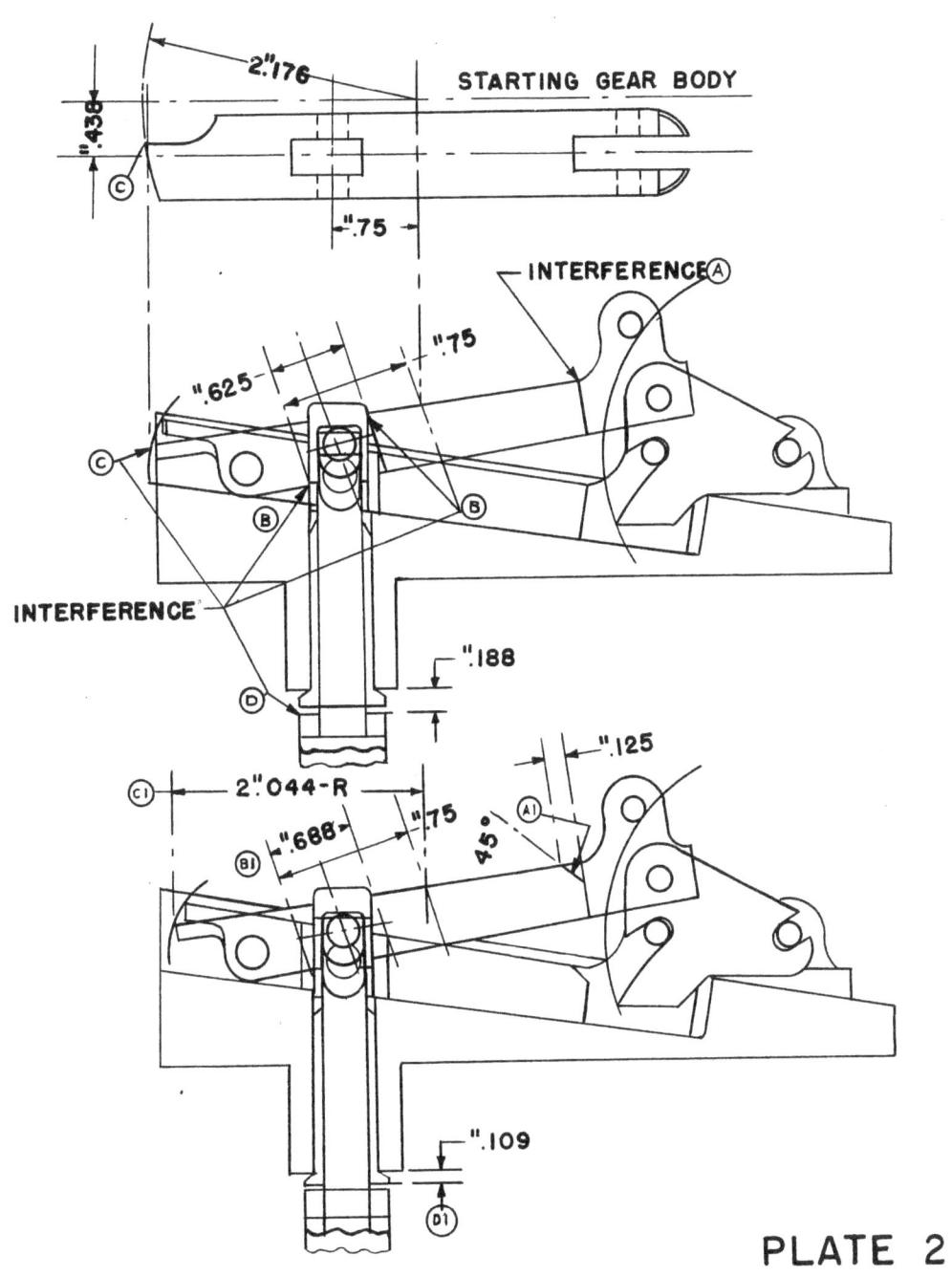

PLATE 2

TS-5 XII-29A

BUREAU OF ORDNANCE CIRCULAR LETTER NO. T-145.

From: The Chief of the Bureau of Ordnance.
To: All vessels carrying torpedoes, destroyer and submarine bases and tenders, torpedo stations, Battle Force Torpedo School, San Diego, Calif., Navy Yards; Torpedo School, Naval Torpedo Station, Newport, R.I.; Submarine School, Submarine Base, New London, Conn.; Naval Mine Depot, Yorktown, Va.; and the Naval Gun Factory.

Subject: Mark 13 and Mods. Torpedoes - Depth Setting Spindle Gear and Change in test weight.

1. The following described condition has been found to exist in the depth setting spindle gear of torpedoes Mark 13 and Mods:

   (a) When the depth is set below "O": Gear, Dr. No. 173749-1 rises, pressing against washer, Dr. No. 99708-12; idler, Dr. No. 99708-11, tends to rise. Washer, Dr. No. 173749-2 should prevent idler from rising, but is so thin that washer, Dr. No. 173749-2 is bent and subsequently damaged by the teeth of spindle gear, Dr. No. 99708-10.

   (b) This results in general jamming and mal-positioning of gear Dr. No. 173749-1 which changes spring tension.

2. This condition exists only in torpedoes Mark 13 and Mods.

3. Pending modification of the design of the spindle gear, Dr. No. 99708-10, S.P. 7113, to eliminate cases of such damage, care must be exercised to prevent setting the depth below zero.

4. The use of a sixteen pound test weight for testing the depth mechanism of the Mark 13 Mod. 1 torpedo in place of a ten pound weight has been found to give better depth performance.

5. The Bureau has therefore requested the Naval Torpedo Station, Newport, to provide the necessary 16 pound test weight as a replacement for the 10 pound test weight and to correct instruction pamphlets accordingly.

BUREAU OF ORDNANCE CIRCULAR LETTER NO. T-150.

From: The Chief of the Bureau of Ordnance.
To: All vessels carrying torpedoes, destroyer and submarine bases and tenders, torpedo stations; Battle Force Torpedo School, San Diego, Calif; Navy Yards; Torpedo School, Naval Torpedo Station, Newport, R.I.;

Submarine School, Submarine Base, New London, Conn.; Naval Mine Depot, Yorktown, Va.; Naval Mine Depot, New London, Conn.; and the Naval Gun Factory.

Subject: Immersion Gear, Torpedoes Mark 7-4B, 13 and 13-1 - Defect in.

1. The following condition which exists in the immersion gear of some Mark 7-4B, 13 and 13-1 torpedoes has recently come to the attention of this Bureau:

In certain torpedoes, the dimension in the spring chamber of the immersion gear casing between the upper extremity of guide slots and the top of the chamber exceeds the height of the spring guide. This condition results either from an adverse combination of tolerances or from actual deviations from drawing dimensions. When a depth at or near the maximum is set on such a torpedo, the spring guide rises above its guide slots and rotates out of alignment with them. Subsequent reduction of the depth setting is then impossible, and an attempt to make a reduced depth setting may result in damage to immersion gear parts. In Mark 7-4B torpedoes, the drawing numbers of the immersion gear casing and of the spring guide are 225514-1 and 173749-3 respectively. In Mark 13 and 13-1 torpedoes these numbers are 173750-1 and 173749-3 respectively.

2. The existence of these unfavorable conditions can be determined only by examination of individual immersion gears. If the spring guide can rise out of engagement with its guide slots at any depth setting, special precautions or corrective action are necessary to prevent the misalignment mentioned in paragraph (1). Pending modifications of parts in torpedoes having such conditions, no depth setting sufficiently great to produce disengagement should be made on those torpedoes.

3. The Naval Torpedo Station, Newport, Rhode Island, will issue a conversion list for the modification of this mechanism by the forces afloat.

BUREAU OF ORDNANCE CIRCULAR LETTER NO. T-155.

From: The Chief of the Bureau of Ordnance.
To: All vessels carrying Mark 13 and Mark 13-1 torpedoes, destroyer and submarine bases and tenders, torpedo stations; Battle Force Torpedo School, San Diego, Calif.; navy yards; naval ammunition depots; Torpedo School, Naval Torpedo Station, Newport, R.I.; Submarine School, Submarine Base, New London, Conn.; Naval Mine Depot, Yorktown, Virginia, and the Naval Gun Factory.
Subject: Mark 13 and Mark 13-1 torpedoes, Instructions for Rendering Governor Cut-off Inoperative; Precautions in Handling with Governor Cut-off Inoperative; Additional Safety Precautions for.

Reference: (a) BuOrd conf. dispatch 152230 of July 15, 1941.

1. Broaching of Mark 13 and Mark 13-1 torpedoes dropped from aircraft has resulted in the cutting of the governor link with loss of the torpedo due to negative buoyancy. Reference (a) directed that the governor be rendered inoperative and removed prior to any further launchings from aircraft. The specific steps necessary to accomplish this are listed herewith:

(a) Remove:

Name	Drawing No.
Tripping lever spring	44318-8
Stud for tripping lever spring	44318-6
Pin for connecting arm.	173670-4
Spacer for connecting arm pin	173670-5

(b) Remove:

Clamp screw	44318-24

(c) Remove as a unit:

Adjusting block	173670-6
Pin for governor link	44318-27
Split pins (2)	44318-34
Governor link	44318-28
Pin for connecting link	173671-1
Connecting link.	173671-2

(d) Remove:

Journal cap screws (4)	95466-5
Journal caps (2)	95466-2

(e) Remove as a unit:

Connecting rod.	173670-1
Shaft coupling	173670-3
Pin for shaft coupling.	46196-23
Split pin for shaft coupling pin	46196-26
Governor shaft	173669-3
Split pin for governor shaft.	63781-1
Governor spring case.	46195-1
Governor casing	46195-2
Governor casing head.	79820-2
Key for governor shaft.	46195-7
Pin for governor shaft.	46195-8
Pin for governor spring casing.	46195-4
Governor spring.	46195-11
Washer for governor spring.	46195-31
Governor spring thimble.	79820-5
Governor weight arms.	79320-3
Link cutters.	79820-6
Pins for cutters.	46195-10

(f) Remove:

Name	Drawing No.
Gear for driving distance mechanism	79103-2
Screw for gear	78050-2

(g) Leave:

Stop pin for tripping lever.	173670-2
Tripping lever.	173672-3

2. Shipment of the parts removed from subject torpedoes will be made to either the Naval Torpedo Station, Newport, or Keyport, whichever is the nearer.

3. Until the Naval Torpedo Station, Newport, prepares a change to O.P. 629 the procedure on pages 81 and 84 of that pamphlet should be corrected to read as follows:

6. Put on wire propeller lock and standard propeller lock.

60. Open stop valve wide. (NOTE: This step applies only to planes where installation is such that stop valve is not accessible with torpedo in position on plane). For all other types of plane see step 66.

61. Hoist torpedo to rack.

62. Insure torpedo is properly positioned in the rack and the rack is properly cocked.

63. Lock the rack or provide safety strap so that torpedo cannot be accidentally released. (On PBY's a safety strap is used. On TBD's a safety pin is used to lock the rack).

64. Install tripping lanyard and starting toggle. (NOTE: Starting spindle arm must be left in down position (flush with starting gear body) and starting toggle inserted by removing pin (206254-5) from bracket 206254-2. When replaced, pin should be secured by cotter pin.

65. Install igniter.

66. Open stop valve wide (if not done, as step 60).

67. Connect igniter.

68. Remove standard propeller lock.

69. At the last practical moment remove safety strap or safety pin.

4. The starting gear should be tested with full flask pressure instead of a reduced charge, valves having been found to function satisfactorily under the reduced charge but not under the full. If in tripping the starting lever the cam toggle operates satisfactorily use the same toggle when launching from the aircraft.

5. Exercise heads should be given a two pound internal hydrostatic test, the pressure being applied through the drain valve.

TS-5   XII-33   Original Page.

6. Until the stronger propellers and upper guide vane braces are received from the Naval Torpedo Station, Newport, the limitation on the altitude from which drops may be made is seventy-five (75) feet.

BUREAU OF ORDNANCE CIRCULAR LETTER NO. T-162.

From: The Chief of the Bureau of Ordnance.
To: All destroyer, submarine and PT boat tenders and bases; all naval air stations with torpedo workshop equipment; all cruisers, aircraft carriers and tenders with torpedo workshop equipment; all torpedo stations and torpedo overhaul activities.

Subject: Exhaust Valve Springs for Mark 11, 12, 13, 14, 15 and Mods. Torpedoes - Drawing 63878-4 SP8741.

Reference: (a) NTS, Newport, ltr. S75-1(12912)(F-RS) of Nov. 6, 1941.

1. A review of the history of the design of subject springs shows that they were originally designed in 1920 of drawn spring steel; in 1927 the material was changed to stainless steel wire; in 1931 the material was changed to corrosion resisting steel. In July 1939, the material was changed to "Inconel", O.S. 651; in February 1941, the material was changed to (47-N-1) "Inconel". Broken springs received at Naval Torpedo Station, Newport, for examination and test have proved to be the stainless steel type. No cases of broken "Inconel" springs have been reported.

2. All exhaust valve springs except the "Inconel" may be attracted by a magnet; this test should be made on all s rings installed and in stock. Springs found to be magnetic should be replaced by the "Inconel" type. Another test consists in touching the end of the spring against a fine grained high speed grinding wheel and observing the sparks; a short orange-red spark, free from yellow bursts, indicates "Inconel", while an abundant spark, yellow-white in color and containing a number of bright carbon bursts or stars, is characteristic of the obsolete CRS.

BUREAU OF ORDNANCE CIRCULAR LETTER NO. T-165.

From: The Chief of the Bureau of Ordnance.
To: All vessels carrying aircraft torpedoes; destroyer, submarine, and aircraft tenders; air stations, destroyer and submarine bases, and mine depots servicing aircraft torpedoes; torpedo stations and schools; division, squadron, flotilla, force, type, and fleet commanders.

Subject: Aircraft Torpedoes - Mark 13 Type Depths of Initial Dive that may be expected when using biplane stabilizers.

Reference: (a) BuOrd ltr. A5-2(2926)(Mn3a) of December 9, 1941.
(b) BuOrd ltr. S75-1 (S5/251)(Mn-4n) of May 1, 1941.

1. With reference (b), instructions were forwarded to certain units in the Naval Service regarding the use of biplane stabilizers on Mark 13 type aircraft torpedoes when they are to be dropped from PBY-3, PBY-4, PBY-5, PBY-5A, PB2Y-1, and PBM-1 planes; reference (a) forwarded instructions for their use with TBD-1 planes. The Bureau has adopted the biplane stabilizer as being the simplest and most effective device for insuring that the axis of the torpedo remains tangent to its trajectory from the time it is released by the plane until it strikes the water.

2. It is particularly to be noted that these stabilizers are useable only with the Mark 13 type torpedo and from the planes indicated in paragraph 1. For any other adaptation it will be necessary to conduct tests to determine the suitability of these particular attachments. The Bureau's instructions should be obtained before such tests are made.

3. As a result of about 200 drops of Mark 13 Mod. 1 torpedoes fitted with biplane stabilizers, conducted at plane speeds of 120 knots with height of drop of 80-120 feet, the depth of the initial dive may be expected to conform to the following table. It was the experience of the Naval Torpedo Station, Newport, that the depth of this initial dive was independent of the actual depth setting on the torpedo. The exploder mechanism will have armed after a torpedo run of approximately 300 yards.

DEPTH OF INITIAL DIVE	PERCENTAGE OF DROPS EXPECTED TO MAKE INITIAL DIVE TO DEPTH INDICATED
10 feet or less	12
20 feet or less	30.5
30 feet or less	49
40 feet or less	73
50 feet or less	89
60 feet or less	94
70 feet or less	98
80 feet or less	98.5
90 feet or less	99.5
100 feet or less	100

4. Reference (a) and the enclosures thereto, consisting of instruction sheets and Ordnance Sketches 96468 and 96469, give the procedure for the use of biplane stabilizers on Mark 13 type torpedoes when dropped from TBD-1 planes. Reference (b) and its enclosures consisting of instruction sheets and Ordnance Sketches 96037, 96060, 96063, and 96079 give the pro-

cedure for the use of biplane stabilizers on Mark 13 type torpedoes when dropped from PBY-3, PBY-4, PBY-5, PBY-5A, PB2Y-1, and PBM-1 planes. The enclosures to reference (a) and (b) have been distributed to airplane carriers, seaplane tenders, patrol wing commanders, and to the higher air commands. Activities desiring these instructions may obtain them from the Naval Torpedo Station, Newport.

5. The information in paragraph 3 concerning depths of initial dive to be expected has been distributed to type and fleet commanders by dispatch.

BUREAU OF ORDNANCE CIRCULAR LETTER NO. T-173.

From: The Chief of the Bureau of Ordnance.
To: All aircraft carriers and tenders; naval air stations and mine depots servicing aircraft torpedoes; torpedo schools and stations; Chief of Army Ordnance and Air Corps.

Subject: Use of Mark 13 Mod. 1 torpedoes in high speed torpedo planes.

Reference: (a) NavTorpSta., Newport, publication on the Mark 13 Mod. 1 Torpedo of January 3, 1942.
(b) BuOrd Circular letter No. T-155 of August 14, 1941.

1. Prior to the production of U.S. Army plane B-26, the fastest torpedo-carrying planes in the United States services made about 130 knots. Reference (a) imposed restrictions on the dropping of the Mark 13 type torpedo to ground speeds of 132 miles per hour (115 knots) and altitudes of release of 75 to 150 feet. Reference (b) stated that until the stronger propellers and upper guide vane braces were received the limitation on the altitude from which drops could be made was 75 feet. These restrictions have now been removed, and torpedoes may be dropped at ground speeds up to 200 knots, altitude 200 feet, subject to the following conditions:

  (a) Torpedo to be Mark 13 Mod. 1 with stabilizer.
  (b) Torpedoes to have strengthened propellers (BuOrd Stock Parts 14810 and 14811).
  (c) Radial braces to be fitted to support the upper guide vane in the same manner as they now support the other three guide vanes.
  (d) War heads to be used. (Exercise heads will not withstand the shock of high speed dropping in excess of 150 knots).

2. Mark 13 Mod. 1 torpedoes having serial numbers higher than 19498 have the strengthened propellers and the requisite radial braces. Material is now being issued to the

forces afloat to permit the local installation in lower serial numbered torpedoes of strengthened propellers and the radial braces.

3. Whereas the strengthened torpedoes can be dropped at the speeds and from the altitudes mentioned in paragraph 1 somewhat better results may be expected if lower speeds are used. The B-26 can drop torpedoes at a speed as low as 160 m.p.h. (139 knots).

4. At air speeds in excess of 270 m.p.h. (234 knots) there is danger of the exploder arming; this when the impeller shaft packing is set up to give the impeller test pressure of 12 to 20 ounces now prescribed. A mechanical stop for the exploder impeller has been designed and will be issued to the service in the near future. This will prevent arming of the impeller while the torpedo is attached to the plane but will not prevent the exploder from arming once the torpedo has been released if the air speed is greater than 234 knots.

5. The Mark 26 Mod. 2 exercise head is the only one that can be used with torpedoes dropped by the B-26; this head can be used at a maximum ground speed of 150 knots. Mark 26 and 26 Mod. 1 exercise heads are too weak to be used on torpedoes dropped by the B-26.

6. Torpedo stabilizers as indicated below have been designed for and will be used on the type of plane indicated:

Plane	Mark of Torpedo	BuOrd Dwg. of air stabilizer	Aircraft installation outline dwg.
PBY-3,4,5,5A PB2Y-1 PBM-1 B-26	Mk. 13 Mk. 13-1	226107 226108	226096 226097
TBD-1	Mk. 13 Mk. 13-1	226109 226110	226098 226099

BUREAU OF ORDNANCE CIRCULAR LETTER NO. T 2-42.

Subject: Aircraft Torpedoes.

Reference: (a) O.D. No. 3816 entitled "Aircraft Torpedoes - General Information".

1. Attention is invited to reference (a), which has been prepared by the Naval Torpedo Station, Newport, Rhode Island. Distribution of reference (a) has been initiated by the Bureau of Ordnance for all aviation activities.

2. The purpose of reference (a) is to present a general picture of the aircraft torpedo problem and to give all available information which has been collected to date which may be of use in dropping torpedoes.

3. Although reference (a) is classed as CONFIDENTIAL, it is considered highly desirable that all pilots, aviation ordnance men, and others directly concerned with aircraft torpedoes be thoroughly familiar with its contents.

4. Sheet No. 39, paragraph 4 of reference (a) states that "It is possible for the impeller to turn and thus arm the knots". This is in error and should read "270 statute miles per hour (234 knots).

BUREAU OF ORDNANCE CIRCULAR LETTER NO. T 8-42.

Subject: Firing Springs for Exploder Mechanisms - Torpedoes.

Reference: (a) BuOrd Conf. Ltr. S75-1 (26) (Mn3a) of 25 May 1942.

1. From time to time the Bureau receives reports indicating that not all firing springs in torpedo exploder mechanisms are in accordance with the latest drawings. Reference (a) cited cases where failures of torpedoes to fire on war shots may probably be charged to the use of improper springs. There appears to be good reason for feeling that most failures of torpedoes to fire during war operations may logically be chargeable to the use of firing springs not in accordance with current drawings.

2. In order that old and new firing springs may be quickly identified the following information is given:

(a) <u>Mark 6 Mod. 1 and Mark 4 Mod. 1 Exploder Mechanisms.</u> New springs are 2.4" long (free length) as against 1.688" for the old springs. Number of <u>actual</u> coils is twelve in the case of new springs and <u>ten</u> in the case of the old springs. The number of <u>active</u> coils has been increased from seven and one-half to ten in the case of the new springs.

(b) <u>Mark 3 Mod. 1 and Mark 3 Mod. 2 Exploder Springs.</u> Present free length approximately 2.3" against 1.6" for the replaced springs. The new springs have eleven actual coils and nine active coils against nine and one-half <u>actual</u> coils in the case of the replaced springs.

3. Proper firing springs should be installed at the earliest practicable time. The Bureau should be informed by despatch of the number of springs required if not available.

©2014 Periscope Film LLC
All Rights Reserved
ISBN #9781940453293
www.PeriscopeFilm.com

www.ingramcontent.com/pod-product-compliance
Lightning Source LLC
Chambersburg PA
CBHW080402300426
44113CB00015B/2379